IRN

Computational Methods in Engineering and Science

COMPUTATIONAL METHODS IN ENGINEERING AND SCIENCE

With Applications to Fluid Dynamics and Nuclear Systems

Shoichiro Nakamura

The Ohio State University
Columbus, Ohio

A Wiley-Interscience Publication

JOHN WILEY & SONS, New York · London · Sydney · Toronto

Library of Congress Cataloging in Publication Data

Nakamura, Shoichiro, 1935—
    Computational methods in engineering and science.

    "A Wiley-Interscience publication."
    Includes bibliographical references and index.
    1. Engineering mathematics. 2. Differential
equations—Numerical solutions. I. Title.

TA330.N35   620'.001'51535   77-5471

ISBN 0-471-01800-7

Printed in the United States of America

10 9 8 7 6 5 4 3 2 1

To
   Akiko
      Tadashi
        Machi

Preface

This book covers the computational methods for differential equations in science and engineering. The computational methods included in this book may be largely divided into three categories: finite difference method, finite element method, and the statistical method (Monte Carlo). From the viewpoint of types of equations, the book includes eigenvalue problems of ordinary differential equations, elliptic partial differential equations, and parabolic and hyperbolic partial differential equations.

The book is intended to be a self-contained text for senior or graduate courses in engineering colleges. More emphasis is placed on comprehensiveness of the materials than on mathematical rigorousness. It is assumed that readers have some preliminary knowledge of linear algebra and manipulations of matrix and vectors. It is also desirable for the reader to have some background in the basic equations in the area of his interest. For example, in teaching the finite difference equation for fluid flow, this book does not describe the derivation of the Navier–Stokes equation, so the reader must know the derivation of the Navier–Stokes equation, and the nature of the equation before he can learn the numerical methods for it.

Chapter 1 covers the fundamentals of the numerical methods. Although it is assumed that the reader has some knowledge of linear algebra and matrix treatments, this chapter reviews and summarizes what is needed in later chapters. Chapters 2 through 4 describe finite difference methods for Sturm-Liouville eigenvalue problems, and elliptic and parabolic partial differential equations. Chapter 5 describes the finite difference methods for fluid flow. Chapters 6 through 8 describe weighted residual methods and their application to computational techniques. The weighted residual method and the variational principle are introduced in Chapter 6. Chapter 7 describes the principle of the finite element method and its application to partial differential equations. In Chapter 8, the coarse-mesh rebalancing method is introduced as a technique to accelerate the convergence rate of the iterative techniques described in Chapter 3. As a statistical approach to the numerical calculation, the Monte Carlo method for neutron transport and heat transfer is described in Chapter 9. Finite difference techniques for transonic aerodynamic analysis are discussed in Chapter 10.

Although various specific applications of the methods are introduced in each chapter, procedures such as derivation of numerical equations and their solutions leading to the specific applications are described as generally as possible; consequently, the text should be useful to those who are interested in the computational methods in general. This book is also designed to be used in mechanical engineering courses as well as nuclear engineering courses.

A reader who is interested in computational fluid dynamics should follow Chapters 1 through 8 and 10. The general aspect of the finite element method can be obtained by reading Chapters 1, 6, and 7. The computational techniques for steady-state and transient heat conduction may be studied in Chapters 1, 2, 3, 4, 7, and a part of Chapter 9.

Until now, numerical analysis courses taught in nuclear engineering have been oriented toward reactor physics calculations. However, as design techniques for nuclear reactors become mature, more emphasis is being placed on computational techniques for safety analyses and economical operation of power plants. In this regard, computational methods other than neutronic calculations described in this book are all closely related to plant safety analyses and important to nuclear engineers except for Chapter 10. For this reason, the author believes that this book is a suitable text for nuclear engineering.

Two important approaches to numerical methods for neutron transport are the finite difference method and the Monte Carlo method. Only the latter is included in this book. This is because (1) there are several comprehensive books and other literatures available on finite difference neutron transport calculations and (2) a short comprehensive text on the Monte Carlo method that covers the recent techniques is more difficult to find.

I am indebted to my previous department chairman, Dr. D. D. Glower, for his helpful advice and administrative support. Early discussions with Dr. R. F. Redmond of The Ohio State University and his encouragement throughout the work were valuable. I benefitted much from the suggestions and discussions of Dr. E. Gelbard of Argonne National Laboratory. The reference materials provided by the NASA Ames Research Center were valuable in writing the last chapter. The assistance of my students, Messrs. P. P. Su, R. H. Lubbers, and R. M. Lell, among others, who proofread at the final stage of the manuscript, is gratefully acknowledged. Most of the numerical examples were directly or indirectly calculated by using IBM 370/168 of the Instruction and Research Computer Center of The Ohio State University. Finally, I express my sincere thanks and appreciation to Miss Carol Edger, who patiently typed the whole manuscript over the several years of my work on this book.

SHOICHIRO NAKAMURA
Columbus, Ohio, January 1977

Contents

Computational Methods in
Engineering and Science

Chapter 1   Fundamentals of Numerical Analysis

1.1 NUMERICAL METHODS AND THEIR BASIC DIFFERENCES FROM ANALYTICAL SOLUTIONS

Deriving the equation that describes a physical phenomenon or system does not need any hardware, although it is almost hopeless to solve the equation analytically in a closed form unless the equation is extremely simple or simplified. The difficulties encountered in finding analytical solutions may be classified into the following cases:

1. The algorithm is simple, but the order of equation or the amount of calculation is too large.
2. The equation is multidimensional.
3. Complexity of geometry.
4. No analytical solution or procedure is known.
5. An analytical solution is available, but complicated and uneconomical.

It is indeed surprising to realize what a small fraction of mathematical equations have their solutions in the straightforward and analytical form. Engineers and scientists historically abandoned mathematical analysis for most of the realistically complicated systems and pursued experimental approaches. Obviously, there are several limitations to experimental approaches, such as experimental errors and the gross nature of the results. Furthermore, experiments that involve dangers or large capital investments can be done only when the risk of failure is acceptably low. As the computer was improved and became cheaper, the gap between analytical solutions and experimental approaches began to be bridged. Today, it is impossible to separate the computer from system design and analysis in most advanced technologies. For example, a substantial amount of computer analysis is done to study the static and dynamic nature of the objectives and to select optimal design parameters, whenever large-scale experiments or plant

1

constructions are involved. The optimal route of a spaceship can be found only by computational analysis. The detailed analysis of hypothetical accidents in a nuclear plant is possible only computationally. Furthermore, numerical analyses are gradually replacing experiments so far as they are more economical and accurate. This trend will continue as the capability of the numerical method increases. In effect, substantial fluid dynamics experiments are expected to be replaced by computational fluid analysis.

Numerical methods are the mathematical procedures based on arithmetic operations by which computers calculate the solution of mathematical equations. Because of digital nature of the computations, there are several differences between numerical methods and analytical approaches. The basic difference is that continuous spaces cannot be represented by a finite core memory. This implies that a function on a continuous coordinate must be approximated by the values of the function on a finite number of discrete points or a set of finite numbers of given functions. Here the numerical methods may be divided into two different approaches: (1) a deterministic approach such as the finite difference and finite element methods and (2) a probabilistic approach such as the Monte Carlo method.

The numerical solution for a differential equation on fixed grid points starts with finding how to express the solution on the discrete coordinate and how to approximate the differential and integral operators on the discrete space. A finite number of dependent variables may be expressed by a vector. Numerical approximations of the differential or integral operators may be expressed by matrices. For this reason, linear algebra in the matrix and vector notations is particularly important in numerical analysis.

It is assumed that the reader has studied basic numerical analysis and linear algebra such as presented in References 1 and 2. However, it is worthwhile to lay a common background. For this reason, the fundamental numerical methods and linear algebra are summarized in the following sections.

1.2 INTERPOLATION METHODS

If the values of a function $f(x)$ are known only on the discrete points $\{x_i\}$, $i = 0, 1, 2, \ldots I$, where $\{\ \}$ denotes a set of numbers but the formula of $f(x)$ is not known, then we may guess the value of $f(x)$ on other points by using an interpolation formula. We restrict our discussions to the polynomial interpolation.[3,4] In case only two discrete points are considered ($I = 1$), the interpolation formula is a linear function of x as

$$g(x) = \frac{x - x_0}{x_1 - x_0}(f_1 - f_0) + f_0 = \frac{x_1 - x}{x_1 - x_0}f_0 + \frac{x - x_0}{x_1 - x_0}f_1 \qquad (1.2.1)$$

where f_i is the known value of $f(x)$ at $x = x_i$.

In general, if $f_i \equiv f(x_i)$, $i = 0, 1, 2, \ldots I$, are given, the interpolation formula can be obtained in the form of a polynomial of order I as

$$g(x) = \sum_{j=0}^{I} a_j x^j \qquad (1.2.2)$$

where a_j are coefficients. Since $g(x)$ can be made to agree with $f(x)$ at $I + 1$ points of x, we have

$$f_i = \sum_{j=0}^{I} a_j x_i^j, \qquad i = 0, 1, 2, \ldots I \qquad (1.2.2a)$$

Solving the above equation determines a_j. The polynomial $g(x)$ thus obtained can be written as

$$g(x) = \frac{(x - x_1)(x - x_2)\ldots(x - x_I)}{(x_0 - x_1)(x_0 - x_2)\ldots(x_0 - x_I)}f_0$$

$$+ \frac{(x - x_0)(x - x_2)\ldots(x - x_I)}{(x_1 - x_0)(x_1 - x_2)\ldots(x_1 - x_I)}f_1 \ldots$$

$$+ \frac{(x - x_0)(x - x_1)\ldots(x - x_{I-1})}{(x_I - x_0)(x_I - x_1)\ldots(x_I - x_{I-1})}f_I \qquad (1.2.3)$$

The above formula is called the *Lagrange interpolation formula*. It is easy to see that Eq. (1.2.3) is equivalent to Eq. (1.2.2): $g(x)$ is a Ith order polynomial of x and is exact at the $I + 1$ points (x_i, f_i), $i = 0, 1, 2, \ldots I$. Since $g(x)$ is exact at $I + 1$ points, $g(x)$ becomes the exact function if $f(x)$ is a polynomial of order I or less. If $f(x)$ is not a polynomial of x or is a polynomial of order greater than I, then $g(x) \neq f(x)$ except at the points x_i, $i = 0, 1, \ldots I$.

If in addition to $\{f_i\}$ the derivatives of $f(x)$ at x_i, $\{f_i'\}$, are given, the polynomial that fits both sequences at $\{x_i\}$ becomes the $(2I + 1)$th order polynomial. This polynomial agrees with $f(x)$ with respect to the values and the derivatives on the discrete points $\{x_i\}$. The polynomial fitting can be extended to include higher order derivatives. These polynomials are called osculating polynomials or Hermite interpolation polynomials.[5] Fitting a function with a high-order polynomial (or equivalently Lagrange formula) is undesirable. This is because the computational time increases rapidly as the order increases and yet there is no guarantee that a given function is well fitted by a polynomial of high order. The Legendre formula should not be applied to a large interval of the independent variable. A poor example with the Lagrange formula is shown in Fig. 1.1, where the error of Lagrange formulae fitted to $y = x^{-2}$ in the range, $0.5 \leq x \leq 5$, is plotted. Figure 1.1 shows also that it is extremely dangerous if the Lagrange formula is used for extrapolation. Fitting $y = x^{-2}$ with the Lagrange formula would be much more successful if a smaller range of x than $[0.5, 5.0]$ was considered and the Lagrange formula was used repeatedly.

The above consideration leads us to the piecewise polynomial method in which a low-order polynomial is repeatedly used in each interval of $x_i \leq x \leq$

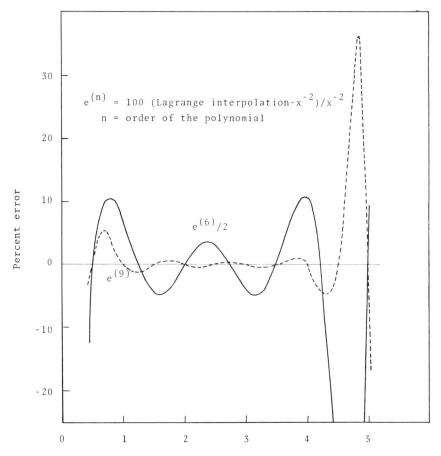

Figure 1.1   Error of Lagrange interpolation for $y = x^{-2}$ in $0.5 \leq x \leq 5$.

x_{i+1}. For instance, if the linear interpolation is used the given function is approximated in each interval of $x_i \le x \le x_{i+1}$ by

$$g_i(x) = \frac{x_{i+1} - x}{x_{i+1} - x_i} f_i + \frac{x - x_i}{x_{i+1} - x_i} f_{i+1} \qquad (1.2.4)$$

The piecewise linear functions, Eq. (1.2.4), that fit $\{f_i\}$, $i = 0, 1, \ldots I$, in the entire region of interest $x_0 \le x \le x_I$ can be written as

$$g(x) = \sum_{i=0}^{I} f_i \eta_i(x) \qquad (1.2.5)$$

where $\eta_i(x)$ is called a shape function and given by

$$\eta_i(x) = \frac{x - x_{i-1}}{x_i - x_{i-1}}, \quad x_{i-1} \le x \le x_i$$

$$= \frac{x_{i+1} - x}{x_{i+1} - x_i}, \quad x_i \le x \le x_{i+1} \qquad (1.2.6)$$

$$= 0, \qquad x < x_{i-1} \quad \text{or} \quad x > x_{i+1}$$

The function $\eta_i(x)$ and its derivative are shown in Fig. 1.2.

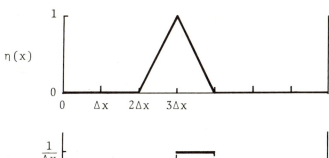

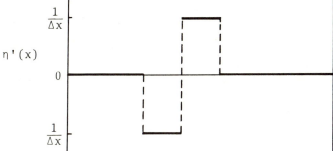

Figure 1.2 A pyramid function and its derivative.

If, in addition to f_i, the first derivative f'_i is given, the polynomial that fits f_i, f_{i+1}, f'_i, and f'_{i+1} in the interval $x_i \le x \le x_{i+1}$ is of order 3 and designated as a cubic polynomial.[6,7] It is given by

$$g_i(x) = f_i u_i^{(0+)}(x) + f_{i+1} u_{i+1}^{(0-)}(x) + f'_i u_i^{(1+)}(x) + f'_{i+1} u_{i+1}^{(1-)}(x) \quad (1.2.7)$$

where

$$u_i^{(0+)}(x) = 3\left(\frac{x_{i+1}-x}{x_{i+1}-x_i}\right)^2 - 2\left(\frac{x_{i+1}-x}{x_{i+1}-x_i}\right)^3 \quad (1.2.8a)$$

$$u_{i+1}^{(0-)}(x) = 3\left(\frac{x-x_i}{x_{i+1}-x_i}\right)^2 - 2\left(\frac{x-x_i}{x_{i+1}-x_i}\right)^3 \quad (1.2.8b)$$

$$u_i^{(1+)}(x) = \left[\left(\frac{x_{i+1}-x}{x_{i+1}-x_i}\right)^2 - \left(\frac{x_{i+1}-x}{x_{i+1}-x_i}\right)^3\right](x_{i+1}-x_i) \quad (1.2.8c)$$

$$u_{i+1}^{(1-)}(x) = \left[-\left(\frac{x-x_i}{x_{i+1}-x_i}\right)^2 + \left(\frac{x-x_i}{x_{i+1}-x_i}\right)^3\right](x_{i+1}-x_i) \quad (1.2.8d)$$

The piecewise cubic polynomials that fit $\{f_i\}$ and $\{f'_i\}$, $i = 0, 1, \ldots I$, in $x_0 \le x \le x_I$ may be written as

$$g(x) = \sum_{i=0}^{I} f_i \eta_i(x) + \sum_{i=0}^{I} f'_i \zeta_i(x) \quad (1.2.9)$$

where

$$\eta_i(x) = u_i^{(0-)}(x), \quad x_{i-1} \le x \le x_i$$
$$= u_i^{(0+)}(x), \quad x_i \le x \le x_{i+1} \quad (1.2.10a)$$
$$= 0, \quad \text{otherwise}$$

$$\zeta_i(x) = u_i^{(1-)}(x), \quad x_{i-1} \le x \le x_i$$
$$= u_i^{(1+)}(x), \quad x_i \le x \le x_{i+1} \quad (1.2.10b)$$
$$= 0, \quad \text{otherwise}$$

The functions, η_i, ζ_i, η'_i, and ζ'_i are plotted in Fig. 1.3. Note that $\eta_i(x_i) = \zeta'_i(x_i) = 1$ and $\zeta_i(x_i) = \eta'_i(x_i) = 0$.

In general, the piecewise polynomial that fits up to the pth derivative at the two ends of an interval is written as

$$g(x) = \sum_{n=0}^{p} [f_i^{(n)} u_i^{(n+)}(x) + f_{i+1}^{(n)} u_{i+1}^{(n-)}(x)] \quad (1.2.11)$$

where $u^{(n+)}(x)$ and $u^{(n-)}(x)$, $n = 0, 1, \ldots p$, are all polynomials of order $2p + 1$. Equation (1.2.11) is called the piecewise Hermite polynomial.

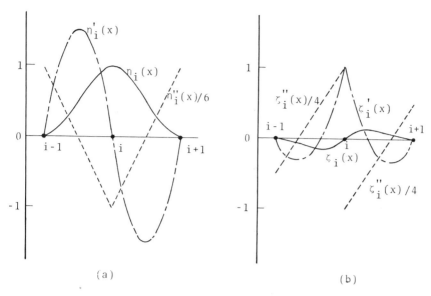

Figure 1.3 Plot of piecewise cubic polynomials ($h_i = h_{i+1} = 1$).

The piecewise cubic polynomial may be used to fit $\{f_i\}$, $i = 0, 1, 2, \ldots I$, even when $\{f_i'\}$ are not specified. $\{f_i'\}$ may be determined so that the second derivative of the piecewise polynomials become continuous at $\{x_i\}$, $i = 1, 2, \ldots I-1$. The piecewise cubic polynomials thus obtained are called *cubic spline*[8-10] and are the smoothest piecewise polynomials of order 3.

The first derivatives of the spline function at x_i are determined as follows. The second derivative of Eq. (1.2.9) at $x_i \pm \epsilon (\epsilon \to 0)$ except for $i = 0$ and I are

$$g''(x_i + \epsilon) = [u_i^{(0+)}(x_i + \epsilon)]''f_i + [u_{i+1}^{(0-)}(x_i + \epsilon)]''f_{i+1}$$
$$+ [u_i^{(1+)}(x_i + \epsilon)]''f_i' + [u_{i+1}^{(1-)}(x_i + \epsilon)]''f_{i+1}' \qquad (1.2.12)$$
$$= 6\frac{-f_i + f_{i+1}}{h_{i+1}^2} - \frac{4}{h_{i+1}}f_i' - \frac{2}{h_{i+1}}f_{i+1}'$$

and

$$g''(x_i - \epsilon) = [u_{i-1}^{(0+)}(x_i - \epsilon)]''f_{i-1} + [u_i^{(0-)}(x_i - \epsilon)]''f_i$$
$$+ [u_{i-1}^{(1+)}(x_i - \epsilon)]''f_{i-1}' + [u_i^{(1-)}(x_i - \epsilon)]''f_i' \qquad (1.2.13)$$
$$= 6\frac{f_{i-1} - f_i}{h_i^2} + \frac{2}{h_i}f_{i-1}' + \frac{4}{h_i}f_i'$$

respectively, where $h_i = x_i - x_{i-1}$. Equating the above two equations and dividing by 2 yield

$$\frac{1}{h_i}f'_{i-1}+\left(\frac{2}{h_i}+\frac{2}{h_{i+1}}\right)f'_i+\frac{1}{h_{i+1}}f'_{i+1}=-3\frac{f_{i-1}-f_i}{h_i^2}+3\frac{f_{i+1}-f_i}{h_{i+1}^2} \quad (1.2.14)$$

At $i=0$, either f'_0 or f''_0 must be specified. If f''_0 is prescribed, f'_0 and Eq. (1.2.12) for $i=0$ are equated to yield

$$\frac{2}{h_1}f'_0+\frac{1}{h_1}f'_1=3\frac{f_1-f_0}{h_1^2}-\frac{1}{2}f''_0 \quad (1.2.15)$$

Similarly, if f''_I is specified for $i=I$, we have

$$\frac{1}{h_I}f'_{I-1}+\frac{2}{h_I}f'_I=3\frac{f_I-f_{I-1}}{h_I^2}+\frac{1}{2}f''_I \quad (1.2.16)$$

When f''_0 and f''_I are both specified, $\{f'_i\}$, $i=0, 1, 2, \ldots I$, are determined by solving Eqs. (1.2.14) to (1.2.16) simultaneously. If f'_0 and f'_I are specified, Eqs. (1.2.15) and (1.2.16) are unnecessary, and Eq. (1.2.14) is solved for $\{f'_i\}$, $i=1, 2, \ldots I-1$. In case any derivative at $i=0$ and I are unknown, the end conditions are arbitrary. The only criterion for the choice of the end conditions is the physical requirement. The following approaches are most commonly used:

a. $f''_0 = f''_I = 0$
b. $f''_0 = f''_1, f''_I = f''_{I-1}$
c. $f''_0 = \lambda_a f''_1, f''_I = \lambda_b f''_{I-1}$ (λs are adjusting parameters)

Among all functions $f(x)$ having a continuous second derivative and passing $\{y_i, x_i\}$, $i=0, 1, 2, \ldots I$, the cubic spline function $g(x)$ with the end conditions $g''(x_0)=g''(x_I)=0$ has the minimal value of the integral

$$\int_{x_0}^{x_I} |f''(x)|^2 \, dx \quad (1.2.17)$$

This is called Holladay's theorem[9] and often called the minimum curvature property. The spline function is not trouble free, however. In fitting points $\{y_i, x_i\}$, oscillation may occur if the grid intervals are too large or if the end conditions are not appropriate. When there is interest in interpolation of points and the derivatives at the grid points are specified, the Hermite interpolation has more accuracy than the spline function. However, the second derivative of the Hermite function is discontinuous, so it cannot be used if continuity of the second derivative of the interpolating function is important.

Example 1.1

Fit the following four points by the cubic splines (Case 1):

i	0	1	2	3
x_i	1	2	3	4
y_i	1	4	9	16

Since the end conditions are given, we first try with $y_0'' = y_3'' = 0$. Eqs. (1.2.14) to (1.2.16) for this problem become

$$
\begin{aligned}
2y_0' + y_1' &= 3(y_1 - y_0) = 9 \\
y_0' + 4y_1' + y_2' &= 3(y_2 - y_0) = 21 \\
y_1' + 4y_2' + y_3' &= 3(y_3 - y_1) = 36 \\
y_2' + 2y_3' &= 3(y_3 - y_2) = 21
\end{aligned}
\qquad (1.2.18)
$$

The solution of the above equation is $y_0' = 3.06667$, $y_1' = 2.86666$, $y_2' = 6.46666$, $y_3' = 7.26667$. Obviously, the given points were taken from $y = x^2$. The values of y interpolated by the cubic splines are compared with the exact values in Table 1.1.

If the end conditions are specified as $y_0' = 2$ and $y_3' = 6$ (Case 2), the second and third equations above become

$$
\begin{aligned}
4y_1' + y_2' &= 21 - 2 = 19 \\
y_1' + 4y_2' &= 36 - 6 = 30
\end{aligned}
\qquad (1.2.19)
$$

the solution is $y_1' = 3.06668$, $y_2' = 6.73333$. The error with this approach is also shown in Table 1.1.

Example 1.2

Suppose $f_i = x_i^{-2}$ and $f_i' = -2x_i^{-3}$, where $x_i = 0.5 + 0.5i$, $i = 0, 1, \ldots 9$, are given. Fit those values by the piecewise cubic Hermite polynomial and cubic spline functions, and check the maximum error in $0.5 \le x \le 5.0$ (This problem was previously used to illustrate a poor application of the Lagrange interpolation as shown in Fig. 1.1).

The piecewise Hermite polynomial is given by Eq. (1.2.9). The cubic spline function does not use f_i' given above except for $i = 0$ and 9. The derivatives f_i' for $i = 1, 2, \ldots, 8$ are determined so that the second derivative of the fitting function is continuous at $i = 1$ through 8.

Equation (1.2.9) is used straightforwardly for the piecewise cubic Hermite interpolation. The error is plotted in Fig. 1.4. We use f_0' and f_I' given previously as the end conditions for the cubic spline interpolation. The derivatives at other grids are determined by Eq. (1.2.14). The error of the cubic spline interpolation is also plotted in Fig. 1.4. By comparing Fig. 1.1

Table 1.1 Relative error of spline functions for $y = x^2$ in $1 \leq x \leq 4$.

	Percent error	
x	Case 1	Case 2
1.0	0	0
1.2	12.00	2.07
1.4	13.39	4.57
1.6	10.38	5.24
1.8	5.53	3.68
2.0	0	0
2.2	- 3.31	-2.95
2.4	- 3.61	- 3.55
2.6	- 2.60	-2.89
2.8	- 1.22	-1.58
3.0	0	0
3.2	0.81	1.54
3.4	1.19	2.57
3.6	1.16	2.76
3.8	0.75	1.94
4.0	0	0

Case 1: $y_0'' = y_3'' = 0$

Case 2: $y_0' = 2, y_3' = 6$

with Fig. 1.4, it is observed that the error of the cubic Hermite and cubic spline interpolations is much smaller than the Lagrange interpolation of 9th order. If we compare the cubic Hermite interpolation with the cubic spline interpolation, the error of the former is smaller than the latter. This tendency is generally true. However, if the second derivative of the inter-polating functions is checked, the former has discontinuity at every grid, while the latter has no discontinuity.

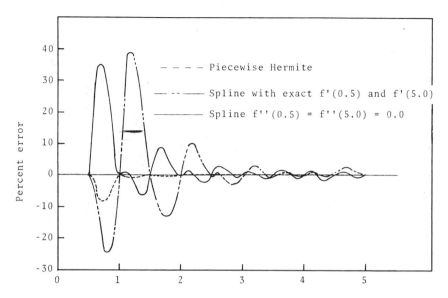

Figure 1.4 Errors of cubic spline and cubic Hermite polynomial fitting for Example 1.2.

1.3 NUMERICAL INTEGRATION

Numerical integration methods become necessary when the values of $f(x)$ are known on the discrete points $\{x_i\}$ but the formula of $f(x)$ is not known, or when the formula of $f(x)$ is known but the analytical integration is difficult. Numerical integration formulas are obtained by integrating the interpolating polynomials described in the previous section.

If we have only one interval, on both ends of which the values of $f(x)$ are given as f_0 and f_1, then the interpolation formula is given by Eq. (1.2.1). Integrating Eq. (1.2.1) in the interval, we obtain

$$\int_{x_0}^{x_1} g(x)\, dx = \frac{\Delta x}{2}[f_0 + f_1] \tag{1.3.1}$$

$$\Delta x = x_1 - x_0$$

which is called the *trapezoidal rule*. If we have I intervals of Δx and apply the trapezoidal rule repeatedly in each interval, the integration formula is

$$\int_{x_0}^{x_I} g(x)\, dx = \frac{\Delta x}{2}[f_0 + 2f_1 + 2f_2 \ldots 2f_{I-1} + f_I] \tag{1.3.2}$$

When we have two intervals for which f_0, f_1, f_2 are known, the interpolation formula of the second order polynomial may be obtained. Integrating the polynomial in those two intervals of Δx, we obtain the integration formula

$$\int_{x_0}^{x_2} g(x)\,dx = \frac{\Delta x}{3}[f_0 + 4f_1 + f_2] \qquad (1.3.3)$$

which is called *Simpson's rule*. If we have $2I$ intervals and apply Simpson's rule repeatedly I times, the integration formula is

$$\int_{x_0}^{x_1} g(x)\,dx = \frac{\Delta x}{3}[f_0 + 4f_1 + 2f_2 + 4f_3 + 2f_4 + \cdots + 4f_{I-1} + f_I] \quad (1.3.4)$$

This formula is called the *extended Simpson's rule*. This derivation can be extended to basic intervals of three as

$$\int_{x_0}^{x_3} g(x)\,dx = \frac{3\Delta x}{8}[f_0 + 3f_1 + 3f_2 + f_3] \qquad (1.3.5)$$

where the right side is obtained by integrating the polynomial of order 3.

The series of integration formulas represented by Eqs. (1.3.1), (1.3.3), and (1.3.5) can be extended to a higher order. The series thus obtained are called the *Newton Cotes formulas*. The Newton Cotes formulas of a large order have some undesirable properties as follows. First, the sequence of integrals does not converge in general to the true value even if $f(x)$ is analytic. In fact, there is no explicit description that a function can be readily approximated by a polynomial. Second, the coefficients in these formulas become large and alternate in sign. This is undesirable for computations because of the round-off error. Therefore, the integration formulas of a low order are usually applied successively to a multiple of the basic number of intervals as Eqs. (1.3.2) and (1.3.4).

Another important formula in the numerical integration is the *Gauss quadrature*.[11,14] By a choice of the interpolating points x_i, one can construct an integrating formula, using $I+1$ interpolating points, that gives the accurate value of the integral if $f(x)$ is a polynomial of degree $2I+1$ or less. We consider a polynomial of order $2I+1$ and try to integrate it in the interval, $-1 \le x \le 1$. By using the Legendre polynomial[11-14] of order $I+1$, $P_{I+1}(x), f(x)$ can be written

$$f(x) = b(x)P_{I+1}(x) + r(x) \qquad (1.3.6)$$

where $b(x)$ and $r(x)$ are polynomials of order I or less. The integral of $f(x)$ becomes

$$\int_{-1}^{+1} f(x)\,dx = \int_{-1}^{+1} b(x)P_{I+1}(x)\,dx + \int_{-1}^{+1} r(x)\,dx \qquad (1.3.7)$$

Since $P_{I+1}(x)$ is orthogonal to any polynomial of order I or less, the first term of Eq. (1.3.7) becomes zero, and we have

$$\int_{-1}^{+1} f(x)\, dx = \int_{-1}^{+1} r(x)\, dx \tag{1.3.8}$$

Suppose the values of $f(x)$ are given for $\{x_i\}$, $i = 0, 1, 2, \ldots I$, which are roots of $P_{I+1}(x) = 0$. referring to Eq. (1.2.3), the polynomial that fits those $I+1$ points may be written

$$g(x) = \sum_{i=0}^{I} \frac{S(x)}{(x - x_i)S'(x_i)} f(x_i) \tag{1.3.9}$$

where

$$S(x) = \prod_{i} (x - x_i) \tag{1.3.10}$$

In the above equation, $\prod$ denotes a product taken over all subscripts i. Substituting Eq. (1.3.6) into Eq. (1.3.9) yields

$$g(x) = \sum_{i=0}^{I} \frac{S(x)}{(x - x_i)S'(x_i)} b(x_i)P_{I+1}(x_i) + \sum_{i=0}^{I} \frac{S(x)}{(x - x_i)S'(x_i)} r(x_i) \tag{1.3.11}$$

The first term of Eq. (1.3.11) becomes zero because $P_{I+1}(x_i) = 0$, while the second term is exactly equal to $r(x)$ because it is the Lagrange interpolation of order I that fits exactly the polynomial of order I or less, namely, $r(x)$. Thus we have $g(x) = r(x)$. By using Eqs. (1.3.8) and (1.3.9), we obtain

$$\int_{-1}^{+1} f(x)\, dx = \int_{-1}^{+1} g(x)\, dx = \int_{-1}^{+1} \sum_{i} \frac{S(x)}{(x - x_i)S'(x_i)} f(x_i)$$

or

$$\int_{-1}^{+1} f(x)\, dx = \sum_{i} w_i f(x_i) \tag{1.3.12}$$

where

$$w_i = \int_{-1}^{+1} \frac{S(x)}{(x - x_i)S'(x_i)}\, dx \tag{1.3.13}$$

Equation (1.3.12) is called the *Gauss quadrature formula*. The values of x_i and w_i are tabulated in Table 1.2 and in various references.

As can be seen from this derivation, the Gauss quadrature formula, Eq. (1.3.12), is exact for any polynomial $f(x)$ of order $2I+1$ or less. If $f(x)$ is a polynomial of order greater than $2I+1$, the error in integration is due to the

Table 1.2 Abscissas and weight Factors for Gaussian
Integration.[a]

$$\int_{-1}^{+1} f(x)\,dx \approx \sum_{i=1}^{n} w_i f(x_i)$$

n	± x_i	w_i
2	0.57735 02691	1.00000 00000
3	0.00000 00000	0.88888 88888
	0.77459 66692	0.55555 55555
4	0.33998 10435	0.65214 51548
	0.86113 63115	0.34785 48451
5	0.00000 00000	0.56888 88888
	0.53846 93101	0.47862 86704
	0.90617 98459	0.23692 68850
6	0.23861 91860	0.46791 39345
	0.66120 93864	0.36076 15730
	0.93246 95142	0.17132 44923
8	0.18343 46424	0.36268 37833
	0.52553 24099	0.31370 66458
	0.79666 64774	0.22238 10344
	0.96028 98564	0.10122 85362
10	0.14887 43389	0.29552 42247
	0.43339 53941	0.26926 67193
	0.67940 95682	0.21908 63625
	0.86506 33666	0.14945 13491
	0.97390 65285	0.06667 13443

[a] Abscissas $= \pm x_i$ (zeros of Legendre polynomials).
Weight factors $= w_i$.

terms of order higher than $2I+1$ in $f(x)$. The coefficients of the Gauss
quadrature, w_i, are always positive, so that the round-off error accumulation
is small. The Gauss formula will give generally better results than the
corresponding equal-interval formulas. The Gauss quadrature is successfully used in various numerical schemes. The disadvantage of the formula is,
however, that the interpolating points and the corresponding weights are
irregular numbers. The Gauss formula can be applied to any interval of x,
$a < x < b$, after the transformation of the variable:

$$x' = \frac{2x - a - b}{b - a}$$
(1.3.14)

Example 1.3

We integrate $y = \sqrt{x}$ in $[0, 1]$ numerically by various methods and compare the result with the exact value:

$$S_{\text{exact}} = \int_0^1 \sqrt{x}\, dx = \tfrac{2}{3} = 0.6666$$

1. Trapezoidal rule [Eq. (1.3.2)] with four intervals ($\Delta x = 0.25$)

$$S = \frac{0.25}{2}[0 + 2\sqrt{0.25} + 2\sqrt{0.5} + 2\sqrt{0.75} + \sqrt{1}] = 0.6432,$$

error $= -3.5\%$

2. Extended Simpson's rule

$$S = \frac{0.25}{3}[0 + 4\sqrt{0.25} + 2\sqrt{0.5} + 4\sqrt{0.75} + \sqrt{1}] = 0.6565,$$

error $= -1.5\%$

3. Gauss quadrature. Since $a = 0$ and $b = 1$ in Eq. (1.3.14), we transform x into x' by $x' = 2x - 1$. In terms of x', the integral becomes

$$\frac{1}{2}\int_{-1}^{+1} \sqrt{\frac{x'+1}{2}}\, dx'$$

If we use the fourth order Gauss quadrature, the answer is

$$S = \sum_{i=1}^{4} w_i f(x_i) = 0.6678, \quad \text{error} = 0.18\%$$

1.4 DIFFERENCE APPROXIMATIONS AND DIFFERENCE OPERATORS

The Taylor series may be used to evaluate a function in the neighborhood of x_0 where the derivatives of all order are known. It can also be used to evaluate approximately the derivatives of $f(x)$ if the values of $f(x)$ are known at a number of discrete points, $\{x_i\}$, $i = 1, 2, \ldots I$.

Suppose we know the values of $f(x_0)$ and $f(x_0 + h)$, where h is a small value. The Taylor series expansion of $f(x_0 + h)$ is

$$f(x_0 + h) = f(x_0) + hf'(x_0) + \frac{h^2}{2}f''(x_0) + \cdots \qquad (1.4.1)$$

If we truncate after the first derivative and solve for $f'(x_0)$, we obtain an approximation

$$f'(x_0) \approx \frac{1}{h}[f(x_0+h)-f(x_0)] \tag{1.4.2}$$

Thus the first derivative at $x = x_0$ is approximately evaluated by $f(x_0+h)$ and $f(x_0)$. By defining the forward differences as

$$\Delta f(x_0) \equiv f(x_0+h)-f(x_0) \tag{1.4.3}$$

and

$$\Delta x_0 \equiv (x_0+h)-x_0 = h$$

the right side of Eq. (1.4.3) is expressed by

$$\frac{f(x_0+h)-f(x_0)}{h} = \frac{\Delta f(x_0)}{\Delta x_0} \tag{1.4.4}$$

which is termed the *forward difference formula*. The *backward difference formula* is also obtained by starting with $f(x_0-h)$:

$$\frac{f(x_0)-f(x_0-h)}{h} = \frac{\nabla f(x_0)}{\nabla x_0} \tag{1.4.5}$$

where ∇ is used to denote backward differences.

The central difference may be defined by

$$\delta f(x_0) \equiv f\left(x_0+\frac{h}{2}\right)-f\left(x_0-\frac{h}{2}\right) \tag{1.4.6}$$

By applying the Taylor series expansion to both terms on the right hand side, we obtain

$$\delta f(x_0) = 2\left[f'(x_0)\frac{h}{2}+\frac{1}{3!}f'''(x_0)\left(\frac{h}{2}\right)^3+\cdots\right] \tag{1.4.7}$$

where the even order derivative terms are zero. Truncating after the second order term, the *central difference formula* is obtained

$$\frac{f\left(x_0+\frac{h}{2}\right)-f\left(x_0-\frac{h}{2}\right)}{h} = \frac{\delta f(x_0)}{\delta x_0} \tag{1.4.8}$$

where δx_0 is defined by

$$\delta x_0 \equiv \left(x_0+\frac{h}{2}\right)-\left(x_0-\frac{h}{2}\right) = h \tag{1.4.9}$$

The errors of the forward difference and backward difference formulas are due to truncating the second order term, while the error of the central difference formula is due to truncating the third order term.

Higher derivatives can be evaluated by using more values of $f(x)$. The central difference approximation for $f''(x_0)$ is obtained as follows. The Taylor series expansion for $f(x_0+h)-2f(x_0)+f(x_0-h)$ is

$$f(x_0+h)-2f(x_0)+f(x_0-h)=h^2f''(x_0)+\tfrac{1}{12}h^4f''''(x_0)+\cdots \quad (1.4.10)$$

Therefore, truncating after the third order term, we obtain

$$f''(x_0)\approx\frac{f(x_0+h)-2f(x_0)+f(x_0-h)}{h^2}\equiv\frac{\delta^2f(x_0)}{\delta x_0^2} \quad (1.4.11)$$

where $\delta x_0 = h$.

Example 1.4

The three difference approximations for the first order ordinary differential equation $y'=f(y,x)$ are

$$\frac{y_{i+1}-y_i}{h}=f(y_{i+1},x_{i+1}) \qquad \text{Backward difference}$$

$$\frac{y_{i+1}-y_i}{h}=\frac{f(y_{i+1},x_{i+1})+f(y_i,x_i)}{2} \qquad \begin{array}{l}\text{Central difference} \quad (1.4.12)\\ \text{(modified)}\end{array}$$

$$\frac{y_{i+1}-y_i}{h}=f(y_i,x_i) \qquad \text{Forward difference}$$

where $x_i = ih$ and $y_i = y(x_i)$.

Generally, the higher order difference formulas are given by

$$\Delta^n f(x_0) = \Delta^{n-1}f(x_0+h)-\Delta^{n-1}f(x_0) \quad (1.4.13)$$

$$\nabla^n f(x_0) = \nabla^{n-1}f(x_0)-\nabla^{n-1}f(x_0-h) \quad (1.4.14)$$

$$\delta^n f(x_0) = \delta^{n-1}f\left(x_0+\frac{h}{2}\right)-\delta^{n-1}f\left(x_0-\frac{h}{2}\right) \quad (1.4.15)$$

In approximating partial derivatives of a two-dimensional function $f(x,y)$ by difference formulas the procedure is essentially the same as for one-dimensional functions. The first order partial derivative, for example,

$$\frac{\partial}{\partial x}f(x,y) \quad \text{at } x=x_0 \quad \text{and} \quad y=y_0$$

can be evaluated by fixing the value of y as $y = y_0$ and considering $f(x, y)$ as a function of only x. The forward, backward, and central difference difference formulas are given, respectively, by

$$f_x \approx \frac{f(x_0+h, y_0)-f(x_0, y_0)}{h} \equiv \frac{\Delta_x f(x_0, y_0)}{\Delta x_0} \tag{1.4.16}$$

$$f_x \approx \frac{f(x_0, y_0)-f(x_0-h, y_0)}{h} \equiv \frac{\nabla_x f(x_0, y_0)}{\nabla x_0} \tag{1.4.17}$$

$$f_x \approx \frac{f\left(x_0+\dfrac{h}{2}, y_0\right)-f\left(x_0-\dfrac{h}{2}, y_0\right)}{h} \equiv \frac{\delta_x f(x_0, y_0)}{\delta x_0} \tag{1.4.18}$$

The central difference approximations for the second order partial derivatives are

$$f_{xx} \approx \frac{1}{h^2}[f(x_0+h, y_0)-2f(x_0, y_0)+f(x_0-h, y_0)] \tag{1.4.19}$$

$$f_{yy} \approx \frac{1}{h^2}[f(x_0, y_0+h)-2f(x_0, y_0)+f(x_0, y_0-h)] \tag{1.4.20}$$

$$f_{xy} \approx \frac{1}{h^2}\left[f\left(x_0+\frac{h}{2}, y_0+\frac{h}{2}\right)-f\left(x_0+\frac{h}{2}, y_0-\frac{h}{2}\right)\right.$$
$$\left.-f\left(x_0-\frac{h}{2}, y_0+\frac{h}{2}\right)+f\left(x_0-\frac{h}{2}, y_0-\frac{h}{2}\right)\right] \tag{1.4.21}$$

Example 1.5

The central difference approximation to

$$-\left[\frac{\partial^2}{\partial x^2}\phi(x, y)+\frac{\partial^2}{\partial y^2}\phi(x, y)\right]=f(x, y)$$

is

$$\frac{-\phi_{i-1,j}+2\phi_{i,j}-\phi_{i+1,j}}{\Delta x^2}+\frac{-\phi_{i,j-1}+2\phi_{i,j}-\phi_{i,j+1}}{\Delta y^2}=f(x_i, y_j)$$

where

$$\phi_{i,j} \equiv \phi(x_i, y_j)$$
$$\Delta x = x_{i+1}-x_i = x_i-x_{i-1}$$
$$\Delta y = y_{j+1}-y_i = y_j-y_{j-1}$$

1.5 NUMERICAL INTEGRATION FOR INITIAL VALUE PROBLEMS OF ORDINARY DIFFERENTIAL EQUATIONS

This section outlines the Runge–Kutta method and the predictor-corrector method, which are the most popular methods for integrating ordinary differential equations. The last subsection summarizes other methods of interest.

RUNGE–KUTTA METHODS

Consider the single equation

$$y' = f(y, x) \tag{1.5.1}$$

with the initial condition $y(0) = y_0$, where f is an arbitrary function of y and x. Numerical schemes to the ordinary differential equation provide the numerical values of the solution on the discrete points of the independent variable, $x_n = n\Delta x$, where Δx is a small interval of x. Since $y'_0 = f(y_0, 0)$ is given by Eq. (1.5.1), an approximation to $y(\Delta x)$ may be calculated by using the forward difference formula as

$$y(\Delta x) = y_0 + \Delta x f(y_0, 0) \tag{1.5.2}$$

Denoting $y(n\Delta x)$ by y_n, the subsequent points can be successively treated as

$$y_{n+1} = y_n + \Delta x f(y_n, x_n) \tag{1.5.3}$$

This scheme is called the *Euler method*. Since the Taylor expansion of y_{n+1} about y_n is

$$y_{n+1} = y_n + \Delta x y'_n + \tfrac{1}{2}(\Delta x)^2 y''_n + \cdots \tag{1.5.4}$$

where y'_n and y''_n are the derivatives of y at $x_n = n\Delta x$, Eq. (1.5.3) is equivalent to approximating y_{n+1} with the first two terms of the Taylor expansion.

In order to increase the accuracy, we consider the numerical scheme that retains the first three terms of Eq. (1.5.4). Although y''_n may be obtained by directly taking the derivatives of Eq. (1.5.1), it requires evaluation of $\partial f / \partial y$ and $\partial f / \partial x$. Another approach is to use the Taylor expansion of $y'[x_n + (\Delta x/2)]$

$$y'\left(x_n + \frac{\Delta x}{2}\right) = y'_n + \frac{\Delta x}{2} y''_n + O(\Delta x^2) \tag{1.5.5}$$

Multiplying Eq. (1.5.5) by Δx and rewriting yield

$$\frac{\Delta x^2}{2} y''_n = \Delta x \left[y'\left(x_n + \frac{\Delta x}{2}\right) - y'_n \right] = \Delta x [f(y_{n+1/2}, x_{n+1/2}) - y'_n] \tag{1.5.6}$$

where $y_{n+1/2} = y(x_n + \Delta x/2)$ and $x_{n+1/2} = x_n + \Delta x/2$. Introducing Eq. (1.5.6) into Eq. (1.5.4), we have

$$y_{n+1} = y_n + \Delta x f(y_{n+1/2}, x_{n+1/2}) + O(\Delta x^3) \qquad (1.5.7)$$

The value of $y_{n+1/2}$ is not known but can be approximated by the Euler's method as

$$y_{n+1/2} = y_n + \frac{\Delta x}{2} y_n' \qquad (1.5.8)$$

With Eq. (1.5.8), Eq. (1.5.7) becomes

$$y_{n+1} = y_n + \Delta x f\left(y_n + \frac{\Delta x}{2} y_n', x_{n+1/2}\right) + O(\Delta x^3) \qquad (1.5.9)$$

This scheme is called the *three-points Runge–Kutta method* and the simplest case of the Runge–Kutta method. This method retains the higher order terms in the Taylor expansion by using the intermediate points between x_n and x_{n+1}.

A higher order method can be obtained by increasing the additional intermediate points of approximating the solution and matching more terms in the Taylor expansion. The rth order Runge–Kutta method requires $r+1$ points ($r-1$ intermediate points). The usual Runge–Kutta method is a five-point method of the fourth order, which is written by

$$\begin{aligned}
y_{n+1} &= y_n + \tfrac{1}{6}[k_1 + 2k_2 + 2k_3 + k_4] \\
k_1 &= \Delta x f(y_n, x_n) \\
k_2 &= \Delta x f(y_n + \tfrac{1}{2}k_1, x_n + \tfrac{1}{2}\Delta x) \\
k_3 &= \Delta x f(y_n + \tfrac{1}{2}k_2, x_n + \tfrac{1}{2}\Delta x) \\
k_4 &= \Delta x f(y_n + k_3, x_n + \Delta x)
\end{aligned} \qquad (1.5.10)$$

or

$$\begin{aligned}
y_{n+1} &= y_n + \tfrac{1}{8}[k_1 + 3k_2 + 3k_3 + k_4] \\
k_1 &= \Delta x f(y_n, x_n) \\
k_2 &= \Delta x f(y_n + \tfrac{1}{3}k_1, x_n + \tfrac{1}{3}\Delta x) \\
k_3 &= \Delta x f(y_n + \tfrac{1}{3}k_1 + k_2, x_n + \tfrac{2}{3}\Delta x) \\
k_4 &= \Delta x f(y_n + k_1 - k_2 + k_3, x_n + \Delta x)
\end{aligned} \qquad (1.5.11)$$

The Runge–Kutta methods are applicable to a higher order ordinary differential equations. For illustration, we apply the Runge–Kutta method to a second order ordinary differential equation:

$$y'' + q(x)y' + r(x)y = s(x) \qquad (1.5.12)$$

with initial values $y(0) = y_0$ and $y'(0) = y_0'$. By defining $z = y'$, the second order differential equation is transformed to coupled first order differential equations as

$$\left. \begin{array}{l} z' = -ry - qz + s \equiv g(y, z, x) \\ y' = z \equiv f(z) \end{array} \right\} \qquad (1.5.13)$$

If we write the second equation of Eq. (1.5.13) more generally as

$$y' = f(y, z, x)$$

the fourth order Runge–Kutta scheme for Eq. (1.5.13) is

$$\left. \begin{array}{l} y_{n+1} = y_n + \frac{1}{6}[k_1 + 2k_2 + 2k_3 + k_4] \\ z_{n+1} = z_{n+1} + \frac{1}{6}[l_1 + 2l_2 + 2l_3 + l_4] \end{array} \right\} \qquad (1.5.14)$$

where

$$k_1 = \Delta x f(y_n, z_n, x_n)$$

$$k_2 = \Delta x f(y_n + \tfrac{1}{2}k_1, z_n + \tfrac{1}{2}l_1, x_n + \tfrac{1}{2}\Delta x)$$

$$k_3 = \Delta x f(y_n + \tfrac{1}{2}k_2, z_n + \tfrac{1}{2}l_2, x_n + \tfrac{1}{2}\Delta x)$$

$$k_4 = \Delta x f(y_n + k_3, z_n + l_3, x_n + \Delta x)$$

The l_i are identical to k_i except that f is replaced by g.

Major concerns with numerical integrating schemes are error and instability. The error of the scheme is due to truncating the higher terms in the Taylor series. The instability is also due to the error of the scheme. If as x increases an error increases unboundedly and behaves in an erratic manner, it is called instability. In order to study instability of a numerical scheme, let us consider the scheme

$$y_{n+1} = y_n + \Delta x y_n' = y_n + \lambda \Delta x y_n \qquad (1.5.15)$$

which is the Euler's method applied to $y' = \lambda y$. With the initial condition, $y_0 = 1$, Eq. (1.5.15) becomes

$$y_{n+1} = \mu y_n = \cdots = \mu^{n+1} y_0 \qquad (1.5.16)$$

where $\mu = 1 + \lambda \Delta x$. The true solution of $y' = \lambda y$ at the grid points is

$$y_{n+1} = (e^{\lambda \Delta x})^{n+1} y_0 \qquad (1.5.17)$$

By comparing Eq. (1.5.16) with Eq. (1.5.17), we find that $\mu = 1 + \lambda \Delta x$ is playing the role of $e^{\lambda \Delta x}$. In fact, μ is the first two terms of the Taylor expansion of $e^{\lambda \Delta x}$:

$$e^{\lambda \Delta x} = 1 + \lambda \Delta x + \tfrac{1}{2}(\lambda \Delta x)^2 + \tfrac{1}{6}(\lambda \Delta x)^3 + \cdots \qquad (1.5.18)$$

If $\lambda \Delta x \to 0$, then μ approaches $e^{\lambda \Delta x}$, so Eq. (1.5.16) converges to the exact solution. On the other hand, if $1 + \lambda \Delta x < -1$, Eq. (1.5.16) will oscillate and diverge as $n \to \infty$, while Eq. (1.5.17) approaches zero. This erratic oscillation is the instability. If $\lambda \Delta x > -2$, no such oscillation appears, although accuracy of the approximation is deteriorated as $|\lambda \Delta x|$ increases.

Generally, instability of a numerical scheme for ordinary differential equations depends on the nature of the differential equation to be integrated. However, general trend of instability of a scheme may be studied by the simplest equation, $y' = \lambda y$. Instability of Eq. (1.5.10) for $y' = \lambda y$ may be studied as follows. By seeking the solution of Eq. (1.5.10) in the form of $y_n = \mu^n y_0$, where μ is undetermined, each step of Eq. (1.5.10) is written as

$$k_1 = \lambda \Delta x y_n$$

$$k_2 = \lambda \Delta x \left(y_n + \frac{k_1}{2} \right) = \lambda \Delta x (1 + \tfrac{1}{2}\lambda \Delta x) y_n$$

$$k_3 = \lambda \Delta x \left(y_n + \frac{k_2}{2} \right) = \lambda \Delta x (1 + \tfrac{1}{2}\lambda \Delta x (1 + \tfrac{1}{2}\lambda \Delta x)) y_n$$

$$k_4 = \lambda \Delta x (y_n + k_3) = \lambda \Delta x (1 + \lambda \Delta x (1 + \tfrac{1}{2}\lambda \Delta x (1 + \tfrac{1}{2}\lambda \Delta x))) y_n$$

and finally

$$y_{n+1} = [1 + \lambda \Delta x + \tfrac{1}{2}(\lambda \Delta x)^2 + \tfrac{1}{6}(\lambda \Delta x)^3 + \tfrac{1}{24}(\lambda \Delta x)^4] y_n \qquad (1.5.19)$$

Therefore, we find

$$\mu = 1 + \lambda \Delta x + \tfrac{1}{2}(\lambda \Delta x)^2 + \tfrac{1}{6}(\lambda \Delta x)^3 + \tfrac{1}{24}(\lambda \Delta x)^4 \qquad (1.5.20)$$

Instability occurs if $\lambda \Delta x < -2.785$.

PREDICTOR-CORRECTOR METHODS

In the Runge–Kutta method, matching higher terms of the Taylor series for y_{n+1} is performed by using additional steps, and y_{n+1} is computed whenever y_n is given, but no previous information other than y_n is used. In contrast to this, the predictor-corrector method uses the previous information, y_n, $y_{n-1}, \ldots, y_{n-k}$ and $y'_n, y'_{n-1}, \ldots, y'_{n-k}$, instead of the additional steps.

The simplest predictor-corrector equation is the Nystrom-trapezoidal rule predictor-corrector method given by

$$\bar{y}_{n+1} = y_{n-1} + 2\Delta x y'_n \qquad\qquad \text{Predictor} \qquad (1.5.21)$$

$$y_{n+1} = y_n + \frac{\Delta x}{2}(\bar{y}'_{n+1} + y'_n) \qquad \text{Corrector} \qquad (1.5.22)$$

The predictor is first evaluated, and $\bar{y}'_{n+1}$ is calculated by

$$\bar{y}'_{n+1} = f(\bar{y}_{n+1}) \qquad\qquad (1.5.23)$$

Then a corrected value for y_{n+1} is calculated by Eq. (1.5.22). This completes one cycle of calculations, and now one can proceed to the next interval. The value of y_{n+1} thus obtained is not satisfying the implicit equation

$$y_{n+1} = y_n + \frac{\Delta x}{2}(y'_{n+1} + y'_n) \qquad\qquad (1.5.24)$$

The value of y_{n+1} that satisfies the above equation may be obtained by additional times of iterations. Namely, whenever an approximation for y_{n+1} is obtained, y'_{n+1} is reevaluated by $y'_{n+1} = f(y_{n+1})$. The next approximation is obtained by introducing y'_{n+1} into the right side of Eq. (1.5.22). This process converges after a few iterations.

Another form of the extrapolation formula for a predictor is the osculating polynomial, which we express by

$$y(x) = \sum_{k=0}^{K} a_{kn}(x_{n+1} - x)^k \qquad\qquad (1.5.25)$$

where a_{kn} is a coefficient and determined so that $y(x)$ and $y'(x)$ fits $k+1$ of $y_n, y_{n-1}, y_{n-2}, \ldots, y'_n, y'_{n-1} \ldots$. Depending upon how y_{n-k} and y'_{n-k} are chosen, various predictor formulas are generated. For example, if we choose $y_n, y'_n, y'_{n-1}, y'_{n-2}$, and y'_{n-3}, Eq. (1.5.25) for $x = x_{n+1}$ may be expressed as

$$y_{n+1} = \alpha_1 y_n + \Delta x(\beta_1 y'_n + \beta_2 y'_{n-1} + \beta_3 y'_{n-2} + \beta_4 y'_{n-3}) \qquad (1.5.26)$$

If we make Eq. (1.5.26) exact for $y(x) = 1, x, x^2, x^3$, and x^4, we obtain

$$y_{n+1} = y_n + \frac{\Delta x}{24}(55y'_n - 59y'_{n-1} + 37y'_{n-2} - 9y'_{n-3}) \qquad (1.5.27)$$

This predictor is known as one of the Adams–Bashforth formulas and is correct if $y(x)$ is a polynomial of order 4.

A corrector based on the osculating polynomial may be derived in the same way as for a predictor except that $y(x)$ is fitted to y'_{n+1} also. For example, if $y_n, y'_{n+1}, y'_n, y'_{n-1}$, and y'_{n-2} are selected and the polynomial is made exact for $y(x) = 1, x, x^2, x^3$, and x^4, we obtain

$$y_{n+1} = y_n + \frac{\Delta x}{24}(9y'_{n+1} + 19y'_n - 5y'_{n-1} + y'_{n-2}) \qquad (1.5.28)$$

This is known as one of the Adams–Moulton formulas.

A few popularly known predictor-corrector methods are summarized in Table 1.3.

One of the problems of the predictor-corrector methods is that the initial condition to the given ordinary differential equation is not enough to start the numerical integration. An additional set of points $x_1, x_2 \ldots$ is required, which must be treated by some other means, for example, a Runge–Kutta method.

Stability of a predictor-corrector method is studied in a similar manner as the Runge–Kutta method. Let us study the stability of a predictor

$$y_{n+1} = y_{n-1} + 2\Delta x y'_n \qquad (1.5.29)$$

applied to $y' = \lambda y$ and $y(0) = 1$ without a corrector. Seeking the solution in the form $y_n = \mu^n y(0)$, μ is found to satisfy

$$\mu^2 = 1 + 2\lambda\,\Delta x\mu \qquad (1.5.30)$$

There are two roots for the above equations.

$$\mu_1 = \lambda\,\Delta x + \sqrt{1 + (\lambda\,\Delta x)^2} = 1 + \lambda\,\Delta x + \tfrac{1}{2}(\lambda\,\Delta x)^2 + O(\Delta x^4) \qquad (1.5.31)$$

$$\mu_2 = \lambda\,\Delta x - \sqrt{1 + (\lambda\,\Delta x)^2} = -1 + \lambda\,\Delta x - \tfrac{1}{2}(\lambda\,\Delta x^2) + O(\Delta x^4) \qquad (1.5.32)$$

Since μ_1^n and μ_2^n are both independent solutions, y_n is a linear combination of those and given by

$$y_n = a\mu_1^n + b\mu_2^n \qquad (1.5.33)$$

where a and b are coefficients determined by the initial condition $y(0) = 1$ and the approximation to y'_1 to start the scheme. Comparing Eq. (1.5.33) with the true solution given by Eq. (1.5.17), we find that the first term is an approximation to the true solution, while the second term is an error and irrelevant to the true solution. If $b \neq 0$ and $\lambda < 0$, then $|\mu_2^n| \to \infty$ as n increases. As a result, Eq. (1.5.33) will oscillate and diverge even if the true solution approaches zero. μ_2 is called a parasitic root.

We now analyze the stability of a simple predictor-corrector set:

$$\bar{y}_{n+1} = y_{n-1} + 2\Delta x y'_n \qquad (1.5.34)$$

$$y_{n+1} = y_n + \frac{\Delta x}{2}(\bar{y}'_{n+1} + y'_n) \qquad (1.5.35)$$

Table 1.3 Commonly Used Predictor-Corrector Methods

Combinations of the predictor and corrector	Formulas	Instability range $\lambda\Delta x$ (Real)[a]
(1) Milne predictor	$\bar{y}_{n+1} = y_{n-3} + \dfrac{4\Delta x}{3}(2y'_n - y'_{n-1} + 2y'_{n-2})$	$-0.3 < \lambda\Delta x \leq 0$ and
Milne corrector	$y_{n+1} = y_{n-1} + \dfrac{\Delta x}{3}(\bar{y}'_{n+1} + 4y'_n + y'_{n-1})$	$-\infty < \lambda\Delta x < -0.8$
(2) Hermite predictor	$\bar{y}_{n+1} = -4y_n + 5y_{n-1} + \Delta x(4y'_n + 2y'_{n-1})$	$\lambda\Delta x < -1$
Milne corrector	$y_{n+1} = y_{n-1} + \dfrac{\Delta x}{3}(\bar{y}'_{n+1} + 4y'_n + y'_{n-1})$	
(3) Milne predictor	$\bar{y}_{n+1} = y_{n-3} + \dfrac{4\Delta x}{3}(2y'_n - y'_{n-1} + 2y'_{n-2})$	$\lambda\Delta x < -0.5 \ (s=1)$
Hamming corrector	$y_{n+1} = \dfrac{1}{8}(9y_n - y_{n-2}) + \dfrac{3\Delta x}{8}(\bar{y}'_{n+1} + 2y'_n - y'_{n-1})$	$\lambda\Delta x < -0.9 \ (s=2)$
(4) Third order Adams	$\bar{y}_n + 1 = y_n + \dfrac{\Delta x}{12}(23\bar{y}'_n - 16y'_{n-1} + 5y'_{n-2})$	$\lambda\Delta x < -1.8 \ (s=1)$
Third order Adams	$y_{n+1} = y_n + \dfrac{\Delta x}{12}(5\bar{y}'_{n+1} + 8y'_n - y'_{n-1})$	$\lambda\Delta x < -1.3 \ (s=2)$

[a] $s = 1$, corrector used once. $s = 2$, corrector used twice in an iterative manner.

This set is called the Nystrom-trapezoidal-rule predictor-corrector method. Applying this scheme to $y' = \lambda y$ yields

$$y_{n+1} = \left[1 + \frac{\lambda \Delta x}{2} + (\lambda \Delta x)^2 \right] y_n + \frac{\lambda \Delta x}{2} y_{n-1} \tag{1.5.36}$$

Seeking the solution of Eq. (1.5.36) in the form $y_n = \mu^n$, the characteristic equation becomes

$$\mu^2 - \left[1 + \frac{\lambda \Delta x}{2} + (\lambda \Delta x)^2 \right] \mu - \frac{\lambda \Delta x}{2} = 0 \tag{1.5.37}$$

The roots are

$$\mu_1 = 1 + \lambda \Delta x + \frac{\lambda^2 \Delta x^2}{2} + \frac{3\lambda^3 \Delta x^3}{12} + O(\Delta x^4) \tag{1.5.38}$$

$$\mu_2 = -\frac{\lambda \Delta x}{2} + O(\Delta x^2) \tag{1.5.39}$$

We again see that $\mu_2 > 1$ if $\lambda < -2/\Delta x$.

Stability limits of the popularly known corrector-predictor methods are shown in the third column of Table 1.3.

OTHER METHODS OF INTEREST

The two methods described in this section are the most popular methods for integrating ordinary differential equations. There are, however, other methods or means for integrating ordinary differential equations. A few useful and interesting methods are briefly introduced next.

1. Burlisch–Store method.[20,21] This method is based on extrapolation of the solution by means of rational functions.
2. Lie series method.[22,23] This method uses as many terms of the Taylor series as required to obtain an accurate numerical solution on discrete points.
3. Weighted residual method.[24] The solution is expressed in the form of a piecewise polynomial, and the coefficients are determined by the weighted residual method.
4. Analog computer. Many engineers once abandoned the use of analog computers because of their low accuracy. However, the accuracy of analog computers has significantly improved in recent years. They can handle nonlinear terms easily and economically for certain class of problems. The applications of analog computers as analog-digital hybrid

computers to initial and coupled initial value problems is attracting a growing interest.[32–34]

5. CSMP.[35,36] This is the computer code package based on the Runge–Kutta method that solves initial value problems with a minimal effort of programming.

1.6 VECTOR AND MATRIX

An array of N numbers may be expressed either as a column or row vector of order N. In this book a column vector representing a columnwise array of N numbers is defined by

$$\mathbf{x} = \begin{bmatrix} x_1 \\ x_2 \\ \vdots \\ x_N \end{bmatrix} \tag{1.6.1}$$

In order to save space, the above equation may be written as

$$\mathbf{x} = \text{col}\,[x_1, x_2, \ldots, x_N] \tag{1.6.2}$$

The row wise array of the same numbers is expressed by $\mathbf{x}^T$, where the superscript T denotes the transpose of the vector $\mathbf{x}$ and means

$$\mathbf{x}^T = [x_1, x_2, \ldots, x_N] \tag{1.6.3}$$

The null vector, the unit vector, and the sum vector are defined, respectively, by

NULL VECTOR

$$\mathbf{0} = \text{col}\,[0, 0, \ldots, 0] \tag{1.6.4}$$

UNIT VECTORS

$$\mathbf{e}_1 = \text{col}\,[1, 0, \ldots, 0] \tag{1.6.5}$$

$$\mathbf{e}_k = \text{col}\,[\underbrace{0, 0, \ldots, 1, 0, \ldots, 0}_{k}] \tag{1.6.6}$$

SUM VECTOR

$$\mathbf{1} = \text{col}\,[1, 1, \ldots, 1] \tag{1.6.7}$$

Some basic operations and properties of vectors are as follows:

a. Vector addition and subtraction. The addition and subtraction of two
   vectors are defined by

$$\mathbf{x} \pm \mathbf{y} = \begin{bmatrix} x_1 \pm y_1 \\ x_2 \pm y_2 \\ \vdots \\ x_N \pm y_N \end{bmatrix} \tag{1.6.8}$$

b. Scalar multiplication.

$$c\mathbf{x} = \mathbf{x}c = \text{col}\,[cx_1, \ldots, cx_N] \tag{1.6.9}$$

c. Equality. Two vectors $\mathbf{x}$ and $\mathbf{y}$ are equal only if $x_k = y_k$ for all of $k = 1$,
   $2, \ldots, N$.

d. Scalar product of two vectors. The scalar product of two vectors $\mathbf{x}$ and $\mathbf{y}$ is
   written and defined by

$$\mathbf{x}^T \cdot \mathbf{y} = \mathbf{y}^T \cdot \mathbf{x} = \sum_{k=1}^{N} x_k y_k \tag{1.6.10}$$

The transpose of a column vector (namely, a row vector) is premultiplied.
The dot between two vectors is often omitted. The scalar product of two
vectors are defined by

$$\langle \mathbf{x}, \mathbf{y} \rangle \equiv \sum_{k=1}^{N} x_k y_k \tag{1.6.11}$$

e. Orthogonality. Two vectors, $\mathbf{x}$ and $\mathbf{y}$, are orthogonal if the scalar product
   is zero, namely $\langle \mathbf{x}, \mathbf{y} \rangle = 0$.

f. Inner product. If $\mathbf{x}$ and $\mathbf{y}$ are allowed to include complex elements, the
   inner product of $\mathbf{x}$ and $\mathbf{y}$ is defined by

$$(\mathbf{x}, \mathbf{y}) = \sum_n \bar{x}_n y_n$$

where $\bar{x}$ is the complex conjugate of x. The inner product of two real
vectors is identical with the scalar product.

g. Norm. Euclidian norm of a real vector $\mathbf{x}$ is defined as

$$\|\mathbf{x}\| = \langle \mathbf{x}, \mathbf{x} \rangle^{1/2} = \left(\sum_{k=1}^{N} x_k^2 \right)^{1/2} \tag{1.6.12}$$

It is seen that the Euclidian norm can be physically interpreted as the "length" of the vector.

Matrix $\mathbf{A}$ is a rectangular array of numbers and defined by

$$\mathbf{A} \equiv \begin{bmatrix} a_{11}a_{12}\ldots a_{1N} \\ a_{21}a_{22}\ldots \\ \vdots \\ a_{M1} \qquad a_{MN} \end{bmatrix} \tag{1.6.13}$$

where a_{ij} is called the element in the ith row and jth column. The right side of Eq. (1.6.13) may be abbreviated by $[a_{ij}]$ or (a_{ij}). Frequently, the elements of $\mathbf{A}$ are denoted by A_{ij} or $(\mathbf{A})_{ij}$: $A_{ij} = (\mathbf{A})_{ij} = a_{ij}$. A comma may be used between subscripts i and j as, for example, $A_{i,j}$ or $a_{i,j}$. A column vector or row vector can be considered as special cases of matrices. When $N = M$, the matrix is called a square matrix of order N. In the remainder of this book, a matrix always means a square matrix.

A list of special forms of matrices of great importance follows:

$\mathbf{A}^T =$ TRANSPOSE OF $\mathbf{A}$

The transpose of $\mathbf{A}$ is defined by $(\mathbf{A}^T)_{ij} = (\mathbf{A})_{ji}$.

$\mathbf{I} =$ IDENTITY MATRIX

All the diagonal elements of a unit matrix are 1, and off-diagonal elements are 0.

$\mathbf{A}^{-1} =$ INVERSE OF $\mathbf{A}$

$$\mathbf{A}^{-1}\mathbf{A} = \mathbf{A}\mathbf{A}^{-1} = \mathbf{I} \tag{1.6.14}$$

$\mathbf{0} =$ NULL MATRIX

Matrix $\mathbf{A}$ is a null matrix if all the elements are zero.

$\mathbf{A}^* =$ CONJUGATE TRANSPOSE OF $\mathbf{A}$

The conjugate transpose of $\mathbf{A}$ is defined by

$$\mathbf{A}^* = [\bar{a}_{ji}] \tag{1.6.15}$$

where $\bar{a}_{ji}$ is the complex conjugate of a_{ji}.

SYMMETRIC MATRIX

Matrix A is symmetric if

$$a_{ij} = a_{ji} \text{ or } \mathbf{A}^T = \mathbf{A}$$

DIAGONAL MATRIX

Matrix $\mathbf{A}$ is diagonal if all the off-diagonal elements are zero: $a_{ij} = 0$ for $i \neq j$.

a. Matrix addition and subtraction. Matrix addition and subtraction are defined by

$$\mathbf{C} = \mathbf{A} + \mathbf{B} = [(a_{ij} + b_{ij})], \; c_{ij} = a_{ij} + b_{ij}$$

$$\mathbf{C} = \mathbf{A} - \mathbf{B}, \; c_{ij} = a_{ij} - b_{ij}$$

b. Scalar multiplication.

$$\mathbf{C} = \alpha \mathbf{A} = \mathbf{A} \alpha$$

$$c_{ij} = \alpha a_{ij}$$

c. Equality. Two matrices $\mathbf{A}$ and $\mathbf{B}$ are equal only if $a_{ij} = b_{ij}$.

d. Product of two matrices. The product of two matrices $\mathbf{A}$ and $\mathbf{B}$ are defined by

$$\mathbf{C} = \mathbf{AB} \text{ where } c_{ij} = \sum_{k=1}^{N} a_{ik} b_{kj}$$

or

$$\mathbf{D} = \mathbf{BA} \text{ where } d_{ij} = \sum_{k=1}^{N} b_{ik} a_{kj}$$

In general, $\mathbf{AB}$ is not equal to $\mathbf{BA}$. However, if $\mathbf{AB} = \mathbf{BA}$ the two matrices $\mathbf{A}$ and $\mathbf{B}$ are said to be commutable.

e. Product of a matrix and a vector. The product of a matrix $\mathbf{A}$ and a vector $\mathbf{x}$ is a vector,

$$\mathbf{y} = \mathbf{Ax} \text{ where } y_i = \sum_{j=1}^{N} a_{ij} x_j$$

f. Determinant. The determinant of a matrix $\mathbf{A}$ of order N is written as $\det [\mathbf{A}]$ or $|\mathbf{A}|$ and defined by

$$\det [\mathbf{A}] \equiv |\mathbf{A}| \equiv \sum (\pm) a_{1i} a_{2j} \ldots a_{Nr}$$

where the sum is taken over all permutations of the second subscripts. $(\pm)$ takes a plus sign if $(i, j, \ldots, r)$ is an even permutation of $(1, 2, \ldots, N)$ and a minus sign if odd.

g. Rank of a matrix. The rank of a square matrix is the maximum number of independent columns. The rank of a matrix $\mathbf{A}$ is equal to that of $\mathbf{A}^T$: in other words, the number of independent columns of $\mathbf{A}$ is always identical with the number of independent rows. If the rank of $\mathbf{A}$ is equal to the order of $\mathbf{A}$, then the determinant of $\mathbf{A}$ is nonzero, $|\mathbf{A}| \neq 0$. If $|\mathbf{A}| = 0$, the rank of $\mathbf{A}$ is less than N and $\mathbf{A}$ is said to be singular. In order for the inverse of $\mathbf{A}$ to exist, $\mathbf{A}$ must not be singular: $|\mathbf{A}| \neq 0$.

h. Positive definite matrix. A real matrix $\mathbf{A}$ is called positive definite if

$$(\mathbf{x}, \mathbf{Ax}) > 0$$

for any nonnull vector $\mathbf{x}$. The real matrix $\mathbf{A}$ is positive definite if and only if it is symmetric and all its eigenvalues are positive (see Theorem 1.1 in Section 1.7).

1.7 LINEAR EQUATIONS

A set of linear equations with N unknowns may be written

$$\sum_{j=1}^{N} a_{ij}x_j = y_i, \qquad i = 1, 2, \ldots, N \tag{1.7.1}$$

where a_{ij} are known coefficients and y_i are known values. If all of y_i are zero, Eq. (1.7.1) is said to be homogeneous, and if at least one of y_i is not zero, it is said to be inhomogeneous. Only inhomogeneous systems are considered in this section. An inhomogeneous system can have a solution if the system is consistent. There is no solution if the system is inconsistent.

Example 1.6

The following system is inconsistent and has no solution:

$$\begin{cases} 2x + 2y = 1 \\ 2x + 2y = 3 \end{cases}$$

Provided that the linear system is consistent, a unique solution of Eq. (1.7.1) exists if and only if all of the N equations in Eq. (1.7.1) are linearly independent (none of the N equations can be obtained by additions and subtractions of the rest of equations). The unique solution of Eq. (1.7.1) can be directly obtained by the Gaussian elimination method as follows. Dividing the first equation by a_{11} and eliminating x_1 in the remaining $N-1$ equations

$$x_1 + \frac{a_{12}}{a_{11}}x_2 + \cdots + \frac{a_{1N}}{a_{11}}x_N = \frac{y_1}{a_{11}}$$

$$\left[a_{22} - a_{21}\left(\frac{a_{12}}{a_{11}}\right)\right]x_2 + \cdots + \left[a_{2N} - a_{21}\left(\frac{a_{1N}}{a_{11}}\right)\right]x_N = y_2 - a_{21}\frac{y_1}{a_{11}} \tag{1.7.2}$$

$$\left[a_{N2} - a_{N1}\left(\frac{a_{12}}{a_{11}}\right)\right]x_2 + \cdots + \left[a_{NN} - a_{N1}\left(\frac{a_{1N}}{a_{11}}\right)\right]x_N = y_N - a_{N1}\frac{y_1}{a_{11}}$$

or

$$x_1 + a'_{12}x_2 + \cdots + a'_{1N}x_N = y'_1$$
$$a'_{22}x_2 + \cdots + a'_{2N}x_N = y'_2$$
$$\vdots \tag{1.7.3}$$
$$a'_{N2}x_2 + \cdots + a'_{NN}x_N = y'_N$$

The second equation is divided by a'_{22}, and x_2 is eliminated in the third through the Nth equations. Then x_3 is eliminated in the fourth through the Nth equations. Finally, we obtain the system in the form:

$$x_1 + h_{12}x_2 + \cdots + h_{1N}x_N = g_1$$
$$x_2 + \cdots + h_{2N}x_N = g_2$$
$$\vdots \tag{1.7.4}$$
$$x_{N-1} + h_{N-1,N}x_N = g_{N-1}$$
$$x_N = g_N$$

The value of x_N is obtained immediately, $x_N = g_N$. This value of x_N is substituted to the $(N-1)$th equation to obtain

$$x_{N-1} = g_{N-1} - h_{N-1,N}g_N \tag{1.7.5}$$

This substitution continues until all of x_i are obtained.

$$x_j = g_j - \sum_{k=j+1}^{N} h_{jk}x_k, \quad j = N-2, N-3, \cdots, 1 \tag{1.7.6}$$

Equation (1.7.1) can be more simply expressed if matrix and vector notations are used as

$$\mathbf{Ax} = \mathbf{y} \tag{1.7.7}$$

where $\mathbf{A}$ is a matrix, which is a square array of coefficients:

$$\mathbf{A} = \begin{bmatrix} a_{11}a_{12} \ldots a_{1N} \\ a_{21} \\ \vdots \\ a_{N1} \qquad\qquad a_{NN} \end{bmatrix} \tag{1.7.8}$$

and, $\mathbf{x}$ and $\mathbf{y}$ are column vectors,

$$\mathbf{x} = \begin{bmatrix} x_1 \\ x_2 \\ \vdots \\ x_N \end{bmatrix}, \qquad \mathbf{y} = \begin{bmatrix} y_1 \\ y_2 \\ \vdots \\ y_N \end{bmatrix} \tag{1.7.9}$$

In terms of rank, Eq. (1.7.7) has a unique solution if the rank of $\mathbf{A}$ is equal to the order of matrix $\mathbf{A}$: $\det \mathbf{A} \neq 0$. We assume consistency of Eq. (1.7.7). The solution of Eq. (1.7.7) can be written as

$$\mathbf{x} = \mathbf{A}^{-1}\mathbf{y}$$

where $\mathbf{A}^{-1}$ is the inverse of $\mathbf{A}$.

When $\mathbf{A}$ is a *tridiagonal matrix*, that is,

$$A = \begin{bmatrix} b_1, c_1 & & & \\ a_2, b_2, c_2 & & & \\ & & \ddots & \\ & & & a_N, b_N \end{bmatrix} \tag{1.7.10}$$

the inversion operation is very simple, but it is important in numerical analyses of boundary value problems of differential equations. The following is the Gaussian elimination for a tridiagonal matrix.

The first equation is divided by b_1:

$$\begin{bmatrix} 1 & c_1/b_1 & & \\ a_2 & b_2 & c_2 & \\ & a_3 & b_3 & c_3 \\ & & & \ddots \end{bmatrix} \begin{bmatrix} x_1 \\ x_2 \\ x_3 \\ \vdots \end{bmatrix} = \begin{bmatrix} y_1/b_1 \\ y_2 \\ y_3 \\ \vdots \end{bmatrix} \tag{1.7.11}$$

The first equation times a_2 is subtracted from the second equation:

$$\begin{bmatrix} 1 & c_1/b_1 & 0 & \\ 0 & (b_2 - a_2c_1/b_1) & c_2 & \\ 0 & a_3 & b_3 & c_3 \\ & & & \ddots \end{bmatrix} \begin{bmatrix} x_1 \\ x_2 \\ x_3 \\ \vdots \end{bmatrix} = \begin{bmatrix} y_1/b_1 \\ y_2 - a_2y_1/b_1 \\ y_3 \\ \vdots \end{bmatrix} \tag{1.7.12}$$

The second equation is divided by the second diagonal element:

$$\begin{bmatrix} 1 & c_1/b_1 & 0 & & \\ 0 & 1 & c_2/(b_2-a_2c_1/b_1) & & \\ 0 & a_3 & b_3 & c_3 & \\ & \diagdown & \diagdown & \diagdown & \diagdown \end{bmatrix} \begin{bmatrix} x_1 \\ x_2 \\ x_3 \\ \vdots \end{bmatrix} = \begin{bmatrix} y_1/b_1 \\ \dfrac{y_2-a_2y_1/b_1}{b_2-a_2c_1/b_1} \\ y_3 \\ \vdots \end{bmatrix} \qquad (1.7.13)$$

By repeating this process, the matrix equation is reduced to

$$\begin{bmatrix} 1 & h_1 & & & & \\ & 1 & h_2 & & & \\ & & 1 & h_3 & & \\ & & & \diagdown & \diagdown & \\ & & & & h_{N-1} \\ & & & & & 1 \end{bmatrix} \begin{bmatrix} x_1 \\ x_2 \\ x_3 \\ \vdots \\ x_{N-1} \\ x_N \end{bmatrix} = \begin{bmatrix} p_1 \\ p_2 \\ p_3 \\ \vdots \\ p_{N-1} \\ p_N \end{bmatrix} \qquad (1.7.14)$$

where h_j and p_j satisfy the recursion formulas:

$$\begin{aligned} h_1 &= \frac{c_1}{b_1}, & h_j &= \frac{c_j}{b_j - a_jh_{j-1}}, \\ p_1 &= \frac{y_1}{b_1}, & p_j &= \frac{y_j - a_jp_{j-1}}{b_j - a_jh_{j-1}} \end{aligned} \qquad (1.7.15)$$

Then backward substitution gives the solution for x_j:

$$\begin{aligned} x_N &= p_N \\ x_j &= p_j - h_jx_{j+1}, \qquad j = N-1, \ldots, 3, 2, 1 \end{aligned} \qquad (1.7.16)$$

If not all of the equations in Eq. (1.7.1) are linearly independent, there is no unique solution, and the number of solutions is infinite. Let us consider the meaning of this. Suppose k among N equations can be obtained by additions and subtractions of the others and $N-k$ equations are linearly dependent. $N-k$ equations are not sufficient to determine N unknowns as is obvious from the procedure of Eq. (1.7.2) through Eq. (1.7.6). However, if k among N unknowns are specified, then $N-k$ equations are sufficient to determine uniquely the rest of the $N-k$ unknowns. Since the specifications of k values are arbitrary, the total number of solutions of Eq. (1.7.1) is infinite.

We have presumed in this section that a_{ij}, x_i, and y_i in Eq. (1.7.1) are all numbers. However, occasionally it happens that a_{ij} are $\kappa \times \kappa$ matrices and x_i and y_i are vectors of order κ (κ is an integer). If such linear equations are expressed by Eq. (1.7.7), then $\mathbf{A}$ is called a block matrix; $\mathbf{x}$ and $\mathbf{y}$ are called block vectors. For example, if a_i, b_i, and c_i in Eq. (1.7.10) are blocks ($\kappa \times \kappa$ submatrices), $\mathbf{A}$ is called a tridiagonal-block matrix. The solution of Eq. (1.7.7) with a block matrix $\mathbf{A}$ is exactly the same as for an ordinary equation except that the arithmetic of numbers is replaced by the arithmetic for submatrices and subvectors. The recursion formula for the solution of a block-tridiagonal equation is given by

$$\mathbf{h}_1 = \mathbf{b}_1^{-1}\mathbf{c}_1, \qquad \mathbf{h}_j = (\mathbf{b}_j - \mathbf{a}_j\mathbf{h}_{j-1})^{-1}\mathbf{c}_j, \qquad\qquad j > 1$$

$$\mathbf{p}_1 = \mathbf{b}_1^{-1}\mathbf{y}_1, \qquad \mathbf{p}_j = (\mathbf{b}_j - \mathbf{a}_j\mathbf{h}_{j-1})^{-1}(\mathbf{y}_j - \mathbf{a}_j\mathbf{p}_{j-1}), \quad j > 1 \qquad (1.7.17)$$

$$\mathbf{x}_N = \mathbf{p}_N, \qquad \mathbf{x}_j = \mathbf{p}_j - \mathbf{h}_j\mathbf{x}_{j-1}, \qquad\qquad\qquad j < N$$

where $\mathbf{h}_j$ is a $\kappa \times \kappa$ square matrix, $\mathbf{p}_j$ is a vector of order κ, and $\mathbf{b}^{-1}$ denotes the inverse of matrix $\mathbf{b}$.

1.8 EIGENVALUES AND EIGENVECTORS

When the right side of Eq. (1.7.1) is zero the linear system is said to be homogeneous. Suppose we have a solution for the homogeneous system. The solution $\{x_i\}$, $i = 1, 2, \ldots, N$ is arbitrary by a constant. In fact, if $\{x_i\}$, $i = 1, 2, \ldots, N$ are solutions, $\{cx_i\}$, $i = 1, 2, \ldots, N$ are also solutions. This means that we can arbitrarily specify at least one of x_is, say x_N, and solve the rest of unknowns in terms of x_N. If we specify the value of x_N, then we have N equations for $N-1$ unknowns, so that one equation becomes abundant. In order that the system be consistent and at least one solution exist, one of the N equations must be linearly dependent to others: in terms of rank, the rank of the coefficient matrix must be lower than N, and accordingly the determinant of the coefficient matrix must be zero:

$$\det[\mathbf{A}] \equiv |\mathbf{A}| = 0 \qquad\qquad (1.8.1)$$

where

$$\mathbf{A} = \begin{bmatrix} a_{11}, a_{12}, \ldots, a_{1N} \\ a_{21}, \\ \vdots \\ a_{N1}, \ldots \qquad a_{NN} \end{bmatrix} \qquad\qquad (1.8.2)$$

When the right side of Eq. (1.7.1) is zero, a solution exists only if Eq. (1.8.1) is satisfied.

Let us consider a homogeneous system of equations

$$(a_{11} - \lambda)x_1 + a_{12}x_2 + \cdots + a_{1N}x_n = 0$$

$$a_{21}x_1 + (a_{22} - \lambda)x_2 + \cdots + a_{2N}x_n = 0 \qquad (1.8.3)$$

$$\vdots$$

$$a_{N1}x_1 + \qquad \cdots \qquad (a_{NN} - \lambda)x_N = 0$$

where λ is an undetermined parameter called *eigenvalue*. Since λ is undetermined, one can adjust λ so that the determinant of the coefficient matrix becomes zero. Equation (1.8.3) is known as an eigenvalue problem. In the matrix and vector notations, Eq. (1.8.3) may be expressed by

$$(\mathbf{A} - \lambda \mathbf{I})\mathbf{x} = 0 \qquad (1.8.4)$$

where $\mathbf{A}$ is defined by Eq. (1.8.2), $\mathbf{I}$ is the identity matrix, and

$$\mathbf{x} = \mathrm{col}\,[x_1, x_2, \ldots, x_N]$$

In order for Eq. (1.8.4) to have a nontrivial solution the determinant of the matrix $[\mathbf{A} - \lambda \mathbf{I}]$ must be zero:

$$\det [\mathbf{A} - \lambda \mathbf{I}] \equiv |\mathbf{A} - \lambda \mathbf{I}| = 0 \qquad (1.8.5)$$

If we define the characteristic function as

$$f(\lambda) = |\mathbf{A} - \lambda \mathbf{I}| \qquad (1.8.6)$$

then $f(\lambda)$ is a polynomial of λ of order N. The characteristic equation $f(\lambda) = 0$ has N roots, $\lambda_1, \lambda_2, \ldots, \lambda_N$, which are all eigenvalues of $\mathbf{A}$.

If the number λ is chosen to be one of the roots λ_n of $f(\lambda) = 0$, there is at least one $\mathbf{x} \neq \mathbf{0}$ that satisfies Eq. (1.8.4). If λ_n is a single root, there exists only one $\mathbf{x}$ corresponding to λ_n that is called the *eigenvector* corresponding to λ_n and is denoted by $\mathbf{u}_n$.

1.7.1 Symmetric Real Matrix

For many eigenvalue problems of physical interest, the matrix $\mathbf{A}$ is real and symmetric. The theory for a symmetric matrix is much simpler than for a nonsymmetric matrix, and, in fact, the theory of eigenvalue problems involving symmetric matrices is very important.

The eigenvalues of a real matrix may include complex eigenvalues since a polynomial with real coefficients can have pairs of complex roots. However, if $\mathbf{A}$ is a symmetric matrix, all the eigenvalues of $\mathbf{A}$ are real as shown later.

When $\mathbf{A}$ is symmetric, the eigenvectors corresponding to different eigenvalues are orthogonal:

$$\langle \mathbf{u}_n, \mathbf{u}_m \rangle = 0 \quad \text{for } n \neq m \tag{1.8.7}$$

If all the eigenvalues of $\mathbf{A}$ are different, the set of orthogonal eigenvectors span E^N (N-dimensional Euclidian space). In other words, the eigenvectors make a complete set so that any arbitrary vector, $\mathbf{p}$, in the N-dimensional space can be expanded in the eigenvectors:

$$\mathbf{p} = \sum_{n=1}^{N} b_n \mathbf{u}_n \tag{1.8.8}$$

where b_n is the coefficient determined by $b_n = \langle \mathbf{u}_n, \mathbf{p} \rangle / \langle \mathbf{u}_n, \mathbf{u}_n \rangle$.

Since each eigenvector is arbitrary by a constant, it can be normalized as

$$\langle \mathbf{u}_n, \mathbf{u}_n \rangle = 1 \tag{1.8.9}$$

Equations (1.8.7) and (1.8.9) can be expressed by a single equation as

$$\langle \mathbf{u}_n, \mathbf{u}_m \rangle = \delta_{nm} \tag{1.8.10}$$

where δ_{nm} is Kronecker's delta. The set of eigenvectors satisfying Eq. (1.8.10) is said to be orthonormal (orthogonal and normalized).

If the eigenvalues of $\mathbf{A}$ are not all distinct, the eigenvectors have the following properties:

a. If an eigenvalue λ_j of $\mathbf{A}$ has multiplicity $k \geq 2$, there exists k orthonormal eigenvectors for λ_j.
b. The k eigenvectors with λ_j span a k-dimensional subspace in E^N. An infinite number of different ways of selecting orthogonal sets of k eigenvectors with λ_j is possible.
c. There exists at least one orthonormal set of eigenvectors of $\mathbf{A}$ that span E^N. However, if one or more eigenvalues has multiplicity $k \geq 2$, the number of different sets of orthonormal eigenvectors of $\mathbf{A}$ that span E^N is infinite because of b.

Example 1.7

$$\mathbf{A} = \begin{bmatrix} 2 & 0 & 0 \\ 0 & 1 & 0 \\ 0 & 0 & 1 \end{bmatrix}$$

Eigenvalues are $\lambda_1 = 2$ and $\lambda_2 = 1$ with multiplicity 2. The eigenvector for

$\lambda_1 = 2$ is $\mathbf{u}_1 = \text{col } (1, 0, 0)$. The two orthogonal vectors for λ_2 are apparently

$$\mathbf{u}_2 = \text{col } (0, 1, 0)$$

$$\mathbf{u}_3 = \text{col } (0, 0, 1)$$

Consider now a new normalized vector that is a linear combination of $\mathbf{u}_2$ and $\mathbf{u}_3$:

$$\mathbf{z}_2 = a\mathbf{u}_2 + \sqrt{1-a^2}\,\mathbf{u}_3$$

where a is arbitrary except for $0 < a < 1$. It is seen that $\mathbf{z}_2$ is an eigenvector for λ_2. Let us consider another normalized eigenvector:

$$\mathbf{z}_3 = b\mathbf{u}_2 + \sqrt{1-b^2}\,\mathbf{u}_3$$

where $0 < |b| < 1$. $\mathbf{z}_3$ can be made orthogonal to $\mathbf{z}_2$ by $b = -\sqrt{1-a^2}$. The new set of $\mathbf{u}_1$, $\mathbf{z}_2$, and $\mathbf{z}_3$ thus obtained is orthonormal and complete. The number of such new sets is infinite because a is arbitrary in $[0, 1]$.

If a real symmetric matrix has only positive (real) eigenvalue it is called the positive definite matrix. The following theorem is useful in later chapters.

Theorem 1.1.   A real symmetric matrix $\mathbf{A}$ is positive definite if and only if all its eigenvalues, λ_n, are positive.

Proof.   If $\mathbf{A}$ is real symmetric, all the eigenvalues and eigenvectors are real. Any vector $\mathbf{x}$ including complex elements can be expanded into the orthonormalized eigenvectors of $\mathbf{A}$ as

$$\mathbf{x} = \sum_{n=1}^{N} b_n \mathbf{u}_n$$

where $\mathbf{u}_n$ are orthonormal eigenvectors of $\mathbf{A}$ and $b_n = \langle \mathbf{u}_n, \mathbf{x} \rangle$. If all the eigenvalues are positive, the inner product of $\mathbf{x}$ and $\mathbf{A}\mathbf{x}$ becomes

$$(\mathbf{x}, \mathbf{A}\mathbf{x}) = \sum_m \sum_n \bar{b}_m b_n \lambda_n \langle \mathbf{u}_m, \mathbf{u}_n \rangle = \sum_n |b_n|^2 \lambda_n > 0$$

where $\bar{b}$ is the complex conjugate of b. Thus $\mathbf{A}$ is positive definite by the definition given in Section 1.6. If $(\mathbf{x}, \mathbf{A}\mathbf{x}) > 0$ for any nonnull $\mathbf{x}$, then setting $\mathbf{x}$ to any eigenvector of $\mathbf{A}$ yields

$$(\mathbf{x}, \mathbf{A}\mathbf{x}) = \langle \mathbf{u}_m, \mathbf{A}\mathbf{u}_m \rangle = \lambda_m > 0$$

Thus all the eigenvalues are positive.

1.7.2 Nonsymmetric Real Matrix

A matrix $\mathbf{A}$ with all real elements may have complex eigenvalues if $\mathbf{A}$ is nonsymmetric. The eigenvalues are again the roots of Eq. (1.8.5). For a real matrix the coefficients of $f(\lambda)$ are all real, so that if λ_n is a complex eigenvalue there exists another eigenvalue, λ_k, that is complex conjugate to λ_n: $\lambda_k = \bar{\lambda}_n$. The eigenvector corresponding to a complex eigenvalue always includes complex elements because the elements of $\mathbf{A}$ are all real. If $\lambda_k = \bar{\lambda}_n$, then it follows that

$$\mathbf{u}_k = \bar{\mathbf{u}}_n$$

In the above equation, each element of $\mathbf{u}_k$ is complex conjugate of the corresponding element of $\mathbf{u}_n$.

The properties of eigenvectors of a nonsymmetric matrix are more complicated than a symmetric matrix. We first consider the case where all the eigenvalues are distinct. Unlike the eigenvectors of a symmetric matrix, the eigenvectors of a nonsymmetric matrix are not in general orthogonal to each other. However, the eigenvectors of $\mathbf{A}$ are orthogonal to the adjoint eigenvectors of $\mathbf{A}$, which are defined by the following equation:

$$(\mathbf{A}^T - \lambda\mathbf{I})\mathbf{v} = 0 \qquad (1.8.11)$$

Since the characteristic equation for Eq. (1.8.11) becomes identical with that of Eq. (1.8.4), the eigenvalues of Eq. (1.8.11) are identical to those of Eq. (1.8.4). We denote the eigenvector of Eq. (1.8.11) corresponding to λ_n by $\mathbf{v}_n$. The property of $\mathbf{u}_n$ as an eigenvector is unchanged even if all the elements are multiplied by a constant, and so is about $\mathbf{v}_n$. Therefore, it is possible for us to adjust the magnitude of $\mathbf{u}_n$ or $\mathbf{v}_n$ so that the product becomes unity:

$$\langle \mathbf{v}_n, \mathbf{u}_n \rangle = 1 \qquad (1.8.12)$$

The orthogonality between $\mathbf{u}_n$ and $\mathbf{v}_n$ is shown as follows. Consider two equations

$$\mathbf{A}\mathbf{u}_n = \lambda_n \mathbf{u}_n \qquad (1.8.13)$$

$$\mathbf{A}^T\mathbf{v}_m = \lambda_m \mathbf{v}_m \qquad (1.8.14)$$

where $n \neq m$ and $\lambda_n \neq \lambda_m$ because we assumed all the eigenvectors are distinct. Premultiplying Eq. (1.8.13) and Eq. (1.8.14) by $\mathbf{v}_m^T$ and $\mathbf{u}_n$, respectively, yields

$$\langle \mathbf{v}_m, \mathbf{A}\mathbf{u}_n \rangle = \lambda_n \langle \mathbf{v}_m, \mathbf{u}_n \rangle \qquad (1.8.15)$$

$$\langle \mathbf{u}_n, \mathbf{A}^T\mathbf{v}_m \rangle = \lambda_m \langle \mathbf{u}_n, \mathbf{v}_m \rangle \qquad (1.8.16)$$

Equation (1.8.16) is equivalent to

$$\langle \mathbf{v}_m, \mathbf{A}\mathbf{u}_n \rangle = \lambda_m \langle \mathbf{v}_m, \mathbf{u}_n \rangle \tag{1.8.16a}$$

so that equating the left sides of Eq. (1.8.15) and Eq. (1.8.16a) yields

$$(\lambda_n - \lambda_m)\langle \mathbf{v}_m, \mathbf{u}_n \rangle = 0 \tag{1.8.17}$$

Since $\lambda_m \neq \lambda_n$ if $m \neq n$ by assumption, we must have

$$\langle \mathbf{v}_m, \mathbf{u}_n \rangle = \delta_{mn} \tag{1.8.18}$$

where δ is Kronecker's delta.

Example 1.8

Consider the matrix

$$A = \begin{bmatrix} 2 & 3 \\ 1 & 2 \end{bmatrix}$$

The eigenvalues and the corresponding eigenvectors are

$$\lambda_1 = 2 + \sqrt{3}, \qquad \mathbf{u}_1 = \begin{bmatrix} \sqrt{3} \\ 1 \end{bmatrix}, \quad \mathbf{v}_1 = \frac{1}{2\sqrt{3}}\begin{bmatrix} 1 \\ \sqrt{3} \end{bmatrix}$$

$$\lambda_2 = 2 - \sqrt{3}, \qquad \mathbf{u}_2 = \begin{bmatrix} \sqrt{3} \\ -1 \end{bmatrix}, \quad \mathbf{v}_2 = \frac{1}{2\sqrt{3}}\begin{bmatrix} 1 \\ -\sqrt{3} \end{bmatrix}$$

The two eigenvectors $\mathbf{u}_1$ and $\mathbf{u}_2$ are not orthogonal to each other, but $\mathbf{u}_1$ and $\mathbf{u}_2$ are, respectively, orthogonal to $\mathbf{v}_2$ and $\mathbf{v}_1$. Furthermore, $\langle \mathbf{v}_n, \mathbf{u}_n \rangle = 1$ for $n = 1$ and 2.

When all the eigenvalues of a nonsymmetric matrix $\mathbf{A}$ are distinct, any arbitrary N-dimensional vector $\mathbf{p}$ can be expressed as a linear combination of N linearly independent eigenvectors $\mathbf{u}_n$:

$$\mathbf{p} = \sum_{n=1}^{N} b_n \mathbf{u}_n \tag{1.8.19}$$

In Eq. (1.8.19) b_n is a coefficient and can be determined by using the biorthogonality as

$$b_n = \langle \mathbf{v}_n, \mathbf{p} \rangle \tag{1.8.20}$$

If an arbitrary vector can be expanded in terms of $\mathbf{u}_n$, the set of $\mathbf{u}_n$ is said to make a basis in the N-dimensional space. When the matrix $\mathbf{A}$ is symmetric, then $\mathbf{u}_n$ and $\mathbf{v}_n$ become identical. Therefore, the previous discussion about a symmetric matrix with distinct eigenvalues can be thought as a special case of nonsymmetry matrices.

Next we consider a nonsymmetric matrix having some eigenvalues with multiplicity. There is at least one eigenvector corresponding to each eigenvalue. The eigenvectors corresponding to different eigenvalues are all independent. It is not always possible to find k independent eigenvectors for the eigenvalue with multiplicity k (*eigenvector deficiency*).

Example 1.9

$$\mathbf{A} = \begin{bmatrix} 1 & 1 \\ 0 & 1 \end{bmatrix}$$

The only eigenvalue of $\mathbf{A}$ is $\lambda = 1$ with multiplicity 2. There is no other eigenvector of $\mathbf{A}$ than

$$\mathbf{u} = \begin{bmatrix} 1 \\ 0 \end{bmatrix}$$

Compare the above matrix with

$$\mathbf{A} = \begin{bmatrix} 1 & 0 \\ 0 & 1 \end{bmatrix}$$

which has $\lambda = 1$ as the only eigenvalue with multiplicity 2 and has two independent eigenvectors:

$$\mathbf{u}_1 = \begin{bmatrix} 1 \\ 0 \end{bmatrix} \quad \text{and} \quad \mathbf{u}_2 = \begin{bmatrix} 0 \\ 1 \end{bmatrix}$$

If eigenvector deficiency occurs, the eigenvectors are not sufficient to express an arbitrary vector in the N-dimensional space. We can, however, find supplementary vectors with which the eigenvectors can form a complete set. Suppose λ_1 is an eigenvalue of multiplicity k. As stated before, at least one eigenvector corresponding to λ_1 can be found:

$$\mathbf{A}\mathbf{u}_1 = \lambda_1 \mathbf{u}_1 \tag{1.8.21}$$

Assuming $\mathbf{u}_1$ is the only eigenvector for λ_1, it is possible to find the following sequence of auxiliary vectors

$$(\mathbf{A} - \lambda_1 \mathbf{I})\mathbf{Z}_1 = \mathbf{u}_1 \tag{1.8.22}$$

$$(\mathbf{A} - \lambda_1 \mathbf{I})\mathbf{Z}_l = \mathbf{Z}_{l-1} \quad \text{for } 1 < l < k \tag{1.8.23}$$

The solution of $\mathbf{Z}_1, \ldots, \mathbf{Z}_{k-1}$ is not unique, but it is possible to make each $\mathbf{Z}$ orthogonal to $\mathbf{u}_1$ and other $\mathbf{Z}$s.

Example 1.10

$$A = \begin{bmatrix} 1 & 1 & 0 \\ 0 & 1 & 1 \\ 0 & 0 & 1 \end{bmatrix}$$

The only eigenvalue of A is $\lambda = 1$ with multiplicity 3. A has only one eigenvector, $u_1 = \text{col}\,[1, 0, 0]$, satisfying

$$Au_1 = u_1$$

The solution for Z_1 of Eq. (1.8.22) is

$$Z_1 = \text{col}\,[\text{arbitrary}, 1, 0]$$

Since Z_1 must be orthogonal to u_1, the first element of Z_1 must be 0, thus $Z_1 = \text{col}\,[0, 1, 0]$. Similarly $Z_2 = \text{col}\,[0, 0, 1]$. The set of u_1, Z_1, and Z_2 form a complete set in the three-dimensional space.

It can be shown that vectors thus found are independent of other eigenvectors as follows. Let us assume λ_1 is the only multiple eigenvalue. If we assume any Z_j is not independent of other eigenvectors, Z_j must be expressed by a linear combination of other eigenvectors:

$$Z_j = \sum_{n=k+1}^{N} d_n u_n \qquad (1.8.24)$$

From Eq. (1.8.23) we have

$$u_1 = (A - \lambda_1 I)^j Z_j = \sum_{n=k+1}^{N} d_n (A - \lambda_1 I)^j u_n$$

so that u_1 cannot be independent of other eigenvectors, thus contradicting the linear independence of eigenvectors u_n.

1.9 SIMILARITY TRANSFORMATION AND JORDAN CANONICAL FORM

Two square matrices A and B are said "similar" if there exists a nonsingular matrix P that transforms A to B by

$$B = P^{-1}AP \qquad (1.9.1)$$

The above transformation is called the *similarity transformation*. Since the eigenvalues are not altered through a similarity transformation, an analysis for A may be done equivalently with B. If a similar matrix B that is much simpler than A is found, the analysis for A may become much easier with B.

Any matrix can be reduced to a diagonal form or at least a near-diagonal form called the *Jordan canonical form* by a similarity transformation. If a matrix **A** has a complete set of eigenvectors (assuming no eigenvector deficiency), the Jordan canonical form is a diagonal matrix as shown below. We first consider a symmetric matrix **A**. Let **P** be chosen as

$$\mathbf{P} = \text{row}\,[\mathbf{u}_1, \mathbf{u}_2, \mathbf{u}_3, \ldots, \mathbf{u}_N] \tag{1.9.2}$$

where the jth column of **P** is equal to the jth orthonormal eigenvector $\mathbf{u}_j$. Since the $\mathbf{u}_j$s are orthonormal, it is easily seen that $\mathbf{P}^{-1}$ has the following form

$$\mathbf{P}^{-1} = \text{col}\,[\mathbf{u}_1^T, \mathbf{u}_2^T, \ldots, \mathbf{u}_N^T] \tag{1.9.3}$$

where the jth row of $\mathbf{P}^{-1}$ is the transpose of $\mathbf{u}_j$. Introducing Eqs. (1.9.2) and (1.9.3) into Eq. (1.9.1) yields a diagonal matrix

$$\mathbf{B} = \begin{bmatrix} \lambda_1 & & & \\ & \lambda_2 & & \\ & & \ddots & \\ & & & \lambda_N \end{bmatrix} \tag{1.9.4}$$

The Jordan canonical form of a nonsymmetric matrix is also a diagonal matrix if there is no eigenvector deficiency. In this case, we choose **P** again as Eq. (1.9.2). Assuming $\mathbf{u}_n$ and $\mathbf{v}_n$ are normalized in the sense of Eq. (1.8.18), it is seen that $\mathbf{P}^{-1}$ is given by

$$\mathbf{P}^{-1} = \text{col}\,[\mathbf{v}_1^T, \mathbf{v}_2^T, \ldots, \mathbf{v}_N^T] \tag{1.9.5}$$

Then Eq. (1.9.4) is obtained.

When a nonsymmetric matrix **A** has an eigenvector deficiency, it cannot be transformed to a diagonal matrix. However, we can transform **A** to a near-diagonal form as follows. Suppose only λ_1 has multiplicity k. Since we do not have sufficient eigenvectors to constitute **P**, we use supplementary vectors for λ_1, which are discussed in Section 1.8, and write **P** as

$$\mathbf{P} = \text{row}\,[\mathbf{u}_1, \mathbf{Z}_1, \mathbf{Z}_2, \ldots, \mathbf{Z}_{k-1}, \mathbf{u}_{k+1}, \ldots, \mathbf{u}_N] \tag{1.9.6}$$

where $\mathbf{Z}_1, \mathbf{Z}_2, \ldots, \mathbf{Z}_{k-1}$ are orthonormal supplementary vectors. As easily seen from Eq. (1.8.23), premultiplication of $\mathbf{Z}_j$ by **A** yields

$$\mathbf{A}\mathbf{Z}_1 = \lambda_1 \mathbf{Z}_1 + \mathbf{u}_1 \tag{1.9.7}$$

and

$$\mathbf{A}\mathbf{Z}_j = \lambda_1 \mathbf{Z}_j + \mathbf{Z}_{j-1}, \qquad j = 2, 3, \ldots, k-1 \tag{1.9.8}$$

Therefore, the product **AP** becomes

$$\mathbf{A}\mathbf{P} = [\lambda_1 \mathbf{u}_1, \lambda_1 \mathbf{Z}_1 + \mathbf{u}_1, \ldots, \lambda_1 \mathbf{Z}_j + \mathbf{Z}_{j-1}, \ldots, \lambda_{k+1} \mathbf{u}_{k+1}, \ldots] = \mathbf{P}\mathbf{B} \tag{1.9.9}$$

where

$$
\mathbf{B} = \begin{bmatrix}
\begin{array}{cccc|cc}
\lambda_1 & 1 & & & & \\
& \lambda_1 & 1 & & & \\
& & \lambda_1 & \ddots & 1 & \\
& & & \ddots & \ddots & \\
& & & & \lambda_1 & \\
\hline
& & & & & \lambda_{k+1} & \ddots \\
& & & & & & \ddots & \lambda_N
\end{array}
\end{bmatrix}
\tag{1.9.10}
$$

The $k \times k$ submatrix enclosed by a dotted line is called a *canonical box*, in which all the diagonal elements are λ_1 and the upper off-diagonal line has 1 for all the elements. If there is more than one eigenvalue with deficiency, **B** will have the corresponding number of canonical boxes.

Eigenvectors of a matrix **B** in Jordan canonical form are simple. If **B** is diagonal and expressed by Eq. (1.9.4), the orthonormal eigenvector corresponding to λ_j is

$$
\mathbf{e}_j = \text{col} \, [\underbrace{0, 0, \ldots, 1, 0, \ldots, 0}_{j \text{ elements}}]
\tag{1.9.11}
$$

In effect, $\mathbf{e}_j$ is related to $\mathbf{u}_j$ by

$$
\mathbf{e}_j = \mathbf{P}^{-1} \mathbf{u}_j
\tag{1.9.12}
$$

When the Jordan canonical form is in the form of Eq. (1.9.10), the eigenvectors for **B** are

$$
\begin{aligned}
\mathbf{e}_1 \quad &= \text{col} \, [1, 0, \ldots, 0] \\
\mathbf{e}_{k+1} &= \text{col} \, [\underbrace{0, 0, \ldots, 1, 0 \ldots 0}_{k+1 \text{ elements}}] \\
\vdots \\
\mathbf{e}_N \quad &= \text{col} \, [0, \ldots, 1]
\end{aligned}
$$

and supplementary vectors are

$$
\begin{aligned}
\mathbf{f}_1 \quad &= \text{col} \, [0, 1, \ldots, 0] \\
\mathbf{f}_{k-1} &= \text{col} \, [\underbrace{0, 0, \ldots, 1, \ldots, 0}_{k \text{ elements}}]
\end{aligned}
$$

PROBLEMS

1. Determine the second order polynomial that passes the three points (x_0, f_0), $(x_0 + \Delta x, f_1)$, and $(x_0 + 2\Delta x, f_2)$. Show that integrating the second order polynomial in $[x_0, x_0 + 2\Delta x]$ results in Eq. (1.3.3).

2. Show that when $y_{n\pm 1}$ and $y'_{n\pm 1}$ are known, the Hermite interpolation formula for y_n becomes

$$y_n = \frac{1}{2}[y_{n-1} + y_{n+1}] + \frac{h}{4}[y'_{n-1} - y'_{n+1}]$$

where $h = x_n - x_{n-1} = x_{n+1} - x_n$.

3. Numerically calculate the integral

$$\int_0^{\pi/2} \sin x \, dx$$

by using four equally divided intervals in accordance with (1) the trapezoidal rule, (2) the extended Simpson's rule, and (3) the Newton Cotes formula of 4th order (integrate the Lagrange interpolation formula with five points).

4. Show that the cubic spline function for equally spaced grids has the following properties:[9]

   a. The second derivative of $g(x)$ given in Eq. (1.2.9) becomes

$$g''(x) = \frac{x - x_i}{x_{i+1} - x_i} g''_{i+1} + \frac{x_{i+1} - x}{x_{i+1} - x_i} g''_i$$

   in the interval, $x_i < x < x_{i+1}$, where $g''_i \equiv g''(x_i)$.

   b. The second derivatives are related to $\{f_i\}$ by

$$\frac{1}{6}g''_{i-1} + \frac{2}{3}g''_i + \frac{1}{6}g''_{i+1} = \frac{f_{i-1} - 2f_i + f_{i+1}}{\Delta x^2}$$

5. Apply the Gauss quadrature with four interpolating points to integrating $\sin x$ in $[0, \pi/2]$, and compare it with the analytical result.

6. Prove that Eq. (1.2.3) and Eq. (1.3.9) are identical.

7. Prove that Eq. (1.2.7) fits f_{i+1}, f'_{i+1} at x_{i+1} and f_i, f'_i at x_i, respectively.

8. Show that

$$f(x + nh) = f(x) + n\Delta f(x) + \frac{n(n-1)}{2!}\Delta^2 f(x) + \cdots + \Delta^n f(x)$$

where Δ is the forward difference operator defined by Eq. (1.4.3).

9. Prove by induction the following formulas:

$$\Delta^n y(x) = \sum_{k=0}^{n} (-1)^k \frac{n!}{k!(n-k)!} y(x+nh-kh)$$

$$\nabla^n y(x) = \sum_{k=0}^{n} (-1)^k \frac{n!}{k!(n-k)!} y(x-kh)$$

$$\delta^{2n} y(x) = \sum_{k=0}^{2n} (-1)^k \frac{(2n)!}{k!(2n-k)!} y(x+nh-kh)$$

10. a. Write the backward difference and the forward difference approximations to

$$\frac{d}{dt} y(t) = y(t), \qquad y(0) = y_0$$

b. Show that the solutions to the two difference equations become, respectively,

$$y(t+nh) = (1+h)^n y_0 \qquad\qquad \text{Backward}$$

$$y(t+nh) = (1-h)^{-n} y_0 \qquad\qquad \text{Forward}$$

11. a. Repeat Problem 10 by using the central difference approximation.
   b. Calculate the error of the central difference approximation for Problem 10 by comparing the numerical result with the analytical solution.

12. Numerically integrate the differential equation

$$-\frac{d^2 y}{dx^2} + 0.1 y = 0$$

$$y(0) = 1, \qquad y'(0) = 0$$

for $0 \le x \le 10$ by using the backward difference formula with $\Delta x = 1$. What is the value of $y(10)$?

13. Integrate the following ordinary differential equations by using the fourth order Runge–Kutta method with $\Delta x = 0.1$:

$$y' = y - 1.5 e^{-0.5x}, \qquad y(0) = 1$$

$$y' = -xy^2, \qquad y(0) = 1$$

14. Integrate the following ordinary differential equation by using the fourth order Runge–Kutta method:

$$y'' = -xy, \qquad y(0) = 1, \qquad y'(0) = 0$$

15. Find the stability criterion for the second order Runge–Kutta formula:

$$y_{n+1} = y_n + \tfrac{1}{2}(k_1 + k_2)$$
$$k_1 = \Delta x f(y_n, x_n)$$
$$k_2 = \Delta x f(y_n + k_1, x_n + \Delta x)$$

which is applied to $y' = \lambda y$.

16. Repeat Problem 15 with the third order Runge–Kutta formula:

$$y_{n+1} = y_n + \tfrac{1}{6}[k_1 + 4k_2 + k_3]$$
$$k_1 = \Delta x f(y_n, x_n)$$
$$k_2 = \Delta x f(y_n + \tfrac{1}{2}k_1, x_n + \tfrac{1}{2}\Delta x)$$
$$k_3 = \Delta x f(y_n - k_1 + 2k_2, x_n + \Delta x)$$

17. Integrate the following ordinary differential equations by using each of the predictor and corrector methods in Table 1.3.

a. $y' = 1 - y$, $y(0) = 2$
b. $y' = y + \sin x$, $y(0) = 1$

18. What should the value of $y'(0)$ in Problem 12 be if the final condition for y is $y(10) = 0$? Find the answer by trial and error for $y'(0)$ with one of the numerical methods in Section 1.5.

19. Find the rank of each matrix:

$$\begin{bmatrix} 2 & -2 & -1 \\ 1 & -1 & 4 \\ 0 & 3 & -2 \end{bmatrix}, \quad \begin{bmatrix} 1 & -2 & 4 & 1 \\ 2 & -3 & 9 & -1 \\ 1 & 0 & 6 & -5 \\ 2 & -5 & 7 & 5 \end{bmatrix}$$

20. Prove that $\mathbf{A}^T\mathbf{A}$ is a square symmetric matrix for any matrix $\mathbf{A}$.

21. Solve the following sets of equations by the Gaussian elimination method:

$$\begin{cases} 3x + y = 2 \\ x + 2y = 3, \end{cases} \qquad \begin{cases} x + y + 2z = 1 \\ x + 2y - z = -2 \\ x + 3y + z = 5 \end{cases}$$

$$\begin{cases} x + y + 2z = 1 \\ x + 3y + z = 5 \\ 2x + 4y + 3z = 6 \end{cases}$$

22. Solve the following equation by using the recursion formula Eqs. (1.7.15) and (1.7.16):

$$\begin{bmatrix} 1 & -1 & 0 \\ -1 & 2 & -1 \\ 0 & -1 & 2 \end{bmatrix} \begin{bmatrix} \phi_1 \\ \phi_2 \\ \phi_3 \end{bmatrix} = \begin{bmatrix} 1 \\ 1 \\ 0 \end{bmatrix}$$

23. A set of inhomogeneous linear equations has no solution if it is inconsistent. Explain the reason geometrically by using an example with two variables.

24. A set of inhomogeneous linear equations has no unique solutions if the equations are not linearly independent. Explain the reason geometrically by using an example with three variables.

25. Write a FORTRAN subroutine to solve inhomogeneous linear equations in accordance with the algorithm of Eqs. (1.7.2) through (1.7.6).

26. Write a FORTRAN subroutine to solve a linear system with a tridiagonal matrix according to the algorithm of Eqs. (1.7.15) and (1.7.16).

27. Develop a Gaussian elimination algorithm for a pentadiagonal matrix and n-diagonal matrix.

28. Find all the independent eigenvectors of the matrix:

$$\mathbf{A} = \begin{bmatrix} 1 & -3 \\ 2 & 1 \end{bmatrix}$$

29. a. Find all the independent eigenvectors and adjoint eigenvectors of

$$\begin{bmatrix} 3 & 4 \\ 1 & 2 \end{bmatrix}$$

b. Show that the normal eigenvectors are orthogonal to the adjoint eigenvectors.

30. a. Find two independent eigenvectors of the matrix

$$\mathbf{A} = \begin{bmatrix} 0 & 1 \\ 1 & 0 \end{bmatrix}$$

b. Check if there are infinite different ways of selecting orthogonal sets of the eigenvectors of $\mathbf{A}$.

31. Find all the eigenvectors and supplemental vectors of the matrix

$$\mathbf{A} = \begin{bmatrix} 1 & 1 & 0 \\ 0 & 1 & 1 \\ 0 & 0 & 1 \end{bmatrix}$$

32. Calculate all the eigenvalues and eigenvectors of

a.
$$\begin{bmatrix} 1 & -1 & 0 \\ -1 & 2 & -1 \\ 0 & -1 & 2 \end{bmatrix} \begin{bmatrix} \phi_1 \\ \phi_2 \\ \phi_3 \end{bmatrix} = \lambda \begin{bmatrix} 1 & 0 & 0 \\ 0 & 1 & 0 \\ 0 & 0 & 1 \end{bmatrix} \begin{bmatrix} \phi_1 \\ \phi_2 \\ \phi_3 \end{bmatrix}$$

b.
$$\begin{bmatrix} 1 & -1 & 0 \\ -1 & 2 & -1 \\ 0 & -1 & 2 \end{bmatrix} \begin{bmatrix} \phi_1 \\ \phi_2 \\ \phi_3 \end{bmatrix} = \lambda \begin{bmatrix} 1 & 0 & 0 \\ 0 & 1 & 0 \\ 0 & 0 & 0 \end{bmatrix} \begin{bmatrix} \phi_1 \\ \phi_2 \\ \phi_3 \end{bmatrix}$$

33. a. Calculate the eigenvalues and the corresponding eigenvectors of

$$\mathbf{A} = \begin{bmatrix} 1 & 1 \\ -1 & 1 \end{bmatrix}$$

b. Expand the following two vectors in the eigenvectors of $\mathbf{A}$:

$$\mathbf{x} = \begin{bmatrix} 1 \\ 0 \end{bmatrix}, \qquad \mathbf{y} = \begin{bmatrix} 0 \\ 1 \end{bmatrix}$$

c. Calculate $\mathbf{A}^n \mathbf{x}$, $n = 1, 2, \ldots, 10$, where n is power.

34. a. Find the normalized eigenvectors of the matrix

$$\mathbf{A} = \begin{bmatrix} 4 & 1 \\ 1 & 1 \end{bmatrix}$$

b. Prove that the matrix

$$\mathbf{B} = [\mathbf{u}_1, \mathbf{u}_2]$$

satisfies

$$\mathbf{B}^T = \mathbf{B}^{-1}$$

where $\mathbf{u}_1$ and $\mathbf{u}_2$ are normalized eigenvectors of $\mathbf{A}$.
c. Calculate $\mathbf{BAB}^{-1}$.

35. Suppose a matrix $\mathbf{A}$ is real and symmetric.
a. Show that the solution of the equation

$$(\mathbf{A} - \lambda \mathbf{I})\psi = \mathbf{f}$$

where $\mathbf{I}$ is an identity matrix, can be written in the form

$$\psi = \sum_n \frac{a_n}{\nu_n - \lambda} \mathbf{u}_n$$

where $\mathbf{u}_n$ is the eigenvectors, ν_n is the corresponding eigenvalues of $\mathbf{A}, \mathbf{f}$ is a known vector, and $\lambda \neq \nu_n$.
b. Determine a_n.

36. Prove that, if matrices $\mathbf{A}$ and $\mathbf{B}$ are symmetric, $\mathbf{C} = \mathbf{BAB}$ is also symmetric.

REFERENCES

1. R. W. Hornbeck, *Numerical methods*, Quantum, New York (1975).
2. V. N. Faddeeva, *Computational methods of linear algebra*, Dover, New York (1959).
3. P. J. Davis, *Interpolation and approximation*, Blaisdell, New York (1963).
4. I. D. Mysovskin, *Lecture on numerical methods*, Walters-Noordhoff, Groningen (1969).
5. L. Fox and D. F. Mayers, *Computing methods for scientists and engineers*, Clarendon, Oxford (1968).
6. P. G. Ciarlet, M. H. Schultz, and R. S. Varga, "Numerical methods of high-order accuracy for non-linear boundary value problems I, one-dimensional problems," *Numer. Math.*, **9**, 394–430 (1967); also see ibid., **13**, 51–77 (1969).
7. G. Birkhoff, M. H. Schultz, and R. S. Varga, "Piecewise Hermite interpolation in one and two variables with applications to partial differential equations," *Numer. Math.*, **11**, 232–256 (1968).
8. T. N. E. Greville, *Theory and applications of spline functions*, Academic, New York (1969).
9. J. H. Ahlberg, E. N. Nilson, and J. L. Walsh, *The theory of splines and their applications*, Academic, New York (1967).
10. P. M. Prenter, *Splines and variational methods*, Wiley-Interscience, New York (1975).
11. A. H. Stroud and D. Secrest, *Gaussian quadrature formulas*, Prentice-Hall, Englewood Cliffs, N.J. (1966).
12. P. M. Morse and H. Feshbach, *Methods of theoretical physics*, McGraw-Hill, New York (1953).
13. D. Courant and D. Hilbert, *Methods of mathematical physics*, Wiley-Interscience, New York (1953).
14. M. Abramowitz and A. Stegun, *Handbook of mathematical functions*, U.S. Government Printing Office, Washington, D.C. (1964).
15. L. Lapidus and J. H. Seinfeld, *Numerical solution ordinary differential equations*, Academic, New York (1971).
16. R. W. Hamming, *Numerical methods for scientists and engineers*, McGraw-Hill, New York (1962).
17. G. L. Kelly, *Handbook of numerical methods and applications*, Addison-Wesley, Reading, Mass. (1967).
18. I. S. Berezin and N. P. Zhidkov, *Computing methods*, Addison-Wesley, Reading, Mass. (1965).
19. A. K. MacPherson and R. E. Kelly, "Global weather forecasting using cubic splines," in *Advances in computer methods for partial differential equations*, AICA, Rutgers University, Dept. of Computer Science, New Brunswick, N.J. (1975).
20. R. Bulirsch and J. Store, "Numerical treatment of ordinary differential equations by extrapolation methods," *Numer. Math.* **8**, 1–13 (1966).
21. N. Clark, "A study of some numerical methods," ANL-7428, Argonne National Laboratory (1968).
22. J. Mennig and J. Auerbach, "The application of Lie series to reactor theory," *Nucl. Sci. Eng.*, **28**, 159 (1967).

23. J. Mennig, T. Auerbach, J. Brunner, J. Haelg, and J. Halin, "Integration of differential equations by means of Lie series and various numerical methods: comparison of speed and reliability," in *Numerical reactor calculations*, IAEA, Vienna (1972).

24. B. L. Hulme, "One-step piecewise polynomial Galerkin method for initial value problems," *Math. Computations*, **26**, 415–436 (1972).

25. D. L. Hetrick, *Dynamics of nuclear reactors*, p. 126, University of Chicago, Chicago (1971).

26. H. G. Campbell, *Matrices with applications*, Appleton-Century-Crofts, New York (1968).

27. J. N. Franklin, *Matrix theory*, Prentice-Hall, Englewood Cliffs, N.J. (1968).

28. R. J. Painter and R. P. Yantis, *Elementary matrix algebra with linear programming*, Prindle, Weber and Schmidt, Boston, Mass. (1971).

29. L. Fox, *Numerical linear algebra*, Oxford U. P., Oxford, England (1965).

30. B. C. Tetra, *Basic linear algebra*, Harper and Row, New York (1969).

31. W. G. Bickley and R. S. H. G. Thompson, *Matrices, their meaning and manipulation*, Van Nostrand, Princeton, N.J. (1964).

32. G. A. Bekey and W. J. Karplus, *Hybrid computation*, Wiley, New York (1968).

33. R. Vichnevetsky, *State of the art in hybrid methods for partial differential equations*, Proc. AICA-IFIP International Conference on Hybrid Computation, Munich, Germany, Aug. 31–Sept. 4, 1970.

34. R. Vichnevetsky, *Advances in computer methods for partial differential equations*, AICA, Dept. of Computer Science, Rutgers Univ., New Brunswick, N.J. (1975).

35. *System/360 continuous system modeling program user's manual*, 5th edit., GH20-0367-4, IBM Corporation, New York (1972).

36. *Continuous system modeling program III program reference manual*, SH19-7001-2, 3rd edit., IBM Corporation, New York (1972).

Chapter 2   Computer Solutions for One-Dimensional Eigenvalue Problems

2.1 BOUNDARY VALUE AND EIGENVALUE PROBLEMS ASSOCIATED WITH ORDINARY DIFFERENTIAL EQUATIONS

The second order ordinary differential equation,

$$-\frac{d}{dx}p(x)\frac{d}{dx}\phi(x)+q(x)\phi(x)=S(x) \tag{2.1.1}$$

defined in $\alpha \leq x \leq \beta$ is called a boundary value problem when ϕ or its derivative is specified at both of $x = \alpha$ and $x = \beta$. The boundary conditions are expressed by a general form as

$$\frac{d}{dx}\phi+\gamma\phi=\sigma \tag{2.1.2}$$

where γ and σ are prescribed values at each boundary. Equation (2.1.2) is called *mixed boundary condition* if $\gamma \neq 0$. If $\gamma = 0$, Eq. (2.1.2) is called the *Neuman boundary condition*. If $\gamma \to \infty$ with $\sigma/\gamma = \text{const}$,* Eq. (2.1.2) is reduced to $\phi = \text{const}$ and called the *Dirichlet boundary condition*. The boundary conditions are said to be homogeneous if $\sigma = 0$ and nonhomogenous if $\sigma \neq 0$. When $S(x)$ is nonzero at least at a point in $\alpha < x < \beta$, a unique solution exists even if all the boundary conditions are homogeneous.

The homogeneous equation

$$-\frac{d}{dx}p(x)\frac{d}{dx}\phi(x)+q(x)\phi(x)=\lambda f(x)\phi(x) \tag{2.1.3}$$

---

* Practically, a sufficiently large number such as $|\gamma| = 10^5$ should be used. However, a too large number for $|\gamma|$ will cause an overflow during the computation.

is called the *Sturm–Liouville eigenvalue problem*,[1,2] where $p(x) > 0, f(x) \geq 0$, λ is an eigenvalue, and

$$\frac{d}{dx}\phi(x) - \gamma_L\phi(x) = 0 \quad \text{at } x = \alpha \tag{2.1.4}$$

$$\frac{d}{dx}\phi(x) + \gamma_R\phi(x) = 0 \quad \text{at } x = \beta \tag{2.1.5}$$

In the above equation, we assume $\gamma_L \geq 0$ and $\gamma_R \geq 0$. If $\gamma_L = \gamma_R = \infty$, the conditions become $\phi(\alpha) = \phi(\beta) = 0$.

The Sturm–Liouville eigenvalue problem has an infinite number of eigenfunctions, $\phi_0(x), \phi_1(x), \ldots, \phi_n(x), \ldots$, with the corresponding real and distinct eigenvalues, $\lambda_0, \lambda_1, \ldots, \lambda_n, \ldots$. The eigenfunctions and the eigenvalues are numbered in the sequence,

$$\lambda_0 < \lambda_1 < \lambda_2 \cdots < \lambda_\infty \tag{2.1.6}$$

The eigenfunction ϕ_0 corresponding to the lowest eigenvalue λ_0 is called the *fundamental eigenfunction* (or *fundamental mode*) and has no root of $\phi_0(x) = 0$ in $\alpha < x < \beta$. Generally, $\phi_n(x)$ has n roots in $\alpha < x < \beta$, as schematically shown in Fig. 2.1.

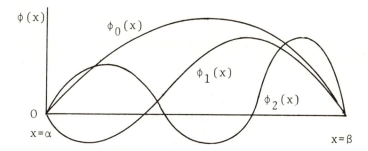

Figure 2.1 Schematic illustration of a few Sturm–Liouville eigenfunctions ($\gamma_\alpha = \gamma_\beta = \infty$).

The eigenvalues of a Sturm–Liouville problem are all positive if $q(x) \geq 0$. A finite number of the eigenvalues may become negative if $q(x)$ is negative for some interval of x. If a number of eigenvalues are negative, the problem can be reduced to an auxiliary eigenvalue problem that is positive definite. By defining

$$\nu = \lambda - \lambda' \tag{2.1.7}$$

where λ' is an arbitrary value satisfying the inequality, $\lambda' < \lambda_0$, Eq. (2.1.3) may be written as

$$-\frac{d}{dx}p(x)\frac{d}{dx}\phi(x)+[q(x)-\lambda'f(x)]\phi(x)=\nu f(x)\phi(x) \qquad (2.1.8)$$

In Eq. (2.1.8), ν is positive for all λ_n.

2.2 FINITE DIFFERENCE APPROXIMATIONS

We assume first that both left and right boundaries are at finite distances from the origin. Therefore, without loss of generality, we assume that the left boundary is at the origin, $\alpha = 0$. The boundaries at infinite distances are considered in Section 2.6.

In order to derive the finite difference approximation for Eq. (2.1.3), we consider the mesh system shown in Fig. 2.2. The mesh intervals Δx_i are arbitrary, but we assume that p, q, and f are constant in each mesh interval. There are two approaches in deriving the finite difference equations for Eq. (2.1.3) as follows.

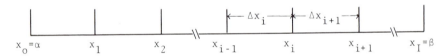

Figure 2.2   One-dimensional grid system.

POINT SCHEME[3-5]

The solution $\phi(x)$ is approximated by discrete values ϕ_i at mesh points. We first consider an internal grid, $0 < i < I$, and integrate Eq. (2.1.3) from $x_A = (x_{i-1}+x_i)/2$ to $x_B = (x_i + x_{i+1})/2$ (Fig. 2.3). The first term of Eq. (2.1.3) becomes then

$$-\int_i \frac{d}{dx}p\frac{d}{dx}\phi\,dx = -p\frac{d}{dx}\phi\bigg|_{x_B} + p\frac{d}{dx}\phi\bigg|_{x_A} \qquad (2.2.1)$$

where i at the integral sign denotes that the integral is performed from x_A to x_B. The derivatives at x_A and x_B are approximated by

$$\frac{d}{dx}\phi\bigg|_{x_A} = \frac{\phi_i - \phi_{i-1}}{\Delta x_i} \qquad (2.2.2)$$

$$\frac{d}{dx}\phi\bigg|_{x_B} = \frac{\phi_{i+1} - \phi_i}{\Delta x_{i+1}} \qquad (2.2.3)$$

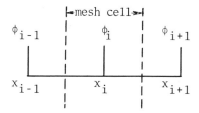

Figure 2.3 Mesh cell for the point scheme.

where $\Delta x_i = x_i - x_{i-1}$. The second term and right side of Eq. (2.1.3) are treated, respectively, as

$$\int_i q(x)\phi(x)\,dx = \frac{q_i\Delta x_i + q_{i+1}\Delta x_{i+1}}{2}\phi_i \qquad (2.2.4)$$

$$\int_i f(x)\phi(x)\,dx = \frac{f_i\Delta x_i + f_{i+1}\Delta x_{i+1}}{2}\phi_i \qquad (2.2.5)$$

where a_i and f_i are the values of q and f in Δx_i, respectively. Collecting terms, we obtain

$$a_i\phi_{i-1} + b_i\phi_i + c_i\phi_{i+1} = \lambda d_i\phi_i, \qquad 0 < i < I \qquad (2.2.6)$$

where

$$a_i = \frac{-p_i}{\Delta x_i} \qquad (2.2.7)$$

$$b_i = \int_i q(x)\,dx - a_i - c_i \qquad (2.2.8)$$

$$c_i = \frac{-p_{i+1}}{\Delta x_{i+1}} \qquad (2.2.9)$$

$$d_i = \int_i f(x)\,dx \qquad (2.2.10)$$

and where p_i is the value of p in Δx_i.

For $i = 0$, Eq. (2.1.3) is integrated from $x_A = \alpha = 0$ to $x_B = (x_0 + x_1)/2$. The integral of the first term becomes the same as Eq. (2.2.1). However, eliminating the first derivative of ϕ at $x_A = 0$ by using Eq. (2.1.4), we have

$$-\int_0^{x_B} \frac{d}{dx}p\frac{d}{dx}\phi\,dx = -p\frac{d}{dx}\phi\bigg|_{x_B} + p_1\gamma_L\phi_0 \qquad (2.2.11)$$

By using Eq. (2.2.3), the right side of the above equation becomes

$$-p_1\frac{\phi_1-\phi_0}{\Delta x_1}+p_1\gamma_L\phi_0 \tag{2.2.12}$$

Collecting terms, the finite difference equation becomes

$$b_0\phi_0+c_0\phi_1=\lambda\,d_0\phi_0 \tag{2.2.13}$$

where

$$b_0=\int_0^{x_B}q(x)\,dx+p_1\gamma_L-c_0 \tag{2.2.14}$$

$$c_0=\frac{-p_1}{\Delta x_1} \tag{2.2.15}$$

$$d_0=\int_0^{x_B}f(x)\,dx \tag{2.2.15'}$$

At $i=I$, integration of Eq. (2.1.2) is performed from $x_A=(x_{I-1}+x_I)/2$ to $x_B=x_I=\beta$. The right boundary condition in the form of Eq. (2.1.5) is used to eliminate the derivative at x_I. The difference equation obtained is

$$a_I\phi_{I-1}+b_I\phi_I=\lambda d_I\phi_I \tag{2.2.16}$$

where

$$a_I=\frac{-p_I}{\Delta x_I} \tag{2.2.17}$$

$$b_I=\int_I q(x)\,dx+p_I\gamma_R-a_I \tag{2.2.18}$$

$$d_I=\int_I f(x)\,dx \tag{2.2.19}$$

Equations (2.2.13), (2.2.6), and (2.2.16) can be expressed in matrix form as

$$\mathbf{A}\boldsymbol{\phi}=\lambda\mathbf{F}\boldsymbol{\phi} \tag{2.2.20}$$

where

$$\boldsymbol{\phi}=\text{col}\,[\phi_0,\phi_1,\phi_2,\ldots,\phi_I]$$

$$\mathbf{A}=\begin{bmatrix}b_0 & c_0 & & & \\ a_1 & b_1 & c_1 & & \\ & a_2 & b_2 & c_2 & \\ & & \ddots & \ddots & \\ & & & a_I & b_I\end{bmatrix} \tag{2.2.21}$$

$$\mathbf{F} = \begin{bmatrix} d_0 & & & & \\ & d_1 & & & \\ & & d_2 & & \\ & & & \cdot & \\ & & & & \cdot \\ & & & & \cdot \, d_I \end{bmatrix} \tag{2.2.22}$$

BOX SCHEME

In this method $\bar{\phi}_i$ represents the average of $\phi(x)$ in the ith mesh interval rather than the value at a grid point as illustrated in Fig. 2.4. We assume again that $p(x)$ is constant in the mesh interval but can be discontinuous across a grid point.

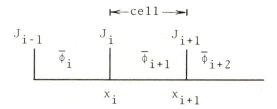

Figure 2.4 Mesh cell for the box scheme.

By defining

$$J(x) = -p\frac{d}{dx}\phi(x) \tag{2.2.23}$$

Eq. (2.1.3) may be written as

$$\frac{d}{dx}J(x) + q(x)\phi(x) = \lambda f(x)\phi(x) \tag{2.2.24}$$

Integrating Eq. (2.2.24) in the ith interval $[x_{i-1}, x_i]$ yields

$$J_i - J_{i-1} + \int_i q\phi \, dx = \lambda \int_i f\phi \, dx \tag{2.2.25}$$

where $J_i = J(x_i)$. Approximating ϕ in the integrals by $\bar{\phi}_i$, Eq. (2.2.25) becomes

$$J_i - J_{i-1} + q_i \Delta x_i \bar{\phi}_i = \lambda f_i \Delta x_i \bar{\phi}_i \tag{2.2.26}$$

The values of J at $x_i + \epsilon$ (ϵ is an infinitesimally small number) and at $x_i - \epsilon$ can be approximated, respectively, by

$$J(x_i + \epsilon) = -p_{i+1}\frac{\bar{\phi}_{i+1} - \phi(x_i)}{\Delta x_{i+1}/2} \tag{2.2.27}$$

$$J(x_i - \epsilon) = -p_i\frac{\phi(x_i) - \bar{\phi}_i}{\Delta x_i/2} \tag{2.2.28}$$

We equate Eq. (2.2.27) and Eq. (2.2.28) for continuity of J at x_i. Solving the resulting equation for $\phi(x_i)$, we have

$$\phi(x_i) = \frac{(p_i/\Delta x_i)\bar{\phi}_i + (p_{i+1}/\Delta x_{i+1})\bar{\phi}_{i+1}}{p_i/\Delta x_i + p_{i+1}/\Delta x_{i+1}} \tag{2.2.29}$$

Substituting Eq. (2.2.29) into Eq. (2.2.27), $J_i \equiv J(x_i \pm \epsilon)$ becomes

$$J_i = \frac{2p_i p_{i+1}}{p_i \Delta x_{i+1} + p_{i+1}\Delta x_i}(\bar{\phi}_i - \bar{\phi}_{i+1}) \tag{2.2.30}$$

Equation (2.2.30) applies at every grid except at the external boundaries. Introducing Eq. (2.2.30) into Eq. (2.2.26), we obtain the difference equation as

$$a_i\bar{\phi}_{i-1} + b_i\bar{\phi}_i + c_i\bar{\phi}_{i+1} = \lambda d_i\bar{\phi}_i \tag{2.2.31}$$

where

$$a_i = -2\frac{p_{i-1}p_i}{p_{i-1}\Delta x_i + p_i\Delta x_{i-1}} \tag{2.2.32}$$

$$b_i = q_i\Delta x_i - a_i - c_i \tag{2.2.33}$$

$$c_i = a_{i+1} \tag{2.2.34}$$

$$d_i = f_i\Delta x_i \tag{2.2.35}$$

To deal with Eq. (2.2.25) for $i = 1$, $J(0)$ must be evaluated. For this purpose, we first eliminate $\phi'(x)$ in Eq. (2.2.23) for $x = 0$ by using Eq. (2.1.4) to obtain

$$J(0) = -p_1\gamma_L\phi(0) \tag{2.2.36}$$

On the other hand, Eq. (2.2.27) for $i = 0$ becomes

$$J(0 + \epsilon) = -p_1\frac{\bar{\phi}_1 - \phi(0)}{\Delta x_1/2} \tag{2.2.37}$$

By equating Eq. (2.2.36) and Eq. (2.2.37), $\phi(0)$ is obtained as

$$\phi(0) = \frac{\bar{\phi}_1}{1 + \gamma_L \Delta x_1/2} \qquad (2.2.38)$$

Introducing Eq. (2.2.38) into Eq. (2.2.36) yields

$$J(0) = J_0 = -p_1 \gamma_L \frac{\bar{\phi}_1}{1 + \gamma_L \Delta x_1/2} \qquad (2.2.39)$$

The difference equation for $i = 1$ becomes, therefore,

$$b_1 \bar{\phi}_1 + c_1 \bar{\phi}_2 = \lambda d_1 \bar{\phi}_1 \qquad (2.2.40)$$

where

$$b_1 = q_1 \Delta x_1 + \frac{p_1 \gamma_L}{1 + \gamma_L \Delta x_1/2} - c_1 \qquad (2.2.41)$$

$$c_1 = a_2 \qquad (2.2.42)$$

By evaluating $J_I = J(\beta)$ similarly to J_0, the finite difference equation for $i = I$ is obtained as

$$a_I \bar{\phi}_{I-1} + b_I \bar{\phi}_I = \lambda d_I \bar{\phi}_I \qquad (2.2.43)$$

where $a_I = c_{I-1}$ and

$$b_I = q_I \Delta x_I + \frac{p_I \gamma_R}{1 + \gamma_R \Delta x_I/2} - a_I \qquad (2.2.44)$$

Thus Eqs. (2.2.40), (2.2.31), and (2.2.43) constitute a matrix eigenvalue problem expressed in the same form as Eq. (2.2.20) except that the subscripts in the first rows of A and F are 1 rather than 0.

2.3 ANALYTICAL EXPRESSION OF EIGENVECTORS AND EIGENVALUES FOR A SIMPLE PROBLEM

Before discussing any detail of solutions of Eq. (2.2.20) for general problems, it may be worthwhile to study analytical expressions[8] for eigenvectors and eigenvalues of Eq. (2.2.20) for a simple geometry. If p, q, and f are constant in Eq. (2.1.3) and the mesh intervals are constant, Eq. (2.2.20) may be equivalently expressed by

$$\tau \frac{2\phi_i - \phi_{i-1} - \phi_{i+1}}{h^2} + \phi_i = \lambda g \phi_i, \qquad i = 1, 2, \ldots, I-1 \qquad (2.3.1)$$

where $\tau = p/q$, $g = f/q$, $h = H/I$, and we assume that boundary conditions are $\phi_0 = \phi_I = 0$ ($H = \beta$, $\alpha = 0$).

For the above equation, there are $I - 1$ independent eigensolutions, which are

$$\phi_{i,n} = \sin(B_n ih), \qquad n = 0, 1, 2, \ldots, I-2 \tag{2.3.2}$$

where

$$B_n = \frac{(n+1)\pi}{H}$$

The corresponding eigenvalues are

$$\lambda_n = \frac{(2\tau/h^2)[1 - \cos(B_n h)] + 1}{g} \tag{2.3.3}$$

Eq. (2.3.2) and Eq. (2.3.3) are comparable to the analytical solutions for Eq. (2.1.2) with the same constants and boundary conditions:

$$\phi_n(x) = \sin(B_n x) \tag{2.3.4}$$

$$\lambda_n = \frac{\tau B_n^2 + 1}{g}, \qquad n = 0, 1 \ldots, \infty \tag{2.3.5}$$

For a small n, Eq. (2.3.3) may be approximated by

$$\lambda_n = \frac{\tau B_n^2 + 1}{g} + O(h^4) \tag{2.3.6}$$

where the following expansion is used:

$$\cos(B_n h) = 1 - \tfrac{1}{2}(B_n h)^2 + O(h^4) \tag{2.3.7}$$

The difference between Eq. (2.3.5) and Eq. (2.3.6) is due to the error introduced by the finite difference approximation. The order of the error in the eigenvalue is smallest for $n = 0$ and is rapidly increased as n is increased.

2.4 THE POWER METHOD

The solution for the minimum eigenvalue of Eq. (2.2.20) and the corresponding eigenvector is important in many areas of science and engineering. The power method is not practically useful for solving the present problem because of its inefficiency. However, the theoretical aspects of the power method is very important since more efficient iterative methods are all based on the power method.

By multiplying both sides of Eq. (2.2.20) by $\mathbf{FA}^{-1}$, we obtain

$$\mathbf{y} = \lambda \mathbf{Ry} \tag{2.4.1}$$

where

$$\mathbf{y} = \mathbf{F}\boldsymbol{\phi} \qquad (2.4.2)$$

$$\mathbf{R} = \mathbf{F}\mathbf{A}^{-1} \qquad (2.4.3)$$

We assume that the diagonal elements of $\mathbf{F}$ are all nonzero. This is the case if $f(x) > 0$ in Eq. (2.1.3). Therefore, the rank of $\mathbf{R}$ is equal to the order of $\mathbf{R}$.

The power method for Eq. (2.4.1) is written as

$$\mathbf{y}^{(t)} = \lambda^{(t-1)}\mathbf{R}\mathbf{y}^{(t-1)}, \qquad t > 0 \qquad (2.4.4)$$

$$\lambda^{(t)} = \lambda^{(t-1)}\frac{\langle \mathbf{w}, \mathbf{y}^{(t-1)}\rangle}{\langle \mathbf{w}, \mathbf{y}^{(t)}\rangle}, \qquad t > 0 \qquad (2.4.5)$$

where t is the iteration number and $\mathbf{w}$ is an arbitrary weighting vector. The vector $\mathbf{y}^{(0)}$ is an arbitrary non-zero vector called the *initial vector*. $\lambda^{(0)}$ is set to unity. The numerical operation of each iteration involves the inversion procedure as $\boldsymbol{\phi}^{(t)} = \mathbf{A}^{-1}\mathbf{s}$ where $\mathbf{s} = \lambda^{(t-1)}\mathbf{F}\boldsymbol{\phi}^{(t-1)}$, so that the present iteration scheme may be called the inverse power method.[11]

For analyzing the convergence rate of an iterative solution method, it is convenient to expand iterative vectors into eigenvectors of the iterative matrix. The eigenvectors $\mathbf{u}_n$ and the corresponding eigenvalues λ_n of $\mathbf{R}$ are defined by

$$\mathbf{u}_n = \lambda_n \mathbf{R}\mathbf{u}_n \qquad (2.4.6)$$

It may be proven that, if the eigenvalues of Eq. (2.1.3) are all real, positive, and distinct, then the eigenvalues of $\mathbf{R}$ are also real, positive, and distinct. The total number of eigenvectors is equal to the order of $\mathbf{R}$ since we assumed the rank of $\mathbf{R}$ is equal to its order. The eigenvalues are numbered in the sequence:

$$\lambda_0 < \lambda_1 < \lambda_2 < \cdots < \lambda_I \qquad (2.4.7)$$

The eigenvectors have orthogonality relations:

$$\langle \mathbf{u}_n, \mathbf{u}_m\rangle = 0 \quad \text{for } n \neq m \qquad (2.4.8)$$

where $\langle\ \rangle$ means the scalar product of the two vectors.

We assume the eigenvectors are normalized:

$$\langle \mathbf{u}_n, \mathbf{u}_n\rangle = 1 \quad \text{for all } n \qquad (2.4.9)$$

The eigenvectors of $\mathbf{R}$ make a complete set.

The initial vector $\mathbf{y}^{(0)}$ is expanded as

$$\mathbf{y}^{(0)} = \sum_n c_n^{(0)}\mathbf{u}_n \qquad (2.4.10)$$

where $c_n^{(0)}$ is a coefficient that can be obtained for a given $\mathbf{y}^{(0)}$ by using the orthogonality of $\mathbf{u}_n$:

$$c_n^{(0)} = \langle \mathbf{u}_n, \mathbf{y}^{(0)} \rangle \qquad (2.4.11)$$

Equation (2.4.4) can be written as

$$\mathbf{y}^{(t)} = \lambda^{(t-1)} \lambda^{(t-2)} \cdots \lambda^{(0)} \mathbf{R}^t \mathbf{y}^{(0)} \qquad (2.4.12)$$

Introducing Eq. (2.4.10) into Eq. (2.4.12) yields

$$
\begin{aligned}
\mathbf{y}^{(t)} &= \lambda^{(t-1)} \cdots \lambda^{(0)} \sum_n c_n^{(0)} \mathbf{R}^t \mathbf{u}_n = \left[\prod_{p=0}^{t-1} \lambda^{(p)} \right] \sum_n c_n^{(0)} \frac{1}{\lambda_n^t} \mathbf{u}_n \\
&= \left[\prod_{p=0}^{t-1} \frac{\lambda^{(p)}}{\lambda_0} \right] c_0^{(0)} \left[\mathbf{u}_0 + \sum_{n=1}^{I} \frac{c_n^{(0)}}{c_0^{(0)}} \left(\frac{\lambda_0}{\lambda_n} \right)^t \mathbf{u}_n \right] \\
&\approx \text{constant} \cdot \left[\mathbf{u}_0 + \sum_{n=1}^{I} \frac{c_n^{(0)}}{c_0^{(0)}} \left(\frac{\lambda_0}{\lambda_n} \right)^t \mathbf{u}_n \right]
\end{aligned}
\qquad (2.4.13)
$$

Here Eq. (2.4.6) is used to get the second equality. Considering $\lambda_0/\lambda_n < 1$ for $n = 1, 2, \ldots, I$, it is easily seen that as t increases $\mathbf{y}^{(t)}$ converges to a constant times $\mathbf{u}_0$. The convergence rate is determined by the *dominance ratio* defined by

$$\sigma \equiv \max_{n>0} \left(\frac{\lambda_0}{\lambda_n} \right) = \frac{\lambda_0}{\lambda_1} \qquad (2.4.14)$$

The dominance ratio approaches unity as the size of the domain increases.

The convergence of $\lambda^{(t)}$ to λ_0 is seen as follows. Introducing Eq. (2.4.13) into Eq. (2.4.5) yields

$$\lambda^{(t)} = \lambda_0 \frac{1 + \sum_{n=1}^{I} \frac{c_n^{(0)}}{c_0^{(0)}} \left(\frac{\lambda_0}{\lambda_n} \right)^{t-1} G_n}{1 + \sum_{n=1}^{I} \frac{c_n^{(0)}}{c_0^{(0)}} \left(\frac{\lambda_0}{\lambda_n} \right)^{t} G_n} \xrightarrow[t\to\infty]{} \lambda_0 \qquad (2.4.15)$$

where

$$G_n = \langle \mathbf{w}, \mathbf{u}_n \rangle / \langle \mathbf{w}, \mathbf{u}_0 \rangle \qquad (2.4.16)$$

In Eq. (2.4.16) $\mathbf{w}$ is a weighting vector.

The selection of the weighting vector is quite arbitrary, although the convergence rate of $\lambda^{(t)}$ is affected to some extent by the choice of $\mathbf{w}$. For example, $\mathbf{w}$ may be set to $\mathbf{1}$ (sumvector). Another practical and frequently used method is to set $\mathbf{w} = \mathbf{y}^{(t)}$ at each iteration, whence the convergence of $\lambda^{(t)}$ becomes two times faster than with the sum vector[7] (see Problem 11).

The iteration is terminated when one or more convergence criteria are satisfied. The simplest convergence criterion is

$$|\lambda^{(t)} - \lambda^{(t-1)}| < \epsilon \qquad (2.4.17)$$

where ϵ is a small prescribed value. If the convergence rate is very slow, the left side of Eq. (2.4.17) becomes small even before true convergence is reached. Furthermore, usually the convergence of the eigenvalue is faster than the convergence of vector elements, so Eq. (2.4.17) is not sufficient by itself to judge the true convergence. Another convergence criterion for vector elements may be written as

$$\max_i \left| \frac{y_i^{(t)} - y_i^{(t-1)}}{y_i^{(t)}} \right| < \epsilon' \qquad (2.4.18)$$

Equations (2.4.17) and (2.4.18) may be used in parallel.

So far we have assumed there are no zero elements in the diagonal matrix $\mathbf{F}$. This is true if $f(x) > 0$ in Eq. (2.1.3). However, $f(x)$ can be zero in any portion of the domain. So we suppose that $\mathbf{F}$ with order I has m zero diagonal elements. Then $\mathbf{y}$ will have m zero elements by definition of Eq. (2.4.2), and $\mathbf{R}$ will have m rows that are all zero. If we remove such zero elements from $\mathbf{y}$, and if we remove all the columns and rows corresponding to the zero diagonal elements of $\mathbf{F}$ from $\mathbf{R}$, we obtain an auxiliary linear problem

$$\mathbf{y'} = \lambda \mathbf{R'} \mathbf{y'} \qquad (2.4.19)$$

where the order and rank of $\mathbf{R'}$ are both $I - m$. Equation (2.4.19) has $(I - m)$ eigenvectors that are orthogonal and complete in the $(I - m)$ dimensional subspace. It is now apparent that the original equation $\mathbf{y} = \lambda \mathbf{R} \mathbf{y}$ has an orthogonal set of $(I - m)$ eigenvectors, each having m zero elements. This set is incomplete in the I-dimensional vector space but is sufficient to expand any $\mathbf{y}$ with the m zero elements. Therefore, all the discussions previously described in this section apply even when $f(x) = 0$ for some intervals of x.

Example 2.1

The power method to solve the fundamental eigenvector of Eq. (2.3.1) with $I = 20$, $h = 10$, $\phi_0 = \phi_{20} = 0$, $\tau = 20$, $g = 1.05$ may be written as

$$-0.2\phi_{i-1}^{(t)} + 1.4\phi_i^{(t)} - 0.2\phi_{i+1}^{(t)} = 1.05\lambda^{(t-1)}\phi^{(t-1)} \qquad (2.4.20)$$

$$\lambda^{(t)} = \lambda^{(t-1)} \frac{\sum\limits_i \phi_i^{(t)} \phi_i^{(t-1)}}{\sum\limits_i [\phi_i^{(t)}]^2}$$

where $\mathbf{w}$ in Eq. (2.4.5) is set to $\boldsymbol{\phi}^{(t)}$ Since the initial vector is arbitrary, we set $\phi_i^{(0)} = 1$, $i = 1, 2, \ldots, I-1$, and $\lambda^{(0)} = 1$.

The right side of Eq. (2.4.20) is calculated with the initial vector or the previous iterative solution, and the new $\phi_i^{(t)}$ is solved by using Eqs. (1.7.15) and (1.7.16). The successive approximations for $\lambda^{(t)}$ are shown below.

t	$\lambda^{(t)}$
1	0.961026
2	0.958631
5	0.956638
10	0.955511
20	0.954718
50	0.954047
100	0.953758
200	0.953636
∞	0.953616

2.5 WIELANDT METHOD

The present method, proposed by Wielandt[9,10] in 1944, has a remarkable convergence speed and is capable of calculating higher eigenvectors and eigenvalues. We first consider the solution for the fundamental eigenvector by this method.

The Wielandt method is characterized by rewriting Eq. (2.2.20) as

$$(\mathbf{A} - \lambda_e \mathbf{F})\boldsymbol{\phi} = \delta\lambda \, \mathbf{F}\boldsymbol{\phi} \qquad (2.5.1)$$

where λ_e is an estimate for the fundamental eigenvalue and

$$\delta\lambda = \lambda - \lambda_e \qquad (2.5.2)$$

In Eq. (2.5.1), $\delta\lambda$ plays the role of eigenvalue. The power method can be applied to Eq. (2.5.1) as

$$\boldsymbol{\phi}^{(t)} = \delta\lambda^{(t-1)}(\mathbf{A} - \lambda_e \mathbf{F})^{-1}\mathbf{F}\boldsymbol{\phi}^{(t-1)}$$

or equivalently

$$\mathbf{y}^{(t)} = \delta\lambda^{(t-1)}\mathbf{R}'\mathbf{y}^{(t-1)} \qquad (2.5.3)$$

where

$$\mathbf{y}^{(t)} = \mathbf{F}\boldsymbol{\phi}^{(t)} \qquad (2.5.4)$$

$$\mathbf{R}' = \mathbf{F}(\mathbf{A} - \lambda_e \mathbf{F})^{-1} \tag{2.5.5}$$

$$\delta\lambda^{(t)} = \delta\lambda^{(t-1)} \frac{\langle \mathbf{w}, \mathbf{y}^{(t-1)} \rangle}{\langle \mathbf{w}, \mathbf{y}^{(t)} \rangle} \tag{2.5.6}$$

and where $\mathbf{w}$ is an arbitrary weighting vector.

The analysis of convergence rate of the power method described in Section 2.4 applies if the eigenvectors and eigenvalues of $\mathbf{R}'$ are used in place of those for $\mathbf{R}$. We denote the eigenvectors and eigenvalues of $\mathbf{R}'$ by $\mathbf{u}_n'$ and $\delta\lambda_n$, respectively. They satisfy

$$\mathbf{u}_n' = \delta\lambda_n \mathbf{R}' \mathbf{u}_n' \tag{2.5.7}$$

It is easily seen that $\mathbf{u}_n'$ and $\delta\lambda_n$ are related to $\mathbf{u}_n$ and λ_n by

$$\mathbf{u}_n' = \mathbf{u}_n \tag{2.5.8}$$

$$\delta\lambda_n = \lambda_n - \lambda_e \tag{2.5.9}$$

The convergence rate of Eq. (2.5.3) is given as

$$\eta = -\log \sigma' \tag{2.5.10}$$

where σ' is the dominance ratio of $\mathbf{R}'$ given by

$$\sigma' \equiv \frac{\delta\lambda_0}{\delta\lambda_1} = \frac{\lambda_0 - \lambda_e}{\lambda_1 - \lambda_e} \tag{2.5.11}$$

Therefore, as λ_e becomes closer to λ_0, the dominance ratio approaches zero, and accordingly the convergence rate becomes very large.

If no estimate for the fundamental eigenvalue is available at the beginning of iterations, one can start with $\lambda_e = 0$, namely, the power method with $\mathbf{R}$. After a few iterations, a reasonably good estimate is obtained. λ_e can be improved after every iteration by using the formula

$$\lambda_e^{(t)} = \lambda_e^{(t-1)} + \delta\lambda^{(t-1)} \tag{2.5.12}$$

It should be noted, however, that if $\lambda_e = \lambda_0$ the matrix $(\mathbf{A} - \lambda_e \mathbf{F})$ becomes singular so that the Wielandt algorithm is broken. Even when λ_e is too close to λ_0, a difficulty still remains because of a serious round-off error in calculating Eq. (2.5.3). To prevent this, using the following formula rather than Eq. (2.5.12) is recommended:

$$\lambda_e^{(t)} = \lambda_e^{(t-1)} + \delta\lambda^{(t-1)} - (\text{a small positive constant}) \tag{2.5.13}$$

Calculations of higher eigenvectors by the present method can be done by setting λ_e to an estimate for the corresponding eigenvalue λ_n. When there is no estimate for the eigenvalue, λ_e is changed by trial and error. The identification of the eigenvector can be done counting the number of nodes.

Example 2.2

The Wielandt method version of Eq. (2.4.20) is

$$-0.2\phi_{i-1}^{(t)}+(1.4-\lambda_e^{(t)})\phi_i^{(t)}-0.2\phi_{i+1}^{(t)}=\delta\lambda^{(t-1)}\phi_i^{(t-1)}$$

$$\delta\lambda^{(t)}=\frac{\delta\lambda^{(t-1)}\sum_i\phi_i^{(t)}\phi_i^{(t-1)}}{\sum_i(\phi_i^{(t)})^2}$$

$$\lambda_e^{(t)}=\lambda_e^{(t-1)}+\delta\lambda^{(t-1)}-0.01$$

The successive approximations for λ are listed below.

t	$\lambda_e^{(t)}+\delta\lambda^{(t)}$
1	0.954778
2	0.955563
3	0.953769
5	0.953605
10	0.953616
∞	0.953616

2.6 ONE-DIMENSIONAL SCHROEDINGER EQUATIONS

The numerical solution methods for Sturm–Liouville equations can be applied to the Schroedinger equations of the quantum mechanics. The one-dimensional Schroedinger equation can be solved by either the initial value method (matching in the middle)[11,12] or the boundary value method.[13,19] We consider here the latter method. The one-dimensional radial form of the Schroedinger equation is

$$-\frac{\hbar^2}{2m}\frac{1}{r^2}\frac{d}{dr}r^2\frac{d}{dr}\psi(r)+\frac{\hbar^2}{2m}\frac{l(l+1)}{r^2}\psi(r)+V(r)\psi(r)=E\psi(r) \quad (2.6.1)$$

where $\psi(r)$ is the wave function to be solved, l is an integer 0, 1, 2, ... called the azimuthal quantum number, m is the mass of the particle, $\hbar$ is Planck's constant, $V(r)$ is the potential function, and E is the energy of the system that is mathematically the eigenvalue of Eq. (2.6.1). The potential function is negative in molecular physics, for example, $V=-e^2/r$ (the coulomb potential). The solutions of Eq. (2.6.1) for the ground state and several excited states are of great interest. For a bound system of a free atom Eq. (2.6.1) is defined in an infinite space ($0 \leq r < \infty$), while it is confined in a finite space for a Wigner–Seitz cell problem.[20]

Derivation of finite difference equations for Eq. (2.6.1) is straightforward except at the two boundaries. Suppose we impose a mesh system to the r coordinate just as in Fig. 2.2, except that x is replaced by r. The mesh intervals can be arbitrary. We denote the midpoints of (r_{i-1}, r_i) and (r_i, r_{i+1}) by r_A and r_B, respectively. Integrating Eq. (2.6.1) in the volume element about an internal grid point i bounded by $r_A \leq r \leq r_B$ and applying the central difference approximation, we obtain

$$a_i \psi_{i-1} + b_i \psi_i + c_i \psi_{i+1} = E d_i \psi \qquad (2.6.2)$$

where

$$a_i = (-2\pi\hbar^2/m) \frac{[(r_{i-1} + r_i)/2]^2}{r_i - r_{i-1}} \qquad (2.6.3)$$

$$b_i = -a_i - c_i + \int_{r_A}^{r_B} \left[\frac{\hbar^2}{2m} \frac{l(l+1)}{r^2} + V(r) \right] 4\pi r^2 \, dr \qquad (2.6.4)$$

$$c_i = (-2\pi\hbar^2/m) \frac{[(r_i + r_{i+1})/2]^2}{r_{i+1} - r_i} \qquad (2.6.5)$$

$$d_i = \int_{r_A}^{r_B} 4\pi r^2 \, dr \qquad (2.6.6)$$

The boundary conditions for a free atom at $r = 0$ and $r = \infty$ are given by

a. $\psi(r)$ is finite at $r = 0$, and
b. $\psi(r)$ is finite at $r = \infty$.

These boundary conditions can be specified more in detail if we consider the asymptotic forms of the solution about the boundaries. For further discussions we assume $V(r) = -e^2/r$.

BOUNDARY CONDITION AT $r = 0$

Let us first consider the case with $l = 0$. It can be easily seen that, as r approaches 0, Eq. (2.6.1) approaches

$$-\frac{\hbar^2}{m} \frac{d}{dr} \psi(r) - e^2 \psi(r) = 0 \qquad (2.6.7)$$

where Eq. (2.6.1) is multiplied by r. Therefore, the finite difference approximation is obtained as

$$\frac{\hbar^2}{m}\frac{\psi_1-\psi_0}{\Delta r_1}+e^2\psi_0=0 \tag{2.6.8}$$

which may be cast into the form of Eq. (2.6.2) if

$$a_0=0 \tag{2.6.9}$$

$$b_0=-\frac{\hbar^2}{m\,\Delta r_1}+e^2 \tag{2.6.10}$$

$$c_0=\frac{\hbar^2}{m\,\Delta r_1} \tag{2.6.11}$$

$$d_0=0 \tag{2.6.12}$$

Next we consider the case with $l\geq1$. As r approaches 0, Eq. (2.6.1) approaches

$$-\frac{d^2}{dr^2}\psi(r)-\frac{2}{r}\frac{d}{dr}\psi(r)+\frac{l(l+1)}{r^2}\psi(r)=0 \tag{2.6.13}$$

In order that $\psi(r)$ be finite at $r=0$, the asymptotic form of $\psi(r)$ about $r=0$ must be in the form,

$$\psi(r)=r^l\sum_{n=0}^{\infty}a_nr^n \tag{2.6.14}$$

Then we find that $\psi(r)=0$ at $r=0$. The numerical form of the boundary condition is $\psi_0=0$. This result can be anticipated directly from Eq. (2.6.13) for the following reason. If $\psi(0)\neq0$, the third term becomes dominant as r approaches 0 and Eq. (2.6.13) is not satisfied.

A more straightforward way of obtaining the finite difference approximation at $r=0$ is directly to integrate Eq. (2.6.1) in the spherical volume element defined by $0\leq r\leq0.5r_1$ and write in the form of Eq. (2.6.2), whence

$$a_0=0 \tag{2.6.15}$$

$$b_0=-c_0+\int_0^{0.5r_1}\left[\frac{\hbar^2}{2m}\frac{l(l+1)}{r^2}-\frac{e^2}{r}\right]4\pi r^2\,dr \tag{2.6.16}$$

$$c_0=\frac{(-2\pi\hbar^2/m)(r_1/2)^2}{r_1} \tag{2.6.17}$$

$$d_0=\int_0^{0.5r_1}4\pi r^2\,dr \tag{2.6.18}$$

Even though the terms of order $1/r^2$ and $1/r$ are involved, the volume integral of those functions become finite. Therefore, b_0 is finite.

BOUNDARY CONDITION AT $r = \infty$

It is advantageous to note that boundary condition (b) can be replaced by b'.

$$\psi(r) = 0 \quad \text{at} \quad r = \infty.$$

The proof of this is obtained easily as follows. As r increases, Eq. (2.6.1) approaches

$$-\frac{\hbar^2}{2m}\left(\frac{d^2}{dr^2} + \frac{2}{r}\frac{d}{dr}\right)\psi(r) = E\psi(r) \tag{2.6.19}$$

($V(r)$ becomes zero as $r \to \infty$). The solution of Eq. (2.6.19) for a negative E is either

$$A\frac{e^{\alpha r}}{r} \quad \text{or} \quad B\frac{e^{-\alpha r}}{r}$$

where A and B are constant and $\alpha = \sqrt{2mE/\hbar^2}$. However, condition (b) restricts us to the latter, so condition (b') is obtained.

Practically, it is impossible to apply the mesh system from $r = 0$ to $r = \infty$. There are two approaches to the outer boundary condition as follows:

1. Assume ψ vanishes at a sufficiently large radius r, and set $\psi_I = 0$ where I is the last mesh index.
2. Suppose we have a good estimate for E and accordingly α. Then assuming the solution at large r is given by

$$\psi(r) \propto \frac{e^{-\alpha r}}{r} \tag{2.6.20}$$

the mixed boundary condition at a sufficiently large r may be written as

$$\psi'(r) = -\alpha\psi(r) \tag{2.6.21}$$

The finite difference form of Eq. (2.6.2) for the last mesh point, $i = I$, is then

$$\frac{\psi_I - \psi_{I-1}}{\Delta r} = -\alpha\psi_I \tag{2.6.22}$$

or

$$a_I\psi_I + b_I\psi_I = 0 \tag{2.6.23}$$

where

$$a_I = -\frac{1}{\Delta r}, \qquad b_I = \frac{1}{\Delta r} + \alpha \qquad (2.6.24)$$

The value of E or α can be improved during the iterative solution.

2.7   ONE DIMENSIONAL MULTIGROUP NEUTRON DIFFUSION EQUATIONS

Multigroup neutron diffusion equations are the most frequently solved equations in nuclear reactor design and analysis. This section describes the numerical solution for a two-group neutron diffusion equation, since it is the simplest case of multigroup equations while it retains most of the mathematical properties of multigroup equations.

The two-group equation may be written

$$\left\{ -\left[\frac{d}{dx} D_1(x) \frac{d}{dx} + \zeta \frac{D_1(x)}{x} \frac{d}{dx} \right] + \Sigma_1(x) \right\} \phi_1(x)$$

$$= \frac{1}{k} [\nu \Sigma_{f1}(x) \phi_1(x) + \nu \Sigma_{f2}(x) \phi_2(x)] \qquad (2.7.1)$$

$$\left\{ -\left[\frac{d}{dx} D_2(x) \frac{d}{dx} + \zeta \frac{D_2(x)}{x} \frac{d}{dx} \right] + \Sigma_2(x) \right\} \phi_2(x) = \Sigma_{sl}(x) \phi_1(x)$$

where ϕ_1 and ϕ_2 are the fast and thermal neutron flux distributions, respectively; D_1, D_2, Σ_1, Σ_2, $\nu \Sigma_{f1}$, and $\nu \Sigma_{f2}$ are coefficients; $\zeta = 0$, 1, and 2 represent slab, cylindrical, and spherical geometries, respectively; and k is the effective multiplication factor (eigenvalue). Both ϕ_1 and ϕ_2 are subject to the left and right boundary conditions in the form of Eq. (2.1.2) with $\sigma = 0$. We are interested in the largest eigenvalue and the corresponding solution of Eq. (2.7.1) here.

Imagine a mesh system that covers the whole one-dimensional space of interest. Assuming the coefficients of Eq. (2.7.1) are all constant in each mesh interval, the difference equation for Eq. (2.7.1) can be written as

$$\mathbf{A}_1 \boldsymbol{\phi}_1 = \frac{1}{k} \left[\mathbf{F}_1 \boldsymbol{\phi}_1 + \mathbf{F}_2 \boldsymbol{\phi}_2 \right]$$

$$\mathbf{A}_2 \boldsymbol{\phi}_2 = \mathbf{Q} \boldsymbol{\phi}_1 \qquad (2.7.2)$$

If the power method is used to solve Eq. (2.7.2), the iterative procedure consists of the following steps:

a. Introduce any initial guess for $\boldsymbol{\phi}_1$ and $\boldsymbol{\phi}_2$ to the right side of the first equation of Eq. (2.7.2) and assume $k = 1$.
b. Solve the first equation for $\boldsymbol{\phi}_1$ on the left side.
c. Introduce $\boldsymbol{\phi}_1$ thus obtained to the right side of the second equation. Solve it for $\boldsymbol{\phi}_2$.
d. Calculate an approximate value for k by using the previous guess for $\boldsymbol{\phi}_1$ and $\boldsymbol{\phi}_2$ as well as the new solutions for $\boldsymbol{\phi}_1$ and $\boldsymbol{\phi}_2$.
e. Introduce the new values of $\boldsymbol{\phi}_1$, $\boldsymbol{\phi}_2$, and k to the right side of the first equation.
f. Repat steps (b) through (e) until $\boldsymbol{\phi}_1$, $\boldsymbol{\phi}_2$, and k converge.

The above iterative procedure may be expressed in the matrix notations as

$$\begin{bmatrix} \boldsymbol{\phi}_1^{(t)} \\ \boldsymbol{\phi}_2^{(t)} \end{bmatrix} = \frac{1}{k^{(t-1)}} \begin{bmatrix} \mathbf{A}_1 & 0 \\ -\mathbf{Q} & \mathbf{A}_2 \end{bmatrix}^{-1} \begin{bmatrix} \mathbf{F}_1 & \mathbf{F}_2 \\ 0 & 0 \end{bmatrix} \begin{bmatrix} \boldsymbol{\phi}_1^{(t-1)} \\ \boldsymbol{\phi}_2^{(t-1)} \end{bmatrix} \qquad (2.7.3)$$

where $\boldsymbol{\phi}_1^{(t)}$ and $\boldsymbol{\phi}_2^{(t)}$ are the vectors after the tth iteration and $k^{(t)}$ is the tth approximation for k. $k^{(t)}$ may be calculated by

$$k^{(t)} = k^{(t-1)} \frac{\mathbf{w}^T [\mathbf{F}_1 \boldsymbol{\phi}_1^{(t)} + \mathbf{F}_2 \boldsymbol{\phi}_2^{(t)}]}{\mathbf{w}^T [\mathbf{F}_1 \boldsymbol{\phi}_1^{(t-1)} + \mathbf{F}_2 \boldsymbol{\phi}_2^{(t-1)}]} \qquad (2.7.4)$$

where $\mathbf{w}^T$ is a weighting vector. Selection of $\mathbf{w}^T$ is rather arbitrary. The sum vector and $\boldsymbol{\phi}_1^{(t)}$ are popularly used, although the latter yields generally more accurate $k^{(t)}$ than the former. The convergence rate of Eq. (2.7.4) becomes slower as the reactor size increases.

By using Eq. (2.7.2) and defining

$$\mathbf{y} = \mathbf{F}_1 \boldsymbol{\phi}_1 + \mathbf{F}_2 \boldsymbol{\phi}_2 \qquad (2.7.5)$$

$\boldsymbol{\phi}_1$ and $\boldsymbol{\phi}_2$ may be expressed as

$$\boldsymbol{\phi}_1 = \frac{1}{k} \mathbf{A}_1^{-1} \mathbf{y} \qquad (2.7.6)$$

$$\boldsymbol{\phi}_2 = \frac{1}{k} \mathbf{A}_2^{-1} \mathbf{Q} \mathbf{A}_1^{-1} \mathbf{y} \qquad (2.7.7)$$

Introducing the above two equations into Eq. (2.7.5) yields

$$\mathbf{y} = \frac{1}{k} \mathbf{R} \mathbf{y} \qquad (2.7.8)$$

where

$$\mathbf{R} = \mathbf{F}_1 \mathbf{A}_1^{-1} + \mathbf{F}_2 \mathbf{A}_2^{-1} \mathbf{Q} \mathbf{A}_1^{-1} \qquad (2.7.9)$$

The iterative solution given by Eq. (2.7.3) may be written as

$$\mathbf{y}^{(t)} = \frac{1}{k^{(t-1)}} \mathbf{R} \mathbf{y}^{(t-1)} \qquad (2.7.10)$$

The properties of the matrices involved in the two-group difference equation are discussed next. By examination, $\mathbf{A}_1$ and $\mathbf{A}_2$ both are found to have the following properties:[10,14]

1. $\mathbf{A}$ is a real symmetric matrix,
2. the diagonal elements of $\mathbf{A}$ are all positive,
3. all the off-diagonal elements are nonpositive (negative or zero),
4. $A_{ii} \geq \sum_{j \neq i} |A_{ij}|$ (diagonal dominance),
5. $\mathbf{A}$ is irreducible.*

When $\mathbf{A}$ has all of the above properties, $\mathbf{A}$ is called an S-matrix,[16] and its inverse, $\mathbf{A}^{-1}$, has all positive elements (positive matrix). Therefore, $\mathbf{A}_1^{-1}$ and $\mathbf{A}_2^{-1}$ are positive symmetric matrices.

Notice that $\mathbf{Q}$, $\mathbf{F}_1$, and $\mathbf{F}_2$ are diagonal matrices. The diagonal elements of $\mathbf{Q}$ are all positive. The diagonal elements of $\mathbf{F}_1$ and $\mathbf{F}_2$ are positive or zero. It now follows that $\mathbf{F}_1 \mathbf{A}_1^{-1}$ is a symmetric, nonnegative matrix and $\mathbf{F}_2 \mathbf{A}_2^{-1} \mathbf{Q} \mathbf{A}_1^{-1}$ is a nonsymmetric, nonnegative matrix. As a result, $\mathbf{R}$ is a nonsymmetric, nonnegative matrix. If $\mathbf{F}_1$ and $\mathbf{F}_2$ have zero diagonal elements at the same positions, then all the elements in the corresponding rows of $\mathbf{R}$ are zero. Consequently, the corresponding elements of $\mathbf{y}$ also become zero. The physical meaning of the ith element of $\mathbf{y}$ is the fission source at the ith mesh point (or mesh interval). If there is no fissionable material at the ith mesh point, the ith diagonal element of $\mathbf{F}_1$ as well as $\mathbf{F}_2$ and the ith element of $\mathbf{y}$ are all zero. When $\mathbf{y}$ includes zeros in this manner, the order of Eq. (2.7.8) can be reduced by removing all such zeroes from $\mathbf{y}$ and by removing the corresponding columns and rows from $\mathbf{R}$. The reduced $\mathbf{R}$ thus obtained is a nonsymmetric positive matrix. Therefore, we assume that $\mathbf{R}$ is a nonsymmetric positive matrix.

---

* Suppose a linear equation $\mathbf{Ax} = \mathbf{s}$, where $\mathbf{A}$ is a matrix, $\mathbf{x}$ is the unknown vector, and $\mathbf{s}$ is a known vector. If any number of the unknown elements in $\mathbf{x}$ can be solved independently of the remaining elements in $\mathbf{x}$, then $\mathbf{A}$ is said to be *reducible*. If no part of $\mathbf{x}$ is solvable independently, the matrix $\mathbf{A}$ is *irreducible*. The physical meaning of irreducibility is discussed in Section 3.5.

A positive matrix has a unique positive eigenvector with a single positive eigenvalue greater in value than the modulus of any other eigenvalue of the matrix (Perron's theorem).[16] It is this positive eigenvalue and the corresponding, unique positive eigenvector in which we are interested. Since $\mathbf{R}$ is nonsymmetric, other eigenvalues may include pairs of complex conjugate values. By using the Perron's theorem, it can be easily proven that the iterative scheme, Eq. (2.7.10), converges to the largest eigenvalue and the unique positive eigenvector. If $\mathbf{R}$ is assumed to have only positive and distinct eigenvalues, the proof of convergence for Eq. (2.7.10) is reduced to the proof of the iterative convergence for Eq. (2.4.4).

The power method for the two-group diffusion equation has been discussed for its basic nature and theoretical interest. However, it is seldom used by itself because of its slow convergence rate. For the one-dimensional few-group equations, the Wielandt method is recommended. In applying the Wielandt method it is advantageous to express the finite difference equation for g-group diffusion equation by a single matrix equation (see Problem 21) as

$$\mathbf{A}\boldsymbol{\phi} = \lambda \mathbf{F}\boldsymbol{\phi} \qquad (2.7.11)$$

where $\lambda = 1/k$, $\mathbf{A}$ is a block-tridiagonal matrix, $\mathbf{F}$ is a block-diagonal matrix, and $\boldsymbol{\phi}$ is a block vector (the order of each block is g). In Eq. (2.7.11) the ith block (subvector) of $\boldsymbol{\phi}$ is col $[\phi_1(x_i), \phi_2(x_i), \ldots, \phi_g(x_i)]$. Equation (2.7.11) may be transformed into the form of Eq. (2.5.1). The major difference in Eq. (2.5.1) between the one-group problem and the g-group problem is that the matrix elements in Eq. (2.5.1) are numbers for a one-group problem but become blocks (submatrices) of order g for a g-group problem. All the procedures after Eq. (2.5.1) applies to the g-group problem if the arithmetics for numbers is replaced by that for $g \times g$ submatrices and subvectors of order g[see Eq. (1.7.17)]. As the number of groups increases, the Wielandt method becomes less attractive because the inversion and multiplication of $g \times g$ submatrices become time consuming, and the total computing time significantly increases even if the total iteration number is small.

For multigroup diffusion equations, the Chebyshev polynomial method described in Section 2.8 is recommended.

2.8 CHEBYSHEV POLYNOMIAL METHODS

In this section, the methods for accelerating iterative convergence of eigenvalue problems by using linear combinations of the previous iterative vectors are described. The Chebyshev polynomials are used to obtain the best linear combinations when there is no knowledge of higher eigenvalues.

The Wielandt method discussed in Section 2.5 becomes less efficient as the number of energy groups increases, when the Chebyshev polynomial method becomes important. They are also important to two- and three-dimensional eigenvalue problems. There are two types of Chebyshev polynomial methods: a one-parameter and a two-parameter method.

2.8.1   One-Parameter Method[15]

The Chebyshev polynomial method to accelerate the iterative convergence of Eq. (2.7.10) is derived with the assumption that the eigenvalues of $\mathbf{R}$ are all positive and real. In general, we do not know if the eigenvalues of a given multigroup equation are all positive and real or include complex values. However, the Chebyshev polynomial method is found to work well even when complex eigenvalues may be included. The Chebyshev polynomial method with complex eigenvalues are discussed in Reference 17.

Let us consider the iterative scheme:

$$\mathbf{y}^{(t)} = \frac{1}{k^{(t-1)}} \theta^{(t)} \mathbf{R} \mathbf{y}^{(t-1)} + (1 - \theta^{(t)}) \mathbf{y}^{(t-1)} \qquad (2.8.1)$$

where $\mathbf{y}^{(t)}$ is the tth iterative vector, $\theta^{(t)}$ are extrapolation parameters, and $k^{(t)}$ is calculated by

$$k^{(t)} = \frac{\langle \mathbf{w}, \mathbf{R} \mathbf{y}^{(t-1)} \rangle}{\langle \mathbf{w}, \mathbf{y}^{(t-1)} \rangle}$$

If $\theta^{(t)}$ is set constant, $1 < \theta < 2$, Eq. (2.8.1) is designated as the fixed-parameter extrapolation method. Our present objective is to determine $\theta^{(t)}$ so that the maximum convergence rate is obtained. In deriving numerical schemes we pretend as if we know all the eigenvalues of $\mathbf{R}$. The eigenvectors of $\mathbf{R}$ are now denoted by $\mathbf{u}_n$, which satisfy

$$\mathbf{R} \mathbf{u}_n = \mu_n \mathbf{u}_n, \qquad n = 0, 1, \dots, N \qquad (2.8.2)$$

where μ_n is the eigenvalue corresponding to $\mathbf{u}_n$. It is assumed that all the eigenvalues are real and positive. The eigenvalues are ordered in the sequence,

$$0 < \mu_N < \mu_{N-1} < \cdots < \mu_1 < \mu_0 = k$$

The largest eigenvalue, μ_0, is equal to k, and we want to solve $\mathbf{u}_0$ by Eq. (2.8.1). It is more convenient to transform μ_n to γ_n by

$$\gamma_n = 2 \frac{\mu_n - \mu_N}{\mu_1 - \mu_N} - 1 \qquad (2.8.3)$$

When the number of grid points is sufficiently large, μ_N is very close to zero, so Eq. (2.8.3) can be approximated by

$$\gamma_n = 2\frac{\mu_n}{\mu_1} - 1 \tag{2.8.4}$$

The range of γ_n is $-1 < \gamma_n < \gamma_1 = 1$ for $n > 1$. γ_0 is larger than 1.

Expanding the intial vector in the eigenvectors as

$$\mathbf{y}^{(0)} = \sum_{n=0}^{N} a_n^{(0)} \mathbf{u}_n \tag{2.8.5}$$

and introducing it in Eq. (2.8.1), the iterative vectors become

$$\mathbf{y}^{(t)} = \eta^{(t)}(\gamma_0) a_0^{(0)} \mathbf{u}_0 + \sum_{n=1}^{N} \eta^{(t)}(\gamma_n) a_n^{(0)} \mathbf{u}_n \tag{2.8.6}$$

In Eq. (2.8.6), $\eta^{(t)}(\gamma_n)$ is a tth order polynomial of γ_n and is defined by

$$\eta^{(t)}(\gamma_n) = \prod_{p=1}^{t} \left[\theta^{(p)} \frac{\mu_1(\gamma_n + 1)}{2k^{(p-1)}} + 1 - \theta^{(p)} \right] \tag{2.8.7}$$

where Eq. (2.8.4) is used. It is assumed in Eq. (2.8.7) that $k^{(p)}$ is very close to $k = \mu_0$. Then Eq. (2.8.7) may be written as

$$\eta^{(t)}(\gamma_n) = \prod_{p=1}^{t} \left[\theta^{(p)} \sigma \frac{\gamma_n + 1}{2} + 1 - \theta^{(p)} \right] \tag{2.8.8}$$

where σ is the dominance ratio defined by $\sigma \equiv \mu_1/\mu_0 < 1$.

The first term of Eq. (2.8.6) is the fundamental eigenvector we seek, and the second term is the residual. Therefore, convergence means that all the ratios, $\eta^{(t)}(\gamma_n)/\eta^{(t)}(\gamma_0)$ for $n > 0$, become zero. Let us try to minimize

$$\max_{n>0} \left| \frac{\eta^{(K)}(\gamma_n)}{\eta^{(K)}(\gamma_0)} \right| \tag{2.8.9}$$

during a fixed number, K, of iterations. This may be done by selecting an appropriate set of $\theta^{(t)}$, $t = 1, 2, \ldots, K$. Generally it is impossible to know individual values of μ_ns or γ_ns, so that we assume γ_n, $n > 0$, are continuously distributed in $[-1, 1]$. In other words, γ is considered to be a continuous variable in the range, $-1 < \gamma \le +1$. Finding the best set of $\theta^{(t)}$ may be restated as follows. The errors at $t = K$ in Eq. (2.8.6) become smallest if $\theta^{(t)}$ are so chosen that

$$\max_{-1<\gamma\le1} \left| \frac{\eta^{(K)}(\gamma)}{\eta^{(K)}(\gamma_0)} \right| \tag{2.8.10}$$

becomes minimum among all possible sets of $\theta^{(t)}$.

The problem of finding such a set of $\theta^{(t)}$ is classic. The answer is to make $\eta^{(K)}(\gamma)$ the Chebyshev polynomial[18] of order K, that is,

$$T^{(K)}(\gamma) = \cos\,(K\,\cos^{-1}\gamma) \qquad (2.8.11)$$

The two polynomials, $\eta^{(K)}(\gamma)$ and $T^{(K)}(\gamma)$, of order K are equivalent if both have the same roots. The roots of $T^{(K)}(\gamma)$ are

$$\xi_t = \cos\frac{\pi}{2K}(2t-1),\ t = 1, 2, \ldots, K \qquad (2.8.12)$$

Forcing $\eta^{(K)}(\gamma)$ to have the same roots, we have the equations

$$\theta^{(t)}\sigma\frac{\xi_t+1}{2}+1-\theta^{(t)}=0,\quad t = 1, 2, \ldots, K \qquad (2.8.13)$$

By solving Eq. (2.8.13), $\theta^{(t)}$ are given by

$$\theta^{(t)} = \frac{1}{1-(\sigma/2)[\cos\,(\pi/2K)(2t-1)+1]} \qquad (2.8.14)$$

The sequence of $\theta^{(t)}$ may be changed arbitrarily. This method assumes that the dominance ratio is known before solving the problem (see Problem 16).

Example 2.3

If the same θ is used for every iteration in Eq. (2.8.1), the scheme is called the fixed extrapolation method. The optimal θ is determined by Eq. (2.8.12) by setting $K = 1$:

$$\theta = \frac{1}{1-\sigma/2}$$

Figure 2.5 shows $\eta^{(t)}(\gamma)/\eta^{(t)}(\gamma_0)$ with the optimal θ for $t = 8$.

Example 2.4

If $\sigma = 0.98$ and $K = 8$, the values of $\theta^{(t)}$ are 33.995, 9.749, 4.206, 2.413, 1.651, 1.278, 1.090, 1.010. For this set of θs, $\eta^{(8)}(\gamma)/\eta^{(8)}(\gamma_0)$ is plotted in Fig. 2.5.

Example 2.5

Suppose the exact dominance ratio is $\mu_1/\mu_0 = 0.98$. If a lower estimate for σ such as $\sigma = 0.95$ is used in Eq. (2.8.13), $\theta^{(t)}$ for $K = 8$ become 16.913, 7.689, 3.830, 2.313, 1.619, 1.268, 1.087, 1.009. $\eta^{(8)}(\gamma)/\eta^{(8)}(\gamma_0)$ is plotted in Fig. 2.5. The decay of the eigenvectors with $\sigma = 0.95$ as an estimate become

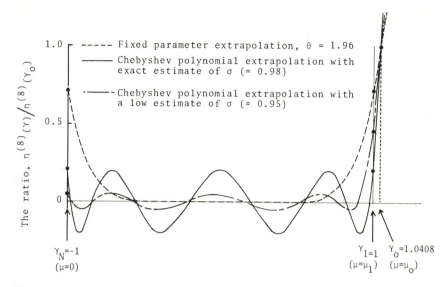

Figure 2.5 Decay rate of error vectors after eight iterations (the dominance ratio is assumed to be $\sigma = \mu_1/\mu_0 = 0.98$).

faster for $\gamma \ll 1$, but those close to 1 become slower compared to when the exact σ is used.

There are two drawbacks associated with the one-parameter Chebyshev polynomial method. One is that the number of iterations to achieve a Chebyshev polynomial must be preset. The other is that if K is large, some values of $\theta^{(t)}$ are very large numbers, thus $\theta^{(t)}$ and $1 - \theta^{(t)}$ will have similar magnitudes with different signs. As the iterative solution approaches the exact solution, two consecutive vectors, $\mathbf{y}^{(t-1)}$ and $\mathbf{y}^{(t)}$ become very close. Consequently, subtraction of two similar vectors, namely, $\theta^{(t)}\mathbf{y}^{(t)} - (\theta^{(t)} - 1)\mathbf{y}^{(t-1)}$, will involve a serious round-off error. Generally, a small K is selected and the set of K values of $\theta^{(t)}$s is repeatedly used. The two-parameter Chebyshev polynomial method is free from those problems.

Another problem with the Chebyshev polynomial method is that the successive approximations of the eigenvalue, $k^{(t)}$, are assumed to be sufficiently close to the exact eigenvalue μ_0. This is a poor assumption, especially in an early stage of iteration. Therefore, it is safer to perform a few power iterations before the Chebyshev polynomial scheme is used. During this period, an initial estimate for the dominance ratio, σ, can also be obtained, and furthermore the accuracy of $k^{(p)}$, which appear in Eq. (2.8.7), particularly for the first few ts, is significantly improved.

2.8.2 Two-Parameter Method[10,17]

In the two-parameter method, a linear combination of three past iterative vectors is considered:

$$\mathbf{y}^{(t)} = \frac{\alpha^{(t)}}{k^{(t-1)}}\mathbf{R}\mathbf{y}^{(t-1)} + (1 - \alpha^{(t)} + \beta^{(t)})\mathbf{y}^{(t-1)} - \beta^{(t)}\mathbf{y}^{(t-2)} \qquad (2.8.15)$$

where $\alpha^{(t)}$ and $\beta^{(t)}$ are the two parameters for extrapolation, and $\alpha^{(0)} = 1$, $\beta^{(0)} = \beta^{(1)} = 0$. Assuming that the initial vector $\mathbf{y}^{(0)}$ is expressed as Eq. (2.8.5), we expand $\mathbf{y}^{(t)}$ in the form

$$\mathbf{y}^{(t)} = \sum_n \eta^{(t)}(\gamma_n)a_n^{(0)}\mathbf{u}_n \qquad (2.8.16)$$

where $\eta^{(t)}(\gamma_n)$ is a polynomial of γ_n of order t, and $\eta^{(0)} = 1$.
 Introducing Eq. (2.8.16) into Eq. (2.8.15), we find

$$\mathbf{y}^{(t)} = \sum_n \left\{ \left[\alpha^{(t)}\frac{\mu_n}{k^{(t-1)}} + 1 - \alpha^{(t)} + \beta^{(t)} \right] \eta^{(t-1)}(\gamma_n) \right.$$
$$\left. - \beta^{(t)}\eta^{(t-2)}(\gamma_n) \right\}a_n^{(0)}\mathbf{u}_n \qquad (2.8.17)$$

It is seen that the polynomial $\eta^{(t)}(\gamma)$ satisfies

$$\eta^{(t)}(\gamma) = \left[\alpha^{(t)}\frac{1+\gamma}{2}\sigma + 1 - \alpha^{(t)} + \beta^{(t)} \right] \eta^{(t-1)}(\gamma)$$
$$- \beta^{(t)}\eta^{(t-2)}(\gamma) \qquad (2.8.18)$$

where $k^{(t)}$ is set to μ_0, assuming that $k^{(t)}$ is very close to μ_0, Eq. (2.8.4) is used and subscript n is omitted.
 We now wish to determine $\alpha^{(t)}$ and $\beta^{(t)}$ so that while $\eta^{(t)}(\gamma_0) = 1$ is satisfied, the maximum of $|\eta^{(t)}(\gamma)|$ for $-1 \le \gamma \le 1$ is minimized. This can be accomplished if $\eta^{(t)}(\gamma)$ becomes the tth order Chebyshev polynomial normalized to 1 at $\gamma = \gamma_0$. We denote the normalized Chebyshev polynomial by

$$\tilde{T}_t(\gamma) \equiv \frac{T_t(\gamma)}{T_t(\gamma_0)} \qquad (2.8.19)$$

where $T_t(\gamma)$ is the tth order Chebyshev polynomial and satisfy the recursion relations as

$$T_0 = 1, \qquad T_1(\gamma) = \gamma$$
$$T_t(\gamma) = 2\gamma T_{t-1}(\gamma) - T_{t-2}(\gamma), \qquad t \ge 2 \qquad (2.8.20)$$

By using Eq. (2.8.20), $\tilde{T}_t(\gamma)$ is found to have the recursion formula given by

$$\tilde{T}_t(\gamma) = 2\gamma \frac{T_{t-1}(\gamma_0)}{T_t(\gamma_0)} \tilde{T}_{t-1}(\gamma)$$

$$- \frac{T_{t-2}(\gamma_0)}{T_t(\gamma_0)} \tilde{T}_{t-2}(\gamma), \qquad t \geq 2 \qquad (2.8.21)$$

In order to satisfy $\eta^{(t)}(\gamma) = \tilde{T}_t(\gamma)$, we equate the coefficients of Eq. (2.8.18) and Eq. (2.8.21) and obtain

$$\alpha^{(t)} \frac{1+\gamma}{2} \sigma + 1 - \alpha^{(t)} + \beta^{(t)} = 2\gamma \frac{T_{t-1}(\gamma_0)}{T_t(\gamma_0)}, \qquad t \geq 2 \qquad (2.8.22)$$

$$\beta^{(t)} = \frac{T_{t-2}(\gamma_0)}{T_t(\gamma_0)} \qquad (2.8.23)$$

Introducing Eq. (2.8.23) into Eq. (2.8.22) and using Eq. (2.8.20), we find

$$\alpha^{(t)} = \frac{4}{\sigma} \frac{T_{t-1}(\gamma_0)}{T_t(\gamma_0)} \qquad (2.8.24)$$

For $t = 1$, we can easily find $\alpha_1 = 1$ and $\beta_1 = 0$.

This method also requires an estimate for the dominance ratio to generate $\alpha^{(t)}$ and $\beta^{(t)}$. The dominance ratio can be estimated in the same way as in the previous method. When the exact dominance ratio is not known, a low dominance ratio estimate must be used. During the iterations using the low dominance ratio estimate, a more accurate dominance ratio can be estimated (see Problem 19).[10]

PROBLEMS

1. Assuming p, q, and f in Eq. (2.1.3) are constant, derive the finite difference equation for Eq. (2.1.3) by directly applying Eq. (1.4.11).

2. What are the differences between the difference equation obtained in Problem 1 and Eq. (2.2.6) derived in Section 2.2 by the point scheme?

3. Prove that, if p, q, and f in Eq. (2.1.3) are constant, the point scheme and the box scheme results in the same difference equation for internal points.

4. The finite difference equations for Sturm–Liouville equations for cylindrical and spherical geometries can be expressed in the form of Eq. (2.2.20).

   a. Calculate the coefficients for internal grid points for cylindrical and spherical geometries in accordance with the point scheme.

    b. Assuming the external boundary conditions are given by Eqs. (2.1.4) and (2.1.5), calculate the coefficients a_i, b_i, and c_i for boundary grid points.

    c. Repeat part **a** by using the box scheme.

5. Prove Eqs. (2.3.2) and (2.3.3).

6. Explicitly write the coefficients of Eqs. (2.2.41) and (2.2.44) for zero external boundary conditions.

7. Prove that the last eigenvector of Eq. (2.3.1) is related to the fundamental eigenvector by

$$\phi_{i,I-2} = (-1)^i \phi_{i,0}$$

8. Prove that, if h in Eq. (2.3.3) is sufficiently small, the dominance ratio is approximately

$$\sigma \equiv \frac{\lambda_0}{\lambda_1} = \frac{\tau\left(\dfrac{\pi}{H}\right)^2 + 1}{4\tau\left(\dfrac{\pi}{H}\right)^2 + 1}$$

    (Notice that, as H increases, σ approaches 1).

9. Prove the following inequalities

$$\min_i \frac{y_i^{(t-1)}}{y_i^{(t)}} \leq \frac{\lambda_0}{\lambda^{(t-1)}} \leq \max_i \frac{y_i^{(t-1)}}{y_i^{(t)}}$$

    where $y_i^{(t)}$ is an element of $\mathbf{y}^{(t)}$ in Eq. (2.4.13).

10. The second eigenvector λ_1 of Eq. (2.4.1) can be estimated during the power iteration by

$$\lambda^{(t)} \frac{\langle \mathbf{w}, \mathbf{y}^{(t-1)} - \mathbf{y}^{(t-2)} \rangle}{\langle \mathbf{w}, \mathbf{y}^{(t)} - \mathbf{y}^{(t-1)} \rangle}$$

    Prove that, as t increases, the above ratio converges to λ_1.

11. Prove that, if $\mathbf{w}$ in Eq. (2.4.16) is set to $\mathbf{y}^{(t)}$ at each iteration, the convergence rate of $\lambda^{(t)}$ becomes two times faster than with $\mathbf{w} = \mathbf{1}$.

12. Prove that Eq. (2.4.5) converges to λ_0 by only one iteration if $\mathbf{w} = \mathbf{u}_1$.

13. In considering the power method in Section 2.4 replace Eq. (2.4.1) by Eq. (2.8.1) and show that $\lambda^{(t)}$ converges to λ_0 if $\theta^{(t)}$ is fixed to a constant in the range, $1 \leq \theta < 2$. Set $\theta^{(0)} = 1$. (This is referred to as the fixed parameter extrapolation method,)

14. Calculate the Chebyshev polynomial extrapolation parameters, $\theta^{(t)}$, $t = 1, 2, \ldots, K$, by using Eq. (2.8.14) for the following three dominance ratios: (a) $\sigma = 0.7$, (b) $\sigma = 0.9$, (c) $\sigma = 0.98$. Set $K = 4$.

15. If $\theta^{(t)}$ is fixed to a constant, what is the optimum value of θ for a given

dominance ratio? Assume $\lambda_{min} = 0$. (Hint: Consider the Chebyshev polynomial extrapolation with $K = 1$.)

16. The dominance ratio can be estimated by using the iterative vectors of the fixed parameter iterations. Derive the equation to estimate the dominance ratio.

17. Calculate the reduction of the first higher eigenvector component in Eq. (2.8.6) via one cycle of Chebyshev polynomial extrapolations with $K = 5$ for $\sigma = 0.98$: (case 1) the dominance ratio is exactly estimated; (case 2) the dominance ratio is 1% overestimated; (case 3) the dominance ratio is 1% underestimated.

18. Suppose the iterative vector is found to include only the fundamental vector and the second eigenvector. Assume that the dominance ratio has been estimated accurately. Derive the equation for the extrapolation parameter, θ, by which the first higher mode is completely eliminated by the next iteration.

19. In applying the two-parameter Chebyshev extrapolation method, the dominance ratio must be estimated before the extrapolated iteration begins. If the estimated dominance ratio is not accurate, the efficiency of the method becomes poor. Derive a method to judge during the extrapolated iterations if the estimated dominance ratio is exact, or overestimated or underestimated.

20. Write explicitly the finite difference equations for Eq. (2.7.1), assuming zero external boundary conditions.

21. **a.** Write the whole difference equation obtained in the above problem in the matrix form

$$\mathbf{A}\boldsymbol{\phi} = \lambda\,\mathbf{R}\boldsymbol{\phi}$$

where the solution vector is expressed as

$$\boldsymbol{\phi} = \text{col}\,[(\phi_{1,1}, \phi_{2,1}), \ldots, (\phi_{1,i}, \phi_{2,i}) \ldots]$$

b. Show that the matrix $\mathbf{A}$ is block tridiagonal.

22. Apply the Wielandt method to the eigenvalue problem in Problem 21. Write a FORTRAN program.

23. Calculate the dominance ratio of the one-energy-group diffusion equation for an infinite homogeneous cylindrical reactor with radius R and a homogeneous spherical reactor with radius R.

24. Develop an algorithm to calculate iteratively higher eigenvectors of a Sturm–Liouville equation by using the power method. [Hint: (a) calculate the eigenvector sequentially starting with the fundamental eigenvector; (b) use the orthogonality of the eigenvector, and eliminate from the iterative vector the eigenvectors that are known already.]

REFERENCES

1. R. Courant and D. Hilbert, *Methods of mathematical physics*, Wiley-Interscience, New York (1962).
2. S. H. Gould, *Variational methods for eigenvalue problems*, University of Toronto, Toronto, Canada (1966).
3. O. J. Marlow and M. C. Suggs, "WANDA-5," WAPD-TM-241, Bettice Atomic Power Laboratory (1960).
4. F. Todt, "FEVER. A one-dimensional few-group depletion program for reactor analysis," GA-2749, General Atomic, Division of General Dynamics (1962).
5. H. Greenspan, C. N. Kelber, and D. Okrent, Ed. *Computing methods in reactor physics*, Gordon and Breach, New York (1968).
6. T. B. Fowler and D. R. Vondy, "Nuclear reactor core analysis code: CITATION," ORNL-TM-2406, Rev 1, Oak Ridge National Laboratory (1970).
7. A. M. Weinberg and E. P. Wigner, *The physical theory of neutron chain reactors*, University of Chicago, Chicago (1958).
8. M. Clark, Jr. and K. F. Hansen, *Numerical methods of reactor analysis*, Academic, New York (1964).
9. H. Wielandt, "Bestimmung hoeheren eigenwerte durch Gebrochene iteration," *Ber. Aerodyn. Versuchsanst. Goett.*, Report 44/J/37 (1944).
10. E. L. Wachspress, *Iterative solution of elliptic systems*, Prentic-Hall, Englewood Cliffs, N.J. (1966).
11. L. Fox, *Numerical solution of ordinary and partial differential equations*, Addison-Wesley, Reading, Mass. (1962).
12. C. W. Sherwin, *Introduction to quantum mechanics*, Holt, Reinholt and Winston, New York, (1959).
13. L. Pauling and E. B. Wilson, Jr., *Introduction to quantum mechanics*, McGraw-Hill, New York (1935).
14. G. Birkhoff and R. S. Varga, "Neutron criticality and non-negative matrices," WAPD-166, Bettice Atomic Power Laboratory (1957).
15. G. G. Bilodeau, "Extrapolation techniques for real symmetric matrices," WAPD-TM-52, Bettice Atomic Power Laboratory (1957).
16. R. S. Varga, *Matrix iterative analysis*, Prentice-Hall, Englewood Cliffs, N.J. (1962).
17. L. A. Hageman, "The Chebyshev polynomical method of iteration," WAPD-TM-537, Bettice Atomic Power Laboratory, Pittsburgh, Pa. (1967).
18. M. A. Snyder, *Chebyshev methods in numerical approximation*, Prentice-Hall, Englewood Cliffs, N.J. (1966).
19. N. W. Winter, "Study of electron correlation in the hydride ion," *J. Chem. Phys.*, **56**, (1972).
20. C. Kittel, *Introduction to solid state physics*, 4th edit., Wiley, New York (1971).

Chapter 3   Iterative Computational Methods for Solving Partial Differential Equations of the Elliptic Type

3.1   PARTIAL DIFFERENTIAL EQUATIONS OF THE ELLIPTIC TYPE

In this chapter we study numerical solutions based on the finite difference method for two-dimensional partial differential equations of elliptic type represented by

$$-\nabla p(\mathbf{r})\nabla \phi(\mathbf{r}) + q(\mathbf{r})\phi(\mathbf{r}) = S(\mathbf{r}) \qquad (3.1.1)$$

This equation is known as the diffusion equation. It becomes the Poisson equation, if $q(\mathbf{r}) = 0$ and $p(\mathbf{r}) =$ constant. If $S(\mathbf{r}) = q(\mathbf{r}) = 0$, it is known as the Laplace equation. The numerical solutions described in this chapter apply to all of those cases. The first step in numerical solution for Eq. (3.1.1) is to derive the finite difference equations. If the coefficients of Eq. (3.1.1) are constant and the mesh intervals are uniform, the central difference formulas for partial derivatives as shown in Section 1.4 may be applied to derive the finite difference equations (see Example 1.5). We, however, consider more general cases, so the difference equations are derived by integrating Eq. (3.1.1) in accordance with the point scheme or box scheme as described in Section 3.2.

Iterative solutions play the central role in solving the elliptic partial differential equations via the finite difference technique. This is because (a) as the number of unknowns is increased, the Gaussian elimination becomes more inefficient or practically impossible except for simple geometries (see Chapter 10), (b) the matrix that represents the coefficients of the difference equations are sparse and regular, so iterative solutions are easier and more economical. As representative iterative schemes, the Jacobi iterative method, the success-over-relaxation (SOR), the cyclic Chebyshev method, and the strongly implicit method are introduced. The mathematical characteristics and the convergence rate of those methods are studied in Sections

3.3 through 3.9. The finite element method and the Monte Carlo method, which have been of increasing interest in recent years as the numerical solutions for Eq. (3.1.1) for irregular geometry, are introduced in Sections 7.2 and 9.9, respectively.

3.2   DERIVATION OF FINITE DIFFERENCE EQUATIONS

We consider a rectangular geometry on an x-y plane. The mesh system is defined as shown in Fig. 3.1 by vertical and horizontal lines. We assume that $p(x)$, $q(x)$, and $S(x)$ are constant in each rectangular mesh box. Two types of difference approximations are shown: one is the *point scheme* and another is the *box scheme*.

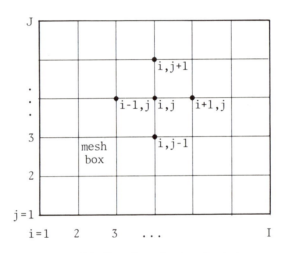

Figure 3.1   Two-dimensional mesh system.

3.2.1   Point Scheme[1-3]

Let us consider the grid point (i, j) and the four adjacent grid points, $(i-1, j)$, $(i+1, j)$, $(i, j-1)$, and $(i, j+1)$ as illustrated in Fig. 3.2. Four rectangular boxes around the point (i, j) are defined by lines intersecting midpoints between (i, j) and the neighboring points. Integrating Eq. (3.1.1) over the four rectangles, we obtain

$$-\int p \frac{\partial}{\partial n} \phi \, ds + \iint q(\mathbf{r}) \phi \, dx \, dy = \iint S(\mathbf{r}) \, dx \, dy \qquad (3.2.1)$$

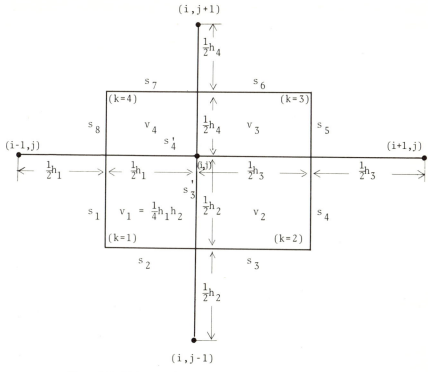

Figure 3.2   Volume and surface elements around a mesh point (i, j).

where the first integral is along the surface of the region of integral and $\partial/\partial n$ is the derivative normal to the surface (*volume* and *surface* are used as generic terms to express an area and the boundary of an area, respectively). Here we approximate ϕ in the second integral by $\phi_{i,j}$ that is, the value of ϕ at (i, j). Then the second integral becomes

$$\iint q(\mathbf{r})\phi \, dx \, dy = (v_1 q_1 + v_2 q_2 + v_3 q_3 + v_4 q_4)\phi_{i,j} \qquad (3.2.2)$$

where

$$v_1 = \frac{h_1 h_2}{4}, \qquad v_2 = \frac{h_2 h_3}{4}, \qquad v_3 = \frac{h_3 h_4}{4}, \qquad v_4 = \frac{h_4 h_1}{4} \qquad (3.2.3)$$

and q_k is the value of $q(\mathbf{r})$ in the volume represented by v_k. Similarly, the right side of Eq. (3.2.1) becomes

$$\iint S \, dx \, dy = v_1 S_1 + v_2 S_2 + v_3 S_3 + v_4 S_4 \qquad (3.2.4)$$

In performing the surface integral of the first term of Eq. (3.2.1), we approximate the derivative by

$$\frac{\partial}{\partial n}\phi = \frac{\phi_{i,j-1}-\phi_{i,j}}{h_2} \qquad \text{on } s_2 \text{ and } s_3$$

$$= \frac{\phi_{i,j+1}-\phi_{i,j}}{h_4} \qquad \text{on } s_6 \text{ and } s_7$$

$$= \frac{\phi_{i-1,j}-\phi_{i,j}}{h_1} \qquad \text{on } s_1 \text{ and } s_8$$

$$= \frac{\phi_{i+1,j}-\phi_{i,j}}{h_3} \qquad \text{on } s_4 \text{ and } s_5$$

(3.2.5)

The surface integral along s_2 and s_3, for example, becomes

$$-\int p\frac{\partial}{\partial n}\phi \, ds = \frac{\phi_{i,j}-\phi_{i,j-1}}{h_2}\left(\frac{p_1 h_1}{2}+\frac{p_2 h_3}{2}\right) \qquad (3.2.6)$$

Collecting all the terms together, we obtain the five-point difference equation,

$$a_{i,j}^L\phi_{i-1,j} + a_{i,j}^R\phi_{i+1,j} + a_{i,j}^B\phi_{i,j-1} + a_{i,j}^T\phi_{i,j+1} + a_{i,j}^C\phi_{i,j} = S_{i,j} \qquad (3.2.7)$$

where the coefficients are defined by

$$a_{i,j}^L = -\frac{p_1 h_2 + p_4 h_4}{2h_1} \qquad (3.2.8)$$

$$a_{i,j}^R = -\frac{p_2 h_2 + p_3 h_4}{2h_3} \qquad (3.2.9)$$

$$a_{i,j}^B = -\frac{p_1 h_1 + p_2 h_3}{2h_2} \qquad (3.2.10)$$

$$a_{i,j}^T = -\frac{p_3 h_3 + p_4 h_1}{2h_4} \qquad (3.2.11)$$

$$a_{i,j}^C = \sum_{k=1}^{4} v_k q_k - (a_{i,j}^L + a_{i,j}^R + a_{i,j}^B + a_{i,j}^T) > 0 \qquad (3.2.12)$$

$$S_{i,j} = \sum_{k=1}^{4} v_k S_k > 0 \qquad (3.2.13)$$

In the preceding equation, a^L, a^R, a^B, a^T are negative and a^C is positive when $p>0$ and $q\geq 0$. Notice also the symmetry properties, $a_{i,j}^L = a_{i-1,j}^R$ and $a_{i,j}^B = a_{i,j-1}^T$.

If the point (i, j) is on the outer boundary, the difference equation is derived by using the boundary condition that is generally expressed by

$$\frac{\partial}{\partial n}\phi + \gamma(s)\phi = \sigma(s) \tag{3.2.14}$$

where s is the coordinate on the outer boundary. The above condition becomes the Neumann boundary condition if $\gamma(s) = 0$, the Dirichlet boundary condition if $\gamma \to \infty$ but $\sigma/\gamma = \text{const}$, and the mixed boundary condition if $\gamma \neq 0$. In the case of the mixed boundary condition with $\sigma = 0$, $1/\gamma$ represents the extrapolation distance.[3]

Suppose v_1 in Fig. 3.2 is interior but s_3' and s_4' are on the outer boundary Integrations in Eq. (3.2.1) apply only to v_1. The second and the right side become, respectively,

$$\iint_{v_1} q(\mathbf{r})\phi \, dx \, dy = v_1 q_1 \phi_{i,j} \tag{3.2.15}$$

$$\iint_{v_1} S(\mathbf{r}) \, dx \, dy = v_1 S_1 \tag{3.2.16}$$

The surface integral for the first term of Eq. (3.1.2) is performed along the surface of v_1, namely, on s_1, s_2, s_3', and s_4'. The derivatives of ϕ normal to the surfaces s_3' and s_4' are evaluated by Eq. (3.2.14), so that the surface integral becomes

$$-\int_{s_1 \sim s_4'} p\frac{\partial}{\partial n}\phi \, ds = \frac{\phi_{i,j} - \phi_{i-1,j}}{h_1}\left(\frac{p_1 h_2}{2}\right) + \frac{\phi_{i,j} - \phi_{i,j-1}}{h_2}\left(\frac{p_1 h_1}{2}\right)$$

$$+ (\gamma_3\phi_{i,j} - \sigma_3)\left(\frac{p_1 h_2}{2}\right) + (\gamma_4\phi_{i,j} - \sigma_4)\left(\frac{p_1 h_1}{2}\right) \tag{3.2.17}$$

The first two terms are due to s_1 and s_2. The third and fourth terms are due to s_3' and s_4', where γ_3 and γ_4 are the values of $\gamma(s)$ on s_3' and s_4', respectively, while σ_3 and σ_4 are for s_3' and s_4', respectively. Collecting all the terms, the difference equation is expressed by Eq. (3.2.7) with the coefficients defined by

$$a_{i,j}^L = -\frac{p_1 h_2}{2h_1} \tag{3.2.18}$$

$$a_{i,j}^R = 0 \tag{3.2.19}$$

$$a_{i,j}^B = -\frac{p_1 h_1}{2h_2} \tag{3.2.20}$$

$$a_{i,j}^T = 0 \qquad (3.2.21)$$

$$a_{i,j}^C = v_1 q_1 - (a_{i,j}^L + a_{i,j}^B) + p_1(\gamma_3 h_2 + \gamma_4 h_1)/2 \qquad (3.2.22)$$

$$S_{i,j} = v_1 S_1 + p_1(\sigma_3 h_2 + \sigma_4 h_1)/2 \qquad (3.2.23)$$

The difference equations for Eq. (3.1.1) on other two-dimensional coordinate systems (r–z and r–θ) can be obtained if the volume and surface integrals are changed properly (see Problem 3).

Eq. (3.2.7) for all the grids may be expressed in the matrix form as

$$\mathbf{A}\boldsymbol{\phi} = \mathbf{S} \qquad (3.2.24)$$

where $\mathbf{A}$ represents $a_{i,j}$s, $\boldsymbol{\phi}$ all of $\phi_{i,j}$s and $\mathbf{S}$ represents all of $S_{i,j}$s. The configuration of $\mathbf{A}$ depends on how the elements of $\boldsymbol{\phi}$ correspond to $\phi_{i,j}$. When $p > 0$ and $q \geq 0$, the matrix $\mathbf{A}$ has positive diagonal elements and nonpositive offdiagonal elements. This matrix is always symmetric.

Example 3.1

Consider the mesh system consisting of nine grids shown in Fig. 3.3a. If we array $\phi_{i,j}$ in the sequence shown in Fig. 3.3b, Eq. (3.2.24) becomes

$$
\begin{bmatrix}
a_{11}^C & a_{11}^R & 0 & a_{11}^T & 0 & 0 & 0 & 0 & 0 \\
a_{21}^L & a_{21}^C & a_{21}^R & 0 & a_{21}^T & 0 & 0 & 0 & 0 \\
0 & a_{31}^L & a_{31}^C & 0 & 0 & a_{31}^T & 0 & 0 & 0 \\
a_{12}^B & 0 & 0 & a_{12}^C & a_{12}^R & 0 & a_{12}^T & 0 & 0 \\
0 & a_{22}^B & 0 & a_{22}^L & a_{22}^C & a_{22}^R & 0 & a_{22}^T & 0 \\
0 & 0 & a_{32}^B & 0 & a_{32}^L & a_{32}^C & 0 & 0 & a_{32}^T \\
0 & 0 & 0 & a_{13}^B & 0 & 0 & a_{13}^C & a_{13}^R & 0 \\
0 & 0 & 0 & 0 & a_{23}^B & 0 & a_{23}^L & a_{23}^C & a_{23}^R \\
0 & 0 & 0 & 0 & 0 & a_{33}^B & 0 & a_{33}^L & a_{33}^C
\end{bmatrix}
\begin{bmatrix}
\phi_{11} \\ \phi_{21} \\ \phi_{31} \\ \phi_{12} \\ \phi_{22} \\ \phi_{32} \\ \phi_{13} \\ \phi_{23} \\ \phi_{33}
\end{bmatrix}
=
\begin{bmatrix}
S_{11} \\ S_{21} \\ S_{31} \\ S_{12} \\ S_{22} \\ S_{32} \\ S_{13} \\ S_{23} \\ S_{33}
\end{bmatrix}
$$

$$(3.2.25)$$

Notice that Eq. (3.2.25) is a pentadiagonal matrix and also a block-tridiagonal matrix. The diagonal blocks are all tridiagonal.

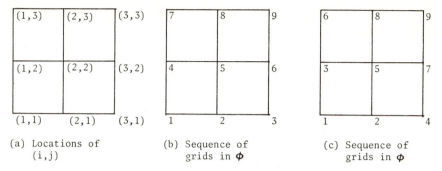

(a) Locations of (i,j) (b) Sequence of grids in φ (c) Sequence of grids in φ

Figure 3.3 A mesh system consisting of nine grids and the two ways of consistent ordering.

If the sequence shown in Fig. 3.3c is used, Eq. (3.2.24) becomes

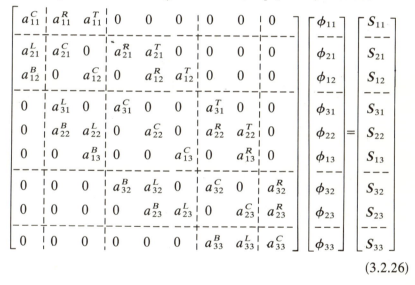

$$(3.2.26)$$

Equation (3.2.26) is a block-tridiagonal matrix, in which the diagonal blocks are diagonal and the nonnull off-diagonal blocks are nonsquare.

3.2.2 Box Scheme[4]

The box scheme was developed particularly for neutron-diffusion equations. The advantage of this scheme is that the computational effort for the second and the right terms of Eq. (3.2.1) becomes much smaller than the point scheme. The discrete values of ϕ in the box scheme represent the average solution in mesh boxes.

Consider the mesh box (i, j) as well as its adjacent mesh boxes as shown in Fig. 3.4. Integrating Eq. (3.1.1) in the box (i, j) yields

$$J^L + J^R + J^B + J^T + \iint_{(i,j)} \phi \, dx \, dy = \iint_{(i,j)} S \, dx \, dy \qquad (3.2.27)$$

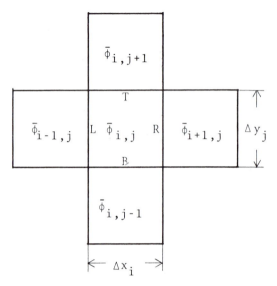

Figure 3.4 Mesh boxes for the box scheme.

where Js represent the particle current across the boundaries of the box:

$$J^L = -\int_L p \frac{\partial}{\partial n} \phi \, ds \qquad (3.2.28)$$

$$J^R = -\int_R p \frac{\partial}{\partial n} \phi \, ds \qquad (3.2.29)$$

$$J^B = -\int_B p \frac{\partial}{\partial n} \phi \, ds \qquad (3.2.30)$$

$$J^T = -\int_T p \frac{\partial}{\partial n} \phi \, ds \qquad (3.2.31)$$

and where $\partial/\partial n$ is a derivative normal to the surface, and T, B, L, and R are top, bottom, left, and right sides of the mesh box (i, j), respectively. Using the average value of the solution in the mesh box, the integrals become

$$\iint_{(i,j)} q \, dx \, dy = q_{i,j}\bar{\phi}_{i,j}\Delta x_i \Delta y_j \qquad (3.2.32)$$

$$\iint_{(i,j)} S \, dx \, dy = S_{i,j} \qquad (3.2.33)$$

where $\bar{\phi}_{i,j}$ denotes the average of ϕ in the mesh box (i, j).

The current terms are approximated in the way similar to the one-dimensional case (see Section 2.2) as

$$J^L = a_{i,j}^L(\bar{\phi}_{i-1,j} - \bar{\phi}_{i,j}) \qquad (3.2.34)$$

$$J^R = a_{i,j}^R(\bar{\phi}_{i+1,j} - \bar{\phi}_{i,j}) \qquad (3.2.35)$$

$$J^B = a_{i,j}^B(\bar{\phi}_{i,j-1} - \bar{\phi}_{i,j}) \qquad (3.2.36)$$

$$J^T = a_{i,j}^T(\bar{\phi}_{i,j+1} - \bar{\phi}_{i,j}) \qquad (3.2.37)$$

where

$$a_{i,j}^L = \frac{-2\Delta y_j}{\Delta x_{i-1}/p_{i-1,j} + \Delta x_i/p_{i,j}} \qquad (3.2.38)$$

$$a_{i,j}^R = \frac{-2\Delta y_j}{\Delta x_{i+1}/p_{i+1,j} + \Delta x_i/p_{i,j}} \qquad (3.2.39)$$

$$a_{i,j}^B = \frac{-2\Delta x_i}{\Delta y_{j-1}/p_{i,j-1} + \Delta y_j/p_{i,j}} \qquad (3.2.40)$$

$$a_{i,j}^T = \frac{-2\Delta x_i}{\Delta y_{j+1}/p_{i,j+1} + \Delta y_j/p_{i,j}} \qquad (3.2.41)$$

Collecting all the terms, the difference equation becomes

$$a_{i,j}^L\bar{\phi}_{i-1,j} + a_{i,j}^R\bar{\phi}_{i+1,j} + a_{i,j}^B\bar{\phi}_{i,j-1} + a_{i,j}^T\bar{\phi}_{i,j+1} + a_{i,j}^C\bar{\phi}_{i,j} = S_{i,j} \qquad (3.2.42)$$

where

$$a_{i,j}^C = q_{i,j}\Delta x_i \Delta y_j - (a_{i,j}^L + a_{i,j}^R + a_{i,j}^B + a_{i,j}^T) \qquad (3.2.43)$$

Notice that a^L, a^R, a^B, and a^T are negative while a^C is positive assuming that $p > 0$ and $q \geq 0$.

If one of the surfaces of the mesh box is on a boundary, the boundary condition given by Eq. (3.2.14) is applied to evaluate the current across that surface. Suppose that the top surface is on the external boundary, for example, then Eq. (3.2.37) is approximated by

$$J_T = -\int_T p\frac{\partial}{\partial n}\phi \, ds = p_{i,j}\gamma_T\bar{\phi}_{i,j}\Delta x_i > 0 \qquad (3.2.44)$$

where γ_T is the value of $\gamma(s)$ on the top boundary and $\sigma(s)$ in Eq. (3.2.14) is assumed to be zero.

3.3   ITERATIVE SOLUTION METHODS AND THEIR REPRESENTATIONS IN THE MATRIX FORM

The two difference schemes, Eqs. (3.2.7) and (3.2.42), that may be expressed in the matrix form as Eq. (3.2.24) have the same form and same mathematical properties, such as the irreducibly diagonal dominance and the symmetry property that are discussed in Section 3.5. Therefore, all the discussions in the following sections of this chapter equally apply to both difference schemes. Equation (3.2.24) could be solved by the Gaussian elimination. However, a large number of unknowns (for example $100 \times 100 = 10,000$) leads to a huge matrix for which the direct Gaussian elimination is very inefficient or practically impossible. Iterative solution methods are more suitable for the set of Eq. (3.2.7). In fact, Eq. (3.2.7) has various favorable properties for iterative solution methods.

The following iteratives schemes are well known:

POINT JACOBI ITERATIVE METHOD

This method is written as

$$\phi_{i,j}^{(t)} = \frac{1}{a^C}[S_{i,j} - a^L\phi_{i-1,j}^{(t-1)} - a^B\phi_{i,j-1}^{(t-1)} - a^R\phi_{i+1,j}^{(t-1)} - a^T\phi_{i,j+1}^{(t-1)}] \quad (3.3.1)$$

where superfix t is the iteration number. This calculation is performed for all of (i, j) in each iteration cycle. When $t = 1$, $\phi_{i,j}^{(0)}$ on the right side is called the initial guess and can be set to any arbitrary value. The order of calculating $\phi_{i,j}$ in each iteration is arbitrary. The Jacobi iterative method is rarely used by itself. However, understanding the mathematical properties of the Jacobi iterative method is fundamental to the study of the successive-over-relaxation method and the cyclic Chebyshev semiiterative method.

LINE-JACOBI METHOD

If the solutions along the $(j-1)$th mesh row and the solutions along the $(j+1)$th mesh row are given by the $(t-1)$th iteration, a set of simultaneous equations for unknowns along the jth row is written as

$$a^L\phi_{i-1,j}^{(t)} + a^C\phi_{i,j}^{(t)} + a^R\phi_{i+1,j}^{(t)} = S_{i,j} - a^B\phi_{i,j-1}^{(t-1)}$$
$$-a^T\phi_{i,j+1}^{(t-1)} \quad \text{for } i = 1, 2, \ldots, I \quad (3.3.2)$$

where unknown terms are on the left side and known terms are on the right side. The left side of Eq. (3.3.2) has the same form as the finite difference equation of the one-dimensional diffusion equation [see Eq. (2.2.6)], so the recursion formula Eqs. (1.7.15) and (1.7.16) can be used to solve the unknowns. Since all the unknowns on a line are solved simultaneously by matrix inversion, this method is called a *line inversion method*. All the lines are inverted in each iteration time. The order of j in each iteration is arbitrary.

GAUSS–SEIDEL ITERATIVE METHODS

When Eq. (3.3.1) is used, one may ask if the convergence rate of iterations is increased by replacing the terms on the right side by the most updated values whenever available. The answer is positive. If the calculation of $\phi_{i,j}$ is in the sequence starting from left bottom and ending at right top of the mesh system, the point Gauss–Seidel method is written as

$$\phi_{i,j}^{(t)} = \frac{1}{a^c}[S_{i,j} - a^L\phi_{i-1,j}^{(t)} - a^B\phi_{i,j-1}^{(t)} - a^R\phi_{i+1,j}^{(t-1)} - a^T\phi_{i,j+1}^{(t-1)}] \quad (3.3.3)$$

This method also saves the core memory, because as soon as a new approximation is obtained it replaces the old value. Although the sequence of calculating $\phi_{i,j}^{(t)}$ in every iteration is not so arbitrary as the point Jacobi iterative method, there are at least three equivalent ways of sweeping grid points as follows:

a. Referring to Fig. 3.1, the calculation in each iteration starts at the left bottom corner, $i=j=1$, and sweeps $i=2, 3, \ldots, I$ in the increasing order. This is repeated for $j=2, 3, \ldots, J$ in the increasing order. The present sweeping scheme corresponds to the sequence shown in Fig. 3.3b.

b. The calculation starts at $i=j=1$ but sweeps $j=2, 3, \ldots, J$ in the increasing order. This is repeated for $i=2, 3, \ldots, I$ in the increasing order.

c. The sweep starts at $i=j=1$. The second step involves the calculations for $\phi_{1,2}^{(t)}$ and $\phi_{2,1}^{(t)}$. The lth step is the calculation for $\phi_{i,l+1-i}$, where i is varied along the diagonal line represented by $(i, l+1-i)$. This corresponds to the sequence shown in Fig. 3.3c.

The line Gauss–Seidel method is derived by replacing $\phi_{i+1,j}^{(t-1)}$ in Eq. (3.3.3) by $\phi_{i+1,j}^{(t)}$. Each grid row is inverted by the tridiagonal solution.

SUCCESSIVE-OVER-RELAXATION (SOR) METHODS

By introducing an extrapolation factor ω that satisfies $1 < \omega < 2$, the Gauss–Seidel method becomes the successive-over-relaxation method. The point successive-over-relaxation method (SOR) is written

$$\phi_{i,j}^{(t)} = \frac{\omega}{a^C}[S_{i,j} - a^L \phi_{i-1,j}^{(t)} - a^B \phi_{i,j-1}^{(t)} - a^R \phi_{i+1,j}^{(t-1)} - a^T \phi_{i,j+1}^{(t-1)}]$$
$$+ (1 - \omega)\phi_{i,j}^{(t-1)}$$

(3.3.4)

The successive-line-over-relaxation method (SLOR) is written

$$\phi_{i,j}^{(t)} = \omega\tilde{\phi}_{i,j}^{(t)} + (1 - \omega)\phi_{i,j}^{(t-1)}$$

(3.3.5)

where $\tilde{\phi}_{ij}$ are the solutions of the set of simultaneous equations,

$$a^L \tilde{\phi}_{i-1,j}^{(t)} + a^C \tilde{\phi}_{i,j}^{(t)} + a^R \tilde{\phi}_{i+1,j}^{(t)} = S_{i,j} - a^B \phi_{i,j-1}^{(t)}$$
$$a^T \phi_{i,j+1}^{(t-1)}, \quad i = 1, 2, \ldots, I$$

(3.3.6)

In the SLOR method, the line inversions start from the bottom horizontal mesh line ($j = 1$; $i = 1, 2, \ldots, I$) and end at the top horizontal mesh line ($j = J$; $i = 1 \sim I$). The same sweep is performed in each iteration.

The convergence rate of the SOR method including SLOR is dependent on the value of ω. If the problem is intrinsically fast convergent, ω may be rather arbitrarily chosen in the range $\omega = 1.4 \sim 1.7$. However, the proper choice of ω is particularly important when the given problem is intrinsically slowly convergent. If the program is to be used repeatedly, it is worthwhile to develop a subprogram to calculate the optimal ω. The effect of ω is discussed in Section 3.7. A very slowly convergent problem can be substantially accelerated by combining the SOR with a coarse mesh rebalancing scheme described in Chapter 8.

ALTERNATING DIRECTION IMPLICIT METHOD[5,6]

The alternating-direction-implicit method consists of two kinds of sweeps: one inverting horizontally from the bottom to top rows, another inverting vertically from the leftmost to the rightmost columns. The present method is expressed by

STEP 1: ROW INVERSIONS

$$a^L\phi_{i-1,j}^{(t-1/2)}+a^R\phi_{i+1,j}^{(t-1/2)}+[a^B+a^T+a^C+\omega^{(t)}(a^L+a^R+a^B+a^T)]\phi_{i,j}^{(t-1/2)}$$

$$=-a^B\phi_{i,j-1}^{(t-1)}-a^T\phi_{i,j+1}^{(t-1)} \qquad\qquad (3.3.7)$$

$$-[a^B+d^T-\omega^{(t)}(a^L+a^R+a^B+a^T)]\phi_{i,j}^{(t-1)}+S_{i,j}$$

STEP 2: COLUMN INVERSION

$$a^B\phi_{i,j-1}^{(t)}+a^T\phi_{i,j+1}^{(t)}+[a^L+a^R+a^C+\omega^{(t)}(a^L+a^R+a^B+a^T)]\phi_{i,j}^{(t)}$$

$$=-a^L\phi_{i-1,j}^{(t-1/2)}-a^R\phi_{i+1,j}^{(t-1/2)} \qquad\qquad (3.3.8)$$

$$-[a^L+a^R-\omega^{(t)}(a^L+a^R+a^B+a^T)]\phi_{i,j}^{(t-1/2)}+S_{i,j}$$

The ADI method has a greater convergence rate than the SOR method for many problems. However, there are several difficulties: (a) The convergence of the scheme has not been proven for general problems. In fact, the convergence of certain problems becomes very slow or nonconvergent. (b) Determination of the extrapolation factor ω is not easy. (c) If an auxiliary core memory such as a large slow core, disc, or drum is to be used, an efficient data transfer between the regular core and the auxiliary is difficult. (d) The programming needs more effort than other methods. For these reasons, the ADI method is not so frequently used in spite of its interesting formula and the fast convergence rate for many problems.

CYCLIC CHEBYSHEV SEMIITERATIVE METHOD

The Jacobi iterative scheme is accelerated by the Chebyshev polynomials. Details are given in Section 3.8.

STRONGLY IMPLICIT METHOD

This method is described in Section 3.9. In spite of its *implicit algorithm*, the scheme is relatively simple.

The alternative-direction-implicit method is called a two-step iterative method, while all other methods in this section are called one-step iterative methods. In studying the mathematical properties of the iterative methods, it is convenient to express the iterative equations in the matrix form. The matrix representations of Eqs. (3.3.1) and (3.3.2) are illustrated in the next example.

Example 3.2

Considering the mesh system consisting of nine grid points of Fig. 3.3a, the point Jacobi iterative method in accordance with the array of $\phi_{i,j}$ as in Eq.

(3.2.25) becomes the following form:

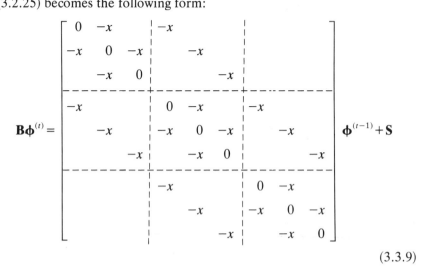

$$\tag{3.3.9}$$

where x corresponds to the nonzero elements in Eq. (3.2.25) and

$$\mathbf{B} = \mathrm{diag}\,[a_{11}^{c}, a_{12}^{c}, a_{13}^{c} \mid a_{21}^{c}, a_{22}^{c}, \ldots, a_{33}^{c}] \tag{3.3.10}$$

$$\boldsymbol{\phi}(t) = \mathrm{col}[\phi_{11}^{(t)}, \phi_{21}^{(t)}, \phi_{31}^{(t)} \mid, \ldots, \phi_{33}^{(t)}] \tag{3.3.11}$$

If we start with Eq. (3.2.26), the point Jacobi iterative method is equivalently written in the form:

$$\tag{3.3.12}$$

where

$$\mathbf{B} = \mathrm{diag}\,[a_{11}^C \;\vdots\; a_{21}^C, a_{12}^C \;\vdots\; a_{31}^C, a_{32}^C, \ldots, a_{33}^C] \qquad (3.3.13)$$

$$\boldsymbol{\phi}^{(t)} = \mathrm{col}\,[\phi_{11}^{(t)} \;\vdots\; \phi_{21}^{(t)}, \phi_{12}^{(t)} \;\vdots\; \phi_{31}^{(t)}, \phi_{32}^{(t)}, \ldots, \phi_{33}^{(t)}] \qquad (3.3.14)$$

The line Jacobi iterative method in accordance with Eq. (3.2.25) is

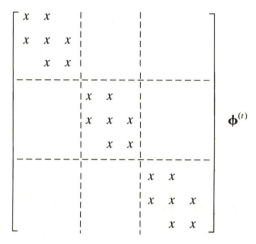

$$\boldsymbol{\phi}^{(t)}$$

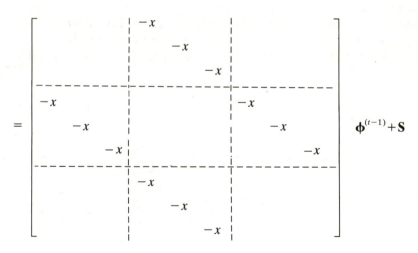

$$= \qquad \boldsymbol{\phi}^{(t-1)} + \mathbf{S}$$

$$(3.3.15)$$

where $\boldsymbol{\phi}^{(t)}$ is the same as Eq. (3.3.11).

Hereafter, we array $\phi_{i,j}$ in $\boldsymbol{\phi}$ for the point iterative methods as illustrated in Fig. 3.5 and assume the calculation is performed in the sequence of $\phi_{i,j}$ in

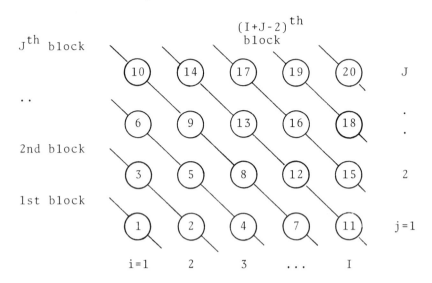

Figure 3.5 Consistent ordering of grid points for a point-iterative scheme (the number in a circle is the position of $\phi_{i,j}$ in the vector $\mathbf{\phi}$).

$\mathbf{\phi}$ (or in the equivalent way in the actual calculations). For the line iterative methods, $\phi_{i,j}$ are arrayed in $\mathbf{\phi}$ as illustrated in Fig. 3.6. The calculations for $\phi_{i,j}$ are performed block by block in the sequence of increasing j. These orderings are called consistent orderings, respectively, for point and line

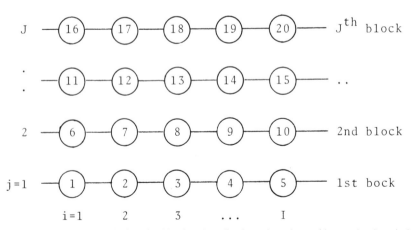

Figure 3.6 Consistent ordering of grid points for a line iterative scheme (the number in a circle is the position of $\phi_{i,j}$ in the vector $\mathbf{\phi}$).

iterative methods. For example, Eq. (3.3.12) is a consistently ordered point iterative scheme, while Eq. (3.3.9) is not. Equation (3.3.15) is a consistently ordered line iterative scheme. With the consistent orderings, matrix $\mathbf{A}$ in Eq. (3.2.25) can be partitioned into

$$\mathbf{A} = \mathbf{B} - \mathbf{L} - \mathbf{U} \tag{3.3.16}$$

where

$$\mathbf{B} = \begin{bmatrix} \boldsymbol{B}_1 & & & & \\ & \boldsymbol{B}_2 & & & \\ & & \boldsymbol{B}_3 & & \\ & & & \ddots & \\ & & & & \boldsymbol{B}_\alpha & \ddots \\ & & & & & \ddots \end{bmatrix} \tag{3.3.17}$$

$$\mathbf{L} = \begin{bmatrix} \mathbf{0} & & & & \\ \mathbf{L}_2 & \mathbf{0} & & & \\ & \mathbf{L}_3 & \mathbf{0} & & \\ & & \ddots & \ddots & \\ & & & \mathbf{L}_\alpha & \mathbf{0} & \ddots \\ & & & & \ddots & \ddots \end{bmatrix} \tag{3.3.18}$$

$$\mathbf{U} = \begin{bmatrix} \mathbf{0} & \mathbf{U}_1 & & & \\ & \mathbf{0} & \mathbf{U}_2 & & \\ & & \mathbf{0} & \mathbf{U}_3 & \\ & & & \ddots & \ddots \\ & & & & \mathbf{0} & \mathbf{U}_\alpha \ddots \\ & & & & & \ddots \end{bmatrix} \tag{3.3.19}$$

In the above equations, $\mathbf{B}_\alpha$, $\mathbf{L}_\alpha$, $\mathbf{U}_\alpha$ are nonnull submatrices and $\mathbf{0}$ are null submatrices. In the case of a consistently ordered *point* scheme, $\mathbf{B}$ is a block diagonal matrix consisting of strictly diagonal submatrices $\mathbf{B}_\alpha$ as illustrated by the diagonal submatrices of Eq. (3.3.13); submatrices of the lower block-triangular matrix $\mathbf{L}$ are all null matrices except $\mathbf{L}_\alpha$, which are not square matrices as seen in Eq. (3.3.12); $\mathbf{U}$ is an upper block-triangular matrix with nonnull $\mathbf{U}_\alpha$ and $\mathbf{U} = \mathbf{L}^T$. In the case of consistently ordered *line* schemes, $\mathbf{B}_\alpha$, $\mathbf{L}_\alpha$, and $\mathbf{U}_\alpha$ are all square submatrices, $\mathbf{B}_\alpha$ is a tridiagonal matrix, and $\mathbf{L}_\alpha$ and $\mathbf{U}_\alpha$ are diagonal matrices as shown in Eq. (3.3.15).

By using $\mathbf{B}$, $\mathbf{L}$, and $\mathbf{U}$ defined in this way, both point and line Jacobi iterative methods, namely, Eqs. (3.3.12) and (3.3.15), can be written as

$$\mathbf{B}\boldsymbol{\phi}^{(t)} = (\mathbf{L}+\mathbf{U})\boldsymbol{\phi}^{(t-1)} + \mathbf{S} \tag{3.3.20}$$

In terms of submatrices of $\mathbf{B}$, $\mathbf{L}$, and $\mathbf{U}$, Eq. (3.3.20) can also be expressed by

$$\mathbf{B}_\alpha \boldsymbol{\phi}_\alpha^{(t)} = \mathbf{L}_\alpha \boldsymbol{\phi}_{\alpha-1}^{(t-1)} + \mathbf{U}_\alpha \boldsymbol{\phi}_{\alpha+1}^{(t-1)} + \mathbf{S}_\alpha \tag{3.3.21}$$

Where $\boldsymbol{\phi}_\alpha^{(t)}$ and $\mathbf{S}_\alpha$ are subvectors of $\boldsymbol{\phi}^{(t)}$ and $\mathbf{S}$, respectively; the order of $\boldsymbol{\phi}_\alpha$ and $\mathbf{S}_\alpha$ is equal to the rank of $\mathbf{B}_\alpha$.

The one-step iterative methods can be expressed in matrix notations as follows:

JACOBI ITERATIVE METHODS

$$\boldsymbol{\phi}^{(t)} = \mathbf{B}^{-1}(\mathbf{L}+\mathbf{U})\boldsymbol{\phi}^{(t-1)} + \mathbf{B}^{-1}\mathbf{S} \tag{3.3.22}$$

GAUSS–SEIDEL ITERATIVE METHODS

$$(\mathbf{B}-\mathbf{L})\boldsymbol{\phi}^{(t)} = \mathbf{U}\boldsymbol{\phi}^{(t-1)} + \mathbf{S}$$

or

$$\boldsymbol{\phi}^{(t)} = (\mathbf{B}-\mathbf{L})^{-1}\mathbf{U}\boldsymbol{\phi}^{(t-1)} + (\mathbf{B}-\mathbf{L})^{-1}\mathbf{S} \tag{3.3.23}$$

SUCCESSIVE-OVER-RELAXATION METHODS

$$(\mathbf{B}-\omega\mathbf{L})\boldsymbol{\phi}^{(t)} = [(1-\omega)\mathbf{B}+\omega\mathbf{U}]\boldsymbol{\phi}^{(t-1)} + \omega\mathbf{S}$$

or

$$\boldsymbol{\phi}^{(t)} = (\mathbf{B}-\omega\mathbf{L})^{-1}[(1-\omega)\mathbf{B}+\omega\mathbf{U}]\boldsymbol{\phi}^{(t-1)} + (\mathbf{B}-\omega\mathbf{L})^{-1}\omega\mathbf{S} \tag{3.3.24}$$

3.4 CONVERGENCE OF ITERATIVE SCHEMES

Any one-step iterative methods introduced in Section 3.3 can be written in the form:

$$\boldsymbol{\phi}^{(t)} = \mathbf{M}\boldsymbol{\phi}^{(t-1)} + \mathbf{g} \tag{3.4.1}$$

where $\mathbf{M}$ is the iterative matrix for a one-step method, and $\mathbf{e}^{(0)}$ is the initial vector that is arbitrary. We define the error vectors $\mathbf{e}^{(t)}$ as

$$\mathbf{e}^{(t)} = \boldsymbol{\phi}^{(t)} - \boldsymbol{\phi} \tag{3.4.2}$$

where $\boldsymbol{\phi}$ is the exact solution satisfying $\boldsymbol{\phi} = \mathbf{M}\boldsymbol{\phi} + \mathbf{g}$. Introducing Eq. (3.4.2) into Eq. (3.4.1), we obtain

$$\mathbf{e}^{(t)} = \mathbf{M}\mathbf{e}^{(t-1)} = \cdots = \mathbf{M}^t\mathbf{e}^{(0)} \qquad (3.4.3)$$

The error vectors $\mathbf{e}^{(t)}$ converge to a null vector for any arbitrary $\mathbf{e}^{(0)}$ if and only if $\mathbf{M}^t$ converges to a null matrix as t increases. $\mathbf{M}^t$ converges to null if the spectral radius μ of $\mathbf{M}$ is less than unity. Convergence rate of the iterative scheme depends on how fast $\mathbf{M}^t$ converges to null.

Example 3.3

Suppose $\mathbf{M}$ and $\mathbf{e}^{(0)}$ are given by

$$\mathbf{M} = \begin{bmatrix} 0.85 & -0.1 \\ -0.1 & 0.85 \end{bmatrix} \quad \mathbf{e}^{(0)} = \begin{bmatrix} 1 \\ 0 \end{bmatrix}$$

The eigenvalues of $\mathbf{M}$ are $\lambda_1 = 0.95$ and $\lambda_2 = 0.75$. The spectral radius is $\mu = 0.95$. $\mathbf{e}^{(t)}$ for various t are shown as follows:

t	0	1	5	10	20	40	100
$e_1^{(t)}$	1.00	0.85	0.51	0.33	0.18	0.064	0.0029
$e_2^{(t)}$	0.00	-0.10	-0.27	-0.27	-0.17	-0.064	-0.0029

If there is no eigenvector deficiency with $\mathbf{M}$ and accordingly the eigenvectors form a complete set, the initial error vector may be expanded in the eigenvectors of $\mathbf{M}$:

$$\mathbf{e}^{(0)} = \sum_n a_n \mathbf{u}_n \qquad (3.4.4)$$

where $\mathbf{u}_n$ is an eigenvector satisfying

$$\mathbf{M}\mathbf{u}_n = \lambda_n \mathbf{u}_n \qquad (3.4.5)$$

Introducing Eq. (3.4.4) into Eq. (3.4.3) and using Eq. (3.4.5), we find

$$\mathbf{e}^{(t)} = \sum_n a_n \lambda_n^t \mathbf{u}_n \qquad (3.4.6)$$

The norms of $\mathbf{e}^{(t)}$ and $\mathbf{e}^{(0)}$ have the relation

$$\|\mathbf{e}^{(t)}\| = \|\sum_n a_n \lambda_n^t \mathbf{u}_n\| \leq \mu^t \|\mathbf{e}^{(0)}\| \qquad (3.4.7)$$

where μ is the spectral radius of $\mathbf{M}$ and is given by

$$\mu = \max_n |\lambda_n| \qquad (3.4.8)$$

Since the initial error vector, $\mathbf{e}^{(0)}$, is arbitrary in practical problems, μ^t serves as a basis for comparing the convergence rate of different iterative methods. We define the convergence rate as

$$R \equiv -\frac{d}{dt} \ln(\mu^t) = -\ln \mu \qquad (3.4.9)$$

where t is treated as if it were a continuous variable.

If the eigenvalue corresponding to μ is a double root and an eigenvector deficiency occurs, the above discussion does not apply.[7] Let us assume that $\mathbf{M}$ has only one double root equal to μ. By using the similarity transformation, the matrix $\mathbf{M}$ may be transformed to the Jordan canonical form,

$$\mathbf{B} = \mathbf{P}^{-1}\mathbf{M}\mathbf{P}$$

where $\mathbf{B}$ may be written as

$$\mathbf{B} = \begin{bmatrix} \mu & 1 & & \\ 0 & \mu & & \\ \hline & & \lambda_3 & \\ & & & \lambda_4 \\ & & & & \ddots \end{bmatrix} \qquad (3.4.10)$$

Premultiplying by $\mathbf{P}^{-1}$, Eq. (3.4.3) becomes

$$\mathbf{P}^{-1}\mathbf{e}^{(t)} = (\mathbf{P}^{-1}\mathbf{M}\mathbf{P})\mathbf{P}^{-1}\mathbf{e}^{(t-1)} = \mathbf{B}^t\mathbf{P}^{-1}\mathbf{e}^{(0)} \qquad (3.4.11)$$

By defining

$$\boldsymbol{\zeta}^{(t)} = \mathbf{P}^{-1}\mathbf{e}^{(t)} \qquad (3.4.12)$$

Eq. (3.4.11) may be equivalently written as

$$\boldsymbol{\zeta}^{(t)} = \mathbf{B}^t\boldsymbol{\zeta}^{(0)} \qquad (3.4.13)$$

The initial vector $\boldsymbol{\zeta}^{(0)}$ may be expressed by

$$\boldsymbol{\zeta}^{(0)} = \sum_{n=1}^{N} a_n \tilde{\mathbf{u}}_n \qquad (3.4.14)$$

where $\tilde{\mathbf{u}}_n$s with $n \neq 2$ are normalized eigenvectors of $\mathbf{B}$ and $\tilde{\mathbf{u}}_2$ is the auxiliary vector. Since $\mathbf{B}$ is in the Jordan canonical form, each of $\tilde{\mathbf{u}}_n$ including $n = 2$ is a unit vector.

By introducing Eq. (3.4.14), Eq. (3.4.13) becomes

$$\zeta^{(t)} = \sum_{n=1}^{N} a_n \mathbf{B}^t \tilde{\mathbf{u}}_n = a_1 \mu^t \tilde{\mathbf{u}}_1 + a_2(t\mu^{t-1}\tilde{\mathbf{u}}_1 + \mu^t \tilde{\mathbf{u}}_2)$$

$$+ \sum_{n=3}^{N} a_n \lambda_n^t \tilde{\mathbf{u}}_n \tag{3.4.15}$$

The last term approaches zero faster than the first two terms. The first term in the parentheses occurs due to the eigenvector deficiency and becomes the dominant term as t increases: therefore, as t increases

$$\zeta^{(t)} \to a_2 t\mu^{t-1}\tilde{\mathbf{u}}_1 \tag{3.4.16}$$

It must be noted that $t\mu^{t-1}$ is an increasing function until t reaches $-1/\ln \mu$. After then, $t\mu^{t-1}$ becomes a decreasing function. The decay rate of $t\mu^{t-1}$ may be defined as

$$R(t\mu^{t-1}) \equiv -\frac{d}{dt}\ln(t\mu^{t-1}) = -\left(\frac{1}{t} + \ln \mu\right) \tag{3.4.17}$$

It can be seen that, as t becomes larger, the decay rate approaches Eq. (3.4.9).

Example 3.4

Suppose $\mathbf{B}$ and $\zeta^{(0)}$ are given by

$$\mathbf{B} = \begin{bmatrix} 0.9 & 1 \\ 0 & 0.9 \end{bmatrix} \qquad \zeta^{(0)} = \begin{bmatrix} 1 \\ 1 \end{bmatrix}$$

The only eigenvalue of $\mathbf{B}$ is $\lambda = 0.9$, and one eigenvector is in deficiency. The $\zeta^{(t)}$ becomes as shown in the following.

t	0	2	5	10	20	30	50	100
$\xi_1^{(t)}$	1.00	2.61	3.87	4.22	2.82	1.46	0.29	.003
$\xi_2^{(t)}$	1.00	0.81	0.59	0.35	0.13	0.042	0.0051	.00003

3.5 BASIC PROPERTIES OF ITERATIVE MATRICES

In applying iterative methods to solving a system of linear equations, the mathematical properties of the matrix are important. We are particularly interested in the sufficient condition that the iterative solution is convergent to the exact solution. As we have learned in Section 3.4, the iterative solution is convergent if the spectral radius of the iterative matrix is less than unity. Of course, it is generally very difficult or impossible to know all the eigenvalues of a matrix if the order of the matrix is large. There are, however, the theorems that give sufficient conditions for the spectral radius to be less than unity. In this section we briefly study a few of the basic properties and theorems related to iterative convergence. Particularly, it is proven that, when q of Eq. (3.1.1) and the boundary condition for it satisfy certain conditions, (i) $\mathbf{A}$ in Eq. (3.2.24) is an irreducibly diagonally dominant matrix with positive diagonal elements, and (ii) $\mathbf{A}^{-1}$ is a positive matrix. Those properties are the basis to prove the convergence of the Jacobi iterative method as discussed in Section 3.6.

A matrix $\mathbf{M}$ is said to be *reducible* if it can be reduced, by permutating rows and columns (permutation transformation), to the form

$$\begin{bmatrix} \mathbf{M}_{11} & \mathbf{0} \\ \mathbf{M}_{12} & \mathbf{M}_{22} \end{bmatrix} \tag{3.5.1}$$

where $\mathbf{M}_{11}$ and $\mathbf{M}_{22}$ are square submatrices. If $\mathbf{M}$ is reducible, the linear equation $\mathbf{Mx} = \mathbf{y}$ may be reduced to

$$\begin{bmatrix} \mathbf{M}_{11} & \mathbf{0} \\ \mathbf{M}_{21} & \mathbf{M}_{22} \end{bmatrix} \begin{bmatrix} \mathbf{x}_1 \\ \mathbf{x}_2 \end{bmatrix} = \begin{bmatrix} \mathbf{y}_1 \\ \mathbf{y}_2 \end{bmatrix} \tag{3.5.2}$$

or

$$\begin{aligned} \mathbf{M}_{11}\mathbf{x}_1 &= \mathbf{y}_1 \\ \mathbf{M}_{21}\mathbf{x}_1 + \mathbf{M}_{22}\mathbf{x}_2 &= \mathbf{y}_2 \end{aligned} \tag{3.5.3}$$

The first equation of Eq. (3.5.3) can be solved independently of the second equation. When a matrix is not reducible, it is said to be *irreducible*.

Example 3.5

Consider the following discrete model of particle transport in equilibrium state

In this system, Ⓐ Ⓑ Ⓒ are the three discrete states where particles can stay. The source particles are supplied to each of the states at the rate of y_1, y_2, and y_3 per second, respectively. Particles can travel between Ⓐ and Ⓑ, as well as between Ⓑ and Ⓒ, and also particles can escape from any of the three states. If the probability that a particle in Ⓙ goes to Ⓘ in one second is P_{ij}, the number of particles staying in each state x_i, satisfies

$$Q_i x_i = \sum_{j \neq i} P_{ij} x_j + y_j \qquad (3.5.4)$$

where y_j is the rate of particles supplied to the state j in one second, and Q_i is the probability that a particle in Ⓘ is removed from Ⓘ in 1 sec. The matrix representation of Eq. (3.5.4) for the three discrete states becomes

$$\begin{bmatrix} Q_A & -P_{AB} & 0 \\ -P_{BA} & Q_B & -P_{BA} \\ 0 & -P_{GB} & Q_C \end{bmatrix} \begin{bmatrix} x_A \\ x_B \\ x_C \end{bmatrix} = \begin{bmatrix} y_A \\ y_B \\ y_C \end{bmatrix} \qquad (3.5.5)$$

The particles can go from any state to any other state. In other words, the three states are *strongly connected*. Eq. (3.5.5) can be easily seen to be irreducible.

Example 3.6

Consider the following system

$$ Ⓐ \rightleftarrows Ⓑ \rightarrow Ⓒ $$

where particles cannot go from Ⓒ to Ⓑ. The equation of the system becomes

$$\begin{bmatrix} Q_A & -P_{AB} & 0 \\ -P_{BA} & Q_B & 0 \\ 0 & -P_{CB} & Q_C \end{bmatrix} \begin{bmatrix} x_A \\ x_B \\ x_C \end{bmatrix} = \begin{bmatrix} y_A \\ y_B \\ y_C \end{bmatrix} \qquad (3.5.6)$$

which is reducible. States Ⓑ and Ⓒ are *weakly connected*.

Definition. A matrix $M = [m_{k,l}]$ is said to be *diagonally dominant* if

$$|m_{k,k}| \geq \sum_{l \neq k} |m_{k,l}|, \quad \text{for all } k \qquad (3.5.7)$$

and *strictly diagonally dominant* if

$$|m_{k,k}| > \sum_{l \neq k} |m_{k,l}|, \quad \text{for all } k \qquad (3.5.8)$$

Definition. A matrix $\mathbf{M}$ is *irreducibly diagonally dominant* if the matrix is irreducible and diagonally dominant and

$$|m_{k,k}| > \sum_{l \neq k} |m_{k,l}| \qquad (3.5.9)$$

for at least one value of k.

Theorem 3.1. The eigenvalues of matrix $\mathbf{M} = [m_{k,l}]$ of order N are all found within the union of the N disks

$$|z - m_{k,k}| \leq \sum_{l \neq k} |m_{k,l}|, \qquad k = 1, 2 \ldots, N \qquad (3.5.10)$$

where z is a complex variable (Fig. 3.7): Gerschgoerin's theorem.[8,9]

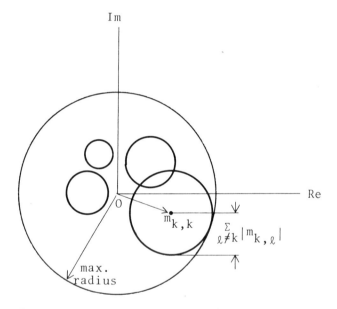

Figure 3.7   Union of disks ($m_{k,k}$ are assumed to be complex numbers).

Proof. Suppose λ is an eigenvalue and $\mathbf{u}$ is the corresponding eigenvector of $\mathbf{M}$. If the kth element of $\mathbf{u}$ has the largest absolute value among other elements, namely, $|u_k| = \max_l |u_l|$, then

$$\sum_l m_{k,l} u_l = \lambda u_k \qquad (3.5.11)$$

which is the kth row of $\mathbf{Mu} = \lambda \mathbf{u}$. Rewriting Eq. (3.5.11) and taking the absolute yield

$$\left|\lambda - m_{k,k}\right|\left|u_k\right| = \left|\sum_{l \neq k} m_{k,l}u_l\right| \leq \sum_{l \neq k} \left|m_{k,l}\right|\left|u_l\right|$$
$$\leq \sum_{l \neq k} \left|m_{k,l}\right|\left|u_k\right| \tag{3.5.12}$$

Dividing the above equation by $\left|u_k\right|$, we obtain

$$\left|\lambda - m_{k,k}\right| \leq \sum_{l \neq k} \left|m_{k,l}\right| \tag{3.5.13}$$

The above equation is true for any eigenvalue, thus completing the proof.

Theorem 3.1 implies that the spectral radius of **M** satisfies the inequality:

$$\mu \leq \max_k \sum_l \left|m_{k,l}\right|$$

Since $\mathbf{M}^T$ has the same eigenvalues as **M**, the right side of Eq. (3.5.10) can be replaced by $\sum_{k \neq l} \left|m_{k,l}\right|$. Therefore, the spectral radius of **M** satisfies

$$\mu(\mathbf{M}) \leq \min\left(\max_l \sum_k \left|m_{k,l}\right|; \max_k \sum_l \left|m_{k,l}\right|\right) \tag{3.5.14}$$

If **M** is real and symmetric, all the eigenvalues are real. With strict diagonal dominance and positive diagonal elements, the minimum eigenvalue satisfies

$$\min_n \lambda_n(\mathbf{M}) \geq \min_k \left(m_{k,k} - \sum_{l \neq k} \left|m_{k,l}\right|\right) > 0 \tag{3.5.15}$$

(see Fig. 3.8). Therefore, all the eigenvalues of **M** are real and positive, and **M** is positive definite. If **M** is a real and symmetric matrix with irreducibly diagonal dominance, at least one circle does not pass the origin of the complex z-plane although all other circles may pass the origin. For an irreducibly diagonally dominant matrix, the $>$ sign before 0 in Eq. (3.5.15) must be replaced by the $\geq$ sign. Nevertheless, the matrix **M** in this case is positive definite because no eigenvalue of **M** can exist at the origin of the z-plane for the following reason. If $z = 0$ is an eigenvalue, it must be on some of the circles. On the other hand, it can be proven (see Theorem 1.7 of Reference 8) that, if an eigenvalue is on a circle, then all the other circles must pass the same point. At least one circle of an irreducibly diagonally dominant matrix cannot pass the origin. Therefore, $z = 0$ cannot be an eigenvalue.

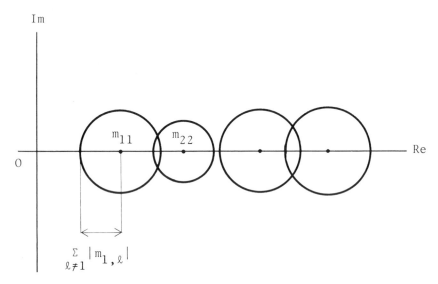

Figure 3.8   Union of disks with positive diagonal elements and strict diagonal dominance.

Let us check the properties of **A** in Eq. (3.2.24). When $p(r)$ of Eq. (3.1.1) is nonzero, the matrix **A** is irreducible. If $q(\mathbf{r})$ in Eq. (3.1.1) and $\gamma(\mathbf{r})$ in Eq. (3.2.14), respectively, satisfy the following conditions (σ in Eq. (3.2.14) is assumed to be zero for simplicity):

$$q(\mathbf{r}) \geq 0$$
$$\gamma(\mathbf{r}) \geq 0 \tag{3.5.16}$$

with strict inequality at least for a portion of **r** for either of q or γ, then the matrix **A** becomes irreducibly diagonal dominant.

The matrix **A** of Eq. (3.2.24) is an S-matrix (see Section 2.7 for definition) if q and γ satisfy Eqs. (3.5.16). The physical meaning of irreducibly diagonal dominance is as follows. If $\gamma = -(\partial/\partial n)\phi/\phi > 0$ the flux of particles (or heat) leaks out of the external boundary, and no leak if $\gamma = 0$. If $q > 0$, particles (or heat) are absorbed in the medium, and if $q = 0$ there is no absorption. If $\gamma = q = 0$, then there is no exit of particles (or heat) and accordingly no solution to Eq. (3.1.1). If a source is supplied continuously to a system that has no exit, there is no steady state. In the discrete system shown in Example 3.5, only one exit, which is either leaking or absorption, is sufficient to guarantee the steady state if all the points are strongly connected (irreducible). In other words, a particle at any state can travel to any other state, so ultimately it can find the exit. Considering Example 3.6 where all the points are not strongly connected, if there is an exit only at Ⓐ but no exit at Ⓒ, then there is no steady state.

Theorem 3.2. The spectral radius of an irreducible nonnegative matrix $\mathbf{A} = [a_{k,l}]$ satisfies either

$$\mu = \sum_l a_{k,l} \quad \text{for all } k \tag{3.5.17}$$

or

$$\mu < \max_k \sum_l a_{k,l} \tag{3.5.18}$$

Proof. If the right side of Eq. (3.5.17) is identical for all k, $\mathbf{u} = \text{col}\,[1, 1, 1, \ldots, 1]$ is an eigenvector of $\mathbf{A}$, and the corresponding eigenvalue is $\lambda = a_{1,1} + a_{1,2} + \cdots + a_{1,N} = Na_{1,1}$. Because of Eq. (3.5.14), λ is equal to the spectral radius. This is the proof for Eq. (3.5.17). Equation (3.5.18) is a special case of Eq. (3.5.14). If the right side of Eq. (3.5.18) is not identical for all k, we can construct a nonnegative irreducible matrix $\mathbf{B} = [b_{ij}]$ by increasing certain positive elements of $\mathbf{A}$ so that for all k,

$$\sum_l b_{k,l} = \alpha = \max_k \left(\sum_l a_{k,l} \right) \tag{3.5.19}$$

Because of Eq. (3.5.17), the spectral radius of $\mathbf{B}$ is α. On the other hand, the spectral radius of a nonnegative irreducible matrix increases if any element of it increases.[9] Therefore, we have $\mu(\mathbf{B}) = \alpha > \mu(\mathbf{A})$, thus proving Eq. (3.5.18)

The following theorem is necessary to prove the convergence of the Jacobi iterative methods and is useful in the proof for Theorems 3.4 and 3.6.

Theorem 3.3. The elements of the inverse of an S-matrix $\mathbf{A}$ of order N are all positive.

Proof. Let

$$\mathbf{A} = \mathbf{B} - (\mathbf{L} + \mathbf{U}) \tag{3.5.20}$$

where $\mathbf{B}$ is the diagonal matrix (not block) with positive diagonal elements of $\mathbf{A}$, namely, $\mathbf{B} = \text{diag}\,[a_{k,k}]$; $\mathbf{L}$ and $\mathbf{U}$ are lower and upper triangular matrices, respectively. By using the Neumann series, the inverse of $\mathbf{A}$ may be written as

$$\mathbf{A}^{-1} = [\mathbf{B} - (\mathbf{L} + \mathbf{U})]^{-1} = \mathbf{B}^{-1}[\mathbf{I} - \mathbf{B}^{-1}(\mathbf{L} + \mathbf{U})]^{-1}$$
$$= \mathbf{B}^{-1} \sum_{n=0}^{\infty} [\mathbf{B}^{-1}(\mathbf{L} + \mathbf{U})]^n \tag{3.5.21}$$

The Neumann series in Eq. (3.5.21) is convergent if the spectral radius of

$\mathbf{M} \equiv \mathbf{B}^{-1}(\mathbf{L} + \mathbf{U})$ is less than 1. The matrix $\mathbf{M} = [m_{k,l}]$ is irreducible, and its diagonal elements are all zero. Since $\mathbf{A}$ is irreducibly diagonally dominant, we have

$$\sum_l m_{k,l} = \frac{1}{a_{k,k}} \sum_{l \neq k} |a_{k,l}| \leq 1 \qquad (3.5.22)$$

with strict inequality for at least one k. Since $\mathbf{M}$ is an irreducible and nonnegative matrix, Eq. (3.5.18) applies to $\mathbf{M}$; thus we have

$$\mu(\mathbf{M}) < \max_k \left(\sum_l m_{k,l} \right) \leq 1 \qquad (3.5.23)$$

It is proven[8] that, if $\mathbf{M}$ is an irreducible and nonnegative matrix, then $\mathbf{M}^{N-1} = [\mathbf{B}^{-1}(\mathbf{L} + \mathbf{U})]^{N-1}$ is a positive matrix (all the elements are positive). Therefore, all the elements of $\mathbf{A}^{-1}$ are positive.

Example 3.7

The following equation is a finite difference approximation with grid points, $x_i = 0, 1, 2$, for $-\phi''(x) = 1$ and $\phi'(0) = \phi(3) = 0$:

$$\begin{bmatrix} 1, & -1, & 0 \\ -1, & 2, & -1 \\ 0, & -1, & 2 \end{bmatrix} \begin{bmatrix} \phi_0 \\ \phi_1 \\ \phi_2 \end{bmatrix} = \begin{bmatrix} 1 \\ 1 \\ 1 \end{bmatrix}$$

The coefficient matrix on the left has the following properties: (a) irreducible, (b) irreducibly diagonally dominant, (c) positive diagonal elements, (d) positive definite, and (e) an S-matrix.

Example 3.8

The diagonal matrices, $\mathbf{B}_\alpha$ of $\mathbf{B}$ in Eq. (3.3.21) for consistently ordered line iterative methods are all S-matrices. Therefore, $\mathbf{B}_\alpha^{-1}$ are all real positive matrices.

3.6 PROPERTIES OF JACOBI ITERATIVE METHODS

The Jacobi iterative method is not practically used by itself because it is not an efficient method compared with other methods. However, when the properties of other iterative matrices including the successive-over-relaxation method and the Chebyshev semiiterative method are studied, properties of the Jacobi iterative matrix become very important.

The Jacobi iterative method defined by Eq. (3.3.22) may be written as

$$\boldsymbol{\phi}^{(t)} = \mathbf{M}_B \boldsymbol{\phi}^{(t-1)} + \mathbf{g},$$ (3.6.1)

where

$$\mathbf{g} = \mathbf{B}^{-1}\mathbf{S}$$ (3.6.2)

$\mathbf{M}_B$ is the Jacobi iterative matrix defined by

$$\mathbf{M}_B = \mathbf{B}^{-1}(\mathbf{L} + \mathbf{U})$$ (3.6.3)

and t in parentheses is an integer denoting the number of iterations. In terms of error vectors the Jacobi iterative method may be expressed as

$$\mathbf{e}^{(t)} = \mathbf{M}_B \mathbf{e}^{(t-1)} = \cdots = \mathbf{M}_B^t \mathbf{e}^{(0)}$$ (3.6.4)

where $\mathbf{e}^{(t)} = \boldsymbol{\phi}^{(t)} - \boldsymbol{\phi}$.

We define the eigenvectors of the Jacobi iterative matrix by

$$\mathbf{M}_B \boldsymbol{\psi}_n = \lambda_n \boldsymbol{\psi}_n$$ (3.6.5)

where $\boldsymbol{\psi}_n$ are the eigenvectors and λ_n are the corresponding eigenvalues, which are numbered by n in the decreasing order of λ_n as

$$\lambda_N < \lambda_{N-1} < \cdots < \lambda_2 < \lambda_1$$ (3.6.6)

We expand the initial error vector $\mathbf{e}^{(0)}$ in the eigenvector series (assuming there is no eigenvector deficiency) as

$$\mathbf{e}^{(0)} = \sum_{n=1}^{N} a_n \boldsymbol{\psi}_n$$ (3.6.7)

Then $\mathbf{e}^{(t)}$ is expressed by

$$\mathbf{e}^{(t)} = \sum_{n=1}^{N} (\lambda_n)^t a_n \boldsymbol{\psi}_n$$ (3.6.8)

where Eq. (3.6.4) is used. As the number of iterations increases, the error vector approaches

$$\mathbf{e}^{(t)} \to \lambda_N^t a_N \boldsymbol{\psi}_N \quad \text{or} \quad \lambda_1^t a_1 \boldsymbol{\psi}_1$$ (3.6.9)

or a combination of both depending on whether $|\lambda_N| > |\lambda_1|$ or $|\lambda_1| > |\lambda_N|$ or $|\lambda_1| = |\lambda_N|$, respectively.

The following two theorems state important properties of the Jacobi iterative matrix:

Theorem 3.4. If $\mathbf{A}$ of Eq. (3.3.24) is an S-matrix, all the eigenvalues of the Jacobi iterative matrix defined by Eq. (3.6.3) are real.

Proof. This can be proven by showing $\mathbf{M}_B$ is a symmetric matrix (see Section 1.7). $\mathbf{B}$ consists of principal square submatrices, $\mathbf{B}_\alpha$, of the positive definite matrix **A** [see Eq. (3.3.17)]. Therefore, each $\mathbf{B}_\alpha$ is also positive definite (see Example 3.8), and accordingly **B** is a positive definite matrix. A real positive definite matrix **B** has a unique square root $\mathbf{B}^{1/2}$ that is a positive definite matrix. By using a similarity transformation, $\mathbf{M}_B$ may be transformed to

$$\mathbf{M}_B' = \mathbf{B}^{1/2}\mathbf{M}_B\mathbf{M}^{-1/2} = \mathbf{B}^{-1/2}(\mathbf{U}+\mathbf{L})\mathbf{B}^{-1/2}$$

Since $\mathbf{B}^{1/2}$ and $(\mathbf{U}+\mathbf{L})$ are both symmetric, $\mathbf{M}_B'$ is also symmetric (see Problem 36, Chapter 1). $\mathbf{M}_B$ is similar to $\mathbf{M}_B'$, so that $\mathbf{M}_B$ is symmetric.

Theorem 3.5.   The eigenvalues of the Jacobi iterative matrix occur in plus and minus pairs.

Proof.   In terms of the submatrices [refer to Eq. (3.3.17) through Eq. (3.3.19)], Eq. (3.6.5) can be expressed by

$$\mathbf{B}_\alpha^{-1}\mathbf{L}_a\boldsymbol{\psi}_{n,\alpha-1}+\mathbf{B}_\alpha^{-1}\mathbf{U}_\alpha\boldsymbol{\psi}_{n,\alpha+1}=\lambda_n\boldsymbol{\psi}_{n,\alpha} \qquad (3.6.10)$$

where $\boldsymbol{\psi}_{n,\alpha}$ is the αth subvector of $\boldsymbol{\psi}_n$. Now, we consider the new vector $\boldsymbol{\phi}_n$, of which the subvectors are related to $\boldsymbol{\psi}_{n,\alpha}$ by

$$\boldsymbol{\phi}_{n,\alpha} = (-1)^\alpha \boldsymbol{\psi}_{n,\alpha} \qquad (3.6.11)$$

It is easily seen that Eq. (3.6.11) is also an eigenvector and satisfies

$$\begin{aligned}\mathbf{B}_\alpha^{-1}[\mathbf{L}_\alpha\boldsymbol{\phi}_{n,\alpha-1}&+\mathbf{U}_\alpha\boldsymbol{\phi}_{n,\alpha+1}]\\ &= -(-1)^\alpha\mathbf{B}_\alpha^{-1}[\mathbf{L}_a\boldsymbol{\psi}_{n,\alpha-1}+\mathbf{U}_\alpha\boldsymbol{\psi}_{n,\alpha+1}] \qquad (3.6.12)\\ &= -(-1)^\alpha\lambda_n\boldsymbol{\psi}_{n,\alpha} = -\lambda_n\boldsymbol{\phi}_{n,\alpha}\end{aligned}$$

or

$$\mathbf{M}_B\boldsymbol{\phi}_n = -\lambda_n\boldsymbol{\phi}_n \qquad (3.6.13)$$

Therefore, if λ_n is an eigenvalue of $\mathbf{M}_B$, $-\lambda_n$ is also an eigenvalue.

It follows that the maximum and the minimum eigenvalues have the same magnitude but different signs. The spectral radius of $\mathbf{M}_B$ is, therefore,

$$\mu = \lambda_1 = -\lambda_N \qquad (3.6.14)$$

Another important property of the Jacobi iterative matrix is that there is a unique positive eigenvector $\boldsymbol{\psi}_1$ corresponding to $\lambda_1 = \mu$ (see Theorem 1.9 of Reference 8). As a consequence of this and Theorem 3.5, the eigenvector $\boldsymbol{\psi}_N$ corresponding to $\lambda_N = -\lambda_1$ is also unique. Note that $\boldsymbol{\psi}_N$ is identical to $\boldsymbol{\psi}_1$ except that signs of elements of $\boldsymbol{\psi}_N$ change from one block to the next block.

Theorem 3.6   The spectral radius of M_B is less than unity.

Proof.   This theorem has already been proven in the proof for Theorem 3.3 in the case of the point Jacobi iterative matrix. The following proof is more general, however, including the case of the line Jacobi iterative matrix.

The positive eigenvector ψ_1 corresponding to λ_1 of M_B satisfies

$$[\mu B - (L + U)]\psi_1 = 0 \tag{3.6.15}$$

Since $B - (L + U) = A$ is positive definite with diagonal dominance and symmetry properties, $[\mu' B - (L + U)]$ must be positive definite for all $\mu' \geq 1$ (nonsingular). In order that Eq. (3.6.15) is satisfied, μ must satisfy det $[\mu B - (L + U)] = 0$; in other words, $[\mu B - (L + U)]$ must be singular. Therefore, $\mu < 1$.

The spectral radius depends on B. If we consider two different partitionings as

$$A = B - (L + U) \tag{3.6.16}$$

$$A = B' - (L' + U') \tag{3.6.17}$$

where all entries of B are greater than or equal to corresponding elements of B', then

$$1 > \mu(M_B) > \mu(M_{B'}) \tag{3.6.18}$$

For example, if the point Jacobi method and line Jacobi method are represented by Eqs. (3.6.16) and (3.6.17), respectively, we find $B \geq B'$. So, the spectral radius of the line Jacobi method is smaller than that of the point Jacobi method. Accordingly, the convergence of the former is faster than the latter.

Example 3.9

Consider a simple one-dimensional difference equation,

$$2\phi_i - \phi_{i-1} - \phi_{i+1} = 1, \qquad i = 1, 2, \ldots, I \tag{3.6.19}$$

with $\phi_0 = \phi_{I+1} = 0$. The associated eigenvalue problem is

$$2\phi_i - \phi_{i-1} - \phi_{i+1} = \nu\phi_i \tag{3.6.20}$$

If we solve Eq. (3.6.19) by the Jacobi iterative method, the eigenvalue problem associated with the Jacobi iterative matrix is

$$\lambda\phi_i = \frac{\phi_{i-1} + \phi_{i+1}}{2} \tag{3.6.21}$$

Eigenfunctions for both Eqs. (3.6.20) and (3.6.21) are given by

$$\phi_i^{(n)} = \sin\left(\frac{n\pi}{I+1}i\right), \qquad n = 1, 2, \ldots, I \tag{3.6.22}$$

where n is the index to identify different eigenfunctions. The corresponding eigenvalues of Eqs. (3.6.20) and (3.6.21) are, respectively,

$$\nu_n = 2 - 2\cos\left(\frac{n\pi}{I+1}\right) \qquad n = 1, 2, \ldots, I \tag{3.6.23}$$

$$\lambda_n = \cos\left(\frac{n\pi}{I+1}\right), \qquad n = 1, 2, \ldots, I \tag{3.6.24}$$

Example 3.10

Consider a set of finite difference equations for an equally spaced two-dimensional mesh system:

$$4\phi_{ij} - \phi_{i-1,j} - \phi_{i+1,j} - \phi_{i,j-1} - \phi_{i,j+1} = 1, \qquad i,j = 1, 2, \ldots, I+1 \tag{3.6.25}$$

with the boundary conditions $\phi_{i,0} = \phi_{i,I+1} = \phi_{0,j} = \phi_{I+1,j} = 0$. This equation may be expressed in the form of Eq. (3.2.24) with matrix notations. The associated eigenvalue problem for Eq. (3.6.25) may be written as

$$4\phi_{ij} - \phi_{i-1,j} - \phi_{i+1,j} - \phi_{i,j-1} - \phi_{i,j+1} = \nu\phi_{i,j} \tag{3.6.26}$$

There are $I \times I$ independent eigenfunctions for Eq. (3.6.26), which are

$$\phi_{i,j}^{(m,n)} = \sin\left(\frac{m\pi}{I+1}i\right)\sin\left(\frac{n\pi}{I+1}j\right), \qquad m = 1, 2, \ldots, I$$
$$n = 1, 2, \ldots, I \tag{3.6.27}$$

The corresponding eigenvalues are

$$\nu_{m,n} = 2\left(2 - \cos\frac{m\pi}{I+1} - \cos\frac{n\pi}{I+1}\right) \tag{3.6.28}$$

3.7 SUCCESSIVE-OVER-RELAXATION (SOR) METHODS

The SOR method is not restricted to the finite difference equation for the elliptic differential equation. The point SOR scheme for a linear equation

$\mathbf{Ax} = \mathbf{S}$ where $\mathbf{A} = [a_{i,j}]$ may be written as

$$x_i^{(t)} = \frac{\omega}{a_{i,i}}\left[S_i - \sum_{j=1}^{i-1} a_{i,j}x_j^{(t)} - \sum_{j=i+1}^{N} a_{i,j}x_j^{(t-1)}\right]$$

$$+ (1-\omega)x_i^{(t-1)}, \qquad 1 < \omega < 2 \tag{3.7.1}$$

where ω is an extrapolation parameter, N is the order of $\mathbf{A}$, and x_i and S_i are elements of $\mathbf{x}$ and $\mathbf{S}$, respectively (i and j refer to the elements in the matrices and vectors but not the grid indices). A sufficient condition for Eq. (3.7.1) to be convergent is that $\mathbf{A}$ is positive definite (p. 77 of Reference 8). This condition is useful, for example, when the SOR method is applied to the finite element method discussed in Chapter 7. However, even when $\mathbf{A}$ is not symmetric nor a positive definite matrix, the SOR method is convergent if all the diagonal elements of $\mathbf{A}$ are all positive and $\mathbf{A}$ is irreducibly diagonally dominant. This case is discussed more in detail in Section 8.3.

In the remainder of this section, we study the eigenvectors and eigen-values of the SOR method applied to large linear systems associated with the elliptic differential equation and derive the optimal extrapolation parameter ω_{opt}.

The successive-over-relaxation method given by Eq. (3.3.24) may be rewritten as

$$\boldsymbol{\phi}^{(t)} = \mathbf{L}_{\omega}\boldsymbol{\phi}^{(t+1)} + \mathbf{g} \tag{3.7.2}$$

where

$$\mathbf{L}_{\omega} = (\mathbf{B} - \omega\mathbf{L})^{-1}[(1-\omega)\mathbf{B} + \omega\mathbf{U}] \tag{3.7.3}$$

$$\mathbf{g} = (\mathbf{B} - \omega\mathbf{L})^{-1}\omega\mathbf{S} \tag{3.7.4}$$

The eigenvectors of $\mathbf{L}_{\omega}$ are defined by

$$\mathbf{L}_{\omega}\boldsymbol{\eta}_n = \gamma_n\boldsymbol{\eta}_n \tag{3.7.5}$$

where $\boldsymbol{\eta}_n$ and γ_n are eigenvectors and corresponding eigenvalues.

In terms of submatrices $\mathbf{B}$, $\mathbf{L}$, and $\mathbf{U}$ defined in Section 3.3, Eq. (3.7.2) can be written as

$$\mathbf{B}_{\alpha}\boldsymbol{\phi}_{\alpha}^{(t)} - \omega\mathbf{L}_{\alpha}\boldsymbol{\phi}_{\alpha-1}^{(t)} = (1-\omega)\mathbf{B}_{\alpha}\boldsymbol{\phi}_{\alpha}^{(t-1)} + \omega\mathbf{U}_{\alpha}\boldsymbol{\phi}_{\alpha+1}^{(t-1)} + \omega\mathbf{S}_{\alpha} \tag{3.7.6}$$

where Eq. (3.7.2) is premultiplied by $(\mathbf{B} - \omega\mathbf{L})$. Similarly, Eq. (3.7.5) can be written as

$$\gamma_n[\mathbf{B}_{\alpha}\boldsymbol{\eta}_{n,\alpha} - \omega\mathbf{L}_{\alpha}\boldsymbol{\eta}_{n,\alpha-1}] = (1-\omega)\mathbf{B}_{\alpha}\boldsymbol{\eta}_{n,\alpha} + \omega\mathbf{U}_{\alpha}\boldsymbol{\eta}_{n,\alpha+1} \tag{3.7.7}$$

where $\boldsymbol{\eta}_{n,\alpha}$ is the αth block (subvector) of $\boldsymbol{\eta}_n$.

The eigenvectors $\boldsymbol{\eta}_n$ are related to the eigenvectors of the Jacobi iterative method, Eq. (3.6.5), as

$$\boldsymbol{\eta}_{n,\alpha} = \gamma_n^{\alpha/2}\boldsymbol{\psi}_{n,\alpha} \tag{3.7.8}$$

Equation (3.7.8) is now proven. Introducing Eq. (3.7.8) into Eq. (3.7.7), we obtain

$$\gamma_n^{\alpha/2}[\gamma_n\mathbf{B}_\alpha\boldsymbol{\psi}_{n,\alpha} - \omega\gamma_n^{1/2}\mathbf{L}_\alpha\boldsymbol{\psi}_{n,\alpha-1}]$$
$$= \gamma_n^{\alpha/2}[(1-\omega)\mathbf{B}_\alpha\boldsymbol{\psi}_{n,\alpha} + \omega\gamma_n^{1/2}\mathbf{U}_\alpha\boldsymbol{\psi}_{n,\alpha+1}] \tag{3.7.9}$$

Dividing both sides by $\gamma_n^{\alpha/2}$ and rearranging terms, Eq. (3.7.9) becomes

$$(\gamma_n + \omega - 1)\mathbf{B}_\alpha\boldsymbol{\psi}_{n,\alpha} = \omega\gamma_n^{1/2}[\mathbf{L}_\alpha\boldsymbol{\psi}_{n,\alpha-1} + \mathbf{U}_\alpha\boldsymbol{\psi}_{n,\alpha+1}] \tag{3.7.10}$$

If we apply Eq. (3.6.10) to the right side of the above equation, we have

$$(\gamma_n + \omega - 1)\boldsymbol{\psi}_{n,\alpha} = \omega\gamma_n^{1/2}\lambda_n\boldsymbol{\psi}_{n,\alpha} \tag{3.7.11}$$

or

$$(\gamma_n + \omega - 1) = \omega\gamma_n^{1/2}\lambda_n \tag{3.7.12}$$

Thus γ_n is the root of Eq. (3.7.12), and Eq. (3.7.8) is proven.

Eigenvalues of the SOR iterative matrix, γ_n, are obtained by solving Eq. (3.7.12) for $\gamma_n^{1/2}$,

$$\gamma_n^{1/2} = \frac{\omega\lambda_n}{2} \pm \sqrt{\frac{\omega^2\lambda_n^2}{4} - \omega + 1} \tag{3.7.13}$$

and by taking the square

$$\gamma_n^\pm = \frac{\omega^2\lambda_n^2}{2} - \omega + 1 \pm \omega|\lambda_n|\sqrt{\frac{\omega^2\lambda_n^2}{4} - \omega + 1} \tag{3.7.14}$$

In the Jacobi iterative matrix, $|\lambda_n|$ and $-|\lambda_n|$ are both eigenvalues, each of which yields an identical pair of γ_ns, schematically shown as

$$+|\lambda_n| \searrow \gamma_n^+ = \frac{\omega^2\lambda_n^2}{2} - \omega + 1 + \omega|\lambda_n|\sqrt{\frac{\omega^2\lambda_n^2}{4} - \omega + 1} \tag{3.7.15}$$

$$-|\lambda_n| \nearrow \gamma_n^- = \frac{\omega^2\lambda_n^2}{2} - \omega + 1 - \omega|\lambda_n|\sqrt{\frac{\omega^2\lambda_n^2}{4} - \omega + 1} \tag{3.7.16}$$

The eigenvalues $\gamma_n^\pm$ become complex if the inside of the square root sign is negative or equivalently

$$\lambda_n^2 < \frac{4(\omega - 1)}{\omega^2} \tag{3.7.17}$$

However, when Eq. (3.7.17) is satisfied, $|\gamma_n^{\pm}|$ is seen to become a constant

$$|\gamma_n^{\pm}| = \sqrt{\gamma_n^+ \gamma_n^-} = \omega - 1 \qquad (3.7.18)$$

The above equation means that, for all λ_n satisfying Eq. (3.7.17), the corresponding γ_ns are found on a circle of radius $\omega - 1$. For those λ_n not satisfying Eq. (3.7.17), γ_n^+ is greater than $\omega - 1$ and γ_n^- is less than $\omega - 1$. The positions of γs in the complex plane are schematically shown in Fig. 3.9,

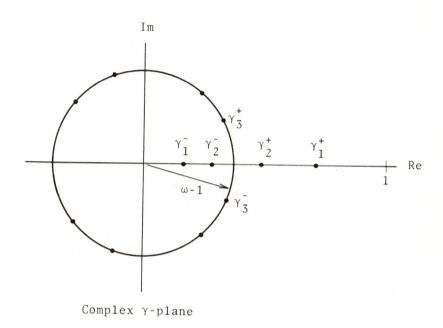

Complex γ-plane

Figure 3.9   Distribution of SOR eigenvalues.

where γ_1^+ and γ_1^- are assumed to correspond to λ_1, and γ_2^+ and γ_2^- to λ_2. It is important to see how the positions of γs change as ω is increased. As ω is gradually increased, the radius of the circle is increased; γ_2^+ and γ_2^- meet at $\gamma = \omega - 1$ and then split into two complex eigenvalues on the circle. If ω is further increased, γ_1^+ is decreased and γ_1^- increased until γ_1^+ and γ_2^- meets at $\omega - 1$, when ω satisfies

$$\frac{\omega^2 \mu^2 (M_B)}{4} - \omega + 1 = 0 \qquad (3.7.19)$$

and

$$\gamma_1^+ = \gamma_1^- = \omega - 1 \tag{3.7.20}$$

where $\mu(\mathbf{M}_B) = \lambda_1$ is the spectral radius of the Jacobi iterative matrix $\mathbf{M}_B$. The minimum spectral radius is attained when Eq. (3.7.19) is satisfied. We denote such ω as ω_{opt}, which is given by

$$\omega_{\text{opt}} = \frac{2}{1 + \sqrt{1 - \mu^2(\mathbf{M}_B)}} \tag{3.7.21}$$

where the right side is the smaller root of Eq. (3.7.19). Therefore, the minimum spectral radius of the SOR method is $\mu(\mathbf{L}_\omega) = \omega_{\text{opt}} - 1$. Three methods to estimate $\mu(\mathbf{M}_B)$ are described in Reference 7.

Example 3.11

If we solve Eq. (3.6.19) by SOR, the associated eigenvalue problem is

$$\gamma \phi_i = \omega \frac{\gamma \phi_{i+1} + \phi_{i+1}}{2} + (1 - \omega)\phi_i$$

The nth eigensolution is

$$\phi_i^{(n)} = (\gamma_n)^{i/2} \sin\left(\frac{n\pi}{I+1} i\right)$$

The corresponding eigenvalues are given by Eq. (3.7.14) in terms of λ_n [see Eq. (3.6.24)].

It must be recognized that when $\omega = \omega_{\text{opt}}$ is used, at least one eigenvector deficiency occurs with the double eigenvalue equal to μ as given by Eq. (3.7.20). As discussed in Section 3.4, the error decay is governed by $t\mu^{t-1}$ rather than μ^t in this case. If a sufficiently large number of iteration is permitted, it can be proven that the use of ω_{opt} gives the fastest convergence. In practice, however, frequently iterations are stopped when a certain convergence criterion is met or a prescribed maximum iteration number is reached. For such a restricted number of iterations, the effect of the eigenvector deficiency is a serious drawback with ω_{opt}. In fact, if the maximum iteration number is restricted, a ω slightly larger than ω_{opt} should give a faster convergence even though the spectral radius becomes larger.

The convergence rate of the SOR iterations can be considerably improved by the coarse-mesh rebalancing scheme discussed in Chapter 8.

3.8 CYCLIC CHEBYSHEV SEMIITERATIVE METHOD

The consistently ordered Jacobi iterative matrix has a two-cyclic property,[12,13] as shown later. It is first shown in this section that the amount of the

Jacobi iterative scheme can be reduced to a half when it has the two-cyclic property. The Chebyshev polynomial method is then applied to the Jacobi iterative scheme. Even though the convergence rate of the Chebyshev polynomial method is half that of the corresponding SOR method, the total computation for 'both methods is about the same because the Chebyshev polynomial method with the cyclic property requires half the number of calculations.

Consider a consistently ordered (point or line) Jacobi iterative scheme expressed in the form of Eq. (3.4.1). With the consistent ordering, the matrix **M** is a block tridiagonal matrix with null diagonal submatrices. Corresponding to the blocks in **M**, ϕ is also divided into blocks. By permutating the rows and columns of the blocks of **M** and ϕ, Eq. (3.4.1) can be transformed to

$$\begin{bmatrix} \phi_1 \\ \phi_2 \end{bmatrix}^{(t)} = \begin{bmatrix} 0 & M_2 \\ M_1 & 0 \end{bmatrix} \begin{bmatrix} \phi_1 \\ \phi_2 \end{bmatrix}^{(t-1)} + \begin{bmatrix} g_1 \\ g_2 \end{bmatrix} \qquad (3.8.1)$$

where ϕ_1 consists of the odd blocks of ϕ, and ϕ_2 consists of the even blocks of ϕ. The transformation that occurred to **M** is called a permutation transformation. When a matrix **M** can be transformed to a 2×2 block matrix with square null diagonal submatrices by a permutation, **M** is called two-cyclic. Any tridiagonal matrix has this property.

Figure 3.10   Mesh system with four points.

Example 3.12

Suppose a point Jacobi iterative scheme for the two-dimensional mesh system with four points as shown in Fig. 3.10. With the consistent ordering, Eq. (3.4.1) becomes

$$\begin{bmatrix} \phi_1 \\ \phi_2 \\ \phi_3 \\ \phi_4 \end{bmatrix} = \begin{bmatrix} 0 & x & x & 0 \\ x & 0 & 0 & x \\ x & 0 & 0 & x \\ 0 & x & x & 0 \end{bmatrix} \begin{bmatrix} \phi_1 \\ \phi_2 \\ \phi_3 \\ \phi_4 \end{bmatrix} + \begin{bmatrix} g_1 \\ g_2 \\ g_3 \\ g_4 \end{bmatrix} \qquad (3.8.2)$$

where x means a nonzero element; ϕ_1 and ϕ_4 belong to the first and the third blocks, respectively; ϕ_2 and ϕ_3 belong to the second block. By changing the order of blocks in $\boldsymbol{\phi}$, Eq. (3.8.2) is transformed to

$$
\begin{bmatrix} \phi_1 \\ \phi_4 \\ \phi_2 \\ \phi_3 \end{bmatrix} = \begin{bmatrix} 0 & 0 & x & x \\ 0 & 0 & x & x \\ x & x & 0 & 0 \\ x & x & 0 & 0 \end{bmatrix} \begin{bmatrix} \phi_1 \\ \phi_4 \\ \phi_2 \\ \phi_3 \end{bmatrix} + \begin{bmatrix} g_1 \\ g_4 \\ g_2 \\ g_3 \end{bmatrix}
\qquad (3.8.3)
$$

Equation (3.8.1) may be divided into two series of calculations

$$
\boldsymbol{\phi}_1^{(t)} = \mathbf{M}_2\boldsymbol{\phi}_2^{(t-1)} + \mathbf{g}_1
\qquad (3.8.4)
$$

$$
\boldsymbol{\phi}_2^{(t)} = \mathbf{M}_1\boldsymbol{\phi}_1^{(t-1)} + \mathbf{g}_2
\qquad (3.8.5)
$$

The sequence of calculations may be schematically shown as

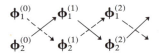

As the iterative vector converges, the difference between $\boldsymbol{\phi}_n^{(t-1)}$ and $\boldsymbol{\phi}_n^{(t)}$ approaches zero. This indicates that we need only one of the series to compute rather than both, thus cutting the amount of calculations by one-half.

We now express Eq. (3.8.1) in the short form as

$$
\boldsymbol{\phi}^{(t)} = \mathbf{M}'\boldsymbol{\phi}^{(t-1)} + \mathbf{g}'
\qquad (3.8.6)
$$

In terms of error vectors defined by $\mathbf{e}^{(t)} = \boldsymbol{\phi}^{(t)} - \boldsymbol{\phi}$, where $\boldsymbol{\phi}$ is the exact solution, Eq. (3.8.6) becomes a homogeneous problem

$$
\mathbf{e}^{(t)} = \mathbf{M}'\mathbf{e}^{(t-1)}
\qquad (3.8.7)
$$

The matrix $\mathbf{M}'$ has the same properties as $\mathbf{M}$ because $\mathbf{M}'$ is a permutation transformation of $\mathbf{M}$, so all the discussions in Section 3.6 apply to Eq. (3.8.7). In essence, all the eigenvalues of $\mathbf{M}'$ are real and the spectral radius is equal to the magnitude of the first and the last eigenvalues.

The Chebyshev polynomial method can be applied to accelerate the convergence of Eq. (3.8.6) or Eq. (3.8.7). As described in Section 2.8, there are two versions of the Chebyshev polynomial method, namely, (1) the one-parameter method, and (2) the two-parameter method. For accelerating the convergence of Eq. (3.8.6), the two-parameter method is much more preferable than the other for the following reasons: (i) the one-parameter method has the truncation problem as described in Section 2.8, and (ii) the second method can take the advantage of using the cyclic nature of $\mathbf{M}'$ to reduce the amount of computations to a half.

Since the largest and the smallest eigenvalues of $\mathbf{M}'$ have the same magnitude equal to the spectral radius μ, the two-parameter Chebyshev polynomial scheme is reduced[7] to

$$\boldsymbol{\phi}^{(t)} = \alpha_t[\mathbf{M}'\boldsymbol{\phi}^{(t-1)} + \mathbf{g}'] - \beta_t\boldsymbol{\phi}^{(t-2)} \tag{3.8.8}$$

where

$$\alpha_1 = 1, \qquad \alpha_t = \frac{2T_{t-1}(1/\mu)}{\mu T_t(1/\mu)}, \qquad t > 1 \tag{3.8.9}$$

$$\beta_1 = 0, \qquad \beta_t = \frac{T_{t-2}(1/\mu)}{T_t(1/\mu)}, \qquad t > 1 \tag{3.8.10}$$

In Eqs. (3.8.9) and (3.8.10), $T_t(z)$ is the Chebyshev polynomial of order t and satisfies the recursion formulas:

$$T_0 = 1, \qquad T_1(z) = z$$
$$T_t(z) = 2zT_{t-1}(z) - T_{t-2}(z), \qquad t \ge 2 \tag{3.8.11}$$

By using Eq. (3.8.11), it is easily seen that

$$\beta_t = \alpha_t - 1 \tag{3.8.12}$$

Because of Eq. (3.8.12), it appears that Eq. (3.8.8) has only one parameter. However, this is the consequence of the special feature of the Jacobi iterative matrix that the maximum and the minimum eigenvalues are equal in magnitude. The distinct difference of the corresponding one-parameter Chebyshev polynomial method from Eq. (3.8.8) is that the last term of the one-parameter method should include $\boldsymbol{\phi}^{(t-1)}$ rather than $\boldsymbol{\phi}^{(t-2)}$. In terms of error vectors, Eq. (3.8.8) becomes

$$\mathbf{e}^{(t)} = \alpha_t\mathbf{M}'\mathbf{e}^{(t-1)} - \beta_t\mathbf{e}^{(t-2)} \tag{3.8.13}$$

With the cyclic property of $\mathbf{M}'$, Eq. (3.8.8) can be divided into two sets of equations

$$\boldsymbol{\phi}_1^{(t)} = \alpha_t[\mathbf{M}_2\boldsymbol{\phi}_2^{(t-1)} + \mathbf{g}_1] + (1 - \alpha_t)\boldsymbol{\phi}_1^{(t-2)} \tag{3.8.14}$$

$$\boldsymbol{\phi}_2^{(t)} = \alpha_t[\mathbf{M}_1\boldsymbol{\phi}_1^{(t-1)} + \mathbf{g}_2] + (1 - \alpha_t)\boldsymbol{\phi}_2^{(t-2)} \tag{3.8.15}$$

Therefore, the series of calculations for $\boldsymbol{\phi}_1^{(0)} \rightarrow \boldsymbol{\phi}_2^{(1)} \rightarrow \boldsymbol{\phi}_1^{(2)} \rightarrow \boldsymbol{\phi}_2^{(3)} \rightarrow \cdots$ is sufficient, thus avoiding the calculations for the other series, similarly to the consistently ordered Jacobi iterative scheme. The present scheme is called the cyclic Chebyshev semiiterative method.

In the remainder of this section, the Chebyshev polynomial acceleration method is shown to be the optimal among all the possible linear acceleration methods applicable to the Jacobi iterative method. For this purpose we first examine the behavior of the tth order polynomial defined by

$$y_t(z) = \frac{T_t(z/\mu)}{T_t(1/\mu)} \qquad (3.8.16)$$

This polynomial satisfies the boundary conditions,

$$\left.\begin{array}{l} y_t(\mu) = 1 \\[2mm] y_t(-\mu) = 1 \quad \text{or} \quad -1 \end{array}\right\} \qquad (3.8.17)$$

It can be also proven that $y_t(z)$ satisfies

$$\max_{-\mu < z < \mu} |y_t(z)| \le \max_{-\mu < z < \mu} |S_t(z)| \qquad (3.8.18)$$

where $S_t(z)$ is any tth order polynomial that is normalized by $|S_t(\pm 1)| = 1$. The Chebyshev polynomial satisfies the recurrence formula

$$T_t\!\left(\frac{z}{\mu}\right) = 2\frac{z}{\mu} T_{t-1}\!\left(\frac{z}{\mu}\right) - T_{t-2}\!\left(\frac{z}{\mu}\right) \qquad (3.8.19)$$

By multiplying Eq. (3.8.16) by $T_t(1/\mu)$ and introducing it in Eq. (3.8.19), we obtain

$$y_t(z) T_t\!\left(\frac{1}{\mu}\right) = 2\frac{z}{\mu} y_{t-1}(z) T_{t-1}\!\left(\frac{1}{\mu}\right) - y_{t-2}(z) T_{t-2}\!\left(\frac{1}{\mu}\right) \qquad (3.8.20)$$

We expand $\mathbf{e}^{(t)}$ in Eq. (3.8.13) as

$$\mathbf{e}^{(t)} = \sum_{n=1}^{N} a_n^{(t)} \boldsymbol{\psi}_n \qquad (3.8.21)$$

where λ_n and $\boldsymbol{\psi}_n$ are the eigenvalue and the eigenvectors of $\mathbf{M}'$:

$$\mathbf{M}' \boldsymbol{\psi}_n = \lambda_n \boldsymbol{\psi}_n \qquad (3.8.22)$$

$$-\mu = \lambda_N < \lambda_{N-1} < \cdots < \lambda_2 < \lambda_1 = \mu \qquad (3.8.23)$$

Introducing Eq. (3.8.21) into Eq. (3.8.13) and using the orthogonality of $\boldsymbol{\psi}_n$, we obtain

$$a_n^{(t)} = \alpha_t \lambda_n a_n^{(t-1)} - \beta_t a_n^{(t-2)} \qquad (3.8.24)$$

If we introduce Eq. (3.8.9) and Eq. (3.8.10) into Eq. (3.8.24) and multiply the results by $T_t(1/\mu)$, we find

$$a_n^{(t)} T_t\!\left(\frac{1}{\mu}\right) = 2\frac{\lambda_n}{\mu} a_n^{(t-1)} T_{t-1}\!\left(\frac{1}{\mu}\right) - a_n^{(t-2)} T_{t-1}\!\left(\frac{1}{\mu}\right) \qquad (3.8.25)$$

By comparing Eq. (3.8.25) with Eq. (3.8.20), $a_n^{(t)}$ is found to be the polynomial of λ_n in the form

$$a_n^{(t)} = a_n^{(0)} y_t(\lambda_n) \tag{3.8.26}$$

Since $a_n^{(t)}$ approaches zero in proportion to $y_t(\lambda_n)$, the convergence rate is the maximum among all the possible linear extrapolation methods for Eq. (3.8.7).

The effective convergence rate of the cyclic Chebyshev semiiterative method is

$$R = -\ln \frac{T_{t-1}(1/\mu)}{T_t(1/\mu)} \xrightarrow[t \to \infty]{} \frac{1}{2}(\omega_{opt} - 1) \tag{3.8.27}$$

where ω_{opt} is the optimal extrapolation parameter for the SOR scheme and is given by Eq. (3.7.21). The convergence rate of the Chebyshev semiiterative method seems half the SOR convergence rate. However, because of the cyclic property of $\mathbf{M}'$, only half the computational effort is necessary in each iteration, so the effective convergence rate is the same as the SOR method. The Chebyshev semiiterative method is preferable to the SOR method when the effect of the eigenvector deficiency of the SOR method with ω_{opt} is serious.

3.9 STRONGLY IMPLICIT METHOD

The strongly implicit iterative method [14-17] has a faster convergence rate than the previously discussed methods such as SOR, the cyclic Chebyshev polynomial method, or ADI. The properties of the iterative matrix and the algorithms with the optimal iteration parameters were studied much later than the previous methods. The essence of the strongly implicit method and its properties are introduced in this section.

3.9.1 A General Form of Iterative Schemes and the Strongly Implicit Method

Consider iterative solutions for Eq. (3.2.24). Although we are not interested in line-inversion schemes, we assume that the grids are numbered in accordance with Fig. 3.6 and $\mathbf{A}$ is a pentadiagonal matrix as illustrated by Eq. (3.2.25).

A general iterative formula for Eq. (3.2.24) may be obtained by adding an auxiliary term $\tilde{\mathbf{A}}$ to each side of Eq. (3.2.24) and putting the iteration numbers to $\boldsymbol{\phi}$ as

$$(\mathbf{A} + \tilde{\mathbf{A}})\boldsymbol{\phi}^{(t)} = \tilde{\mathbf{A}}\boldsymbol{\phi}^{(t-1)} + \mathbf{S} \tag{3.9.1}$$

where t is the number of iterations, and $\tilde{\mathbf{A}}$ is so chosen that the inversion of $(\mathbf{A}+\tilde{\mathbf{A}})$ becomes easy. All the iterative schemes discussed in the earlier sections may be considered as special cases of Eq. (3.9.1). For example, $\tilde{\mathbf{A}}$ for the point Jacobi iterative method is set to $\mathbf{L}+\mathbf{U}$, where $\mathbf{L}$ and $\mathbf{U}$ are lower and upper triangular matrices, respectively. If $(\mathbf{A}+\tilde{\mathbf{A}})$ is set to $(\mathbf{L}+\mathbf{B})$, one obtains the Gauss–Seidel methods, where $\mathbf{B}$ is the diagonal matrix.

In general, the convergence rate becomes faster as $(\mathbf{A}+\tilde{\mathbf{A}})$ becomes closer to $\mathbf{A}$, or in other words, more "implicit". In this regard $\tilde{\mathbf{A}}=0$ gives the fastest convergence rate. In effect the solution is obtained by "one iteration", $\boldsymbol{\phi}=\mathbf{A}^{-1}\mathbf{S}$, although the iterative solutions are necessary when such a direct inversion is difficult or costly.

In the strongly implicit method, $\tilde{\mathbf{A}}$ is so chosen that $\mathbf{A}+\tilde{\mathbf{A}}$ can be factorized as

$$\mathbf{A}+\tilde{\mathbf{A}}=\mathbf{L}\cdot\mathbf{U} \tag{3.9.2}$$

where $\mathbf{L}$ and $\mathbf{U}$ are, respectively, lower and upper triangular matrices, each of which has only three nonzero elements in each row. Inversion of a matrix in the form of $\mathbf{L}\cdot\mathbf{U}$ is very easy. If $\tilde{\mathbf{A}}$ is a minor correction to $\mathbf{A}$, then the iterative scheme represented by Eq. (3.9.1) may be regarded as *strongly implicit*.

3.9.2 Derivation of the Auxiliary Matrix, $\tilde{\mathbf{A}}$

In Eq. (3.9.1), $\mathbf{A}$ is a pentadiagonal matrix as shown in Fig. 3.11. If $\mathbf{L}$ and $\mathbf{U}$ are as shown in Fig. 3.11, the product $\mathbf{L}\cdot\mathbf{U}$ becomes a heptadiagonal matrix as shown in Fig. 3.11. The two additional elements of $\mathbf{L}\cdot\mathbf{U}$ compared with $\mathbf{A}$ correspond to the white circles in Fig. 3.12. In order for $(\mathbf{A}+\tilde{\mathbf{A}})$ to be equal to $\mathbf{L}\cdot\mathbf{U}$, $\tilde{\mathbf{A}}$ must be at most a heptadiagonal matrix. We denote the seven elements in a row of $\tilde{\mathbf{A}}$ sequentially from left by $\tilde{a}_{i,j}^L$, $\tilde{a}_{i,j}^{BR}$, $\tilde{a}_{i,j}^T$, $\tilde{a}_{i,j}^C$, $\tilde{a}_{i,j}^B$, $\tilde{a}_{i,j}^{TL}$, and $\tilde{a}_{i,j}^R$. The elements of $\mathbf{L}\cdot\mathbf{U}$ and $\mathbf{A}+\tilde{\mathbf{A}}$ must satisfy the relations:

$$b_{i,j}=a_{i,j}^B+\tilde{a}_{i,j}^B \tag{3.9.3a}$$

$$b_{i,j}e_{i,j-1}=\tilde{a}_{i,j}^{BR} \tag{3.9.3b}$$

$$c_{i,j}=a_{i,j}^L+\tilde{a}_{i,j}^L \tag{3.9.3c}$$

$$d_{i,j}+b_{i,j}f_{i,j-1}+c_{i,j}e_{i-1,j}=a_{i,j}^C+\tilde{a}_{i,j}^C \tag{3.9.3d}$$

$$d_{i,j}e_{i,j}=a_{i,j}^R+\tilde{a}_{i,j}^R \tag{3.9.3e}$$

$$c_{i,j}f_{i-1,j}=\tilde{a}_{i,j}^{TL} \tag{3.9.3f}$$

$$d_{i,j}f_{i,j}=a_{i,j}^T+\tilde{a}_{i,j}^T \tag{3.9.3g}$$

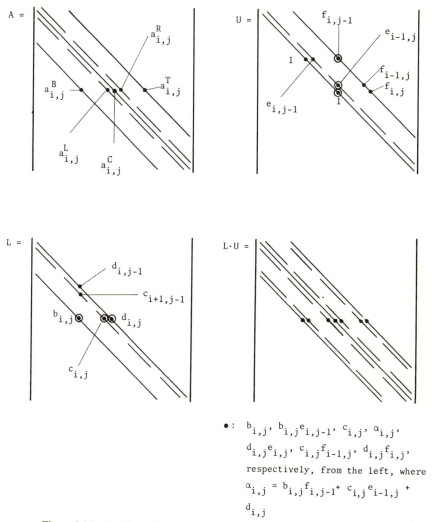

Figure 3.11 Configuration of the matrices in the strongly implicit method.

In these equations, the values with indices for nonexisting points such as $e_{0,1}$, $e_{1,0}, f_{0,1}$, and $f_{1,0}$ are all zero. The question now is whether the elements of $\mathbf{L}$ and $\mathbf{U}$ for any given $\tilde{\mathbf{A}}$ can be determined. The answer is positive if $\tilde{\mathbf{A}}$ has a certain property. In order to see this, we assume that the right sides of Eq. (3.9.3) are all known and also that e and f with indices $(i-1, j)$ or $(i, j-1)$ are all previously calculated. The second assumption is valid if the calculation for b, c, d, e, and f is a recursive procedure starting from $i = j = 1$ with $e_{1,0} = e_{0,1} = f_{0,1} = f_{1,0} = 0$ and proceeding in the increasing order for i and j.

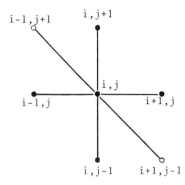

Figure 3.12 Mesh scheme of the strongly implicit method.

Equations (3.9.3) now have five unknowns: $b_{i,j}$, $c_{i,j}$, $d_{i,j}$, $e_{i,j}$, and $f_{i,j}$. Since there are seven equations for five unknowns, there is no solution unless two of Eqs. (3.9.3) are linearly dependent. This is the property that $\tilde{\mathbf{A}}$ must have in order that the elements of $\mathbf{L}$ and $\mathbf{U}$ be uniquely determined.

In the method proposed by Stone,[15] Eqs. (3.9.3b) and (3.9.3f) are omitted in determining the five unknowns, and only $\tilde{a}^L$, $\tilde{a}^T$, $\tilde{a}^C$, $\tilde{a}^B$, and $\tilde{a}^R$ are given. The values of $\tilde{a}^{BR}$ and $\tilde{a}^{TL}$ are calculated by Eqs. (3.9.3b) and (3.9.3f), respectively, only after $b_{i,j}$ and $c_{i,j}$ are determined.

In choosing $\tilde{\mathbf{A}}$ (viz., $\tilde{a}^L$, $\tilde{a}^T$, $\tilde{a}^C$, $\tilde{a}^B$, and $\tilde{a}^R$) we desire the physical properties represented by $\mathbf{A} + \tilde{\mathbf{A}}$ to remain as close to $\mathbf{A}$ as possible. In fact, there are several ways of choosing such an $\tilde{\mathbf{A}}$. Stone obtained $\tilde{\mathbf{A}}$ as follows. $\phi_{i-1,j-1}$ and $\phi_{i+1,j-1}$ can be approximately expressed by a truncated Taylor series as

$$\phi_{i+1,j-1} = -\phi_{i,j} + \phi_{i,j-1} + \phi_{i+1,j} \tag{3.9.4}$$

$$\phi_{i-1,j+1} = -\phi_{i,j} + \phi_{i,j+1} + \phi_{i-1,j} \tag{3.9.5}$$

If the mesh spaces are sufficiently small, Eqs. (3.9.4) and (3.9.5) are good approximations. A constant times Eq. (3.9.4) and another constant times Eq. (3.9.5) can be added to the five-point difference equation Eq. (3.2.8) to yield seven-point equations without disturbing the five-point equations very much. Since Eqs. (3.9.4) and (3.9.5) are not exact relations because of the truncated Taylor series, a parameter α may be introduced to adjust such errors:

$$\phi_{i+1,j-1} = \alpha[-\phi_{i,j} + \phi_{i,j-1} + \phi_{i+1,j}] \tag{3.9.6}$$

$$\phi_{i-1,j+1} = \alpha[-\phi_{i,j} + \phi_{i,j+1} + \phi_{i-1,j}] \tag{3.9.7}$$

We multiply Eq. (3.9.6) by $\tilde{a}_{i,j}^{BR}$ and Eq. (3.9.7) by $\tilde{a}_{i,j}^{TL}$ and add these to both sides of Eq. (3.2.7). The left side of the resulting equation becomes

$$a_{i,j}^{B}\phi_{i,j-1}+a_{i,j}^{L}\phi_{i-1,j}+a_{i,j}^{B}\phi_{i+1,j}+a_{i,j}^{T}\phi_{i,j+1}+a_{i,j}^{C}\phi_{i,j}$$
$$+\tilde{a}_{i,j}^{BR}[\phi_{i+1,j-1}+\alpha(\phi_{i,j}-\phi_{i,j-1}-\phi_{i+1,j})] \qquad (3.9.8)$$
$$+\tilde{a}_{i,j}^{TL}[\phi_{i-1,j+1}+\alpha(\phi_{i,j}-\phi_{i,j+1}-\phi_{i-1,j})]$$

By comparing Eqs. (3.9.3a), (3.9.3c), (3.9.3d), (3.9.3e), and (3.9.3g) with Eq. (3.9.8), we obtain

$$b_{i,j}=a_{i,j}^{B}-\alpha\tilde{a}_{i,j}^{BR} \qquad (3.9.9a)$$
$$c_{i,j}=a_{i,j}^{L}-\alpha\tilde{a}_{i,j}^{TL} \qquad (3.9.9c)$$
$$d_{i,j}+b_{i,j}f_{i,j-1}+c_{i,j}e_{i-1,j}=a_{i,j}^{C}+\alpha(\tilde{a}_{i,j}^{BR}+\tilde{a}_{i,j}^{TL}) \qquad (3.9.9d)$$
$$d_{i,j}e_{i,j}=a_{i,j}^{R}-\alpha\tilde{a}_{i,j}^{BR} \qquad (3.9.9e)$$
$$d_{i,j}f_{i,j}=a_{i,j}^{T}-\alpha\tilde{a}_{i,j}^{TL} \qquad (3.9.9f)$$

By using Eqs. (3.9.3b) and (3.9.3f), $\tilde{a}_{i,j}^{BR}$ and $\tilde{a}_{i,j}^{TL}$ are given in terms of $b_{i,j}$ and $c_{i,j}$. Substituting those into Eq. (3.9.9) yields

$$b_{i,j}+\alpha b_{i,j}e_{i,j-1}=a_{i,j}^{B} \qquad (3.9.10a)$$
$$c_{i,j}+\alpha c_{i,j}f_{i-1,j}=a_{i,j}^{L} \qquad (3.9.10b)$$
$$d_{i,j}+b_{i,j}f_{i,j-1}+c_{i,j}e_{i-1,j}-\alpha(b_{i,j}e_{i,j-1}+c_{i,j}f_{i-1,j})=a_{i,j}^{C} \qquad (3.9.10c)$$
$$d_{i,j}e_{i,j}+\alpha b_{i,j}e_{i,j-1}=a_{i,j}^{R} \qquad (3.9.10e)$$
$$d_{i,j}f_{i,j}+\alpha c_{i,j}f_{i-1,j}=a_{i,j}^{T} \qquad (3.9.10g)$$

The values of $b_{i,j}$, $c_{i,j}$, $d_{i,j}$, $e_{i,j}$ and $f_{i,j}$ are determined by solving Eq. (3.9.10).

The auxiliary matrix $\tilde{\mathbf{A}}$ thus obtained is not symmetric. A symmetric auxiliary matrix would have the advantages in that the eigenvalues of $\mathbf{L}\cdot\mathbf{U}$ are all real so that analysis of iterative convergence becomes easier and that the Chebyshev polynomial method can be applied without difficulties. Stone developed the symmetric auxiliary matrix[16] by modifying Eq. (3.9.10) as

$$b_{i,j}+\alpha c_{i,j-1}f_{i-1,j-1}=a_{i,j}^{B} \qquad (3.9.11a)$$
$$c_{i,j}+\alpha b_{i-1,j}e_{i-1,j-1}=a_{i,j}^{L} \qquad (3.9.11c)$$
$$d_{i,j}+b_{i,j}f_{i,j-1}+c_{i,j}e_{i-1,j}-\alpha(b_{i-1,j}e_{i-1,j-1}+c_{i,j-1}f_{i-1,j-1})=a_{i,j}^{C} \qquad (3.9.11d)$$
$$d_{i,j}e_{i,j}+\alpha b_{i,j}e_{i,j-1}=a_{i,j}^{R} \qquad (3.9.11e)$$
$$d_{i,j}f_{i,j}+\alpha c_{i,j}f_{i-1,j}=a_{i,j}^{T} \qquad (3.9.11g)$$

It is proven next that $\mathbf{A} + \tilde{\mathbf{A}}$ thus determined is symmetric and positive definite. Note that $a_{i,j}^B = a_{i,j-1}^T$ and $a_{i,j}^L = a_{i-1,j}^R$ because $\mathbf{A}$ is symmetric. It is recognized from Eqs. (3.9.11a) and (3.9.11g) that

$$b_{i,j} = d_{i,j-1} f_{i,j-1} \tag{3.9.12}$$

and also from Eqs. (3.9.11c) and (3.9.11e) that

$$c_{i,j} = d_{i-1,j} e_{i-1,j} \tag{3.9.13}$$

With reference to the form of $\mathbf{L}$ and $\mathbf{U}$ shown in Fig. 3.11 (see the values marked by $\odot$), we have the following relation:

$$\mathbf{L} = \mathbf{U}^T \cdot \text{diag}\,(d_{i,j}) \tag{3.9.14}$$

where $\text{diag}\,(d_{i,j})$ is the diagonal matrix with $d_{i,j}$ along the diagonal. By using Eq. (3.9.2), $\mathbf{A} + \tilde{\mathbf{A}}$ becomes

$$\mathbf{A} + \tilde{\mathbf{A}} = \mathbf{L} \cdot \mathbf{U} = \mathbf{U}^T \cdot \text{diag}\,(d_{i,j}) \mathbf{U}$$

$$= [\mathbf{U} \cdot \text{diag}\,(\sqrt{d_{i,j}})]^T \cdot \text{diag}\,(\sqrt{d_{i,j}}) \cdot \mathbf{U} \tag{3.9.15}$$

which is symmetric and positive definite because $\mathbf{M}^T \mathbf{M}$ is a positive definite if $\mathbf{M}$ is a nonsingular matrix.

3.9.3   Iterative Convergence

In Stone's algorithm, the auxiliary matrix $\tilde{\mathbf{A}}$ changes as a set of αs is used cyclically. If $\tilde{\mathbf{A}}$ changes every time, the analysis of convergence rate becomes very difficult. The iterative scheme proposed by Dupont et al. is in the following form:[14]

$$(\mathbf{A} + \tilde{\mathbf{A}})\boldsymbol{\phi}^{(t)} = (\mathbf{A} + \tilde{\mathbf{A}})\boldsymbol{\phi}^{(t-1)} - \omega(\mathbf{A}\boldsymbol{\phi}^{(t-1)} - \mathbf{S}) \tag{3.9.16}$$

where $\tilde{\mathbf{A}}$ is a fixed auxiliary matrix and ω is the iterative parameter that can be changed at every iteration. We assume that $\tilde{\mathbf{A}}$ in Eq. (3.9.16) is defined by Eq. (3.9.11) with $\alpha = 1$.

The exact solution satisfies

$$(\mathbf{A} + \tilde{\mathbf{A}})\boldsymbol{\phi} = (\mathbf{A} + \tilde{\mathbf{A}})\boldsymbol{\phi} - \omega(\mathbf{A}\boldsymbol{\phi} - \mathbf{S}) \tag{3.9.17}$$

Subtracting Eq. (3.9.17) from Eq. (3.9.16) yields

$$\mathbf{e}^{(t)} = \mathbf{M}_\omega \mathbf{e}^{(t-1)} \tag{3.9.18}$$

where

$$\mathbf{e}^{(t)} = \boldsymbol{\phi}^{(t)} - \boldsymbol{\phi} \tag{3.9.19}$$

$$\mathbf{M}_\omega = \mathbf{I} - \omega(\mathbf{A} + \tilde{\mathbf{A}})^{-1}\mathbf{A} \tag{3.9.20}$$

It has been proven in Section 3.9.2 that for a positive definite matrix $\mathbf{A}$ that represents the finite difference operator for an elliptic partial differential equation, $\mathbf{A} + \tilde{\mathbf{A}}$ is also positive definite. Therefore, the Chebyshev polynomial method can be applied to select an optimum set of ωs, provided that the maximum and minimum eigenvalues of $(\mathbf{A} + \tilde{\mathbf{A}})^{-1}\mathbf{A}$ are known. In practice, however, the minimum and maximum eigenvalues of $(\mathbf{A} + \tilde{\mathbf{A}})^{-1}\mathbf{A}$ are not known before the iteration. Diamond[16,17] developed the algorithm of using a fixed ω that can be applied even when little is known about the eigenvalues of $(\mathbf{A} + \tilde{\mathbf{A}})^{-1}\mathbf{A}$. In the remainder of this section, Diamond's method is discussed.

We have the theoretical basis for the convergence of Eq. (3.9.18) as follows:

Theorem 3.7. If $\mathbf{A}$ and $\mathbf{A} + \tilde{\mathbf{A}}$ are positive definite, then $\mathbf{e}^{(t)}$ converges to a null vector if and only if

$$0 < \omega < \frac{2}{\lambda_{\max}[(\mathbf{A} + \mathbf{A})^{-1}\tilde{\mathbf{A}}]} \tag{3.9.21}$$

where $\lambda_{\max}[\mathbf{B}]$ is the maximum eigenvalue of the matrix $\mathbf{B}$.

Proof. If $\mathbf{A}$ and $\mathbf{A} + \tilde{\mathbf{A}}$ are both positive definite, then $(\mathbf{A} + \tilde{\mathbf{A}})^{-1}\mathbf{A}$ is also positive definite (see Problem 25). Therefore, when ω satisfies Eq. (3.9.21), we have $|\lambda_i[\mathbf{I} - \omega(\mathbf{A} + \mathbf{A})^{-1}\mathbf{A}]| < 1.0$ for all i where $\lambda_i[\mathbf{B}]$ represent the ith eigenvalue of $\mathbf{B}$.

The optimal choice for a fixed ω is given by the following lemma.

Lemma 3.1. If $\mathbf{A}$ and $\mathbf{A} + \tilde{\mathbf{A}}$ are positive definite, then the optimal choice for ω is given by

$$\omega_{\mathrm{opt}} = \frac{2}{\lambda_{\max}[(\mathbf{A} + \tilde{\mathbf{A}})^{-1}\mathbf{A}] + \lambda_{\min}[(\mathbf{A} + \tilde{\mathbf{A}})^{-1}\mathbf{A}]} \tag{3.9.22}$$

Proof. Since all the eigenvalues of $(\mathbf{A} + \tilde{\mathbf{A}})^{-1}\mathbf{A}$ are positive, the eigenvalues of the iterative matrix $\mathbf{T}(\omega) = \mathbf{I} - \omega(\mathbf{A} + \tilde{\mathbf{A}})^{-1}\mathbf{A}$ are decreasing functions of ω. Clearly, the spectral radius of $\mathbf{T}(\omega)$ is minimized with respect to ω, when $|\lambda_{\min}[\mathbf{T}(\omega)]| = |\lambda_{\max}[\mathbf{T}(\omega)]|$, or equivalently,

$$1 - \omega\lambda_{\max}[(\mathbf{A} + \tilde{\mathbf{A}})^{-1}\mathbf{A}] = \omega\lambda_{\min}[(\mathbf{A} + \tilde{\mathbf{A}})^{-1}\mathbf{A}] - 1 \tag{3.9.23}$$

Therefore, Eq. (3.9.22) is obtained.

Diamond has proposed a method to find ω_{opt} in which the iterations start with crude estimates for $\mu_{\max}$ and $\mu_{\min}$, for example, $a = 2.5$ and $b = 0.5$ for

μ_{max} and μ_{min}, respectively, and by observing the residue, a and b are revised.

3.9.4 Iterative Procedure

Equation (3.9.16) may be written as

$$(\mathbf{A} + \tilde{\mathbf{A}})\boldsymbol{\delta}^{(t)} = \mathbf{r}^{(t-1)} \qquad (3.9.24)$$

where

$$\boldsymbol{\delta}^{(t)} = \boldsymbol{\phi}^{(t)} - \boldsymbol{\phi}^{(t-1)} \qquad (3.9.25)$$

$$\mathbf{r}^{(t-1)} = -\omega(\mathbf{A}\boldsymbol{\phi}^{(t-1)} - \mathbf{S}) \qquad (3.9.26)$$

As the iterative solution converges, $\boldsymbol{\delta}^{(t)}$ becomes a smaller correction to $\boldsymbol{\phi}^{(t-1)}$, so $\boldsymbol{\delta}^{(t)}$ will not require a large number of significant figures. This is the advantage of the form of Eq. (3.9.24). On the other hand, the drawback is that more arithmetic operations are necessary than for the form of Eq. (3.9.16).

Having the matrix elements of $\mathbf{L}$ and $\mathbf{U}$, we rewrite Eq. (3.9.24) in the form

$$\mathbf{L} \cdot \mathbf{U}\boldsymbol{\delta}^{(t)} = \mathbf{r}^{(t-1)} \qquad (3.9.27)$$

where $\mathbf{r}^{(t-1)}$ is calculated by using Eq. (3.9.26) with the previous iterative solution $\boldsymbol{\phi}^{(t-1)}$. The first step of solving Eq. (3.9.27) is to obtain $\mathbf{v}^{(t-1)} = \mathbf{L}^{-1}\mathbf{r}^{(t-1)}$. Calculations of $\mathbf{v}^{(t-1)}$ is easy because $\mathbf{L}$ is a lower tridiagonal matrix:

$$v_{i,j}^{(t-1)} = \frac{1}{d_{i,j}}[r_{i,j}^{(t-1)} - b_{i,j}v_{i,j-1}^{(t-1)} - c_{i,j}v_{i-1,j}^{(t-1)}] \qquad (3.9.28)$$

This calculation is performed in the increasing order in i and j (forward sweep). The second step is to solve $\boldsymbol{\delta}^{(t)}$ from

$$\mathbf{U}\boldsymbol{\delta}^{(t)} = \mathbf{L}^{-1}\mathbf{r}^{(t-1)} \equiv \mathbf{v}^{(t-1)} \qquad (3.9.29)$$

The solution is also easy because $\mathbf{U}$ is an upper tridiagonal matrix:

$$\delta_{i,j}^{(t)} = v_{i,j}^{(t-1)} - e_{i,j}\delta_{i+1,j}^{(t)} - f_{i,j}\delta_{i,j+1}^{(t)} \qquad (3.9.30)$$

This calculation is performed in the decreasing order in i and j (backward sweep). The values of $\phi_{i,j}^{(t)}$ are obtained by using Eq. (3.9.25).

3.10 TWO-DIMENSIONAL NEUTRON DIFFUSION EQUATIONS

The numerical schemes to solve multigroup, multidimensional neutron diffusion equations are the typical examples of utilizing the iterative schemes described in previous sections. In this section, we first introduce the standard iterative technique for two-dimensional neutron diffusion equations and then introduce a modified scheme.

The numerical method for multigroup, two-dimensional diffusion equations[1,2] can be well illustrated with a two-group equation[3]

$$-\nabla D_1 \nabla \phi_1(x, y) + \Sigma_1 \phi_1 = \frac{1}{k} [\nu \Sigma_{f1} \phi_1 + \nu \Sigma_{f2} \phi_2]$$

$$-\nabla D_2 \nabla \phi_2(x, y) + \Sigma_2 \phi_2 = \Sigma_{sl} \phi_1$$

(3.10.1)

where the standard notations are used (see Section 2.7). The finite difference approximation to Eq. (3.10.1) may be written as

$$\mathbf{A}_1 \boldsymbol{\phi}_1 = \frac{1}{k} [\mathbf{F}_1 \boldsymbol{\phi}_1 + \mathbf{F}_2 \boldsymbol{\phi}_2]$$

$$\mathbf{A}_2 \boldsymbol{\phi}_2 = \mathbf{Q} \boldsymbol{\phi}_1$$

(3.10.2)

where $\boldsymbol{\phi}_1$ and $\boldsymbol{\phi}_2$ are discrete approximations for ϕ_1 and ϕ_2, respectively; $\mathbf{A}_1$ and $\mathbf{A}_2$ are matrices representing the finite difference operators; and $\mathbf{F}_1$ and $\mathbf{F}_2$ and $\mathbf{Q}$ represent $\nu \Sigma_{f1}$, $\nu \Sigma_{f2}$ and Σ_{sl}, respectively.

The standard numerical solution to solve Eq. (3.10.2) involves two kinds of iterative schemes: inner iterations and outer iterations. Equation (3.10.2) is similar to Eq. (2.7.2), except that $\mathbf{A}_1$ and $\mathbf{A}_2$ in Eq. (3.10.2) are pentadiagonal, and the direct inversions of them are not practical.

The power method or a Chebyshev polynomial method for Eq. (2.7.2) is necessary for Eq. (3.10.2) and is called "outer iteration". Each outer iteration needs the solution of the left sides of Eq. (3.10.2) whenever the right sides are given. This problem is reduced to the solution for Eq. (3.2.24). The iterative schemes discussed earlier in this chapter are used for this purpose. The iterative scheme to solve the inhomogeneous problem is called "inner iteration". The flow of iterative schemes for Eq. (3.10.2) is illustrated in Fig. 3.13 (SOR for the inner iteration and the Chebyshev polynomial method for the outer iteration are assumed).

The total computing time for the standard iterative scheme is proportional to (computing time for one inner iteration) × (average number of inner iterations) × (number of energy groups) × (number of outer iterations). Any savings in convergence rate of inner or outer iterations will reduce the computing cost significantly. Experiences show the convergence criterion for inner iterations may not be too strict. The rule of thumb is that the inner iterations during each outer iteration may be terminated when the residue of iteration becomes less than one tenth of the first iteration. If the number of inner iterations per outer iteration is smaller, the number of outer iterations

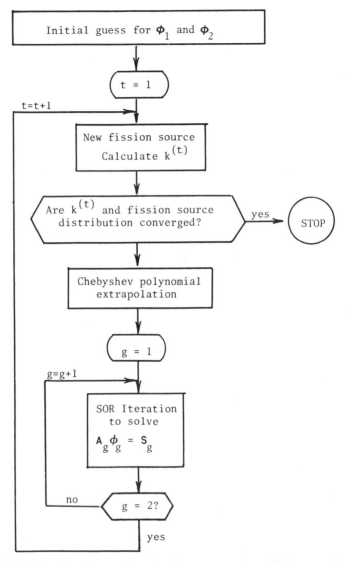

Figure 3.13   Flow chart for two-dimensional two-group neutron diffusion equation code.

is increased because of inaccurate inversion of $\mathbf{A}_1$ and $\mathbf{A}_2$. However, there is no guideline for an optimal convergence criteria with which the over-all computing time is minimized.

The Wielandt method does not work for two- or three-dimensional problems because the direct inversion of a huge matrix including $\mathbf{A}_1$, $\mathbf{A}_2$, $\mathbf{F}_1$,

F_2, and $\mathbf{Q}$ as submatrices is practically impossible. The coarse-mesh rebalancing method described in Chapter 8 has a significant potential to accelerate convergence of both iterative schemes, especially the outer iterations.

In a modified scheme called the Equipoise method[4,18-22] essentially only one iteration for each energy group is done during one outer iteration. For certain problems, this method is more economical than the standard scheme without any sophisticated acceleration technique. The Equipoise method uses the weighted neutron balance equation to estimate the eigenvalue:

$$k = \frac{\mathbf{w}^T[\mathbf{F}_1\boldsymbol{\phi}_1^{(t)}+\mathbf{F}_2\boldsymbol{\phi}_2^{(t)}]}{\mathbf{w}^T[\mathbf{A}_1\boldsymbol{\phi}_1^{(t)}+\mathbf{A}_2\boldsymbol{\phi}_2^{(t)}-\mathbf{Q}\boldsymbol{\phi}_1^{(t)}]}$$

where $\mathbf{w}^T$ is a weighting vector. The calculation of the eigenvalue in the form of Eq. (2.7.4) does not work because the inversions of $\mathbf{A}_1$ and $\mathbf{A}_2$ during each outer iteration is very inaccurate. The eigenvalue estimate is updated only once in five to ten iterations.

The proof for convergence of the Equipoise scheme is an interesting problem, although not any satisfactory study has been done. A simple model for the study is the eigenvalue problem of a one-energy group, two-dimensional diffusion equation. The Equipoise scheme for this equation may be expressed by

$$a\phi_{i-1,j}^{(t)}+b\phi_{i+1,j}^{(t-1)}+c\phi_{i,j-1}^{(t)}+d\phi_{i,j+1}^{(t-1)}+e\phi_{i,j}^{(t)}=\lambda\phi_{i,j}^{(t-1)}$$

where t is the iteration time, the scheme is based on the Gauss–Seidel method, and λ is an estimate for the eigenvalue $(1/k)$. The above scheme can be proven to converge to the true solution if λ is close to the exact fundamental eigenvalue. As a method to accelerate the convergence of the Equipoise method, the coarse-mesh rebalancing acceleration technique has been successfully used (see Section 8.5).

PROBLEMS

1. Assume $p(\mathbf{r})$ and $q(\mathbf{r})$ are constant and $S=0$ in Eq. (3.1.1). Grids are spaced uniformly, that is, $h_2=h_4=\Delta x$ and $h_4=h_3=\Delta y$ in Fig. 3.2. Derive the finite difference equation for Eq. (3.1.1) by directly applying the central difference formulas, Eqs. (1.4.19) and (1.4.20). Compare the result with Eq. (3.2.7), and discuss the differences in the coefficients of unknowns.

2. Derive the finite difference formulas of the equation

$$-\nabla p\nabla\phi(x, y)+q\phi(x, y)=S$$

$$\frac{\partial}{\partial n}\phi+\gamma\phi=0, \qquad \gamma>0 \text{ on } \Gamma$$

for the mesh point (i, j) on the upper boundary Γ of a geometry as shown below.

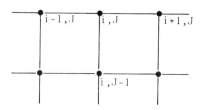

3. a. Derive the finite difference equations for

$$\left[-\frac{1}{r}\frac{\partial}{\partial r}rp\frac{\partial}{\partial r}-\frac{\partial}{\partial z}p\frac{\partial}{\partial z}+q\right]\phi(r, z)=S(r, z)$$

where r is the radial coordinate and z is the axial coordinate.

b. Derive the finite difference equations for

$$\left[-\frac{1}{r}\frac{\partial}{\partial r}rp\frac{\partial}{\partial r}-\frac{1}{r^2}\frac{\partial}{\partial\theta}p\frac{\partial}{\partial\theta}+q\right]\phi(r, \theta)=S(r, \theta)$$

where r is the radial coordinate and θ is the angular coordinate.

4. Consider the one-dimensional difference equation

$$\frac{2\phi_i-\phi_{i-1}-\phi_{i+1}}{h^2}=S_i, \qquad i=1, 2, \ldots, I-1$$

$$\phi_0=\phi_I=0$$

Write the iterative schemes for this equation in accordance with (a) the Jacobi iterative method, (b) the Gauss–Seidel iterative method, and (c) SOR.

5. The equation in Problem 4 can be written in the form of Eq. (3.3.15). Write **B**, **L**, and **U** explicitly.

6. The equation in Problem 4 can be expressed in the form, $\mathbf{A}\phi=\mathbf{S}$. If ϕ is defined as

$$\phi=\text{col}(\phi_1, \phi_2)$$

where

$$\boldsymbol{\phi}_1 = \text{col}\,(\phi_1, \phi_3, \phi_5, \ldots)$$

$$\boldsymbol{\phi}_2 = \text{col}\,(\phi_2, \phi_4, \phi_6, \ldots)$$

then **A** may be written as

$$\mathbf{A} = \begin{bmatrix} \mathbf{A}_{11} & \mathbf{A}_{12} \\ \mathbf{A}_{21} & \mathbf{A}_{22} \end{bmatrix}$$

Show that $\mathbf{A}_{11}$ and $\mathbf{A}_{22}$ are diagonal matrices, while $\mathbf{A}_{21}$ and $\mathbf{A}_{12}$ are bidiagonal matrices and $\mathbf{A}_{12} = \mathbf{A}_{21}^T$.

7. Suppose the elements of $\boldsymbol{\phi}$ in Eq. (3.2.24) are numbered in accordance with the consistent ordering for line inversions. Then a permutation of blocks of **A** will lead to the form

$$\begin{bmatrix} \mathbf{A}_{11} & \mathbf{A}_{12} \\ \mathbf{A}_{21} & \mathbf{A}_{22} \end{bmatrix}\begin{bmatrix} \boldsymbol{\phi}_1 \\ \boldsymbol{\phi}_2 \end{bmatrix} = \begin{bmatrix} \mathbf{S}_1 \\ \mathbf{S}_2 \end{bmatrix}$$

where $\boldsymbol{\phi}_1$ is a block vector involving only odd lines, $\boldsymbol{\phi}_2$ even lines. Show that $\mathbf{A}_{11}$ and $\mathbf{A}_{22}$ are block diagonal matrices while $\mathbf{A}_{12}$ and $\mathbf{A}_{21}$ are block bidiagonal matrices.

8. Repeat the calculation of Eq. (3.4.15) for the case where the matrix **B** has a canonical box of order three for the eigenvalue equal to the spectral radius.

9. Show two examples of physical systems for which the linear equations are reducible.

10. Prove that the difference equations for

$$-\frac{d^2}{dx^2}\phi(x) + q\phi(x) = S, \qquad q > 0$$

are strictly diagonally dominant.

11. Prove that the difference equations for

$$-\frac{d^2}{dx^2}\phi(x) = S, \qquad \alpha < x < \beta$$

$$\phi'(\alpha) = \phi(\beta) = 0$$

are irreducibly diagonally dominant.

12. Show that the matrix representing the difference equations in Problem 10 is an S-matrix.

13. Prove Eq. (3.7.18).

14. Write a FORTRAN program to solve

$$\frac{4\phi_{i,j} - \phi_{i-1,j} - \phi_{i+1,j} - \phi_{i,j-1} - \phi_{i,j+1}}{h^2} = S_{i,j}$$

$$\phi_{1,j} = \phi_{I,j} = \phi_{i,1} = \phi_{i,J} = 0$$

by using (a) the point Jacobi and (b) the point SOR method.

15. Repeat Problem 14 by using (a) the line Jacobi method and (b) SLOR.

16. Show that a half of the eigenvectors of the Gauss–Seidel matrix are in deficiency. [Hint: (1) SOR with $\omega = 1$ is equivalent to the Gauss–Siedel method. (2) Refer to Fig. 3.9.]

17. If the spectral radius of the Jacobi iterative matrix is $\mu(\mathbf{M}_B) = 1 - \epsilon$ where $\epsilon \ll 1$, what is the spectral radius of the Gauss–Seidel matrix?

18. Show that the number of the Gauss–Seidel iterations to meet a certain convergence criterion is approximately half of the Jacobi iterations.

19. When SOR is used, one needs to know $\mu(\mathbf{M}_B)$ to calculate ω_{opt}. The algorithm of Problem 20 may be used for this purpose. Another method to estimate $\mu(\mathbf{M}_B)$ is to perform SOR with ω less than ω_{opt}.
  a. Show that

$$\frac{\langle \boldsymbol{\phi}^{(t)} - \boldsymbol{\phi}^{(t-1)}, \boldsymbol{\phi}^{(t)} - \boldsymbol{\phi}^{(t-1)} \rangle}{\langle \boldsymbol{\phi}^{(t-1)} - \boldsymbol{\phi}^{(t-2)}, \boldsymbol{\phi}^{(t-1)} - \boldsymbol{\phi}^{(t-2)} \rangle}$$

converges to $\mu^2(\mathbf{L}_\omega)$, which is the spectral radius of SOR with ω.
  b. How can $\mu(\mathbf{M}_B)$ be estimated by using an estimate for $\mu^2(\mathbf{L}_\omega)$?

20. The spectral radius $\mu(\mathbf{M}_B)$ of the Jacobi iterative matrix can be obtained by performing iterations given by Eq. (3.6.1) with $\mathbf{g} = \mathbf{0}$ and an arbitrary initial vector $\boldsymbol{\phi}^{(0)}$. Show that

$$\frac{\langle \boldsymbol{\phi}^{(t)}, \boldsymbol{\phi}^{(t)} \rangle}{\langle \boldsymbol{\phi}^{(t-1)}, \boldsymbol{\phi}^{(t-1)} \rangle}$$

converges to $\mu^2(\mathbf{M}_B)$. (Hint: expand $\boldsymbol{\phi}^{(0)}$ into the eigenvectors of $\mathbf{M}_B$.)

21. Suppose we solve the equation in Problem 14 by the Jacobi iterative method. Discuss how the convergence rate is affected if I and J arc doubled while h is reduced to a half.

22. Repeat Problem 21 for the SOR method.

23. Solve the following equation by using the SOR method

$$-\nabla^2\phi + 0.1\phi = 1.0$$

The geometry and the boundary conditions are specified in the following figure. Use 10×10 grid points. Compare the convergence rates for $\omega = 1.5, 1.7, 1.9$.

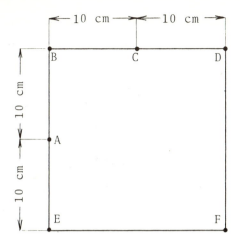

Boundary conditions

$A-B-C \quad \nabla_n \phi = -2\phi$

$C-D-F-E-A \quad \nabla_n \phi = 0$

24. Apply the one-parameter Chebyshev polynomial extrapolation to the Jacobi iterative method, and derive the formula to calculate the extrapolation parameters.

25. Prove that if **A** and **B** are both positive definite, $\mathbf{A}^{-1}\mathbf{B}$ is also positive definite. [Hint: (1) A positive definite matrix has its square root. (2) Use a similarity transformation. (3) See also Theorem 3.4.]

REFERENCES

1. L. A. Hageman and C. J. Pfeifer, *The utilization of the neutron diffusion program, PDQ-5*, WAPD-TM-395, Bettice Atomic Power Laboratory, Pittsburgh, Pa. (1965).

2. W. R. Cadwell, *PDQ-7 reference manual*, WADP-TM-678, Bettice Atomic Power Laboratory, Pittsburgh, Pa. (1969).

3. J. R. Lamarsh, *Nuclear reactor theory*, Addison-Wesley, Reading, Mass. (1966).

4. T. B. Fowler, D. R. Vondy, and G. W. Cunningham, *Nuclear reactor core analysis code: CITATION,"* ORNL-TM-2496, Oak Ridge National Laboratory, Tenn. (1971).

5. E. L. Wachspress, *CURE, a generalized two-space-dimension multigroup coding for IBM-704*, KAPL-1724, Knolls Atomic Power Laboratory, Schenectady, N.Y. (1957).

6. E. L. Wachspress, "Optimum alternating-direction-implicit iteration parameters for a model problem," *J. Soc. Indust. Appl. Math.*, **10**, 994–1016 (1963).

7. E. L. Wachspress, *Iterative solution of elliptic systems*, Prentice-Hall, Englewood Cliffs, N.J. (1966).

8. R. S. Varga, *Matrix iterative analysis*, Prentice-Hall, Englewood Cliffs, N.J. (1962).

9. J. N. Franklin, *Matrix theory*, Prentice-Hall, Englewood Cliffs, N.J. (1968).

10. L. A. Hageman and R. B. Kellogg, "Estimating optimum overrelaxation parameters," *Math. Comp.*, **22**, 60–68 (1968).

11. D. M. Young, *Iterative solution of large linear systems*, Academic, New York (1962).

12. L. A. Hageman, "The estimation of acceleration parameters for the Chebyshev polynomial and successive overrelaxation iteration methods," WAPD-TM-1038, Bettice Atomic Power Laboratory, Pennsylvania (1969).

13. L. A. Hageman, *Block iterative methods for two-cyclic matrix equations with special application to the numerical solution of the second order self-adjoint elliptic partial differential equation in two dimensions,* WAPD-TM-327, Bettice Atomic Power Laboratory, Pennsylvania (1962).

14. T. Dupont, R. P. Kendall, and H. H. Rachford, Jr., "A factorization procedure for the solution of elliptic difference equation," *SIAM J. Numer. Anal.,* **5**, 559–585 (1968).

15. H. L. Stone, "Iterative solution of implicit approximations of multi-dimensional partial differential equations," SIAM J. Numer. Anal., **5**, 530–559 (1968).

16. M. A. Diamond, *An economical algorithm for the solution of finite difference equations,* Ph.D. thesis, University of Illinois at Urbana-Champaign (1972).

17. M. A. Diamond, "An adaptive algorithm for the application of factorization techniques in the solution of elliptic difference equations," Department of Computer Science, University of Illinois, Urbana, Ill. (1974).

18. T. B. Fowler and M. L. Tobias, *Equipois-3: A two-dimensional, two-group, neutron diffusion code for IBM-7090 computer,* ORNL-3199, Oak Ridge National Laboratory, Tenn. (1962).

19. M. L. Tobias and T. B. Fowler, *The TWENTY GRAND program for the numerical solution of few-group neutron diffusion equations in two dimensions,* ORNL-3200, Oak Ridge National Laboratory, Tenn. (1962).

20. T. B. Fowler, M. L. Tobias, and D. R. Vondy, *EXTERMINATOR—a multigroup code for solving neutron diffusion equations in one and two dimensions,* ORNL-TM-842, Oak Ridge National Laboratory, Tenn. (1966).

21. D. L. Delp, D. L. Fisher, J. M. Harriman, and M. J. Stedwell, *FLARE, a three-dimensional boiling water reactor simulator,* GEAP-4598, General Electric Company, Calif. (1964).

22. P. T. Choong and H. Soodak, "Simplified eigenvalue problem solutions in diffusion calculation," *Trans. Amer. Nucl. Soc.,* **13**, 744–745 (1970).

23. M. K. Butler, M. Legan, L. Ranzini, and W. J. Snow, *Compilation of program abstract,* ANL-7411, Argonne National Laboratory. (The original version was published in 1968. Since then, several supplements have been added.)

Chapter 4 Numerical Solution for Partial Differential Equations of the Parabolic Type

4.1 PARTIAL DIFFERENTIAL EQUATION OF THE PARABOLIC TYPE

The partial differential equation in the form

$$\frac{\partial \phi(\mathbf{r}, t)}{\partial t} = \nabla p \nabla \phi(\mathbf{r}, t) - q\phi(\mathbf{r}, t) + S(\mathbf{r}, t) \qquad (4.1.1)$$

is called a parabolic partial differential equation, where p and q are coefficients and are functions of $\mathbf{r}$ and t. For one-dimensional slab geometries, Eq. (4.1.1) is reduced to

$$\frac{\partial \phi(x, t)}{\partial t} = \frac{\partial}{\partial x} p \frac{\partial}{\partial x} \phi(x, t) - q\phi(x, t) + S(x, t) \qquad (4.1.2)$$

Parabolic partial differential equations are encountered in various branches of science. For example, the heat transfer equation in a slab geometry is

$$\frac{\partial T(x, t)}{\partial t} = \frac{\partial}{\partial x} p \frac{\partial}{\partial x} T(x, t) + Q(x, t) \qquad (4.1.3)$$

where T is the temperature at space coordinate x and time t, p is the thermal conductivity, and Q is the heat-generation term. Another example is the time-dependent monoenergetic neutron diffusion equation, given by

$$\frac{1}{v} \frac{\partial \phi(x, t)}{\partial t} = \frac{\partial}{\partial x} D \frac{\partial}{\partial x} \phi(x, t) + (\nu\Sigma_f - \Sigma_a)\phi(x, t) + S(x, t) \qquad (4.1.4)$$

where ϕ is the neutron flux, v is the neutron velocity, D is the diffusion coefficient, Σ_a is the absorption cross section, ν is the average number of neutrons born per fission, Σ_f is the fission cross section, and S is the neutron source.

In order for these parabolic equations to become physically meaningful, boundary conditions for external boundaries and an initial condition must be specified. The treatment of boundary conditions are the same as the steady state problems discussed in Chapters 2 and 3. If we set the initial time at $t = 0$, then the initial condition is

$$\phi(\mathbf{r}, 0) = \phi_0(\mathbf{r}) \tag{4.1.5}$$

where the right side is a prescribed function.

The partial differential equations of parabolic type are often accompanied by the nonlinear feedback effect. For example, the thermal conductivity p in Eq. (4.1.3) may be a function of T. If the change of p in each time step is not significant (this is the case when the feedback effect itself is small or Δt is small), p is recalculated in every time step by using the previous values of T.

4.2   DERIVATION OF DIFFERENCE EQUATIONS

The difference equations for parabolic partial differential equations are closely related to those for steady-state equations discussed in Chapters 2 and 3, since if the time derivative is zero, the parabolic equations are reduced to steady-state problems. The purpose of this section is to show how the time derivatives can be treated. Only the one-dimensional slab case is considered here, since the extension to multidimensional case or other one-dimensional geometries is easy once the one-dimensional slab case is studied.

Let us assume that the width of the slab considered is H and that the boundary and initial conditions for Eq. (4.1.2) are given by

$$\phi(0, t) = \phi(H, t) = 0 \quad \text{Boundary conditions}$$
$$\phi(x, 0) = \phi_0(x) \quad\quad\quad \text{Initial condition} \tag{4.2.1}$$

In order to apply the finite difference approximation to Eq. (4.1.2), we consider the mesh system on the x-t coordinate as shown in Fig. 4.1. The values of p, q, and S are assumed to be constant in each mesh box on the x-t plane.

We first apply the difference approximation on the spatial coordinate. If we use the point scheme of finite differencing discussed in Section 2.2, the semidifference equation becomes

$$\frac{d}{dt}\phi_i(t) = a_i[\phi_{i-1}(t) - \phi_i(t)] + c_i[\phi_{i+1}(t) - \phi_i(t)] - \Gamma_i\phi_i(t) + Q_i(t) \tag{4.2.2}$$

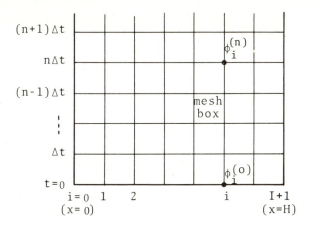

Figure 4.1   Time-space mesh system.

where

$$a_i = \frac{2p_i}{\Delta x_i(\Delta x_i + \Delta x_{i+1})}, \qquad c_i = \frac{2p_{i+1}}{\Delta x_{i+1}(\Delta x_i + \Delta x_{i+1})} \qquad (4.2.3)$$

$$\Gamma_i = \frac{q_i \Delta x_i + q_{i+1} \Delta x_{i+1}}{\Delta x_i + \Delta x_{i+1}} \qquad (4.2.4)$$

$$Q_i = \frac{S_i \Delta x_i + S_{i+1} \Delta x_{i+1}}{\Delta x_i + \Delta x_{i+1}} \qquad (4.2.5)$$

Equation (4.2.2) can also be expressed with matrix and vector notations by

$$\frac{d}{dt}\boldsymbol{\phi}(t) = \mathbf{A}\boldsymbol{\phi}(t) + \mathbf{Q}(t) \qquad (4.2.6)$$

where $\boldsymbol{\phi}$ is the vector consisting of $\{\phi_i(t)\}$, $i = 1, 2, \ldots, I$, and $\mathbf{A}$ is a tridiagonal matrix, of which the diagonal elements are the coefficients of $\phi_i(t)$, the left off-diagonal elements are the coefficients of $\phi_{i-1}(t)$, and the right off-diagonal elements are the coefficients of $\phi_{i+1}(t)$.

Next we apply the finite difference approximation on the time coordinate. Suppose the time t is in the interval, $t_n < t < t_n + \Delta t$, then the left side of Eq. (4.2.2) can be approximated by

$$\frac{d}{dt}\phi_i(t) = \frac{\phi_i^{(n+1)} - \phi_i^{(n)}}{\Delta t} \qquad (4.2.7)$$

where

$$\phi_i^{(n)} = \phi_i(t_n) \qquad (4.2.8)$$

The values of $\phi_i(t)$ on the right side of Eq. (4.2.4) may be approximated in three ways as follows:

$$\phi_i(t) \approx \phi_i^{(n)} \tag{4.2.9}$$

$$\phi_i(t) \approx \phi_i^{(n+1)} \tag{4.2.10}$$

$$\phi_i(t) \approx \alpha\phi_i^{(n+1)} + (1-\alpha)\phi_i^{(n)} \tag{4.2.11}$$

where α is an arbitrary parameter in the range $0 \le \alpha \le 1$.

In the case of Eq. (4.2.9), Eq. (4.2.2) becomes the forward difference formula:

$$\frac{\phi_i^{(n+1)} - \phi_i^{(n)}}{\Delta t} = a_i\phi_{i-1}^{(n)} + b_i\phi_i^{(n)} + c_i\phi_{i+1}^{(n)} + Q_i^{(n)} \tag{4.2.12}$$

or equivalently,

$$\phi_i^{(n+1)} = \Delta t \left[a_i\phi_{i-1}^{(n)} + \left(b_i + \frac{1}{\Delta t} \right)\phi_i^{(n)} + c_i\phi_{i+1}^{(n)} + Q_i^{(n)} \right] \tag{4.2.13}$$

where $Q_i^{(n)}$ is the average of $Q_i(t)$ in the time interval Δt, and

$$b_i = -a_i - c_i - \Gamma_i$$

Therefore, $\phi_i^{(n+1)}$ are calculated if $\phi_i^{(n)}$ are known. In fact, $\boldsymbol{\phi}_i^{(0)}$ are specified as the initial conditions so that the $\boldsymbol{\phi}_i^{(1)}$ at the first time step can be calculated, and then the calculations go on to the next time step. Since the solutions for the new time step are expressed explicitly in terms of the solution for the previous time step, this method is called the *explicit* method.

In the case of Eq. (4.2.10), Eq. (4.2.2) becomes the backward difference formula,

$$-a_i\phi_{i-1}^{(n+1)} + \left(\frac{1}{\Delta t} - b_i \right)\phi_i^{(n+1)} - c_i\phi_{i+1}^{(n+1)} = \frac{1}{\Delta t}\phi_i^{(n)} + Q_i^{(n)} \tag{4.2.14}$$

where the three unknown values are on the left side of the equation. The set of the above equations for all the spatial grid points must be solved simultaneously. This set of equations has the same form as the finite difference equations for the one-group one-dimensional steady-state diffusion equation, so that it may be solved by using the recursion formula, Eqs. (1.7.15) and (1.7.16). Since the field values of the new time steps are described implicitly by Eq. (4.2.14) the present method is called the *implicit* method.

The difference equation based on Eq. (4.2.11) is written

$$
-a_i\alpha\phi_{i-1}^{(n+1)} + \left(\frac{1}{\Delta t} - b_i\alpha\right)\phi_i^{(n+1)} - c_i\alpha\phi_{i+1}^{(n+1)}
$$

$$
= a_i(1-\alpha)\phi_{i-1}^{(n)} + \left[b_i(1-\alpha) + \frac{1}{\Delta t}\right]\phi_i^{(n)} + c_i(1-\alpha)\phi_{i+1}^{(n)} + Q_i^{(n)}
$$

$$(4.2.15)$$

Equation (4.2.15) is a general expression including the former two methods, because $\alpha = 0$ yields Eq. (4.2.13) and $\alpha = 1$ yields Eq. (4.2.14). If $\alpha = 1/2$, Eq. (4.2.15) becomes the central difference formula and is called the Crank–Nicholson method. Except for $\alpha = 0$, $\phi_i^{(n+1)}$ on the left side must be solved simultaneously so that Eq. (4.2.15) is an *implicit* method.

In the matrix notations, Eq. (4.2.15) is expressed by

$$
\frac{\boldsymbol{\phi}^{(n+1)} - \boldsymbol{\phi}^{(n)}}{\Delta t} = \alpha\mathbf{A}\boldsymbol{\phi}^{(n+1)} + (1-\alpha)\mathbf{A}\boldsymbol{\phi}^{(n)} + \mathbf{Q}^{(n)} \qquad (4.2.16)
$$

where

$$
\boldsymbol{\phi}^{(n)} = \boldsymbol{\phi}(t_n) \qquad (4.2.17)
$$

4.3   STABILITY OF THE DIFFERENCE EQUATIONS

The solution of the difference equations may differ from the solution of the differential equations for several reasons. First, the difference equation only approximates the differential equation, and this introduces the truncation error. The truncation error will vanish in the limit as Δx and Δt approach zero. Second, the solution of the difference and the differential equations may differ even in the limit as Δx and Δt approaches zero. This lack of convergence may be interpreted as instability of the difference equations. For example, for the explicit method, it is shown that $\Delta t < (\Delta x)^2/2p$ is the stability condition. Unless this inequality between Δx and Δt is satisfied, the numerical solution is unstable and the solution oscillates in an erratic fashion.[2] Therefore, even if Δx and Δt are both made smaller and smaller, but in such a way that the inequality is not satisfied, then the numerical solution will not converge to the true solution of the difference equation.

The stability of the three numerical methods derived in Section 4.2 is studied by comparing the behavior of the solution of Eq. (4.2.16) with that of Eq. (4.2.6). Equation (4.2.16) has been derived by applying a difference approximation to the time derivative in Eq. (4.2.6): the instability is introduced in this procedure. For simplicity of discussion in this section, we consider the homogeneous slab with $S(x, t) = 0$, for which the coefficients of Eq. (4.1.2) do not change in time. We assume the initial and boundary conditions are given by Eq. (4.2.1). The equally spaced mesh intervals are considered: $\Delta x = H/(I+1)$. We express the solutions of Eqs. (4.2.6) and (4.2.16) in terms of the eigenvectors of the associated eigenvalue problem. The stability of the numerical solution can be seen by observing how the coefficients of eigenvectors change as time increases.

The matrix $\mathbf{A}$ defined in Eq. (4.2.6) is a finite difference operator for a Sturm–Liouville equation (see Section 2.1). Therefore, $\mathbf{A}$ has distinct eigenvalues and corresponding eigenvectors satisfying

$$\omega_k \mathbf{u}_k = \mathbf{A}\mathbf{u}_k, \qquad k = 1, 2, \ldots, I \qquad (4.3.1)$$

where ω_k is an eigenvalue and $\mathbf{u}_k$ is the corresponding eigenvector. In the case of the homogeneous slab with an equally spaced mesh system, Eq. (4.3.1) may be written as

$$\omega_k u_{i,k} = \frac{p}{(\Delta x)^2}(u_{i-1,k} - 2u_{i,k} + u_{i+1,k}) - q u_{i,k} \qquad (4.3.2)$$

where $u_{i,k}$ is the ith element of $\mathbf{u}_k$ ($u_{0,k} = u_{I+1,k} = 0$ due to the boundary conditions). It is easily seen that the eigensolution and the eigenvalue for Eq. (4.3.2) are given by

$$u_{i,k} = \sin(kBi\Delta x) \qquad (4.3.3a)$$

$$\omega_k = \frac{2p}{(\Delta x)^2}[\cos(kB\Delta x) - 1] - q \qquad (4.3.3b)$$

where

$$B = \frac{\pi}{H}$$

Assuming that $\Delta x/H$ is sufficiently small, the maximum and minimum eigenvalues are, respectively,

$$\max_k(\omega_k) = \omega_1 \approx -pB^2 - q \qquad (4.3.4)$$

$$\min_k(\omega_k) = \omega_I \approx -\frac{4p}{(\Delta x)^2} - q \approx -\frac{4p}{(\Delta x)^2} \qquad (4.3.5)$$

It should be noted that ω_1 is related to the size of the slab and ω_I to the mesh size.

Let us investigate the difference between the solutions for Eqs. (4.2.6) and (4.2.16). We know that $\{\mathbf{u}_k\}$, $k = 1, 2, \ldots, I$, form a complete set (see Section 1.7), so that the solution for each equation can be expressed in terms of $\mathbf{u}_k$. The solution for Eq. (4.2.6) with $\mathbf{Q} = \mathbf{0}$ is given by

$$\boldsymbol{\phi}(t) = \sum_{k=1}^{I} g_k \mathbf{u}_k \exp(\omega_k t) \qquad (4.3.6)$$

In Eq. (4.3.6) g_ks are constants determined by the initial condition:

$$g_k = \langle \mathbf{u}_k, \boldsymbol{\phi}_0 \rangle \tag{4.3.7}$$

where $\mathbf{u}_k$ is assumed to be orthonormal. For $t > 0$, the sequence of $\exp(\omega_k t)$ satisfies the inequalities

$$e^{\omega_1 t} > e^{\omega_2 t} > \cdots > e^{\omega_I t} > 0 \tag{4.3.8}$$

Therefore, as time increases, Eq. (4.3.6) approaches the asymptotic form:

$$\boldsymbol{\phi}(t) \rightarrow g_1 \mathbf{u}_1 \exp(\omega_1 t) \tag{4.3.9}$$

Equation (4.2.16) with $\mathbf{Q} = \mathbf{0}$ may be written as

$$\left[\frac{\mathbf{I}}{\Delta t} - \alpha \mathbf{A} \right] \boldsymbol{\phi}^{(n+1)} = \left[\frac{\mathbf{I}}{\Delta t} + (1-\alpha) \mathbf{A} \right] \boldsymbol{\phi}^{(n)} \tag{4.3.10}$$

where $\mathbf{I}$ is the identity matrix. We note that the eigenvectors of $\mathbf{A}$ satisfy the eigenvalue problem associated with Eq. (4.3.10) as

$$\gamma_k \left[\frac{\mathbf{I}}{\Delta t} - \alpha \mathbf{A} \right] \mathbf{u}_k = \left[\frac{\mathbf{I}}{\Delta t} + (1-\alpha) \mathbf{A} \right] \mathbf{u}_k \tag{4.3.11}$$

where γ_k is the eigenvalue. By introducing Eq. (4.3.1) into Eq. (4.3.11) we find that the eigenvalue γ_k is correlated with ω_k by

$$\gamma_k = \frac{1 + (1-\alpha)\omega_k \Delta t}{1 - \alpha \omega_k \Delta t} \tag{4.3.12}$$

Thus the solution of Eq. (4.2.16) with $\mathbf{Q} = \mathbf{0}$ or equivalently Eq. (4.3.10) may be expressed by

$$\boldsymbol{\phi}^{(n)} = \sum_{k=1}^{I} g_k \gamma_k^n \mathbf{u}_k \tag{4.3.13}$$

where g_k is defined by Eq. (4.3.7) and γ_k^n is the nth power of γ_k.

By comparing Eq. (4.3.13) with Eq. (4.3.6), γ_k^n in the former is found to be playing the role of $\exp(n\Delta t \omega_k)$ in the latter. While $\exp(n\Delta t \omega_k)$ is positive for all ω_ks, γ_k^n may become negative depending on the value of α. If any one of γ_k is less than -1, γ_k^n will oscillate and diverge as n increases.

For $\alpha = 0$ (explicit method), Eq. (4.3.12) becomes

$$\gamma_k = 1 + \Delta t \omega_k \tag{4.3.14}$$

so that $\gamma_k < -1$ occurs for $\omega_k < -2/\Delta t$. In order to prevent this, the minimum eigenvalue must satisfy

$$\min_k \gamma_k = \gamma_I \approx 1 - \Delta t \frac{4p}{(\Delta x)^2} > -1 \tag{4.3.15}$$

where Eqs. (4.3.5) and (4.3.14) are used. Equation (4.3.15) may be rewritten as

$$\Delta t < \frac{(\Delta x)^2}{2p} \tag{4.3.16}$$

This is the condition for stability of the explicit method. The above equation indicates that, as the spatial mesh interval is made smaller, the time interval must be made much smaller. Therefore, computational time will be significantly increased.

For $\alpha = 1$ (implicit method), Eq. (4.3.12) becomes

$$\gamma_k = \frac{1}{1 - \Delta t \omega_k} \tag{4.3.17}$$

Since γ_k becomes positive and smaller than unity for all negative ω_k, there is no possibility of instability for any Δt. However, γ_1 becomes negative for a positive ω_1 unless $\Delta t < 1/\omega_k$. As discussed in Section 4.4, γ_1 is an approximation to $\exp(\omega_1 \Delta t)$, so that $\Delta t \ll 1/\omega_1$ must be satisfied in order that γ_1 be a good approximation to $\exp(\omega_1 \Delta t)$. Therefore, as long as $\Delta t \ll 1/\omega_1$ is satisfied, γ_k are all positive and satisfy the inequality

$$0 < \gamma_l < \cdots < \gamma_2 < \gamma_1 \tag{4.3.18}$$

Accordingly, there is no possibility of oscillation.

For $\alpha = 1/2$ (Crank-Nicholson), Eq. (4.3.9) becomes

$$\gamma_k = \frac{1 + 0.5 \Delta t \omega_k}{1 - 0.5 \Delta t \omega_k} \tag{4.3.19}$$

Similarly to the case of Eq. (4.3.17) Δt must satisfy $\Delta t \ll 2/\omega_1$ in order that γ_1 becomes a good approximation to $\exp(\omega_1 \Delta t)$. When this is satisfied, γ_k satisfies the inequality

$$-1 < \gamma_k < \gamma_1, \qquad k > 1 \tag{4.3.20}$$

and there is no possible instability or oscillation.

4.4 TRUNCATION ERROR AND EXPONENTIAL TRANSFORMATION

There are two major sources of error introduced into numerical solutions: one is due to the spatial difference scheme and the other to the time difference scheme. In order to see the effect of spatial differencing, we first consider the analytical solution of Eq. (4.1.2) with $S = 0$ and constant coefficients. We compare Eq. (4.3.6), which is the solution of the semidifference approximation, Eq. (4.2.6), to the analytical solution of Eq. (4.1.2).

With the boundary and initial conditions given by Eq. (4.2.1), the analytical solution can be written as

$$\phi(x, t) = \sum_{k=1}^{\infty} h_k v_k(x) \exp(\eta_k t) \qquad (4.4.1)$$

where η_k and $v_k(x)$ are the kth eigenvalue and the corresponding eigenfunction of the equation,

$$\eta v(x) = \left(p \frac{d^2}{dx^2} - q \right) v(x) \qquad (4.4.2)$$

and h_k is the coefficient. η_k and v_k are given as

$$\eta_k = -pk^2 B^2 - q \qquad (4.4.3)$$

$$v_k(x) = \sqrt{\frac{2}{H}} \sin(kBx) \qquad (4.4.4)$$

where $B = \pi/H$. It is easily seen that $v_k(x)$ are orthonormal eigenfunctions and form a complete set. The coefficients h_k are determined by the initial condition and orthogonality of $v_k(x)$.

Let us compare the solution of the semidifference approximation, Eq. (4.3.6), with the analytical solution, Eq. (4.4.1). It can be seen that (1) terms higher than $k = I$ are missing in Eq. (4.3.6), (2) $g_k = h_k$ since $\mathbf{u}_k$ and $v_k(x)$ are both orthonormal, (3) the terms with $k \le I$ in the two equations differ because of different exponents. Since $\mathbf{u}_k$ is a discrete approximation to $v_k(x)$, it is natural to consider that ω_k is also an approximation to η_k. However, this is true only for small k. If we define the error of ω_k as $\epsilon_k \equiv \omega_k - \eta_k$ it may be written as

$$\epsilon_k = \frac{2p}{(\Delta x)^2}(\cos kB\Delta x - 1) + pk^2 B^2 \qquad (4.4.5)$$

For a small k, ϵ_k can be expressed by

$$\epsilon_k \approx +\frac{p}{12} k^4 B^4 \Delta x^2 \qquad (4.4.6)$$

so that the error for ω_1 is the smallest among other ω_ks, and as k increases the error is rapidly increased. The magnitude of ϵ_1 is more important for us than ϵ_k, $k > 1$, because $\mathbf{u}_1$ is the asymptotic solution. As seen from Eq. (4.4.6) ϵ_1 becomes smaller as Δx decreases.

The error introduced by time-differencing procedure can be studied by comparing Eqs. (4.3.6) and (4.3.13) again. The error of Eqs. (4.3.13) introduced by time differencing is given as the difference betweeen those two equations:

$$\boldsymbol{\epsilon} \equiv \boldsymbol{\phi}^{(n)} - \boldsymbol{\phi}(t_n) = \sum_k g_k \mathbf{u}_k (\gamma_k^n - \exp(\omega_k t_n)) \tag{4.4.7}$$

where γ_k^n is the nth power of γ_k. Equation (4.4.7) shows that the error associated with each eigenvector component is given by

$$\kappa_k^{(n)} \equiv \gamma_k^n - \exp(\omega_k t_n) = \gamma_k^n - \exp(\omega_k n \Delta t) \tag{4.4.8}$$

where $t_n = n \Delta t$ is used. Let us consider Eq. (4.4.8) for $n = 1$. Using Eq. (4.3.12), $\kappa_k^{(1)}$ becomes

$$\kappa_k^{(1)} = \frac{1 + (1-\alpha)\Delta t \omega_k}{1 - \alpha \Delta t \omega_k} - \exp(\omega_k \Delta t) \tag{4.4.9}$$

For a small time step such that $\omega_k \Delta t \ll 1$, Eq. (4.4.9) can be expanded as

$$\kappa_k^{(1)} = (\alpha - 1/2)(\omega_k \Delta t)^2 + 0(\Delta t^3) \tag{4.4.10}$$

Therefore, when $\alpha = 1/2$ (Crank–Nicholson) $\kappa_k^{(1)}$ becomes the third order of Δt, while $\kappa_k^{(1)}$ is the second order of Δt for $\alpha = 0$ and 1.

We now consider a method to decrease the truncation error of time differencing by using the exponential transformation proposed by Hansen.[13] For a short time after a sudden change in the coefficients or the source term, there is a fast transient where the spatial shape of the solution changes significantly. After then the solutions are well represented by the asymptotic exponential term plus small contributions by the next few exponential terms. This suggests that the solution at each mesh point may be expressed as an exponential function in time modified by a correction function:

$$\phi_i(t) = e^{\Omega t} \psi_i(t) \tag{4.4.11}$$

where $\psi_i(t)$ is the correction function. If $e^{\Omega t}$ approximates the time behavior of solution at each point well, the change of $\psi_i(t)$ in time is small and smooth. We can reasonably expect that if the difference equation is written in terms of ψ_i the truncation error introduced by the first order difference approximation for $d\psi_i/dt$ will be small, and accordingly, we may increase the size of the time steps.

Introducing Eq. (4.4.11) into Eq. (4.2.2), the difference equation for the correction function is obtained:

$$\frac{d}{dt}\psi_i(t) = a_i(\psi_{i-1} - \psi_i) + c_i(\psi_{i+1} - \psi_i) + (\Gamma(t) - \Omega)\psi_i + Q_i e^{-\Omega t} \tag{4.4.12}$$

If we apply the backward time differencing to the time derivative of $\psi_i(t)$, the implicit scheme for ψ is obtained:

$$\frac{\psi_i^{(n+1)} - \psi_i^{(n)}}{\Delta t} = a_i(\psi_{i-1}^{(n+1)} - \psi_i^{(n+1)}) + c_i(\psi_{i+1}^{(n+1)} - \psi_i^{(n+1)})$$
$$+ (\Gamma(t) - \Omega)\psi_i^{(n+1)} + \tilde{Q}_i \qquad (4.4.13)$$

where $\tilde{Q}_i$ is the time average of $Q_i(t) \exp(-\Omega_i t)$ in Δt.

This method improves the accuracy of the numerical scheme whenever the change in the spacial shape of the solution in time is small, even when the coefficients of the source term are space- and time-dependent. The value of Ω can be calculated during the previous time step and can be changed from time to time.

The use of such a single parameter Ω is obviously restrictive if the solution in different positions and in different time responds at different rates to a localized perturbation. Equation (4.4.9) can be extended to include time- and space-dependent parameters as

$$\phi_i(t) = \exp(\Omega_i^{(n)} t)\psi_i(t), \qquad t_n \leq t \leq t_{n+1} \qquad (4.4.14)$$

The parameters $\Omega_i^{(n)}$ can be calculated at the previous time step for each mesh point by

$$\Omega_i^{(n)} = \frac{1}{\Delta t} \ln \frac{\phi_i^{(n)}}{\phi_i^{(n-1)}} \qquad (4.4.15)$$

4.5 THE FOURIER STABILITY ANALYSIS

The stability analysis in the previous section using the eigenvectors is rigorous and easy to understand. However, the eigenvectors for many problems cannot be expressed analytically. A more general method to examine the stability of a numerical scheme for a transient problem is the Fourier analysis originally proposed by von Neumann[1-3] in which the solution for each time step is expanded into the Fourier series. If the magnitude of any Fourier component does not increase unboundedly, then the scheme is said to be stable. Usually the Fourier analysis is applied to an infinite geometry in order to avoid the complexity involved with the boundary conditions. However, we apply the Fourier analysis to finite geometries that are subject to boundary conditions.

The difference scheme for a parabolic partial differential equation on an equally spaced grid system may be written as

$$\mathscr{F}\phi_i^{(n)} = \mathscr{B}\phi_i^{(n-1)} + S_i^{(n)} \qquad (4.5.1)$$

where $\mathcal{F}$ and $\mathcal{B}$ are difference operators. We restrict ourselves to a second order difference operator $\mathcal{F}$, and assume that $\phi_i^{(n)}$ is subject to the boundary conditions

$$\phi_0^{(n)} = \phi_{I+1}^{(n)} = 0 \qquad (4.5.2)$$

If $\mathcal{F}$ and $\mathcal{B}$ do not change with n, we call Eq. (4.5.1) a one-step scheme. It is assumed that the exact solution is not expected to grow unboundedly. Generally, $\mathcal{F}$ and $\mathcal{B}$ satisfy the following property:

$$\gamma\mathcal{F} \sin (i\alpha + \theta) = \mathcal{B} \sin (i\alpha) \qquad (4.5.3)$$

which is the imaginary part of

$$\gamma\mathcal{F} \exp (ij\alpha + j\theta) = \mathcal{B} \exp (ij\alpha) \qquad (4.5.4)$$

where $j = \sqrt{-1}$, γ is a real constant, and θ is the phase shift. Both γ and θ are characteristic values of the combination of $\mathcal{F}$ and $\mathcal{B}$.

Numerical errors due to round-off are generated at each time step of Eq. (4.5.1). In order to study how the error generated at a time step will behave in later time steps, the following special situation is assumed: the round-off error occur at $n = 0$ but not in the subsequent time steps. Under this condition, the solution of Eq. (4.5.1) for $n \geq 0$ may be written as

$$\phi_i^{(n)} = \phi_{i,\,\text{exact}}^{(n)} + \epsilon_i^{(n)}, \qquad i = 1, 2, \ldots, I \qquad (4.5.5)$$

where $\phi_{i,\,\text{exact}}$ is the exact solution and ϵ_i is the error originated at $n = 0$. Introducing Eq. (4.5.5) into Eq. (4.5.1), we find

$$\mathcal{F}\epsilon_i^{(n)} = \mathcal{B}\epsilon_i^{(n-1)} \qquad (4.5.6)$$

The errors are also subject to the boundary conditions, $\epsilon_0^{(n)} = \epsilon_{I+1}^{(n)} = 0$, because $\phi_i^{(n)}$ and $\phi_{i,\,\text{exact}}^{(n)}$ both satisfy the same boundary conditions.

The errors are expanded into the Fourier series,

$$\epsilon_i^{(n)} = \sum_{k=1}^{I} b_k^{(n)} \sin (\alpha_k i)$$

$$\alpha_k = \frac{k\pi \Delta x}{H} \qquad (4.5.7)$$

where H is the length of the region, $\Delta x = H/(I+1)$, and $b_k^{(n)}$ is the coefficient that is determined by

$$b_k^{(n)} = \frac{\displaystyle\sum_{i=1}^{I} \epsilon_i^{(n)} \sin (\alpha_k i)}{\displaystyle\sum_{i=1}^{I} \sin^2 (\alpha_k i)} \qquad (4.5.8)$$

It is of interest to see how each of $b_k^{(n)}$ behaves as n increases. The relation between $b_k^{(n)}$ to $b_k^{(n-1)}$ is not always simple because of the phase shift to the Fourier components and boundary conditions. Assuming that right side of Eq. (4.5.6) is known, the general solution to $\epsilon_i^{(n)}$ is very similar to that for an inhomogeneous differential equation. It may be written as

$$\epsilon_i^{(n)} = \epsilon_{i,h}^{(n)} + \epsilon_{i,p}^{(n)} \tag{4.5.9}$$

where $\epsilon_{i,h}^{(n)}$ is the homogeneous solution satisfying

$$\mathscr{F}\epsilon_{i,h}^{(n)} = 0 \tag{4.5.10}$$

and $\epsilon_{i,p}^{(n)}$ is a particular solution.

Since $\mathscr{F}$ is a second order difference operator, the solution for $\epsilon_{i,h}^{(n)}$ is given by

$$\epsilon_{i,h}^{(n)} = c_1 e^{\tau_1 i} + c_2 e^{\tau_2 i} \tag{4.5.11}$$

where τ_1 and τ_2 are the characteristic values of

$$\mathscr{F}e^{\tau i} = 0 \tag{4.5.12}$$

and c_1 and c_2 are undetermined coefficients. The particular solution is found by introducing Eq. (4.5.7) for $n-1$ to the right side of Eq. (4.5.6) and using Eq. (4.5.3) as

$$\epsilon_{i,p}^{(n)} = \mathscr{F}^{-1}\mathscr{B}\epsilon_i^{(n-1)} = \sum_{k=1}^{I} b_k^{(n-1)}\gamma_k \sin(\alpha_k i + \theta_k) \tag{4.5.13}$$

Therefore, Eq. (4.5.9) becomes

$$\epsilon_i^{(n)} = c_1^{(n)}e^{\tau_1 i} + c_2^{(n)}e^{\tau_2 i} + \mathscr{F}^{-1}\mathscr{B}\epsilon_i^{(n-1)} \tag{4.5.14}$$

Constants c_1 and c_2 are determined by using the boundary conditions

$$
\begin{aligned}
\epsilon_0^{(n)} &= 0 = c_1^{(n)} + c_2^{(n)} + \epsilon_{0,p}^{(n)} \\
\epsilon_{I+1}^{(n)} &= 0 = c_1^{(n)}e^{\tau_1(I+1)} + c_2^{(n)}e^{\tau_2(I+1)} + \epsilon_{I+1,p}^{(n)}
\end{aligned}
\tag{4.5.15}
$$

A successive application of Eq. (4.5.14) yields

$$
\begin{aligned}
\epsilon_i^{(n)} &= c_1^{(n)}e^{\tau_1 i} + c_2^{(n)}e^{\tau_2 i} + \mathscr{F}^{-1}\mathscr{B}\epsilon_i^{(n-1)} \\
&= c_1^{(n)}e^{\tau_1 i} + c_2^{(n)}e^{\tau_2 i} + \mathscr{F}^{-1}\mathscr{B}(c_1^{(n-1)}e^{\tau_1 i} + c_2^{(n-1)}e^{\tau_1 i}) + (\mathscr{F}^{-1}\mathscr{B})^2\epsilon_i^{(n-2)} \\
&= \cdots
\end{aligned}
\tag{4.5.16}
$$

$$= \sum_{t=0}^{n-1} c_1^{(n-t)}(\mathscr{F}^{-1}\mathscr{B})^t e^{\tau_1 i} + \sum_{t=0}^{n-1} c_2^{(n-t)}(\mathscr{F}^{-1}\mathscr{B})^t e^{\tau_2 i} + (\mathscr{F}^{-1}\mathscr{B})^n\epsilon_i^{(0)}$$

where t is an integer. The discrete functions, $e^{\tau_1 i}$ and $e^{\tau_2 i}$ for $i = 1$, $2, \ldots, I$ can be expanded into the Fourier series:

$$e^{\tau_g i} = \sum_{k=1}^{I} b_{gk} \sin(\alpha_k i), \qquad g = 1, 2 \tag{4.5.17}$$

The first and second terms after the last equality of Eq. (4.5.16) satisfies the following inequalities

$$
\begin{aligned}
\left| \sum_{t=0}^{n-1} c_g^{(n-t)} (\mathscr{F}^{-1}\mathscr{B})^t \sum_{k=1}^{I} b_{gk} \sin(\alpha_k i) \right| \\
= \left| \sum_{k=1}^{I} b_{gk} \sum_{t=0}^{n-1} c_g^{(n-t)} (\mathscr{F}^{-1}\mathscr{B})^t \sin(\alpha_k i) \right| \\
\leq \sum_{k=1}^{I} |b_{gk}| \max_t |c_g^{(n-t)}| \left| \sum_{t=0}^{n-1} \gamma^t \sin(\alpha_k i + t\theta_k) \right| \\
\leq \sum_{k=1}^{I} G_{gk} \sum_{t=0}^{n-1} |\gamma^t| \leq \sum_{k=1}^{I} G_{gk} \frac{1}{1 - |\gamma_k|}, \qquad g = 1, 2
\end{aligned}
\tag{4.5.18}
$$

where

$$G_{gk} = |b_{gk}| \max_t |c_g^{(n-t)}| \tag{4.5.19}$$

Therefore, the first and second terms of Eq. (4.5.16) do not increase unboundedly if $|\gamma_k| < 1$. The last term of Eq. (4.5.16) becomes

$$(\mathscr{F}^{-1}\mathscr{B})^n \epsilon_1^{(0)} = \sum_{k=1}^{I} b_k^{(0)} \gamma_k^n \sin(\alpha_k i + n\theta_k) \tag{4.5.20}$$

where Eq. (4.5.4) is used. Equation (4.5.20) converges to zero if $|\gamma_k| < 1$. Thus we prove that if $|\gamma_k| < 1$ for all k, the errors generated originally at $n = 0$ do not unboundedly increase as n increases. If $\theta_k = 0$ for all k, the particular solution given by Eq. (4.5.13) satisfies the boundary conditions, $\epsilon_{0,p}^{(n)} = \epsilon_{I+1,p}^{(n)} = 0$, accordingly, Eq. (4.5.15) yields $c_1^{(n)} = c_2^{(n)} = 0$. In this case, the first and second terms of Eq. (4.5.16) vanish, and $\epsilon_i^{(n)} \to 0$ as n increases if $|\gamma_k| < 1$.

Although this analysis is performed assuming that the difference operator $\mathscr{F}$ is second order, we can extend the analysis to the first order or the higher order difference operators. The Fourier stability analysis not only applies to the parabolic difference equations but also to difference equations associated with hyperbolic partial differential equations.

In general, the Fourier stability analysis is applied to the infinite mesh space with uniform Δx, neglecting the effect of boundary conditions. This means $c_1^{(n)} = c_2^{(n)} = 0$ is assumed in Eq. (4.5.14). The initial error in the infinite mesh space can be expressed by the Fourier series (see Problem 9):

$$\epsilon_i^{(0)} = \sum_{k=-\infty}^{+\infty} A_k \exp\left(ijk\,\Delta x\right)$$

where A_k are coefficient of the Fourier components given by

$$A_k = \lim_{I \to \infty} \frac{1}{2I+1} \sum_{i=-I}^{+I} \exp\left(-ikj\Delta x\right)\epsilon_i^{(0)} \qquad (4.5.21)$$

Each Fourier component satisfies Eq. (4.5.4) or equivalently

$$\xi \exp\left(ikj\Delta x\right) = \mathscr{F}^{-1}\mathscr{B} \exp\left(ikj\Delta x\right) \qquad (4.5.22)$$

where $\xi = \gamma e^{i\theta}$. Therefore, the error at any n becomes

$$\epsilon_i^{(n)} = (\mathscr{F}^{-1}\mathscr{B})^n \epsilon_i^{(0)} = \sum_{k=-\infty}^{+\infty} A_k \xi_k^n \exp\left(ikj\Delta x\right) \qquad (4.5.23)$$

The stability condition is given by

$$\max_k |\xi_k| < 1 \qquad (4.5.24)$$

Applying this approach to the forward, backward, and Crank–Nicholson method, the same stability criteria as derived in Section 4.3 are obtained. The Fourier stability analysis is applied to the parabolic equation with advective terms in Section 4.6.

4.6 STABILITY OF PARABOLIC EQUATIONS WITH THE ADVECTIVE TERM

The parabolic equation in the form (see Section 5.6)

$$\frac{\partial \phi}{\partial t} = -\left[u \frac{\partial \phi}{\partial x} + v \frac{\partial \phi}{\partial y} \right] + \nu \nabla^2 \phi \qquad (4.6.1)$$

appears as the vorticity transport equation for incompressible fluid flow, where u and v are coefficients, ϕ is the field function (vorticity), and ν is a constant. The first term on the right side is the advective term,[3] and the second term is the diffusion term. If $\nu = 0$, Eq. (4.6.1) is reduced to the hyperbolic equation that is called the advective equation. Because of the advective term, Eq. (4.6.1) has different properties than the diffusion equation.

In this section, we study the numerical schemes for the one-dimensional model for Eq. (4.6.1), that is,

$$\frac{\partial \phi}{\partial t} = -u\frac{\partial \phi}{\partial x} + \nu\frac{\partial^2 \phi}{\partial x^2} \tag{4.6.2}$$

The numerical schemes for two-dimension are described in Section 4.7. Various difference formulas for Eq. (4.6.2) can be obtained depending on how each term on the right side is treated. Let us study the stability of three schemes by using Fourier analysis.

We first consider the explicit scheme

$$\frac{\phi_i^{(n+1)} - \phi_i^{(n)}}{\Delta t} = -u\frac{\phi_{i+1}^{(n)} - \phi_{i-1}^{(n)}}{2\Delta x} \\ + \nu\frac{\phi_{i+1}^{(n)} - 2\phi_i^{(n)} + \phi_{i-1}^{(n)}}{\Delta x^2} \tag{4.6.3}$$

In applying Fourier analysis, we consider an infinite space and assume that u in Eq. (4.5.3) is a positive constant, although u is space dependent in general. By introducing $\phi_i^{(n)} = \xi_k^n e^{ikj\Delta x}$ into Eq. (4.6.3), we find

$$\xi_k = 1 + 2\nu\frac{\Delta t}{\Delta x^2}[\cos(k\,\Delta x) - 1] - ju\frac{\Delta t}{\Delta x}\sin(k\,\Delta x) \tag{4.6.4}$$

Applying Eq. (4.5.24) to Eq. (4.6.4), two necessary conditions are obtained as

$$d \equiv \nu\frac{\Delta t}{\Delta x^2} < \frac{1}{2} \tag{4.6.5}$$

$$c \equiv u\frac{\Delta t}{\Delta x} < 1 \tag{4.6.6}$$

Equation (4.6.5) is equivalent to Eq. (4.3.16) for the diffusion equation, while Eq. (4.6.6) is the stability condition for the advective equation ($\nu = 0$). The most general criteria for stability is found by calculation of the modulus of ξ as

$$|\xi_k|^2 = \xi_k\bar{\xi}_k = [1 + 2d(\cos k\Delta x - 1)]^2 + c^2(1 - \cos^2 k\Delta x) < 1 \tag{4.6.7}$$

The necessary condition thus obtained[3] is

$$\frac{u\Delta x}{\nu} \le 2 \tag{4.6.8}$$

Equations (4.6.5) and (4.6.8) are the necessary and sufficient conditions for stability.

The fully implicit scheme for Eq. (4.6.2) is

$$\frac{\phi_i^{(n+1)} - \phi_i^{(n)}}{\Delta t} = -u \frac{\phi_{i+1}^{(n+1)} - \phi_{i-1}^{(n+1)}}{2\Delta x}$$
$$+ v \frac{\phi_{i+1}^{(n+1)} - 2\phi_i^{(n+1)} + \phi_{i-1}^{(n+1)}}{\Delta x^2} \tag{4.6.9}$$

The amplitude factor for Eq. (4.6.9) is

$$\xi_k = \frac{1}{2d[1 - \cos(k\Delta x)] + jc\,\sin(k\Delta t)} \tag{4.6.10}$$

Since Eq. (4.6.10) satisfies Eq. (4.5.24), the implicit scheme is unconditionally stable.

Another interesting scheme is obtained by applying the leapfrog method[3] to the advective term:

$$\frac{\phi_i^{(n+1)} - \phi_i^{(n-1)}}{2\Delta t} = -u \frac{\phi_{i+1}^{(n)} - \phi_{i-1}^{(n)}}{2\Delta x} + v \frac{\phi_{i+1}^{(n-1)} - 2\phi_i^{(n-1)} + \phi_{i-1}^{(n-1)}}{\Delta x^2} \tag{4.6.11}$$

The Fourier stability analysis shows that the stability of Eq. (4.6.11) is limited by the smallest of the advection limit, $c = u\Delta t/\Delta x \leq 1$, and the diffusion limit, $d = v\Delta t/\Delta x^2 \leq 1/2$. For the advective equation ($v = 0$), Eq. (4.6.11) is proven unconditionally unstable.

4.7 TWO-DIMENSIONAL PARABOLIC EQUATIONS

Two-dimensional parabolic equations are encountered in transients such as heat transfer, neutron diffusion, and fluid flow. In this section we study the numerical schemes for two-dimensional parabolic equations by using the vorticity equation as an example. However, the schemes can be applied to the diffusion equation if the advective term in Eq. (4.6.1) is set to zero. For simplicity of discussion, the mesh spacings, Δx and Δy, are assumed to be uniform. The cases of nonuniform mesh spacing and a space-dependent coefficient of the diffusion term can be treated by the method discussed in Section 3.2.

The explicit scheme for the two-dimensional vorticity equation is

$$\frac{\phi_{i,j}^{(n+1)} - \phi_{i,j}^{(n)}}{\Delta t} = u \frac{\phi_{i-1,j}^{(n)} - \phi_{i+1,j}^{(n)}}{2\Delta x} + v \frac{\phi_{i,j-1}^{(n)} - \phi_{i,j+1}^{(n)}}{2\Delta y}$$
$$+ v \left[\frac{\phi_{i-1,j}^{(n)} + \phi_{i+1,j}^{(n)} - 2\phi_{i,j}^{(n)}}{(\Delta x)^2} + \frac{\phi_{i,j-1}^{(n)} + \phi_{i,j+1}^{(n)} - 2\phi_{i,j}^{(n)}}{(\Delta y)^2} \right] \tag{4.7.1}$$

where the first and the second terms on the right side represent the advective term. Assuming that u, v, and ν are constant, the Fourier stability criteria are

$$\nu\left[\frac{\Delta t}{(\Delta x)^2}+\frac{\Delta t}{(\Delta y)^2}\right]\le 1 \tag{4.7.2}$$

$$u\frac{\Delta t}{\Delta x}+v\frac{\Delta t}{\Delta y}\le 1 \tag{4.7.3}$$

The fully implicit method is obtained by replacing all superscript ns with $n+1$ on the right side of Eq. (4.7.1). The implicit scheme is unconditionally stable, thus allowing a large time step. The disadvantage, however, is that $\phi_{i,j}^{(n+1)}$ for all the grid points must be solved simultaneously. This can be done either by the Gaussian elimination method or an iterative solution method such as the SOR method. The time saved by using a larger time step can be offset by a long computational time to solve the implicit equation. The application of the implicit method for neutron diffusion equations is briefly described in Section 4.8.

As an alternative to the fully explicit or fully implicit method, the alternative direction implicit (ADI) method is favored for heat conduction and fluid flow problems, since it is unconditionally stable, while no iteration or direct inversion of the whole matrix is required. The ADI method has been also applied for two- and three-dimensional time-dependent neutron diffusion equations.

In the case of the vorticity equation, the ADI scheme is represented by the following two steps:

Step 1.   (Row inversion)

$$\frac{\phi_{i,j}^{(n+1/2)}-\phi_{i,j}^{(n)}}{\Delta t/2}=u\frac{\phi_{i-1,j}^{(n+1/2)}-\phi_{i+1,j}^{(n+1/2)}}{2\Delta x}+v\frac{\phi_{i,j-1}^{(n)}-\phi_{i,j+1}^{(n)}}{2\Delta x}$$

$$+\nu\left[\frac{\phi_{i-1,j}^{(n+1/2)}+\phi_{i+1,j}^{(n+1/2)}-2\phi_{i,j}^{(n+1/2)}}{(\Delta x)^2}\right. \tag{4.7.4}$$

$$\left.+\frac{\phi_{i,j-1}^{(n)}+\phi_{i,j+1}^{(n)}-2\phi_{i,j}^{(n)}}{(\Delta y)^2}\right]$$

Step 2. (Column inversion)

$$\frac{\phi_{i,j}^{(n+1)} - \phi_{i,j}^{(n+1/2)}}{\Delta t/2} = u\frac{\phi_{i-1,j}^{(n+1/2)} - \phi_{i+1,j}^{(n+1/2)}}{2\Delta x} + v\frac{\phi_{i,j-1}^{(n+1)} - \phi_{i,j+1}^{(n+1)}}{2\Delta y}$$

$$+ v\left[\frac{\phi_{i-1,j}^{(n+1/2)} + \phi_{i+1,j}^{(n+1/2)} - 2\phi_{i,j}^{(n+1/2)}}{(\Delta x)^2}\right.$$

$$\left. + \frac{\phi_{i,j-1}^{(n+1)} + \phi_{i,j+1}^{(n+1)} - 2\phi_{i,j}^{(n+1)}}{(\Delta y)^2}\right] \tag{4.7.5}$$

Equation (4.7.4) includes unknowns only on a row of grid points, so that the equations for a row can be solved by using the line inversion technique for a tridiagonal matrix (see Section 1.7). Similarly, Eq. (4.7.5) for Step 2 involves line inversions for columns of grid points. This is essentially the same as the ADI method for elliptic equations introduced in Section 3.3.

4.8 NUMERICAL SOLUTIONS OF TIME-DEPENDENT NEUTRON DIFFUSION EQUATIONS WITH DELAYED NEUTRONS (ONE AND TWO DIMENSIONS)

The purpose of this section is to outline the numerical solution for unsteady state multigroup diffusion equations for nuclear reactors. We first consider the two-group diffusion equation with one delayed neutron group for a slab geometry. This model is simple but includes all the essential features of the multigroup equation with six delayed neutron groups. The change of flux distribution in time will generally affect the group constants through changes in xenon concentration and material density. Therefore, the neutron diffusion equations must be coupled with feedback equations such as for the xenon effect, heat conduction, and coolant flow. However, in discussing the numerical solutions, we assume that the group constants are given functions of time and also that they can be approximated by constants in each time step.

IMPLICIT ONE-SPACE-DIMENSIONAL SCHEME

The one-dimensional two-group neutron diffusion equation with one-delayed group neutrons may be written as

$$\frac{\partial\phi_1(x, t)}{v_1 \partial t} - \left(\frac{\partial}{\partial x}D_1\frac{\partial}{\partial x} - \Sigma_1\right)\phi_1 = (1 - \beta)(\nu\Sigma_{f1}\phi + \Sigma_{f2}\phi_2) + \lambda C$$

$$\frac{\partial\phi_2(x, t)}{v_2 \partial t} - \left(\frac{\partial}{\partial x}D_2\frac{\partial}{\partial x} - \Sigma_2\right)\phi_2 = \Sigma_{sl}\phi_1 \tag{4.8.1}$$

$$\frac{\partial C(x, t)}{\partial t} + \lambda C = \beta(\nu\Sigma_{f1}\phi_1 + \nu\Sigma_{f2}\phi_2)$$

where subscripts 1 and 2 represent the fast and thermal energy groups, respectively, no external source is assumed, and other notations are standard in reactor physics.[9] We make each equation of Eq. (4.8.1) discrete with respect to the space coordinate just as Eq. (4.2.2) was derived from Eq. (4.1.2). The semidiscrete equations are written as

$$\frac{1}{v_1}\frac{d\boldsymbol{\phi}_1}{dt}+(\mathbf{H}_1+\mathbf{R}_1)\boldsymbol{\phi}_1=(1-\beta)(\mathbf{M}_1\boldsymbol{\phi}_1+\mathbf{M}_2\boldsymbol{\phi}_2)+\lambda\,\mathbf{C}$$

$$\frac{1}{v_2}\frac{d\boldsymbol{\phi}_2}{dt}+(\mathbf{H}_2+\mathbf{R}_2)\boldsymbol{\phi}_2=\mathbf{T}\boldsymbol{\phi}_1 \qquad (4.8.2)$$

$$\frac{d\mathbf{C}}{dt}+\lambda\,\mathbf{C}=\beta\,(\mathbf{M}_1\boldsymbol{\phi}_1+\mathbf{M}_2\boldsymbol{\phi}_2)$$

where $\boldsymbol{\phi}$ is a vector representing the flux on the grid points, $\mathbf{H}$ is the tridiagonal matrix representing the difference operator for $-(\partial/\partial x)D(\partial/\partial x)$, $\boldsymbol{R}$ is a diagonal matrix representing the removal cross section, Σ_g, $g = 1, 2$; $\mathbf{T}$ is a diagonal matrix representing the slowing down cross section, Σ_{sl}; $\mathbf{M}$ is a diagonal matrix representing the fission cross section times the average number of neutrons per fission, $\nu\Sigma_{fg}$, $g = 1, 2$; and $\mathbf{C}$ is the vector representing the precursor concentration at the grids.

Equation (4.8.2) may be expressed in a compact form as

$$\frac{d\boldsymbol{\Phi}}{dt}=\mathbf{B}\boldsymbol{\Phi} \qquad (4.8.3)$$

if we define

$$\boldsymbol{\Phi}=\mathrm{col}\,[\boldsymbol{\phi}_1,\ \boldsymbol{\phi}_2,\ \mathbf{C}] \qquad (4.8.4)$$

and

$$\mathbf{B}=\begin{bmatrix} -v_1[\mathbf{H}_1+\mathbf{R}_1-(1-\beta)\mathbf{M}_1], & v_1(1-\beta)\mathbf{M}_2, & v_1\lambda\mathbf{I} \\ v_2\mathbf{T}, & -v_2(\mathbf{H}_2+\mathbf{R}_2), & \mathbf{0} \\ \beta\mathbf{M}_1, & \beta\mathbf{M}_2, & -\lambda\mathbf{I} \end{bmatrix}$$

$$\equiv\begin{bmatrix} \mathbf{B}_{11} & \mathbf{B}_{12} & \mathbf{B}_{13} \\ \mathbf{B}_{21} & \mathbf{B}_{22} & \mathbf{B}_{23} \\ \mathbf{B}_{31} & \mathbf{B}_{32} & \mathbf{B}_{33} \end{bmatrix} \qquad (4.8.5)$$

We now apply a difference scheme to the time derivatives of Eq. (4.8.3) and derive an implicit scheme:

$$\frac{\boldsymbol{\Phi}^{(n+1)}-\boldsymbol{\Phi}^{(n)}}{\Delta t}=\alpha\mathbf{B}\boldsymbol{\Phi}^{(n+1)}+(1-\alpha)\mathbf{B}\boldsymbol{\Phi}^{(n)} \qquad (4.8.6)$$

which corresponds to Eq. (4.2.16) for Eq. (4.2.6). Notice in Eq. (4.8.4) that the third component of $\boldsymbol{\Phi}$ is $\mathbf{C}$. If we explicitly write Eq. (4.8.6) in terms of $\boldsymbol{\phi}_1^{(n+1)}$ and $\boldsymbol{\phi}_2^{(n+1)}$, we obtain the equation in the following form:

$$\begin{bmatrix} \mathbf{E}_{11} & -\mathbf{E}_{12} \\ -\mathbf{E}_{21} & \mathbf{E}_{22} \end{bmatrix} \begin{bmatrix} \boldsymbol{\phi}_1^{(n+1)} \\ \boldsymbol{\phi}_2^{(n+1)} \end{bmatrix} = \begin{bmatrix} \mathbf{d}_1 \\ \mathbf{d}_2 \end{bmatrix} \tag{4.8.7}$$

where $\mathbf{C}^{(n+1)}$ is eliminated and

$$\mathbf{E}_{11} = \frac{\mathbf{I}}{\Delta t} - \alpha \mathbf{B}_{11} - \alpha \eta \mathbf{B}_{31} \tag{4.8.8}$$

$$\mathbf{E}_{12} = \alpha \mathbf{B}_{12} + \alpha \eta \mathbf{B}_{32} \tag{4.8.9}$$

$$\mathbf{E}_{21} = \alpha \mathbf{B}_{21} \tag{4.8.10}$$

$$\mathbf{E}_{22} = \frac{\mathbf{I}}{\Delta t} - \alpha \mathbf{B}_{22} \tag{4.8.11}$$

$$\begin{aligned} \mathbf{d}_1 = \frac{1}{\Delta t} \boldsymbol{\phi}_1^{(n)} + (1-\alpha)(\mathbf{B}_{11}\boldsymbol{\phi}_1^{(n)} + \mathbf{B}_{12}\boldsymbol{\phi}_2^{(n)}) \\ + (1-\alpha)\eta(\mathbf{B}_{31}\boldsymbol{\phi}_1^{(n)} + \mathbf{B}_{32}\boldsymbol{\phi}_2^{(n)}) + \lambda(v_1 - \eta)\mathbf{C}^{(n)}. \end{aligned} \tag{4.8.12}$$

$$\mathbf{d}_2 = \frac{1}{\Delta t} \boldsymbol{\phi}_2^{(n)} + (1-\alpha)(\mathbf{B}_{21}\boldsymbol{\phi}_1^{(n)} + \mathbf{B}_{22}\boldsymbol{\phi}_2^{(n)}) \tag{4.8.13}$$

$$\eta = v_1 \left(1 - \frac{1}{1 + \lambda \alpha \Delta t} \right) \tag{4.8.14}$$

In deriving this equation, we assume that $\boldsymbol{\phi}_1^{(n)}$, $\boldsymbol{\phi}_2^{(n)}$, and $\mathbf{C}^{(n)}$ are known from the previous step of calculation or the initial conditions, so all the terms proportional to those are included in $\mathbf{d}_1$ or $\mathbf{d}_2$. $\mathbf{E}_{11}$ and $\mathbf{E}_{22}$ are tridiagonal matrices; $\mathbf{E}_{22}$ and $\mathbf{E}_{21}$ are diagonal matrices.

If Eq. (4.8.7) is permutated so that the elements of the unknown vector are reordered as

$$\operatorname{col}\left[(\phi_{1,1}^{(n+1)}, \phi_{2,1}^{(n+1)}), \ldots, (\phi_{1,i}^{(n+1)}, \phi_{2,i}^{(n+1)}), \ldots, (\phi_{1,I}^{(n+1)}, \phi_{2,I}^{(n+1)})\right] \tag{4.8.15}$$

then the coefficient matrix of Eq. (4.8.7) may be transformed to a block tridiagonal matrix with 2×2 submatrices. For block tridiagonal matrices a simple recursion formula is available to solve the unknowns [see Eq. (1.7.17)].

STANDARD TWO-DIMENSIONAL IMPLICIT SCHEME—TWIGL[7]

The extension of the one-dimensional implicit method to two dimensions is straightforward, and the scheme can be expressed again by Eq. (4.8.7) if each term is redefined appropriately. Eq. (4.8.1) is altered to a two-dimensional equation if ϕ_1, ϕ_2, and C are redefined as functions of x, y, and t, and $-(\partial/\partial y)D(\partial/\partial y)$ is added to the second term of the first and second equations of Eq. (4.8.1). In two-dimension, tridiagonal matrices $\mathbf{V}_1$ and $\mathbf{V}_2$ each representing the difference operators for $-(\partial/\partial y)D_1(\partial/\partial y)$ and $-(\partial/\partial y)D_2(\partial/\partial y)$, respectively, must be added to the second term of the first and second equation of Eq. (4.8.2). Assuming the elements of ϕ_1 and ϕ_2 correspond to the grid points in accordance with the line scheme described in Section 3.3 and illustrated by Fig. 3.6, $\mathbf{V}_1$ and $\mathbf{V}_2$ are tridiagonal matrices with half-band width equal to the number of grid points on the x direction. $\mathbf{E}_{11}$ and $\mathbf{E}_{22}$ are block-tridiagonal matrices of order equal to the number of mesh rows representing difference formulas for elliptic partial differential equations. The order of each submatrix is equal to the number of columns. $\mathbf{E}_{12}$ and $\mathbf{E}_{21}$ are block-diagonal matrices. For two-dimensional problems, the block matrices on the left side of Eq. (4.8.7) cannot be inverted directly because the submatrices of $\mathbf{E}_{11}$ and $\mathbf{E}_{22}$ represent the difference operators for partial differential operators of elliptic type. So two loops of iterative schemes are used: outer iteration and inner iteration. The basic form of the outer iteration scheme is written as

$$\begin{bmatrix} \mathbf{E}_{11} & \mathbf{0} \\ \mathbf{0} & \mathbf{E}_{22} \end{bmatrix}\begin{bmatrix} \boldsymbol{\phi}_1^{(n+1)} \\ \boldsymbol{\phi}_2^{(n+1)} \end{bmatrix}^{(t)} = \begin{bmatrix} \mathbf{0} & \mathbf{E}_{12} \\ \mathbf{E}_{21} & \mathbf{0} \end{bmatrix}\begin{bmatrix} \boldsymbol{\phi}_1^{(n+1)} \\ \boldsymbol{\phi}_2^{(n+1)} \end{bmatrix}^{(t-1)} + \begin{bmatrix} \mathbf{d}_1 \\ \mathbf{d}_2 \end{bmatrix} \quad (4.8.16)$$

where t is the outer iteration number. The Chebyshev polynomial acceleration method based on the cyclic nature of Eq. (4.7.19) reduces the number of iterations. An inner iteration scheme is used to solve the equations in the form, $\mathbf{E}_{gg}\boldsymbol{\phi}_g^{(n+1)} = \mathbf{S}_g$, $g = 1, 2$. This equation is the difference equation for a partial differential equation of elliptic type, so all the iterative solutions described in Chapter 3 are applicable.

EXPONENTIAL TRANSFORMATION APPLIED TO ONE-DIMENSIONAL NEUTRON DIFFUSION EQUATIONS

The upper limit of the time step interval Δt of the one-dimensional implicit scheme is bounded because of the truncation error as mentioned in Section 4.3. This is a serious penalty when slow transient problems are calculated, since they usually involve a long physical time and accordingly a long computing time. The exponential transformation introduced in Section 4.4 is useful in increasing the time-step interval. In this subsection the numerical method for one-dimensional space–kinetic equations called the GAKIN method[6] is introduced.

In the GAKIN method, matrix **B** given by Eq. (4.8.5) is partitioned into the form

$$\mathbf{B} = \mathbf{L} + \mathbf{U} + \mathbf{H} + \mathbf{\Gamma} \tag{4.8.17}$$

where **L** is strictly block lower triangular and **U** strictly block upper triangular as

$$\mathbf{L} = \begin{bmatrix} \mathbf{0}, & \mathbf{0}, & \mathbf{0} \\ v_2\mathbf{T}, & \mathbf{0}, & \mathbf{0} \\ \beta\mathbf{M}_1, & \beta\mathbf{M}_2, & \mathbf{0} \end{bmatrix} \quad \mathbf{U} = \begin{bmatrix} \mathbf{0}, & (1-\beta)\mathbf{M}_2, & v_1\lambda\mathbf{I} \\ \mathbf{0}, & \mathbf{0}, & \mathbf{0} \\ \mathbf{0}, & \mathbf{0}, & \mathbf{0} \end{bmatrix} \tag{4.8.18}$$

and **H** and **Γ** are strictly block diagonal; **H** is the tridiagonal matrix representing the difference approximation to the diffusion term, and **Γ** represents other terms as

$$\mathbf{\Gamma} = \begin{bmatrix} -v_1[\mathbf{R}_1 - (1-\beta)\mathbf{M}_1], & \mathbf{0}, & \mathbf{0} \\ \mathbf{0}, & -v_2\mathbf{R}_2, & \mathbf{0} \\ \mathbf{0}, & \mathbf{0}, & -\lambda\mathbf{I} \end{bmatrix} \tag{4.8.19}$$

Equation (4.8.3) is now written in the form

$$\frac{d\mathbf{\Phi}}{dt} - \mathbf{\Gamma}\mathbf{\Phi} = (\mathbf{L} + \mathbf{U} + \mathbf{H})\mathbf{\Phi} \tag{4.8.20}$$

Integrating Eq. (4.8.20) in the time interval, $t_n < t < t_{n+1}$, where $t_{n+1} = t_n + \Delta t$, yields

$$\mathbf{\Phi}^{(n+1)} = e^{\mathbf{\Gamma}\Delta t}\mathbf{\Phi}^{(n)} + \int_0^{\Delta t} e^{\mathbf{\Gamma}(\Delta t - \xi)}(\mathbf{L} + \mathbf{U} + \mathbf{H})\mathbf{\Phi}(t_n + \xi)\, d\xi \tag{4.8.21}$$

In Eq. (4.8.21) the exponential of a matrix is interpreted as

$$e^{\mathbf{P}} = \mathbf{I} + \sum_{m=1}^{\infty} \frac{1}{m!}\mathbf{P}^m \equiv \mathbf{Q} \tag{4.8.22}$$

where **I** is the identity matrix.

If **P** is a diagonal matrix the ith diagonal element of **Q** becomes

$$Q_{i,i} = 1 + \sum_{m=1}^{\infty} \frac{1}{m!}P_{i,j}^m = \exp(P_{i,i}) \tag{4.8.23}$$

where $Q_{i,i}$ and $P_{i,i}$ are, respectively, the ith diagonal elements of **Q** and **P**. We assume that the above power series converges. If the time behavior of the flux in the entire domain is well represented by $e^{\Omega \xi}$ in the time interval, $0 < \xi < \Delta t$, where $\xi = t - t_n$ and Ω is a constant, the terms in the integral of Eq. (4.8.21) may be approximated by

$$\mathbf{L}\Phi(t_n + \xi) \approx \mathbf{L}\Phi^{(n)} e^{\Omega \xi} \tag{4.8.24}$$

$$\mathbf{U}\Phi(t_n + \xi) \approx \mathbf{U}\Phi^{(n)} e^{\Omega \xi} \tag{4.8.25}$$

$$\mathbf{H}\Phi(t_n + \xi) \approx \mathbf{H}\Phi^{(n+1)} e^{\Omega(\xi - \Delta t)} \tag{4.8.26}$$

The constant Ω may be found from the previous time interval. A pointwise definition of Ω has also been used. Introducing the approximations of Eqs. (4.8.24) through (4.8.26) into Eq. (4.8.21) and performing the integration yield

$$[\mathbf{I} - (\mathbf{\Omega} - \mathbf{\Gamma})^{-1}(\mathbf{I} - e^{(\mathbf{\Gamma} - \mathbf{\Omega})\Delta t})\mathbf{H}]\Phi^{(n+1)}$$
$$= [e^{\mathbf{\Gamma}\Delta t} + (\mathbf{\Omega} - \mathbf{\Gamma})^{-1}(e^{\mathbf{\Omega}\Delta t} - e^{\mathbf{\Gamma}\Delta t})(\mathbf{\Gamma} + \mathbf{U})]\Phi^{(n)} \tag{4.8.27}$$

where $\mathbf{I}$ is the identity matrix and $\mathbf{\Omega} = \Omega \mathbf{I}$. By expressing the coefficients of $\Phi^{(n+1)}$ and $\Phi^{(n)}$ in Eq. (4.8.27) by $\mathbf{A}_1$ and $\mathbf{A}_2$, respectively, the Eq. (4.8.27) may be written more compactly as

$$\mathbf{A}_1 \Phi^{(n+1)} = \mathbf{A}_2 \Phi^{(n)} \tag{4.8.28}$$

The matrix $\mathbf{H}$ is a block-diagonal matrix with tridiagonal submatrices, so $\mathbf{A}_1$ is also a block-diagonal matrix with tridiagonal submatrices. The solution of Eq. (4.8.28) is easy. The inverse of $\mathbf{A}_1$ can be shown to be positive ($\mathbf{A}_1$ is a positive definite matrix regardless of the value of Δt), so Eq. (4.8.28) is unconditionally stable. If $e^{\Omega \Delta t}$ represents the major variation of the neutron flux in Δt, the error introduced by Eqs. (4.8.24) through (4.8.26) will be small. Therefore, Eq. (4.8.27) is more accurate than the standard implicit scheme, and, an accordingly much larger Δt than for the latter can be used.

The GAKIN method has been extended to two- and three-dimensional space–kinetic equations[11–14,16] in conjunction with the alternative-direction-implicit method as well as alternative-direction-explicit method. The major advantage of the multidimensional versions of the GAKIN method is that it does not require the two loops of iterative schemes as in the standard method. In other words, one time step needs only one sweep of grid points, so the computing time for one time step is shorter than the standard implicit method. The numerical study shows, however, that the multidimensional version of the GAKIN method for fast transient calculations needs much smaller time steps Δt than the standard implicit method.

ADI-B^2 METHOD

Another two-dimensional method designated as the ADI-B^2 method[17] closely resembles the TWIGL scheme but results in block-tridiagonal difference equations that are amenable to noniterative solution techniques. We first express the semidifference equation for a two-group two-dimensional equation in the following form:

$$\frac{1}{v_1}\frac{d\boldsymbol{\phi}_1}{dt} + (\mathbf{H}_1 + \mathbf{V}_1 + \mathbf{R}_1)\boldsymbol{\phi}_1 = (1-\beta)(\mathbf{M}_1\boldsymbol{\phi}_1 + \mathbf{M}_2\boldsymbol{\phi}_2) + \lambda\,\mathbf{C}$$

$$\frac{1}{v_2}\frac{d\boldsymbol{\phi}_2}{dt} + (\mathbf{H}_2 + \mathbf{V}_2 + \mathbf{R}_2)\boldsymbol{\phi}_2 = \mathbf{T}\boldsymbol{\phi}_1 \qquad (4.8.29)$$

$$\frac{d\mathbf{C}}{dt} + \lambda\,\mathbf{C} = \beta\,(\mathbf{M}_1\boldsymbol{\phi}_1 + \mathbf{M}_2\boldsymbol{\phi}_2)$$

where $\boldsymbol{\phi}_1$, $\boldsymbol{\phi}_2$ are the vectors representing the fluxes at the grid points; $\mathbf{H}$ is the matrix representing the difference operator for $-(\partial/\partial x)D(\partial/\partial x)$; $\mathbf{V}$ is the matrix representing the difference operator for $-(\partial/\partial y)D(\partial/\partial y)$; $\mathbf{R}$ is the diagonal matrix representing removal of neutrons; $\mathbf{T}$ is the diagonal matrix representing slowing down; and $\mathbf{M}$ is the diagonal matrix representing the fission cross section times number of neutrons per fission.

The ADI-B^2 method consists of two steps of calculations for each time interval Δt as follows.

ROW INVERSION

$$\left[\frac{2\mathbf{I}}{v_1\Delta t} + \mathbf{H}_1 + (\mathbf{D}_1\mathbf{B}^2)_y + \mathbf{R}_1 - (1-\beta)\mathbf{M}_1\right]\boldsymbol{\phi}_1^{(n+1/2)}$$

$$-(1-\beta)\mathbf{M}_2\boldsymbol{\phi}_2^{(n+1/2)} - \frac{1}{2}\lambda\,\mathbf{C}^{(n+1/2)} = \frac{2}{v_1\Delta t}\boldsymbol{\phi}_1^{(n)} + \frac{1}{2}\lambda\,\mathbf{C}^{(n)} \qquad (4.8.30)$$

$$\left[\frac{2\mathbf{I}}{v_2\Delta t} + \mathbf{H}_2 + (\mathbf{D}_2\mathbf{B}^2)_y + \mathbf{R}_2\right]\boldsymbol{\phi}_2^{(n+1/2)} - \mathbf{T}\boldsymbol{\phi}^{(n+1/2)} = \frac{2}{v_2\Delta t}\boldsymbol{\phi}_2^{(n)} \qquad (4.8.31)$$

$$\mathbf{C}^{(n+1/2)} = \frac{1-\lambda\Delta t/4}{1+\lambda\Delta t/4}\mathbf{C}^{(n)} + \frac{\beta\Delta t/4}{1+\lambda\Delta t/4}$$

$$\times[\mathbf{M}_1(\boldsymbol{\phi}_1^{(n+1/2)} + \boldsymbol{\phi}_1^{(n)}) + \mathbf{M}_2(\boldsymbol{\phi}_2^{(n+1/2)} + \boldsymbol{\phi}_2^{(n)})] \qquad (4.8.32)$$

where $(\mathbf{D}_g\mathbf{B}^2)_y$ is a diagonal matrix defined by the relation

$$(\mathbf{D}_g\mathbf{B}^2)_y\boldsymbol{\phi}_g^{(n)} = \mathbf{V}_g\boldsymbol{\phi}_g^{(n)} \qquad (4.8.33)$$

and physically represents the vertical leakage of neutron flux evaluated with the previous flux distribution $\phi_g^{(n)}$.

COLUMN INVERSION

$$\left[\frac{2\mathbf{I}}{v_1\Delta t}+\mathbf{V}_1+(\mathbf{D}_1\mathbf{B}^2)_x+\mathbf{R}_1-(1-\beta)\mathbf{M}_1\right]\boldsymbol{\phi}_1^{(n+1)}$$

$$-(1-\beta)\mathbf{M}_2\boldsymbol{\phi}_2^{(n+1)}-\frac{1}{2}\lambda\,\mathbf{C}^{(n+1)}$$

$$=\frac{2}{v_1\Delta t}\boldsymbol{\phi}_1^{(n+1/2)}+\frac{1}{2}\lambda\,\mathbf{C}^{(n+1/2)}$$

(4.8.34)

$$\left[\frac{2\mathbf{I}}{v_2\Delta t}+\mathbf{V}_2+(\mathbf{D}_2\mathbf{B}^2)_x+\mathbf{R}_2\right]\boldsymbol{\phi}_2^{(n+1)}-\mathbf{T}\boldsymbol{\phi}^{(n+1)}=\frac{2}{v_2\Delta t}\boldsymbol{\phi}_2^{(n)} \quad (4.8.35)$$

$$\mathbf{C}^{(n+1)}=\frac{1-\lambda\,\Delta t/4}{1+\lambda\,\Delta t/4}\mathbf{C}^{(n+1/2)}+\frac{\beta\,\Delta t/4}{1+\lambda\,\Delta t/4}$$

$$\times[\mathbf{M}_1(\boldsymbol{\phi}_1^{(n+1)}+\boldsymbol{\phi}_1^{(n+1/2)})+\mathbf{M}_2(\boldsymbol{\phi}_2^{(n+1)}+\boldsymbol{\phi}_2^{(n+1/2)})]$$

(4.8.36)

where $(\mathbf{D}_g\mathbf{B}^2)_x$ is a diagonal matrix defined by the relation

$$(\mathbf{D}_g\mathbf{B}^2)_x\boldsymbol{\phi}_g^{(n+1/2)}=\mathbf{H}_g\boldsymbol{\phi}_g^{(n+1/2)}$$

(4.8.37)

The term, $\mathbf{C}^{(n+1/2)}$, on the left side of Eq. (4.8.30) can be eliminated by using Eq. (4.8.32). Then, the set of Eqs. (4.8.30) and (4.8.31) may be written in the form of block-tridiagonal difference equations with 2×2 submatrices. In other words, the unknown in each grid row can be solved independently of the unknown in any other rows by using the recursion formula for a block-tridiagonal matrix given by Eq. (1.7.17). The independent solution for each grid row is made possible because $\mathbf{V}_1$ and $\mathbf{V}_2$ in Eq. (4.8.29) are replaced by diagonal matrices $(\mathbf{D}_g\mathbf{B}^2)_y$, $g=1, 2$, respectively. Similarly to the row inversions, the set of Eqs. (4.8.34) and (4.8.35) may be written in the form of block-tridiagonal difference equations after eliminatinc $\mathbf{C}^{(n+1)}$ with Eq. (4.8.36). Thus unknowns along each grid column can be solved independently of other columns. This scheme is nonlinear in the sense that the solution of the previous half step is introduced to the operator through $\mathbf{B}^2$, so the Fourier stability analysis cannot be applied. The numerical study performed by the author of the method shows that the scheme is stable for all the problems tested and the computing time is three to ten times faster than the TWIGL scheme on the same accuracy basis for a specific class of problems tested. For a more difficult class of problems the running time was equivalent to the TWIGL scheme because the ADI-$\mathbf{B}^2$ scheme requires much smaller Δt if the transient is fast and the flux distribution is complicated.

PROBLEMS

1. Develop the explicit and implicit numerical schemes for

$$\frac{d}{dt}\begin{bmatrix} y_1(t) \\ y_2(t) \end{bmatrix} = \begin{bmatrix} -0.5, & 0.4 \\ 0.4, & -0.5 \end{bmatrix}\begin{bmatrix} y_1(t) \\ y_2(t) \end{bmatrix} + \begin{bmatrix} \sin(\omega t) \\ 0 \end{bmatrix}$$

where the initial conditions are $y_1(0) = y_2(0) = 0$, and find the stability limit for Δt for each scheme.

2. Prove that Eqs. (4.3.3a) and (4.3.3b) satisfy Eq. (4.3.2).

3. Write a FORTRAN program that solves the finite difference equation for Eq. (4.1.2) by using the implicit, explicit, and the Crank–Nicholson schemes. Assume p, q, and S are space dependent but are constant in each mesh interval, and use $\phi = 0$ as left and right boundary conditions.

4. Derive an implicit numerical scheme for the neutron diffusion equation with one-delayed group:

$$\frac{\partial \phi(x, t)}{v\, \partial t} = \frac{\partial}{\partial x} D \frac{\partial}{\partial x} \phi(x, t)$$

$$+ [(1-\beta)v\Sigma_f - \Sigma_a]\phi(x, t) + \lambda C(x, t)$$

$$\frac{\partial}{\partial t} C(x, t) = -\lambda C(x, t) + \beta v\Sigma_f \phi(x, t)$$

In these equations the notations are standard for nuclear reactor physics (see References 8 and 9).

5. Derive the truncation error of Eqs. (4.2.12), (4.2.14), and (4.2.15) as approximations to Eq. (4.1.2). [Hint: Expand $\phi_{i\pm1}^{n\pm1} \equiv \phi((i+1)\Delta x, (n\pm1)\Delta t)$ in the Taylor series and introduce to the difference equations, then compare the results with Eq. (4.1.2).]

6. Find the analytical solution of the difference equation

$$\mathcal{F}\epsilon \equiv -2\epsilon_{i+1} + 3\epsilon_i - \epsilon_{i-1} = 1, \qquad 0 < i < 10$$

Boundary conditions are $\epsilon_0 = \epsilon_{10} = 0$.

7. Express the following one-step schemes in the form of Eq. (4.5.3), and find γ and θ:

   a. $\phi_i^{(n)} - \phi_{i-1}^{(n)} = \phi_i^{(n-1)} - \phi_{i+1}^{(n-1)}$

   b. $\phi_i^{(n)} = \phi_i^{(n-1)} + \frac{\Delta t}{(\Delta x)^2}[\phi_{i-1}^{(n-1)} - 2\phi_i^{(n-1)} + \phi_{i+1}^{(n-1)}]$

   c. $\phi_i^{(n)} - \frac{\Delta t}{(\Delta x)^2}[\phi_{i-1}^{(n)} - 2\phi_i^{(n)} + \phi_{i+1}^{(n)}] = \phi_i^{(n-1)}$

8. Prove the following orthogonality relations

$$\sum_{i=0}^{I} \sin(\beta_k i) \sin(\beta_{k'} i) = 0 \quad \text{if } k \neq k'$$

where $\beta_k = k\pi/I$, k is an integer and $1 \leq k \leq I - 1$.

9. Prove that the array of real values $\{\phi_i\}$, $i = -\infty$ to $+\infty$, on the grids $x_i = i\Delta x$ can be expressed by the Fourier series

$$\phi_i = \sum_{k=-\infty}^{+\infty} A_k e^{ikj\Delta x}$$

where $j = \sqrt{-1}$ and A_k are coefficients ($\bar{A}_k = A_{-k}$).

10. Apply the Fourier stability analysis to

$$\frac{\phi_j^{(n-1)} - 2\phi_j^{(n)} + \phi_j^{(n+1)}}{(\Delta t)^2} = c^2 \frac{\phi_{j-1}^{(n)} - 2\phi_j^{(n)} + \phi_{j+1}^{(n)}}{(\Delta x)^2}$$

for an infinite mesh space, and express the amplitude factor in terms of Δt, Δx, and c. Prove that the scheme is stable if $c\Delta t/\Delta x < 1$.

11. Apply the Fourier analysis to

$$\frac{\phi_j^{(n+1)} - 2\phi_j^{(n)} + \phi_j^{(n-1)}}{(\Delta t)^2} = c^2 \frac{\phi_{j+1}^{(n+1)} - 2\phi_j^{(n+1)} + \phi_{j-1}^{(n+1)}}{(\Delta x)^2}$$

for an infinite mesh system, and prove the above equation is unconditionally stable.

12. Apply the Fourier stability analysis to

$$\frac{\phi_j^{(n)} - \phi_j^{(n-1)}}{\Delta t} = \alpha \frac{\phi_{j-1}^{(n)} - 2\phi_j^{(n)} + \phi_{j+1}^{(n)}}{(\Delta x)^2} + (1-\alpha) \frac{\phi_{j-1}^{(n-1)} - 2\phi_j^{(n-1)} + \phi_{j+1}^{(n-1)}}{(\Delta x)^2}$$

and prove that the amplitude factor satisfies

$$\xi_k = \frac{1 - 4(1-\alpha)(\Delta t/\Delta x^2) \sin^2(k\Delta x/2)}{1 + 4\alpha(\Delta t/\Delta x^2) \sin^2(k\Delta x/2)}$$

and $\Delta t/\Delta x^2$ is unrestricted for $0.5 \leq \alpha \leq 1$.

13. The two-step difference scheme for Eq. (4.1.4)

$$\frac{\phi_i^{(n+1/2)}-\phi_i^{(n)}}{v\Delta t/2}=D\frac{\phi_{i-1}^{(n+1/2)}-\phi_i^{(n+1/2)}-\phi_i^{(n)}+\phi_{i+1}^{(n)}}{(\Delta x)^2}$$

$$+(v\Sigma_f-\Sigma_a)\frac{\phi_i^{(n+1/2)}+\phi_i^{(n)}}{2}$$

$$\frac{\phi_i^{(n+1)}-\phi_i^{(n+1/2)}}{v\Delta t/2}=D\frac{\phi_{i-1}^{(n+1/2)}-\phi_i^{(n+1/2)}-\phi_i^{(n+1)}+\phi_{i+1}^{(n+1)}}{(\Delta x)^2}$$

$$+(v\Sigma_f-\Sigma_a)\frac{\phi_i^{(n+1)}+\phi_i^{(n+1/2)}}{2}$$

is called the ADE (alternative-direction-explicit) method.[15] Assuming an infinite geometry, prove that the ADE method is unconditionally stable.

14. We apply the ADE method to the mesh system consisting of only two grids, $i=1$ and 2, and assume the boundary conditions are $\phi_0=\phi_3=0$. If we express the solution for those two grids by $\boldsymbol{\phi}^{(n)}=\mathrm{col}\,(\phi_1^{(n)},\phi_2^{(n)})$, the ADE scheme may be expressed as

$$\mathbf{A}_1\boldsymbol{\phi}^{(n-1/2)}=\mathbf{B}_1\boldsymbol{\phi}^{(n-1)}$$

$$\mathbf{A}_2\boldsymbol{\phi}^{(n)}=\mathbf{B}_2\boldsymbol{\phi}^{(n-1/2)}$$

or equivalently, one cycle of calculations is

$$\boldsymbol{\phi}^{(n)}=\mathbf{M}\boldsymbol{\phi}^{(n-1)}$$

where

$$\mathbf{M}=\mathbf{A}_2^{-1}\mathbf{B}_2\mathbf{A}_1^{-1}\mathbf{B}_1$$

Calculate the elements of **M** numerically for various combinations of Δx and Δt, and calculate the spectral radius of **M** for each combination.

15. Transform the equations in Problem 1 with

$$y_g(t)=e^{\Omega t}z_g(t),\qquad g=1,2$$

where Ω is a constant, and derive the implicit numerical scheme for $Z_g(t)$.

REFERENCES

1. D. U. Von Rosenberg, *Methods for numerical solution of partial differential equations,* American Elsevier, New York (1969).

2. R. D. Richtmeyer and K. W. Morton, *Difference methods of initial-value problems*, 2nd edit., Wiley, New York (1967).

3. P. J. Roache, *Computational fluid dynamics*, Hermosa, Albuquerque, N.M. (1972).

4. G. G. O'Brien, M. A. Hyman, and S. Kaplan, "A study of the numerical solution of partial differential equations," *J. Math. Phys.*, **29**, 223 (1951).

5. E. Isaacson and H. B. Keller, *Analysis of numerical methods*, Wiley, New York (1966).

6. K. F. Hansen and S. R. Johnson, *GAKIN, a program for the solution of the one-dimensional, multigroup, space-time dependent diffusion equations*, USAEC Rpt. GA-7543 (1967).

7. J. B. Yasinsky, M. Natelson, and L. A. Hageman, *TWIGL—a program to solve the two dimensional, two-group, space-time neutron diffusion equations with temperature feedback*, USAEC, WAPD-TM-743, Bettice Atomic Power Laboratory, Pittsburgh, Pa. (1968).

8. D. K. Hetrick, *Dynamics of nuclear reactors*, University of Chicago, Chicago (1971).

9. J. R. Lamarsh, *Introduction to nuclear reactor theory*, Addison-Wesley, Reading, Mass. (1966).

10. P. E. Rohan, C. M. Kang, R. A. Shober, and S. G. Wagner, "A multi-dimensional space-time kinetics code for PWR transients," *Proc. Conf. Computational Methods in Nuclear Engineering*, CONF-750413, Vol. 2, National Technical Information Service, Springfield, Va. (1975).

11. D. R. Ferguson, "Multidimensional reactor dynamics today: an overview," *Proc. Conf. Computational Methods in Nuclear Engineering*, CONF-750413, Vol. 2, National Technical Information Service, Springfield, Va. (1975).

12. William H. Reed and K. F. Hansen, "Alternating direction methods for the reactor kinetics equations," *Nucl. Sci. Eng.*, **41**, 431–442 (1970).

13. K. F. Hansen, "Finite-difference solutions for space-dependent kinetics equations," *Dynamics of nuclear systems*, D. L. Hetrick, ed., University of Arizona, Tucson, Arizona (1972).

14. A. F. Henry, "Review of computational methods for space-dependent kinetics," *Dynamics of nuclear systems*, D. L. Hetrick, Ed., University of Arizona, Tucson, Arizona (1972).

15. R. S. Denning, R. F. Redmond, and S. S. Iyer, "A stable explicit finite difference technique for spatial kinetics," *Trans. Amer. Nucl. Soc.* **12**, 148 (1969).

16. D. R. Ferguson and K. F. Hansen, "Solution of the space-dependent reactor kinetics equations in three-dimensions," *Nucl. Sci. Eng.*, **51**, 189–205 (1973).

17. L. A. Hageman and J. B. Yasinsky, "Comparison of alternative-direction time-differencing methods with other implicit methods for the solution of the neutron group diffusion equations," *Nucl. Sci. Eng.*, **38**, 8–32 (1969).

Chapter 5   Computational Fluid Dynamics—I

5.1 EXPERIMENTS OR COMPUTATIONS

Computational fluid dynamics is one of the areas of computational methods that is being most dramatically developed. The numerical techniques are developed almost in any branch of fluid dynamics including aircraft design, weather science, civil engineering, oceanography, and nuclear science. Today, computers can solve the fluid dynamic problems that were considered impossible or impractical in the 1960s.[1,2] Traditionally, fluid dynamics engineers have relied heavily on experimental approaches. This trend, however, is expected to change in the next decade for the following reasons. First, the advances in technology and science require more information than can be obtained by experiments. Second, the cost for experiments is increasing almost exponentially as the size of the systems, such as aircraft, increases (see Fig. 5.1). Third, the cost of computers is decreasing

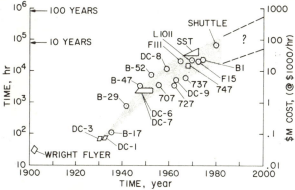

Figure 5.1   Total wind tunnel test hours for development of various aircraft.[1] (Reprinted from the April 1975 *Astronautics & Aeronautics*, a publication of the American Institute of Aeronautics and Astronautics).

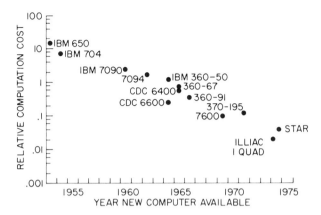

Figure 5.2 The trend of computational cost for computer simulation of a given flow.[1] (Reprinted from the April
1975 *Astronautics & Aeronautics*, a publication
of the American Institute of Aeronautics and
Astronautics).

(see Fig. 5.2). Fourth, numerical techniques for fluid analysis are being
advanced rapidly, thus making computational fluid analysis practical in
many design and analysis purposes.

It should be worthwhile to notice that, with regard to subsonic
aerodynamics, the Aeronautics and Astronautics Association is trying to
determine whether design problems and the aerodynamic equations of
motion can be solved with good mathematical models, thus eliminating the
need to build and test scale models of aircraft in wind tunnels. Chapman,
Mark, and Pirtle[1] point out that many parts of experimental fluid analysis
will be replaced by computer analysis and that a balanced use of some
experiments and many computations will probably be the solution. They
also predict that the future history of the numerical solution to the Navier–
Stokes equation will be similar to the past history of the neutron Boltzmann
equation. It was during 1955 to 1970 that computer solutions to the neutron
diffusion and neutron Boltzmann equation were most vigorously developed.
As the result of this, the calculations for prediction of neutron flux distribution and behavior of chain reactions became so accurate and so economical
that a large part of experimental work was replaced by computer solutions.
In fact, the experimental facilities used to determine nuclear reactor
parameters in the early days were demolished.

The Navier–Stokes equation is the most frequently used approximation to
the molecular Boltzmann equation that precisely describes fluid behavior.
The Navier–Stokes equation is far more difficult to solve than the neutron

Boltzmann equations because of its coupled form, nonlinearity, and a large number of dependent variables. The numerical solutions are subject not only to numerical instabilities but also to hydraulic instabilities. Therefore, it is natural that computer solution is becoming realistic more than one decade later than for the neutron Boltsmann equation. It is interesting, however, that there are numerous similarities between the Navier–Stokes equation and the neutron Boltzmann equation. For this reason, many numerical techniques developed for neutronic calculations are used in computational fluid dynamics.

The purpose of this chapter is to introduce basic aspects of the numerical methods for fluid flow analyses. For this purpose the following three subjects are selected: (a) characteristic methods for one-dimensional compressive flow, (b) control volume method for an arbitrary network system of conduits, and (c) two-dimensional incompressible fluid flow. For two-dimensional incompressible flow analyses, two representative finite difference methods, namely, the vorticity method and the MAC method, are introduced. In both methods, the Navier–Stokes equations are reduced to the finite difference equations for elliptic partial differential equations and parabolic partial differential equations. Therefore, the methods described in Chapters 3 and 4 are necessary. The methods described in this chapter are based on the finite difference approximations, while the finite element approaches for two-dimensional fluid flow are introduced in Chapter 7. The computational methods for aerodynamics are described in Chapter 10.

5.2 DERIVATION OF CHARACTERISTIC EQUATIONS FOR ONE-DIMENSIONAL COMPRESSIBLE FLOW

The basic equations of inviscid one-dimensional unsteady flow are partial differential equations of hyperbolic type. Those equations can be transformed into characteristic forms[3–11] that are ordinary differential equations. One of the most attractive features of the characteristic method is that the numerical schemes based on it conserve the physical properties of the system. Basically, the characteristic method tracks the propagation of waves and calculates their strength. Therefore, it is comparatively easy to simulate a fluid system including fluid discontinuities or shock waves. Another significant point is that the region of influence of any point in the numerical calculations is very close to the physical ones. The purpose of this section is to derive the characteristic equations from the basic partial differential equations for unsteady flow of compressible fluid.

Consider one-dimensional homogeneous flow of a compressible fluid in a uniform conduit. The equation of continuity is

$$\frac{\partial \rho}{\partial t} + v \frac{\partial \rho}{\partial x} + \rho \frac{\partial v}{\partial x} = 0 \tag{5.2.1}$$

where ρ is the density of the fluid and v is the velocity. The equation of motion under the influence of gravity and flow resistance is

$$\frac{\partial \rho v}{\partial t} + \frac{\partial \rho v^2}{\partial x} + \frac{\partial p}{\partial x} = f + \rho g \cos \theta \tag{5.2.2}$$

where f is the flow resistance, p is the pressure, and $g \cos \theta$ is the component of gravity in the axial direction of the conduit. Equation (5.2.2) may be equivalently written by using Eq. (5.2.1) as

$$\rho \left(\frac{\partial v}{\partial t} + v \frac{\partial v}{\partial x} \right) + \frac{\partial p}{\partial x} = f + \rho g \cos \theta \tag{5.2.3}$$

The energy equation in terms of specific enthalpy is

$$\rho \left(\frac{\partial H}{\partial t} + v \frac{\partial H}{\partial x} \right) - \frac{1}{J} \left(\frac{\partial p}{\partial t} + v \frac{\partial p}{\partial x} \right) = q''' \tag{5.2.4}$$

where H is the specific enthalpy, q''' is the heat generation rate, and J is a conversion factor to relate units of work to units of heat.

There are four unknowns, ρ, v, p, and H, for the three equations, Eqs. (5.2.1), (5.2.3), and (5.2.4). In order to make the system of equations complete, an equation of state is necessary. For this purpose, we use the relation among ρ, H, and p:

$$\rho = \pi(H, p) \tag{5.2.5}$$

For a homogeneous two-phase flow, a steam table may be used for π. The partial derivatives of H may be written as

$$\frac{\partial H}{\partial t} = \frac{\partial H}{\partial \rho} \frac{\partial \rho}{\partial t} + \frac{\partial H}{\partial p} \frac{\partial p}{\partial t} \tag{5.2.6}$$

$$\frac{\partial H}{\partial x} = \frac{\partial H}{\partial \rho} \frac{\partial \rho}{\partial x} + \frac{\partial H}{\partial p} \frac{\partial p}{\partial x} \tag{5.2.7}$$

The square of sonic velocity can be defined[12] as

$$c^2 = -\frac{\partial H / \partial \rho}{\partial H / \partial p - 1/J\rho} \tag{5.2.8}$$

By using Eqs. (5.2.6), (5.2.7), and (5.2.8), Eq. (5.2.4) is transformed to

$$-\rho c^2\left(\frac{\partial\rho}{\partial t}+v\frac{\partial\rho}{\partial x}\right)+\rho\left(\frac{\partial p}{\partial t}+v\frac{\partial p}{\partial x}\right)=-\frac{q'''c^2}{\partial H/\partial\rho} \tag{5.2.9}$$

The partial derivative of ρ in Eq. (5.2.9) can be eliminated by using Eq. (5.2.1) to yield

$$\rho c^2\frac{\partial v}{\partial x}+\left(\frac{\partial p}{\partial t}+v\frac{\partial p}{\partial x}\right)=-\frac{q'''c^2}{\rho(\partial H/\partial\rho)} \tag{5.2.10}$$

where both sides are divided by ρ. Therefore, we consider the set of Eqs. (5.2.1), (5.2.3), and (5.2.10) as the basic equations and use them to derive the characteristic equations.

Let us first consider the sets of Eqs. (5.2.3) and (5.2.10). Along an arbitrarily chosen curve on the x–t plane we have the equations

$$dv=\frac{\partial v}{\partial t}dt+\frac{\partial v}{\partial x}dx \tag{5.2.11}$$

$$dp=\frac{\partial p}{\partial t}dt+\frac{\partial p}{\partial x}dx \tag{5.2.12}$$

Equations (5.2.3), (5.2.10), (5.2.11), and (5.2.12) provide sufficient equations to determine $\partial v/\partial t$, $\partial v/\partial x$, $\partial p/\partial t$, and $\partial p/\partial x$ along this curve if the determinant of the coefficient matrix is not zero. However, we are rather interested in those curves for which the determinant becomes zero. To find those curves, we set the determinant to zero:

$$\begin{vmatrix} \rho & \rho v & 0 & 1 \\ 0 & \rho c^2 & 1 & v \\ dt & dx & 0 & 0 \\ 0 & 0 & dt & dx \end{vmatrix}=0 \tag{5.2.13}$$

Expanding the determinant yields

$$(v^2-c^2)\,dt^2-2v\,dx\,dt+dx^2=0 \tag{5.2.14}$$

from which we obtain

$$\frac{dx}{dt}=v+c \tag{5.2.15}$$

$$\frac{dx}{dt}=v-c \tag{5.2.16}$$

Equations (5.2.15) and (5.2.16) are the equations for the characteristic lines, which are referred to as *sonic characteristic lines*.

When Eq. (5.2.15) is satisfied by a curve on the $x–t$ plane, the set of Eqs. (5.2.3), (5.2.10), (5.2.15), and (5.2.12) are consistent only if the following determinant becomes zero:

$$\begin{vmatrix} \rho & \rho v & 0 & \zeta \\ 0 & \rho c^2 & 1 & \xi \\ dt & dx & 0 & dv \\ 0 & 0 & dt & dp \end{vmatrix} = 0 \tag{5.2.17}$$

where ζ and ξ are the right sides of Eq. (5.2.3) and Eq. (5.2.10), respectively. Expanding the determinant and using Eq. (5.2.15) we obtain

$$\rho c \frac{dv}{dt} + \frac{dp}{dt} = c\zeta + \xi \tag{5.2.18}$$

Similarly, along a curve satisfying Eq. (5.2.16), we obtain

$$-\rho c \frac{dv}{dt} + \frac{dp}{dt} = -c\zeta + \xi \tag{5.2.19}$$

Thus summarizing Eqs. (5.2.15), (5.2.16), (5.2.18), and (5.2.19), we obtained two sets of equations:

$$\frac{dx}{dt} = v + c \tag{5.2.20}$$

$$\rho c \frac{dv}{dt} + \frac{dp}{dt} = c(f + \rho g \cos\theta) - \frac{q'''c^2}{\rho(\partial H/\partial\rho)} \tag{5.2.21}$$

and

$$\frac{dx}{dt} = v - c \tag{5.2.22}$$

$$-\rho c \frac{dv}{dt} + \frac{dp}{dt} = -c(f + \rho g \cos\theta) - \frac{q'''c^2}{\rho(\partial H/\partial\rho)} \tag{5.2.23}$$

The above sets of equations can be used to calculate p and v along the sonic characteristic lines.

In order to obtain the equations for H, we start with Eq. (5.2.4). Along any arbitrary curve on the $x–t$ plane, we have Eq. (5.2.12) and

$$dH = \frac{\partial H}{\partial t} dt + \frac{\partial H}{\partial x} dx \tag{5.2.24}$$

Along the same line, Eq. (5.2.4) becomes

$$\rho\left[\frac{dH}{dt}+\frac{\partial H}{\partial x}\left(v-\frac{dx}{dt}\right)\right]-\frac{1}{J}\left[\frac{dp}{dt}+\frac{\partial p}{\partial x}\left(v-\frac{dx}{dt}\right)\right]=q''' \qquad (5.2.25)$$

where Eqs. (5.2.12) and (5.2.24) were introduced into Eq. (5.2.4). There-fore, if the curve is chosen as $dx/dt=v$, we obtain the third set of characteris-tic equations

$$\frac{dx}{dt}=v \qquad (5.2.26)$$

$$\rho\frac{dH}{dt}-\frac{1}{J}\frac{dp}{dt}=q''' \qquad (5.2.27)$$

The curve represented by Eq. (5.2.26) is referred to as a *material charac-teristic line*.

5.3 EXPLICIT CHARACTERISTIC METHOD

The explicit characteristic method is suitable for very rapid transients such as a water hammer of compressible fluid in a network of conduits.[4] The main restriction of the method is that the upper limit of time step Δt is bounded by $\Delta t < \Delta x/(c+|v|)$ for stability of the scheme, where Δx is the mesh spacing, c is the sonic velocity, and v is the fluid velocity. Therefore, if the transient considered is slow and long, the computation may be prohibitively expen-sive. However, the implicit method described in Section 5.4 should provide more suitable means for such cases.

Consider a single uniform conduit with the assumption that there is no fluid discontinuity such as shock or change of fluid type. The time and space coordinates are divided into mesh intervals, and difference equations for p and v are derived. Then the difference equation for H is derived. The superscript n and the subscript j denote time and spatial grids, respectively. Spatial mesh intervals may be chosen arbitrarily within the constraints of the stability condition.

As formulated, the problem is one of an initial and boundary value type, so when a time interval $[t^n, t^{n+1}]$ is considered, p, v, and H are known at t^n, either from initial condition or from previous calculations. In deriving the difference equations, it is further assumed that the property-dependent coefficients are known, even though the properties are the functions of p, v, and H at t^{n+1}.

If a conduit is divided into N mesh intervals, there will be $N+1$ grids on which p and v are considered, so the total number of unknown ps and vs is

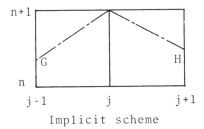
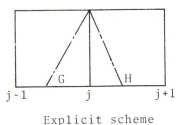

Implicit scheme   Explicit scheme

Figure 5.3 Sonic characteristic lines.

$2N+2$. $2N$ equations among $2N+2$ are difference equations. Two additional equations for each conduit are given as boundary conditions.

The lines represented by Eqs. (5.2.20) and (5.2.22) in a time and space mesh box are referred to as positive and negative characteristic lines, respectively, or simply as *sonic characteristic lines*. They are illustrated in Fig. 5.3 by dashed lines.

The difference approximation to Eqs. (5.2.21) and (5.2.23) along the characteristic lines may be written, respectively, as

$$\frac{p_j^{n+1}-p_G}{\Delta t}+\langle \rho c\rangle_+\frac{v_j^{n+1}-v_G}{\Delta t}=R_+\equiv \text{RHS of Eq. (5.2.21)} \qquad (5.3.1)$$

$$\frac{p_j^{n+1}-p_H}{\Delta t}-\langle \rho c\rangle_-\frac{v_j^{n+1}-v_H}{\Delta t}=R_-\equiv \text{RHS of Eq. (5.2.23)} \qquad (5.3.2)$$

In the above equations, $\langle \ \rangle$ denotes an average along the corresponding characteristic line, and RHS denotes an appropriate representation to the right-hand side of Eq. (5.2.21) or Eq. (5.2.23). Subscripts G and H refer to the intersection of sonic characteristic lines (see Fig. 5.3). When G and H are on the bottom horizontal line as in Fig. 5.3, the difference equations are referred to as *explicit*, while when G and H are on the side vertical lines the difference equation is referred to as *implicit*.

If points G and H are on the bottom line of the mesh box, the values of p and v at those points may be calculated by interpolating the known values at (j, n) and $(j+1, n)$ or at (j, n) and $(j-1, n)$:

$$p_G=\frac{x_j-x_G}{\Delta x}p_{j-1}^n+\frac{x_G-x_{j-1}}{\Delta x}p_j^n \qquad (5.3.3)$$

$$v_G=\frac{x_j-x_G}{\Delta x}v_{j-1}^n+\frac{x_G-x_{j-1}}{\Delta x}v_j^n \qquad (5.3.4)$$

$$p_H=\frac{x_{j+1}-x_H}{\Delta x}p_j^n+\frac{x_H-x_j}{\Delta x}p_{j+1}^n \qquad (5.3.5)$$

$$v_H = \frac{x_{j+1} - x_H}{\Delta x} v_j^n + \frac{x_H - x_j}{\Delta x} v_{j+1}^n \qquad (5.3.6)$$

where $\Delta x = x_j - x_{j-1}$ or $\Delta x = x_{j+1} - x_j$; x_G and x_H are x coordinates at G and H, respectively.

Considering the explicit scheme in the remainder of this section, Eqs. (5.3.1) and (5.3.2) may be written:

$$p_j^{n+1} + \langle \rho c \rangle_+ v_j^{n+1} = \Delta t R_+ + p_G + \langle \rho c \rangle_+ v_G \qquad (5.3.7)$$

$$p_j^{n+1} - \langle \rho c \rangle_- v_j^{n+1} = \Delta t R_- + p_H - \langle \rho c \rangle_- v_H \qquad (5.3.8)$$

Since the right sides of Eqs. (5.3.7) and (5.3.8) are all known, this set of equations can be solved easily at each grid point. If grid point j is a left boundary, Eq. (5.3.7) does not apply, but v_j^{n+1} or p_j^{n+1} is prescribed as a boundary condition, thus allowing the solution of Eq. (5.3.8). The right boundary may be treated in a similar way. If a boundary is a closed end, $v = 0$ is the boundary condition.

The enthalpies are then calculated explicitly by using the third set of characteristic equations, Eqs. (5.2.26) and (5.2.27). Along the material charactertic line respresented by Eq. (5.2.26), the difference approximation for Eq. (5.2.27) is given by

$$\langle \rho \rangle_m \frac{H_j^{n+1} - H_D}{\Delta t} - \frac{1}{J} \frac{p_j^{n+1} - p_D}{\Delta t} = \langle q''' \rangle_m \qquad (5.3.9)$$

where H_D and p_D are the values of H and p on the material line at $t = t_j^n$, and $\langle \ \rangle_m$ is an average along the material line. The position of D falls in (x_{j-1}, x_j) or (x_j, x_{j+1}) depending on the sign of v as schematically shown in Fig. 5.4. The value of H_D is interpolated if $\langle v \rangle_m > 0$ as

$$H_D = \langle v \rangle_m \frac{\Delta t}{\Delta x} H_{j-1}^n + \left(1 - \langle v \rangle_m \frac{\Delta t}{\Delta x}\right) H_j^n \qquad (5.3.10)$$

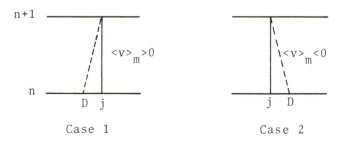

Figure 5.4 Material characteristic line (– – –)

and if $\langle v \rangle_m < 0$ as

$$H_D = |\langle v \rangle_m| \frac{\Delta t}{\Delta x} H_{j+1}^n + \left(1 - |\langle v \rangle_m| \frac{\Delta t}{\Delta x}\right) H_j^n \qquad (5.3.11)$$

The value of p_D is also obtained by a similar interpolation. p_j^{n+1} is already obtained. Therefore, the calculation of H_j^{n+1} is

$$H_j^{n+1} = \frac{1}{\langle \rho \rangle_m}\left[\Delta t \langle q''' \rangle_m + \frac{1}{J}(p_j^{n+1} - p_D)\right] + H_D \qquad (5.3.12)$$

If fluid is entering at an open end, the enthalpy is specified as a boundary condition.

Once H_j^{n+1} and p_j^{n+1} are determined, ρ and other fluid properties are calculated by using equations of state, thus completing the calculation for the time interval.

A network of conduits generally includes junctions of two different conduits or more than two conduits. Double or multiple grids are considered at a junction as illustrated in Fig. 5.5. For a two-way junction where j and $j+1$ are double grids (corresponding to $j = 3$ and 4 in Fig. 5.5), the boundary conditions are

$$p_j = p_{j+1} + \Delta p$$
$$v_j A_j = v_{j+1} A_{j+1} \qquad (5.3.13)$$

where Δp is pressure loss (or gain) at the junction and A is the cross-sectional area. Assuming j, $j+1$, and k are triple grids at a three-way junction, the boundary conditions are

$$p_{j+1} = p_j + \Delta p_{j,j+1}$$
$$p_k = p_j + \Delta p_{j,k}$$
$$v_j A_j + v_{j+1} A_{j+1} + v_k A_k = 0 \qquad (5.3.14)$$

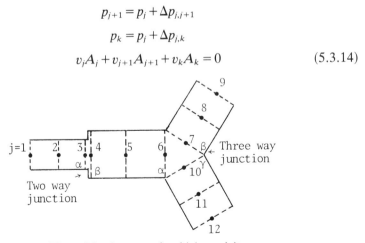

Figure 5.5 A system of multiple conduits.

where $\Delta p_{j,k}$ is pressure loss between j and k, and v is assumed to be positive when the direction of flow is toward the junction.

5.4 IMPLICIT CHARACTERISTIC METHOD

While the explicit scheme is limited by the upper bound at Δt, the implicit scheme is limited by the lower bound of Δt. However, for specified large time steps with the implicit method, the acoustic effects are damped out during the calculations, thus providing a method that is suitable for long transients whenever acoustic effects are not important.[7,9]

Implicit difference schemes for characteristic equations are obtained in accordance with the second scheme in Fig. 5.3, where the points of G and H are on the vertical sides of each mesh box. The values of p and v at those points are interpolated between the values of (j, n) and $(j, n+1)$ for G and between $(j+1, n)$ and $(j+1, n+1)$ for H as

$$p_G = (1-\eta_+)p_j^{n+1} + \eta_+ p_j^n \tag{5.4.1}$$

$$v_G = (1-\eta_+)v_j^{n+1} + \eta_+ v_j^n \tag{5.4.2}$$

$$p_H = (1-\eta_-)p_{j+1}^{n+1} + \eta_- p_{j+1}^n \tag{5.4.3}$$

$$v_H = (1-\eta_-)p_{j+1}^{n+1} + \eta_- p_{j+1}^n \tag{5.4.4}$$

where

$$\eta_+ = \frac{\Delta x}{\Delta t}\frac{1}{\langle c+v\rangle_+} \tag{5.4.5}$$

$$\eta_- = \frac{\Delta x}{\Delta t}\frac{1}{\langle c-v\rangle_-} \tag{5.4.6}$$

Introducing Eqs. (5.4.1) through (5.4.4) into Eqs. (5.3.1) and (5.3.2) yields

$$-(1-\eta_+)p_j^{n+1} - \langle\rho c\rangle_+(1-\eta_+)v_j^{n+1} + p_{j+1}^{n+1} + \langle\rho c\rangle v_{j+1}^{n+1}$$
$$= \frac{\Delta x}{\langle c+v\rangle_+}\langle R\rangle_+ + \langle\rho c\rangle_+\eta_+ v_j^n + \eta_+ p_j^n \tag{5.4.7}$$

$$p_j^{n+1} - \langle\rho c\rangle_- v_j^{n+1} - (1-\eta_-)p_{j+1}^{n+1} + \langle\rho c\rangle_-(1-\eta_-)v_{j+1}^{n+1}$$
$$= -\frac{\Delta x}{\langle c-v\rangle_-}\langle R\rangle_- - \langle\rho c\rangle_-\eta_- v_{j+1}^n + \eta_- p_{j+1}^n \tag{5.4.8}$$

where $\langle\ \rangle_+$ and $\langle\ \rangle_-$ denote averages along the positive and negative sonic characteristic line, respectively, and

$$\langle R \rangle_+ = \left\langle c(f + \rho g \cos \theta) - \frac{q''' c^2}{\rho (\partial H / \partial \rho)} \right\rangle_+ \qquad (5.4.9)$$

$$\langle R \rangle_- = \left\langle c(f + \rho g \cos \theta) + \frac{q''' c^2}{\rho (\partial H / \partial \rho)} \right\rangle_- \qquad (5.4.10)$$

Hereafter, $\langle R \rangle_+$ and $\langle R \rangle_-$ will be referred to as *general force terms* for simplicity. The enthalpy calculation is performed explicitly by Eq. (5.3.12).

With the following change of variables made:

$$\Delta v_j = v_j^{n+1} - v_j^n, \qquad \Delta p_j = p_j^{n+1} - p_j^n$$

Eqs. (5.4.7) and (5.4.8) become

$$-(1 - \eta_+)\Delta p_j - \langle \rho c \rangle_+ (1 - \eta_+)\Delta v_j + \Delta p_{j+1} + \langle \rho c \rangle_+ \Delta v_{j+1}$$
$$= \frac{\Delta x}{\langle c + v \rangle_+} \langle R \rangle_+ + \langle \rho c \rangle_+ (v_j^n - v_{j+1}^n) + (p_j^n - p_{j+1}^n) \qquad (5.4.11)$$

$$\Delta p_j - \langle \rho c \rangle_- \Delta v_j - (1 - \eta_-)\Delta p_{j+1} + \langle \rho c \rangle_- (1 - \eta_-)\Delta v_{j+1}$$
$$= -\frac{\Delta x}{\langle c - v \rangle_-} \langle R \rangle_- + \langle \rho c \rangle_- (v_j^n - v_{j+1}^n) + (p_{j+1}^n - p_j^n) \qquad (5.4.12)$$

At each time step, Eqs. (5.4.11) and (5.4.12) are simultaneously solved for the pressure and velocity fields, by inverting a band diagonal matrix (band width = 4) or by an iterative scheme described in Appendix IV.

The sizes of Δx and Δt are bounded by two conditions. One is the condition for the implicit difference scheme for p and v, given by

$$\Delta t \geq \frac{\Delta x}{c - |v|} \qquad (5.4.13)$$

The other condition due to the explicit formula Eq. (5.3.12) for H is given by

$$\Delta t \leq \frac{\Delta x}{|v|} \qquad (5.4.14)$$

Both of these conditions can be satisfied if the Mach number $|v|/c$ is less than 0.5.

Example 5.1

The implicit characteristic method was used to calculate the transient pressure and velocity response of an intitially stagnant gas in a 10-ft pipe closed at one end and exposed to a step uniform heat source.[7] The equation of state in terms of the sonic velocity was taken to be

$$c^2 = \frac{p}{\rho} \text{ ft}^2 \sec^{-2}$$

The ratio of the heat generated to the isobaric expansion term was set to

$$\frac{q'''}{\rho^2 H_\rho} = -1 \sec^{-1}$$

The initial conditions are

$$v = 0, \qquad p = 2116.8 \text{ lbf ft}^{-2}, \qquad \rho = 0.04 \text{ lbm ft}^{-3}$$

The boundary conditions are

$$v(10) = 0, \qquad p(0) = 2116.8 \text{ lbf ft}^{-2}$$

Figure 5.6 shows the time histories of pressure and density at the closed end of the pipe. As $\alpha \approx \Delta t_{imp}/\Delta t_c$ increases (where $\Delta t_c \approx \Delta x/c$, the upper limit of the explicit method and the lower limit of the implicit method), the pressure oscillation at the closed end decays more. This is expected since in the limit as $\Delta t \to \infty$ the implicit solution approaches the quasisteady solution for a thermally expanding fluid. Where high-frequency oscillation is not

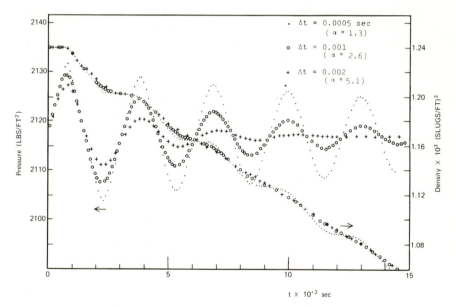

Figure 5.6 Pressure and density at the closed end of the 10-ft pipe.

important this dampening effect is of no consequence. Good agreement between the density time history and the analytical solution[12] is obtained for Δt_{imp} as large as $\alpha = 10$.

These test results indicate that utilization of the implicit method for reactor flow transients, where high-frequency oscillations are not important, should produce satisfactory results and a significant decrease in computational time.

The time step size of the explicit method is bounded by an upper limit, while that of the implicit method is bounded by a lower limit. This is seen from Eq. (5.4.13) and also Fig. 5.3. Since Δx, c, and v vary from grid to grid, the upper bound of the explicit method is the minimum of $\Delta x/(c + |v|)$ among all the mesh boxes, and the lower bound of the implicit method is the maximum of $\Delta x/(c - |v|)$. Therefore, there is a gap of Δt between min $\Delta x/(c + |v|)$ and max $\Delta x/(c - |v|)$, where neither method is applicable. The partially implicit method[11] proposed to bridge this gap uses either of the explicit differencing and implicit differencing for each mesh box whichever satisfies the stability condition as illustrated in Fig. 5.7.

In systems where more than one fluid phase can exist, the model must be capable of handling any boundaries between fluid phases (interfaces) that might occur.[9] The basic assumptions in dealing with such interfaces are that (1) a distinct interface exists and (2) no mixing occurs across any such interface. If the interface is between liquid and two phase, it is a boiling boundary and may move faster or slower than the fluid velocity. On the other hand, the velocity of the interface between liquid and steam is mostly governed by the velocity of the liquid.

In the analysis of hypothetical transient of reactor coolant, the treatment of the interface is particularly important for the following reasons. First, the acoustic velocities are substantially different among the various states of water, and hence the characteristic lines are significantly refracted across an interface. Second, other fluid properties may change significantly across an

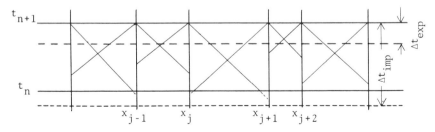

Figure 5.7 Differencing scheme of the partially implicit method (Δt_{exp} is the upper bound of the explicit method, and Δt_{imp} is the lower bound of the implicit method).

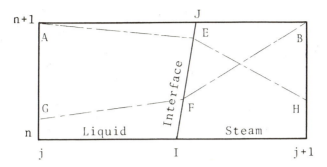

Figure 5.8   Characteristic lines refracted at an interface.

interface, so the difference equation cannot be justified without a proper treatment of interfaces. A typical example of characteristic lines in a mesh cell with such a fluid boundary is shown in Fig. 5.8, where the two characteristic lines representing the sonic fronts are refracted across the boundary due to the difference in sonic velocities.

Given the x coordinate of point I, I_x and the velocity of the interface v, the be all the points can be computed as follows:

$$J_x = I_x + \bar{v}\Delta t \tag{5.4.15}$$

$$E_t = -\frac{I_x - \Delta t \langle c - v \rangle_{-1}}{\bar{v} + \langle c - v \rangle_{-1}} \tag{5.4.16}$$

$$E_x = J_x\left(\frac{\langle c - v \rangle_{-1}}{\bar{v} + \langle c - v \rangle_{-1}}\right) \tag{5.4.17}$$

$$F_t = \frac{I_x - \Delta x + \Delta t \langle c + v \rangle_{+2}}{\langle c + v \rangle_{+2} - \bar{v}} \tag{5.4.18}$$

$$F_x = \frac{J_x \langle c + v \rangle_{+2} - \bar{v}\Delta x}{(c + v)_{+2} - \bar{v}} \tag{5.4.19}$$

$$G_t = F_t - \frac{F_x}{\langle c + v \rangle_{+1}} \tag{5.4.20}$$

$$H_t = E_t - \frac{\Delta x - E_x}{\langle c - v \rangle_{-2}} \tag{5.4.21}$$

where subscripts t and x indicate the t and x coordinates, respectively.

In Eqs. (5.4.15) through (5.4.21) $\langle\ \rangle_{+1}$ denotes an average along that part of the positive characteristic to the left of the interface, $\langle\ \rangle_{+2}$ denotes an average along that part of the positive characteristic to the right of the interface, $\langle\ \rangle_{-1}$ denotes an average along that part of the negative characteristic to the left of the interface, and $\langle\ \rangle_{-2}$ denotes an average along that part of the negative characteristic to the right of the interface.

Now consider the refracted characteristic line *GFB*. Difference approximations to Eq. (5.2.21) may be written for the line *GF*, as well as for *FB*, in terms of p and v at G, F, and B. Adding the two equations thus obtained yields:

$$-(1-\eta_+)\Delta p_j - \langle\rho c\rangle_{+1}(1-\eta_+)\Delta v_j + \Delta p_{j+1} + \langle\rho c\rangle_{+2}\Delta v_{j+1}$$

$$= \frac{F_x}{\langle c+v\rangle_{+1}}\langle R\rangle_{+1} + \frac{\Delta x - F_x}{\langle c+v\rangle_{+2}}\langle R\rangle_{+2} + \langle\rho c\rangle_{+1}v_j^n \qquad (5.4.22)$$

$$-\langle\rho c\rangle_{+2}v_{j+1}^n + p_j^n - p_{j+1}^n + v_F(\langle\rho c\rangle_{+2} - \langle\rho c\rangle_{+1})$$

where Δp_j and Δv_j follow the previous definitions, v_F is the fluid velocity at F (in general, v_F may differ from the velocity of the interface $\bar{v}$), and

$$\eta_+ = 1 - \frac{G_t}{\Delta t} \qquad (5.4.23)$$

A similar equation is obtained for the line *AEH* as

$$\Delta p_j - \langle\rho c\rangle_{-1}\Delta v_j - (1-\eta_-)\Delta p_{j+1} + \langle\rho c\rangle_{-2}(1-\eta_-)\Delta v_{j+1}$$

$$= -\frac{E_x}{\langle c-v\rangle_{-1}}\langle R\rangle_{-1} - \frac{\Delta x - E_x}{\langle c-v\rangle_{-2}}\langle R\rangle_{-2} + \langle\rho c\rangle_{-1}v_j^n \qquad (5.4.24)$$

$$-\langle\rho c\rangle_{-2}v_{j+1}^n - p_j^n + p_{j+1}^n + v_E(\langle\rho c\rangle_{-2} - \langle\rho c\rangle_{-1})$$

where v_E is the fluid velocity at E and

$$\eta_- = 1 - \frac{H_t}{\Delta t} \qquad (5.4.25)$$

Left sides of Eqs. (5.4.22) and (5.4.24) have the same unknown variables as Eqs. (5.4.11) and (5.4.12). The right sides of Eqs. (5.4.22) and (5.4.24) are all known except v_F and v_E.

In dealing with v_F and v_E, it is assumed that the fluid velocity at an interface is equal or nearly equal to the velocity of the less compressible fluid. Thus in the case of Fig. 5.8, for example, the following two equations are obtained:

$$-(1-\eta_+)\Delta p_j - \left[\langle\rho c\rangle_{+1}\left(1-\eta_+ - \frac{F_t}{\Delta t}\right) + \langle\rho c\rangle_{+2}\frac{F_t}{\Delta t}\right]\Delta v_j$$

$$+\Delta p_{j+1} + \langle\rho c\rangle_{+2}\Delta v_{j+1} = \frac{F_x}{\langle c+v\rangle_{+1}}\langle R\rangle_{+1} + \frac{\Delta x - F_x}{\langle c+v\rangle_{+2}}\langle R\rangle_{+2} \quad (5.4.26)$$

$$+\langle\rho c\rangle_{+2}(v_j^n - v_{j+1}^n) + p_j^n - p_{j+1}^n$$

and

$$\Delta p_j - \left[\langle \rho c \rangle_{-1} \left(1 - \frac{E_t}{\Delta t} \right) + \langle \rho c \rangle_{-2} \frac{E_t}{\Delta t} \right] \Delta v_j - (1 - \eta_-) \Delta p_{j+1}$$

$$+ \langle \rho c \rangle_{-2} (1 - \eta_-) \Delta v_{j+1} = -\frac{E_x}{\langle c - v \rangle_{-1}} \langle R \rangle_{-1} - \frac{\Delta x - E_x}{\langle c - v \rangle_{-2}} \langle R \rangle_{-2} \quad (5.4.27)$$

$$+ \langle \rho c \rangle_{-2} (v_j^n - v_{j+1}^n) - p_j^n + p_{j+1}^n$$

Equations (5.4.26) and (5.4.27) have the same unknown variables as Eqs. (5.4.11) and (5.4.12). Therefore, the introduction of interfaces does not change the structure of the simultaneous equations for a homogeneous fluid. Thus the effect of interfaces can be introduced by changing the coefficients and nonhomogeneous terms of the characteristic equations for a homogeneous fluid. The calculations of Δp, Δv, and H for t^{n+1} are done iteratively. During the first iteration, the fluid properties and sonic velocities at t^{n+1} are assumed wherever necessary. The velocity of the interface is also assumed. When p_j^{n+1} and v_j^{n+1} are obtained, H^{n+1} is calculated by using Eq. (5.3.12). Then the fluid properties and sonic velocities are found by using the new values of p and H. The velocity of the interface is recalculated. The second and further iterations are continued in the same way as the first iteration except that updated values of fluid property, sonic velocities, and interface velocity are used to compute the new coefficients.

Example 5.2

Consider a uniform, symmetrical, frictionless U-tube, 26 ft long, partially filled with liquid water over 16 ft, and the remaining part being filled with superheated steam. Both ends of the tube are specified to be open with a pressure of 60 psia. The initial conditions at $t = 0$ are that (1) $v = 0$ in the entire tube, and (2) the left liquid elevation is 3 ft below the top of the U-tube and the right liquid surface is 4 ft below the left liquid surface. Solutions were obtained with several combinations of Δt and Δx.

If the mass of superheated steam is neglected, the problem may be easily solved analytically. Assuming a harmonic oscillation, the equation of motion is

$$l\ddot{y} = 2gy$$

where l is the length of liquid water, g is the gravity constant, and y is the half of the difference in elevation between the left and right surfaces. From the equation of motion, the period of oscillation for $l = 16$ ft and $g = 32.17$ ft sec^2 is $T = 3.13$ sec. For each numerical solution the period obtained by the implicit characteristic method is compared with this analytical result as shown in Table 5.1.

Table 5.1   Period of Manometer Oscillation

Δt (sec)	Δx (ft)	Period (sec)
Implicit characteristic		
0.05	2	2.95
0.025	2	2.95
0.0125	2	2.95
0.05	1	3.05
0.05	0.5	3.10
Analytical		3.13

Example 5.3

Consider a uniform 12-ft-long frictionless pipe. The left half of the pipe is heated at $q''' = 7272$ Btu ft^{-3} sec^{-1}, and the right half is cooled at $q''' = -7272$ Btu ft^{-3} sec^{-1}. Subcooled water at an enthalpy of 500 Btu lbm^{-1}

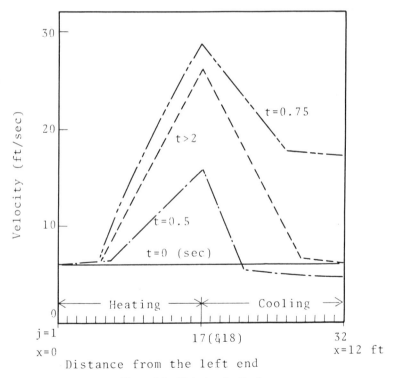

Figure 5.9   Velocity distribution.

enters the left end at a constant velocity of 6 ft sec^{-1}. The right end is open at a fixed pressure of 1000 psia. Initial conditions are $H = 500$ Btu lbm^{-1}, $v = 6$ ft sec^{-1}, and $p = 1000.0$ psia.

Numerical computations were made with $\Delta x = 0.4$ ft and $\Delta t = 0.005$ sec. Fluid velocity distributions at various times are shown in Fig. 5.9. The sharp change of acceleration at $j = 17$ is due to the change from heating to cooling. Sharp changes of acceleration at other locations are due to an interface between subcooled and two-phase water.

5.5 CONTROL VOLUME METHOD[13-19]

This is a method for unsteady flow of a compressible fluid in a network of volumes and links. The control volume method has been successfully used for coolant transient analyses of nuclear power plants, especially in hypothetical loss of coolant analyses. The advantages of the method are that (1) two-phase fluid can be easily treated, and (2) computational cost is low compared with finite difference methods such as the characteristic method. The disadvantages are that (1) calculations are gross and detailed information such as the propagation of waves is lost, and (2) accuracy is less than with finite difference methods. The explicit and implicit schemes are both discussed.

Consider a network consisting of control volumes and links, as illustrated in Fig. 5.10, for example. The links that connect control volumes are called *ordinary links,* while the link that connects a volume and the external atmosphere is called a *boundary link.*

Equations for energy and mass conservations may be written at each control volume as

$$\frac{d}{dt}U_i = \sum_l H_l W_l - \sum_m H_m W_m + Q_i \qquad (5.5.1)$$

$$\frac{d}{dt}M_i = \sum_l W_l - \sum_m W_m \qquad (5.5.2)$$

In these equations U_i and M_i denote the internal energy and mass of the fluid in volume i, Q_i is a known time-dependent heat source, H_l and H_m are the enthalpies, respectively, at inlet and exit of the volume i, and W_l and W_m are the corresponding flow rates. Once the internal energy and mass of the fluid in a volume is calculated, the pressure in that volume is obtained by the equation of state:

$$P_i = \eta(U_i, M_i) \qquad (5.5.3)$$

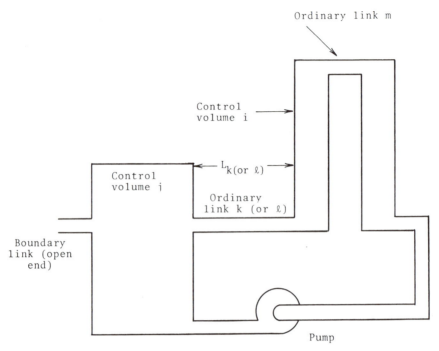

Figure 5.10   An illustration of the system consisting of control volumes and links.

Enthalpies are obtained by using the relation $M_iH_i = U_i + P_iV_i$. For a two-phase fluid, a steam table may be used to obtain P and H.

Assuming the cross sectional area of a link is uniform, the flow at an ordinary link is governed by a one-dimensional conservation of momentum equation and may be written as

$$\frac{d}{dt} W_k = \frac{A_k}{L_k}[P_j - P_i - F_k - \rho_k \Delta z] \equiv Y_k \qquad (5.5.4)$$

where W_k is the mass rate of flow at the link k, P_i and P_j are the pressures, respectively, at the inlet and outlet of the link; F_k is the friction loss in the link; A_k is the link flow area, L_k is the link length; and Δz is the change in elevation of the link. In Eq. (5.5.4), the effect of kinetic energy change in the link is neglected. The friction loss may be approximated by

$$F_k = \frac{f_k L_k W_k |W_k|}{2g_c\rho_k A_k^2} \qquad (5.5.5)$$

where f_k is the friction factor g_c is the conversion factor. An orifice may be included in a link by using an appropriate value for f_k. A pump may also be included in a link by using an equivalent value of Δz in Eq. (5.5.4).

Assuming the external pressure is constant, the momentum equation for a boundary link may be expressed by

$$\frac{d}{dt} W_k = g_k(P_i) \tag{5.5.6}$$

where P_i is the pressure of the volume that is upstream to the boundary link and g_k is a given function of P_i.

EXPLICIT SCHEME[13,14,17]

By using the forward difference formula in the time interval (t_n, t_{n+1}), where n is the index for a time grid, Eqs. (5.5.1) and (5.5.2) are approximated by

$$U_i^{n+1} = U_i^n + \Delta t\left(\sum_l H_l^n W_l^n - \sum_m H_m^n W_m^n + Q_i^n\right) \tag{5.5.7}$$

$$M_i^{n+1} = M_i^n + \Delta t\left(\sum_l W_l^n - \sum_m W_m^n\right) \tag{5.5.8}$$

The pressure in volume i is then

$$P_i^{n+1} = \eta(U_i^{n+1}, M_i^{n+1}) \tag{5.5.9}$$

The enthalpy H_i^{n+1} for volume i is calculated by

$$M_i^{n+1} H_i^{n+1} = U_i^{n+1} + P_i^{n+1} V_i$$

where V_i is the capacity of volume i. Then, the enthalpy in the link k, namely H_k^{n+1}, is set to the enthalpy of the volume from which the flow is coming into link k. W_k^{n+1} is calculated by the difference formula for Eq. (5.5.4) as

$$W_k^{n+1} = W_k^n + \Delta t \frac{A_k}{L_k}[P_j^{n+1} - P_i^{n+1} - F_k^n - \rho_k^{n+1}\Delta z] \tag{5.5.10}$$

Equation (5.5.6) can be treated similarly as

$$W_k^{n+1} = W_k^n + \Delta t g_k(P_i^{n+1}) \tag{5.5.11}$$

When the values of U_i^{n+1}, M_i^{n+1}, W_k^{n+1} and fluid properties at all the volumes and links are calculated, the computation cycle for $[t_n, t_{n+1}]$ is completed, and the calculations for the next time step are initiated.

While the explicit method described in this section is simple and easy to handle, one disadvantage is that the upper bound of Δt is limited by instability of the solution. The main reason for instability is due to the resonance effect of a control volume. For an isolated volume and link, the Helmholtz frequency δ is

$$\delta = \frac{\omega}{2\pi} = \frac{c}{2\pi}\sqrt{\frac{A}{Vl}} \qquad (5.5.12)$$

where ω is the angular velocity, c is the sonic velocity, V is the volume of fluid, l is the length of the volume and A is the cross-sectional area. When the time-step size approaches the natural period, a numerical instability can result. Therefore, an appropriate time step must be 5 to 10 times smaller than the natural period, namely,

$$\Delta t < \frac{2\pi}{5c}\sqrt{\frac{Vl}{A}} \qquad (5.5.13)$$

Another type of numerical instability is due to mass and energy flow. However, the frequencies associated with these instabilities are normally lower than the Helmholtz frequencies.[17]

IMPLICIT METHOD

The implicit method[15,16] requires more computational effort and time per each time step, but there is no upper bound of Δt. It is particularly attractive when the transient is slow and long. We first study the principle of the implicit scheme for a nonlinear system and then develop the implicit control volume method.

For simplicity, let us consider a set of two nonlinear first order equations

$$\frac{d}{dt}y_1(t) = f_1(y_1, y_2, t)$$

$$\frac{d}{dt}y_2(t) = f_2(y_1, y_2, t) \qquad (5.5.14)$$

A numerical scheme for the above set may be written

$$y_k^{n+1} = y_k^n + hf_k[y_1(t_n + h\theta), y_2(t_n + h\theta), t_n + \theta h], \qquad k = 1, 2 \quad (5.5.15)$$

where $h = \Delta t = t_{n+1} - t_n$ and $0 \le \theta \le 1$. If $\theta = 0$, Eq. (5.5.15) is an explicit scheme. If $\theta = 1$, the scheme is fully implicit but cannot be solved because f_k are nonlinear functions of y_k. To obtain a solvable implicit scheme, f_k is linearized by expanding it around t_n with $\theta = 1$:

$$f_k^{n+1} \approx f_k^n + \sum_{j=1}^{2}\frac{\partial f_k}{\partial y_j}(y_j^{n+1} - y_j^n) + \frac{\partial f_k}{\partial t}h \qquad (5.5.16)$$

where f_k^n is defined as

$$f_k^n = f_k(y_1^n, y_2^n, t_n) \qquad (5.5.17)$$

and the partial derivatives are evaluated at t_n. Introducing Eq. (5.5.16) into Eq. (5.5.15) with $\theta = 1$ yields

$$\sum_{j=1}^{2} \left(\delta_{kj} - h \frac{\partial f_k}{\partial y_j} \right) (y_k^{n+1} - y_k^n) = h f_k^n + h^2 \frac{\partial f_k}{\partial t}, \qquad k = 1, 2 \quad (5.5.18)$$

where $\delta_{kj} = 1$ if $k = j$, and $\delta_{kj} = 0$ if $k \neq j$. We neglect the second term on the right side, because it does not contribute to the implicit form and the magnitude is the second order in h. Besides, if f_k does not include t explicitly, the second term is automatically zero.

The Jacobian matrix is defined as

$$\mathbf{J} = \begin{bmatrix} \dfrac{\partial f_1}{\partial y_1}, & \dfrac{\partial f_1}{\partial y_2} \\[2mm] \dfrac{\partial f_2}{\partial y_1}, & \dfrac{\partial f_2}{\partial y_2} \end{bmatrix} \qquad (5.5.19)$$

By using Eq. (5.5.19), Eq. (5.5.18) can be written as

$$(\mathbf{I} - h\mathbf{J})(\mathbf{y}^{n+1} - \mathbf{y}^n) = h\mathbf{f}^n \qquad (5.5.19a)$$

where $\mathbf{I}$ is an identity matrix and

$$\mathbf{y}^n = \mathrm{col}\,(y_1^n, y_2^n) \qquad (5.5.20)$$

$$\mathbf{f}^n = \mathrm{col}\,(f_1^n, f_2^n) \qquad (5.5.21)$$

Inverting $(\mathbf{I} - h\mathbf{J})$, the solution of Eq. (5.5.19a) is obtained as

$$\mathbf{y}^{n+1} = \mathbf{y}^n + (\mathbf{I} - h\mathbf{J})^{-1} h\mathbf{f}^n \qquad (5.5.22)$$

Thus an implicit scheme for Eq. (5.5.14) is obtained. Although the derivation has been done with two variables, Eq. (5.5.22) is valid for any number of variables.

In applying Eq. (5.5.22) to the set of Eqs. (5.5.1) through (5.5.4), we could define $\mathbf{y}$ consisting of U_i, M_i, and W_k for all the volumes and links. However, the order of $\mathbf{y}$ can be significantly reduced by elmininating U_i and M_i as described next.

The backward difference formulas for Eqs. (5.5.1) and (5.5.2) are

$$\Delta U_i \equiv U_i^{n+1} - U_i^n = \Delta t \left(\sum_l H_l W_l^{n+1} - \sum_m H_m W_m^{n+1} + Q_i^{n+1} \right) \quad (5.5.23)$$

$$\Delta M_i \equiv M_i^{n+1} - M_i^n = \Delta t \left(\sum_l W^{n+1} - \sum_m W_m^{n+1} \right) \qquad (5.5.24)$$

where l are inlet links and m are outlet links to volume i. The implicit scheme for Eq. (5.5.4) is

$$\Delta W_k \equiv W_k^{n+1} - W_k^n = \Delta t\left(Y_k^n + \frac{\partial Y_k}{\partial W_k}\Delta W_k \right.$$

$$\left. + \frac{\partial Y_k}{\partial U_i}\Delta U_i + \frac{\partial Y_k}{\partial M_i}\Delta M_i + \frac{\partial Y_k}{\partial U_j}\Delta U_j + \frac{\partial Y_k}{\partial M_j}\Delta M_j \right) \quad (5.5.25)$$

where the right side of Eq. (5.5.4) has been linearized. The partial derivatives in Eq. (5.5.25) are given by

$$\frac{\partial Y_k}{\partial W_k} = -c_k\frac{\partial F_k}{\partial W_k} \quad (5.5.26)$$

$$\frac{\partial Y_k}{\partial U_j} = c_k\frac{\partial P_j}{\partial U_j} = c_k\frac{\partial \eta(U_j, M_j)}{\partial U_j} \quad (5.5.27a)$$

$$\frac{\partial Y_k}{\partial U_i} = -c_k\frac{\partial P_i}{\partial U_i} = -c_k\frac{\partial \eta(U_i, M_i)}{\partial U_i} \quad (5.5.27b)$$

$$\frac{\partial Y_k}{\partial M_j} = c_k\frac{\partial P_j}{\partial M_j} = c_k\frac{\partial \eta(U_j, M_j)}{\partial M_j} \quad (5.5.28a)$$

$$\frac{\partial Y_k}{\partial M_i} = -c_k\frac{\partial P_i}{\partial M_i} = -c_k\frac{\partial \eta(U_i, M_i)}{\partial M_i} \quad (5.5.28b)$$

where Eq. (5.5.3) was used and $c_k = A_k/L_k$. Introducing Eqs. (5.5.23) and (5.5.24) into Eq. (5.5.25) yields

$$W_k^{n+1} - W_k^n = \Delta t\left[Y_k^n + \frac{\partial F_k}{\partial W_k}(W_k^{n+1} - W_k^n) \right.$$

$$+ \Delta t\frac{\partial Y_k}{\partial U_i}\left(\sum_l H_l W_l^{n+1} - \sum_m H_m W_m^{n+1} + Q_i^{n+1}\right)$$

$$+ \Delta t\frac{\partial Y_k}{\partial M_i}\left(\sum_l W_l^{n+1} - \sum_m W_m^{n+1}\right) \quad (5.5.29)$$

$$+ \Delta t\frac{\partial Y_k}{\partial U_j}\left(\sum_q H_q W_q^{n+1} - \sum_r H_r W_r^{n+1} + Q_j^{n+1}\right)$$

$$\left. + \Delta t\frac{\partial Y_k}{\partial M_j}\left(\sum_q W_q^{n+1} - \sum_r W_r^{n+1}\right)\right]$$

where l and m are the inlet and outlet links, respectively, to volume i, and q and r are the inlet and outlet links, respectively, to volume j.

In accordance with Eq. (5.5.16), the difference formula for Eq. (5.5.6) is

$$\Delta W_k \equiv W_k^{n+1} - W_k^n = \Delta t\left(g_k^n + \frac{\partial g_k}{\partial W_k}\Delta W_k + \frac{\partial g_k}{\partial U_i}\Delta U_i + \frac{\partial g_k}{\partial M_i}\Delta M_i\right) \quad (5.5.30)$$

where i is inlet volume to the boundary link k, and

$$\frac{\partial g_k}{\partial U_i} = \frac{\partial g_k}{\partial P_i}\frac{\partial P_i}{\partial U_i} = \frac{\partial g_k}{\partial P_i}\frac{\partial \eta(U_i, M_i)}{\partial U_i} \quad (5.5.31)$$

$$\frac{\partial g_k}{\partial M_i} = \frac{\partial g_k}{\partial P_i}\frac{\partial P_i}{\partial M_i} = \frac{\partial g_k}{\partial P_i}\frac{\partial \eta(U_i, M_i)}{\partial M_i} \quad (5.5.32)$$

Introducing Eqs. (5.5.23) and (5.5.24) into Eq. (5.5.30) yields

$$W_k^{n+1} - W_k^n = \Delta t\left[g_k^n + \frac{\partial g_k}{\partial W_k}(W_k^{n+1} - W_k^n) \right.$$

$$+ \Delta t\frac{\partial g_k}{\partial U_i}\left(\sum_l H_l W_l^{n+1} - \sum_m H_m W_m^{n+1} + Q_i^{n+1}\right) \quad (5.5.33)$$

$$\left. + \Delta t\frac{\partial g_k}{\partial M_i}\left(\sum_l W_l^{n+1} - \sum_m W_m^{n+1}\right)\right]$$

Equation (5.5.29) for all the ordinary links and Eq. (5.5.33) for boundary links are coupled. The set of those equations are linear in W_k^{n+1}s. The flows for all the links are simultaneously solved by the Gaussian elimination technique. For a numerical example, see Reference 16.

5.6 VORTICITY METHOD FOR TWO-DIMENSIONAL INCOMPRESSIBLE FLOW

In the vorticity method discussed in this section, the vorticity function and the stream function are used as field functions rather than velocities and pressure, while the MAC method described in Section 5.7 uses velocities and pressure directly. In both methods, the flow equations are transformed to a set of parabolic and elliptic partial differential equations, so the numerical schemes to solve those types of equations as described in the previous chapters are applied. Another numerical method for two-dimensional incompressible flow based on the finite element method is described in Section 7.5.

The basic equations for two-dimensional incompressible flow of a Newtonian fluid in rectangular geometry are as follows.

MOMENTUM (NAVIER–STOKES) EQUATIONS

$$\frac{\partial u}{\partial t}+u\frac{\partial u}{\partial x}+v\frac{\partial u}{\partial y}=-\frac{1}{\rho}\frac{\partial p}{\partial x}+\nu\nabla^2 u+F_x \tag{5.6.1}$$

$$\frac{\partial v}{\partial t}+u\frac{\partial v}{\partial x}+v\frac{\partial v}{\partial y}=-\frac{1}{\rho}\frac{\partial p}{\partial y}+\nu\nabla^2 v+F_y \tag{5.6.2}$$

CONTINUITY EQUATION

$$\frac{\partial u}{\partial x}+\frac{\partial v}{\partial y}=0 \tag{5.6.3}$$

where u and v are velocity components in x and y directions, respectively; p is the pressure; ν is the kinematic viscosity; ρ is the fluid density; and F_x and F_y are the external forces which are assumed to be zero in this section.

The pressure in Eqs. (5.6.1) and (5.6.2) is eliminated by differentiating Eq. (5.6.1) with respect to y, differentiating Eq. (5.6.2) with respect to x, and subtracting one from the other. Using the voriticity defined by

$$\zeta=\frac{\partial u}{\partial y}-\frac{\partial v}{\partial x} \tag{5.6.4}$$

the result of this operation becomes

$$\frac{\partial \zeta}{\partial t}=-u\frac{\partial \zeta}{\partial x}-v\frac{\partial \zeta}{\partial y}+\nu\nabla^2\zeta \tag{5.6.5}$$

Equation (5.6.5) may be written in a compact form as

$$\frac{\partial \zeta}{\partial t}=-\mathbf{V}\cdot\nabla\zeta+\nu\nabla^2\zeta \tag{5.6.6}$$

where $\mathbf{V}$ represents the vector (u, v). Equation (5.6.5) or (5.6.6) is called the *vorticity equation*. If the *stream function* is defined by

$$\frac{\partial \psi}{\partial y}=u, \qquad \frac{\partial \psi}{\partial x}=-v \tag{5.6.7}$$

then Eq. (5.6.4) becomes the Poisson equation:

$$\nabla^2\psi=\zeta \tag{5.6.8}$$

Since the continuity equation was not used, Eq. (5.6.5) is called the *nonconservative form*. The continuity equation can be incorporated, if desired, in the following manner. Since Eq. (5.6.4) is equivalently written as $\nabla\cdot\mathbf{V}=0$, the term $\mathbf{V}\cdot\nabla\zeta$ in Eq. (5.6.6) becomes

$$\mathbf{V}\cdot\nabla\zeta=\nabla\cdot\mathbf{V}\zeta+\zeta\nabla\cdot\mathbf{V}=\nabla(\mathbf{V}\zeta) \tag{5.6.9}$$

Introducing Eq. (5.6.9) into Eq. (5.6.6) yields

$$\frac{\partial \zeta}{\partial t} = -\boldsymbol{\nabla}(\mathbf{V}\zeta) + \nu\boldsymbol{\nabla}^2\zeta = -\frac{\partial u\zeta}{\partial x} - \frac{\partial v\zeta}{\partial y} + \nu\boldsymbol{\nabla}^2\zeta \qquad (5.6.10)$$

Equation (5.6.10) is called the *conservative form*. The conservative form is not very significant for incompressive flow, so only the nonconservative form is considered in the rest of this section.

In deriving the finite different schemes for the vorticity and Poisson equations, we first discretize those equations on the time coordinate. Applying the forward differencing scheme to Eqs. (5.6.6) and (5.6.8), we have

$$\frac{\zeta^{n+1} - \zeta^n}{\Delta t} = -\mathbf{V}^n \cdot \boldsymbol{\nabla}\zeta^n + \nu\boldsymbol{\nabla}^2\zeta^n \qquad (5.6.11)$$

$$\boldsymbol{\nabla}^2\psi^{n+1} = \zeta^{n+1} \qquad (5.6.12)$$

Basically, ζ^{n+1} and ψ^{n+1} are both solved at each time step according to the diagram in Fig. 5.11. Iterative methods for elliptic partial differential equations discussed in Chapter 3 are used to solve Eq. (5.6.12) in each time

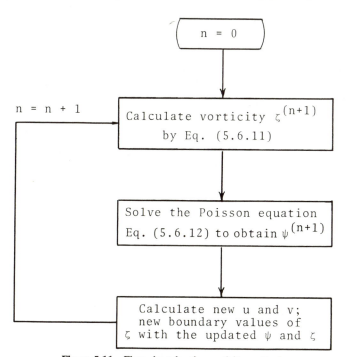

Figure 5.11 Flow chart for the vorticity method.

step. If the geometry is rectangle and the boundary conditions for Eq. (5.6.12) are simple, the fast direct inversion method described in Section 10.5 will be more economical than the iterative schemes. Since Eq. (5.6.11) is explicit, no iterative procedure is required, so computational time for Eq. (5.6.11) is short. The upper bound of Δt is limited to prevent the numerical instability of the explicit scheme. Although the use of a very small Δt is a disadvantage of the explicit method, the stream function changes little in such a small Δt. Therefore, the iterative solution for Eq. (5.6.12) is necessary only once in several time steps, so old values of u and v can be used in Eq. (5.6.11) until they are updated. This technique saves much computing time. Stability analysis for Eq. (5.6.11) is described in Section 4.7.

The implicit scheme may be also derived by applying the backward differencing to Eq. (5.6.6):

$$\frac{\zeta^{n+1}-\zeta^n}{\Delta t} = -\mathbf{V}^n \cdot \nabla\zeta^{n+1} + \nu\nabla^2\zeta^{n+1} \qquad (5.6.13)$$

The solution of Eq. (5.6.13) requires direct inversion of a large matrix or iterative solution. Although the time step for the implicit method is unlimited, two iterative schemes, one for Eq. (5.6.13) and another for the Poisson equation, may be time-consuming and uneconomical compared with the explicit method.

The boundary condition for the vorticity method must be specified in terms of vorticity and the stream function. They are given as follows:

INFLOW BOUNDARY CONDITIONS

There is no unique way of specifying the boundary condition at an inlet, because the flow characteristics are dependent on the upstream flow condition. Assuming the inflow boundary is along the bottom side of the geometry as in Fig. 5.12, one approach is to set

$$v(x, 0) = v_0(x)$$
$$u(x, 0) = 0 \qquad (5.6.14)$$

where $v_0(x)$ is a prescribed function. Since $\partial\psi/\partial x = -v$, the stream function ψ along the inflow boundary is obtained by integrating $-v_0(x)$ along the boundary. We denote the stream function thus obtained along the inflow boundary by ψ_0. The stream function is a continuous function along the whole boundary of the geometry. It is constant along a wall boundary. Any arbitrary constant can be assigned to a point on the boundary. The integrating constant involved in ψ_0 along the inflow boundary is determined so that ψ_0 becomes continuous at the ends of the inflow boundary.

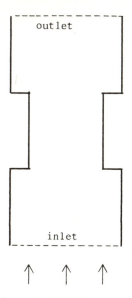

↑   ↑   ↑   **Figure 5.12**   Illustration of a flow channel.

By using Eq. (5.6.4) and Eq. (5.6.14), the vorticity at the inflow boundary is given by

$$\zeta = \frac{\partial u}{\partial y} - \frac{\partial v}{\partial x} = -\frac{\partial v_0}{\partial x} = \frac{\partial^2 \psi_0}{\partial x^2} \tag{5.6.15}$$

OUTFLOW BOUNDARY CONDITIONS

The outflow boundary conditions are treated similarly to the inflow boundary. Assuming the boundary is on the top side of the geometry, the simplest approach is to set $u(x, y) = 0$ and $v(x, y) = $ const. If there is an outlet conduit behind the outflow boundary, an improved approach is to set v to a parabolic distribution. Such complete specifications of velocity at the outflow boundary are favorable for the stability of the numerical scheme. However, they are not appropriate whenever the development of the velocity distributions during the calculation is important.

The less restrictive boundary conditions are

$$u \doteq \frac{\partial \psi}{\partial y} = 0$$

$$\frac{\partial \zeta}{\partial y} = 0 \tag{5.6.16}$$

This set of conditions implies that an outlet conduit behind the outflow boundary is assumed where the parallel flow is established. The vertical velocity at the outflow boundary is developed in the course of computation. With the finite difference approximation, these boundary conditions are approximated by

$$\psi_{i,J} = \psi_{i,J-1}$$

$$\zeta_{i,J} = \zeta_{i,J-1}$$

(5.6.17)

where J is the grid number j at the outflow boundary.

WALL BOUNDARY CONDITIONS

The stream function becomes constant along a wall boundary. Since the value of the stream function can be specified at one stream line, the conventional choice is $\psi = 0$ along one selected wall boundary. The stream function is continuous along all the external boundary. Once the value of ψ along a wall is specified, the value of ψ along the inflow, outflow, and other wall boundaries is determined so that ψ is continuous.

If the wall is a free-slip boundary, the wall boundary condition for vorticity is $\zeta = 0$. The boundary condition for a nonslip wall is rather complicated. Assume the wall is parallel to the y coordinate. The velocities are specified by $u = v = 0$. In order to express the vorticity at a nonslip wall, we expand $\psi_{i,j}$ in a Taylor series about the wall (we consider a left wall and assume the mesh index at the wall is $i = 0$):

$$\psi_{1,j} = \psi_{0,j} + \frac{\partial \psi}{\partial x}\bigg|_{0,j} \Delta x + \frac{1}{2} \frac{\partial^2 \psi}{\partial x^2}\bigg|_{0,j} \Delta x^2 + 0(\Delta x^3)$$

(5.6.18)

Because of the nonslip condition, the second term is zero:

$$\frac{\partial \psi}{\partial x}\bigg|_{0,j} = -v_{0,j} = 0$$

(5.6.19)

Since $u = 0$ along the wall, we have

$$\frac{\partial u}{\partial y}\bigg|_{0,j} = 0$$

(5.6.20)

Introducing this into Eq. (5.6.4) yields $\zeta = -\partial v / \partial x$ along the wall. Referring to Eq. (5.6.7), this equation further becomes

$$\zeta = -\frac{\partial v}{\partial x} = \frac{\partial^2 \psi}{\partial x^2}$$

(5.6.21)

Eliminating the second derivative of ψ from Eq. (5.6.18), Eq. (5.6.21) yields the vorticity at the nonslip wall as

$$\zeta_{0,j} = \frac{2}{\Delta x^2}(\psi_{1,j} - \psi_{0,j})$$

5.7 MAC METHOD[24-27]

While the vorticity and stream functions are used as variables in the vorticity method, the MAC method uses u, v, and p directly. The numerical scheme of the MAC method consists of one Poisson equation and two parabolic equations in the explicit differencing scheme. The only time-consuming part of the MAC method is the implicit solution of the Poisson equation at each time step. Since pressure is an explicit variable, the pressure boundary condition can be easily incorporated. The MAC method is capable of including the free surface fluid boundary and the interface of different fluids in a system, although such boundaries are not discussed here.

The mesh structure used in the MAC method is shown in Fig. 5.13. Pressure is defined at the center of a cell with indicies (i, j). The velocity u is defined at the left and right sides of a cell with indices $(i \pm \frac{1}{2}, j)$, and the velocities v are defined at the top and bottom sides of a cell with indices $(i, j \pm \frac{1}{2})$. The boundaries of the domain coincide with the sides of cells. It is assumed that spatial meshes are equally spaced.

Direct application of central differencing to the spatial differential terms of Eqs. (5.6.1) and (5.6.2) and application of a forward differencing scheme to the time differentiation, except for p, yield the following two equations:*

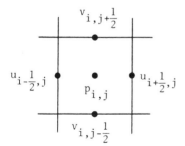

Figure 5.13   Mesh scheme of the MAC method.

---

* Before applying difference approximations, u times Eq. (5.6.3) and v times Eq. (5.6.3) were added to Eqs. (5.6.1) and (5.6.2), respectively. Therefore, the left sides of Eqs. (5.6.1) and (5.6.2), respectively were written,

$$\frac{\partial u}{\partial t} + \frac{\partial u^2}{\partial x} + \frac{\partial uv}{\partial y} \quad \text{and} \quad \frac{\partial v}{\partial t} + \frac{\partial uv}{\partial x} + \frac{\partial v^2}{\partial y}$$

$$u_{i+1/2,j}^{n+1} = u_{i+1/2,j}^n - \frac{\Delta t}{\Delta y}[(uv)_{i+1/2,j+1/2}^n - (uv)_{i+1/2,j-1/2}^n]$$

$$- \frac{\Delta t}{\Delta x}[(u_{i+1,j}^n)^2 - (u_{i,j}^n)^2] - \frac{\Delta t}{\rho \Delta x}(p_{i+1,j}^{n+1} - p_{i,j}^{n+1})$$

$$+ \frac{\nu \Delta t}{(\Delta y)^2}(u_{i+1/2,j+1}^n - 2u_{i+1/2,j}^n + u_{i+1/2,j-1}^n) \qquad (5.7.1)$$

$$+ \frac{\nu \Delta t}{(\Delta x)^2}(u_{i+3/2,j}^n - 2u_{i+1/2,j}^n + u_{i-1/2,j}^n)$$

$$+ \Delta t F_{x,i+1/2,j}^n$$

$$v_{i,j+1/2}^{n+1} = v_{i,j+1/2}^n - \frac{\Delta t}{\Delta x}[(uv)_{i+1/2,j+1/2}^n - (uv)_{i-1/2,j+1/2}^n]$$

$$- \frac{\Delta t}{\Delta y}[(v_{i,j+1}^n)^2 - (v_{i,j}^n)^2] - \frac{\Delta t}{\rho \Delta y}(p_{i,j+1}^{n+1} - p_{i,j}^{n+1})$$

$$+ \frac{\nu \Delta t}{(\Delta y)^2}(v_{i,j+3/2}^n - 2v_{i,j+1/2}^n + v_{i,j-1/2}^n)$$

$$+ \frac{\nu \Delta t}{(\Delta x)^2}(v_{i+1,j+1/2}^n - 2v_{i,j+1/2}^n + v_{i-1,j+1/2}^n) \qquad (5.7.2)$$

$$+ \Delta t F_{y,i,j+1/2}^n$$

where n denotes the time step and the external forces are assumed to be space dependent. In Eqs. (5.7.1) and (5.7.2), the velocities that do not fall at the points specified in Fig. 5.13 are obtained by taking appropriate averages of the values in the neighborhood, for example,

$$v_{i,j} = \frac{v_{i,j+1/2} + v_{i,j-1/2}}{2} \qquad (5.7.3)$$

$$(uv)_{i+1/2,j+1/2} = \left(\frac{v_{i,j+1/2} + v_{i+1,j+1/2}}{2}\right)\left(\frac{u_{i+1/2,j} + u_{i+1/2,j+1}}{2}\right) \qquad (5.7.4)$$

Equations (5.7.1) and (5.7.2) may be written more compactly as

$$u_{i+1/2,j}^{n+1} = \xi_{i+1/2,j}^n - \frac{\Delta t}{\rho \Delta x}(p_{i+1,j}^{n+1} - p_{i,j}^{n+1}) \qquad (5.7.5)$$

$$v_{i,j+1/2}^{n+1} = \eta_{i,j+1/2}^n - \frac{\Delta t}{\rho \Delta y}(p_{i,j+1}^{n+1} - p_{i,j}^{n+1}) \qquad (5.7.6)$$

where ξ and η represent the values with index n on the right sides of Eq. (5.7.1) and (5.7.2), respectively.

The difference approximation for the continuity equation, Eq. (5.6.3), is

$$\frac{u_{i+1/2,j}^{n+1}-u_{i-1/2,j}^{n+1}}{\Delta x}+\frac{v_{i,j+1/2}^{n+1}-v_{i,j-1/2}^{n+1}}{\Delta y}=0 \tag{5.7.7}$$

Introducing Eqs. (5.7.5) and (5.7.6) into Eq. (5.7.7) yields

$$\frac{-p_{i-1,j}^{n+1}+2p_{i,j}^{n+1}-p_{i+1,j}^{n+1}}{\Delta x^2}+\frac{-p_{i,j-1}^{n+1}+2p_{i,j}^{n+1}-p_{i,j+1}^{n+1}}{\Delta y^2}$$

$$=\frac{\rho}{\Delta t}\left\{\frac{\xi_{i+1/2,j}^{n}-\xi_{i-1/2,j}^{n}}{\Delta x}+\frac{\eta_{i,j+1/2}^{n}-\eta_{i,j-1/2}^{n}}{\Delta y}\right\} \tag{5.7.8}$$

which is the difference approximation for the Poisson equation, $-\nabla^2 p = q$, and may be solved iteratively as discussed in Chapter 3. If the geometry is simple, the discrete Poisson equation can be solved economically by using the fast direct method described in Section 10.5. Once $p_{i,j}$ are obtained by solving Eq. (5.7.8), they are back-substituted into Eqs. (5.7.5) and (5.7.6) to calculate u and v for the new time step.

Boundary conditions on u and v are evaluated by considering hypothetical grid points outside the domain. The values of u and v on the hypothetical grids are necessary to evaluate Eqs. (5.7.1) and (5.7.2). For example, if $u = 0$ and $(\partial/\partial x)v = 0$ (free-slip) along the left wall in Fig. 5.14, we set

$$u_{1/2,j}=0, \qquad v_{0,j-1/2}=v_{1,j-1/2} \tag{5.7.9}$$

where $(0, j-\frac{1}{2})$ is a hypothetical grid. In the case of $v = 0$ (nonslip) along the left wall

$$\tfrac{1}{2}(v_{0,j-1/2}+v_{1,j-1/2})=0 \quad \text{or} \quad v_{0,j-1/2}=-v_{1,j-1/2} \tag{5.7.10}$$

When v and u are both specified at a side of a cell, the pressure for the hypothetical cell is not required. If the values of u and v are specified at an inlet or outlet boundary, the treatment is essentially the same as a wall boundary.

The boundary conditions for an inlet or outlet where only p is specified are more complicated. Generally, a supplementary condition is required in addition to p. For example, we consider the outlet in Fig. 5.14 and assume p is specified there. We use $\partial u/\partial y = 0$ as the supplementary condition. The values at the fictitious grids are set to

$$u_{i+1/2,J+1}=u_{i+1/2,J} \tag{5.7.11}$$

$$v_{i,J+3/2}=v_{i,J+1/2}-\frac{\Delta y}{\Delta x}(u_{i+1/2,J+1}-u_{i-1/2,J+1}) \tag{5.7.12}$$

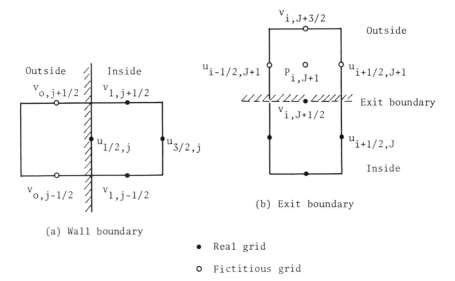

(a) Wall boundary

(b) Exit boundary

• Real grid

○ Fictitious grid

Figure 5.14 Illustration of fictitious grids at the boundaries.

where the second equation is based on the continuity equation, Eq. (5.6.3). The value of $p_{i,J+1}$ in the fictitious cell, which appears in Eq. (5.7.8), is related to the prescribed value of p on the boundary as

$$\frac{p_{i,J+1}+p_{i,J}}{2} = p_{i,J+1/2}: \text{ prescribed value} \qquad (5.7.13)$$

The boundary condition for Eq. (5.7.8) is the Dirichlet type if p on the boundary is prescribed; otherwise, the boundary condition for p is given as the Neumann type:

$$\frac{\partial p}{\partial n} = q(s) \qquad (5.7.14)$$

where n is the normal outward and s is the coordinate along the boundary. The normal derivative of p at the boundary is obtained by introducing the latest results for u and v into the difference form of Eq. (5.6.1) or Eq. (5.6.2). If we consider a vertical nonslip wall on the left side, the normal derivative is equal to $-\partial p/\partial x$ and is obtained from Eq. (5.6.1) as

$$\frac{\partial p}{\partial n} = \rho\left(\nu\frac{\partial^2 u}{\partial x^2}+F_x\right) \qquad (5.7.15)$$

where $u = v = 0$ is used in Eq. (5.6.1). By using the central difference formula for the left side of Eq. (5.7.15) and the forward difference formula for the right side, we have

$$\frac{p_{1,j}^{n+1} - p_{0,j}^{n+1}}{\Delta x} = \rho \left[\nu \frac{u_{1/2,j}^n - 2u_{3/2,j}^n + u_{5/2,j}^n}{(\Delta x)^2} - F_{x,1/2,j} \right] \quad (5.7.16)$$

where $p_{0,j}$ is the hypothetical pressure outside the boundary. The pressure equation, Eq. (5.7.8), for $i = 1$ includes $p_{0,j}$, which is now eliminated by using Eq. (5.7.16).

For a free-slip wall or a symmetry boundary $q(s)$ is zero in Eq. (5.7.14), so $p_{0,j} = p_{1,j}$ is used in Eq. (5.7.8) for $i = 1$.

If the pressure boundary conditions are all Neumann type, the pressure distribution is arbitrary by a constant. In this case, the pressure at one grid should be specified to assure the numerical solution of an iterative scheme. The convergence rate of the iterative solution for Eq. (5.7.8) is slower than that for Eq. (5.6.8) because the former is the Neumann problem while the latter is Dirichlet problem. Tight convergence of the iterative scheme in each time step is very important because errors due to poor convergence will be accumulated as time steps are increased.[28]

The numerical stability depends on the viscosity term. If $\nu = 0$, the numerical scheme is unconditionally unstable. As a useful rule of thumb,[29] the MAC method is considered stable if ν satisfies

$$\nu > \tfrac{1}{2} \Delta t V^2$$
$$\nu > \tfrac{1}{2} \Delta x^2 \dot{V}$$

where V is the average maximum fluid speed and $\dot{V}$ is the average maximum velocity gradient in the direction of flow.

PROBLEMS

1. Derive the characteristic equation for the hyperbolic equation

$$\frac{\partial}{\partial t} \phi(x, t) + c \frac{\partial}{\partial x} \phi(x, t) = 0$$

where c is a constant.

2. a. Solve the characteristic equation obtained in the previous problem with the initial condition:

$$\phi(x, 0) = 1 \quad \text{for } x \leq 0$$
$$= 2 \quad \text{for } x > 0$$

b. Plot the discontinuity originated at $x = 0$ and $t = 0$ on the x–t plane.

3. Apply the Fourier stability analysis to the following difference schemes for the hyperbolic equation in Problem 1 and find the stability criteria for Δt:

    a.

$$\phi_i^{(n+1)} = \eta\phi_i^{(n)} + (1-\eta)\phi_{i-1}^{(n)}$$

$$\eta = 1 - \frac{c\,\Delta t}{\Delta x}$$

    b.

$$\phi_i^{(n+1)} = \xi\phi_{i-1}^{(n+1)} + (1-\xi)\phi_{i-1}^{(n)}$$

$$\xi = 1 - \frac{\Delta x}{c\,\Delta t}$$

4. Write the implicit characteristic difference equations for a homogeneous fluid in the conduit shown by Fig. 5.5, and express the entire system of equations in a matrix form.

5. Apply the explicit and implicit control volume method to the problem of Example 5.1, and study numerically the effect of Δt and the number of control volumes on the accuracy and stability of the schemes.

6. The best way to learn the vorticity method and the MAC method is to develop small computer programs for a simple two-dimensional geometry. Write a FORTRAN program for each method for the geometry shown in Fig. 7.12.

REFERENCES

1. D. R. Chapman, H. Mark, and M. W. Pirtle, "Computers vs. wind tunnels for aerodynamic flow simulations," *Astronaut. Aeronaut.*, **13**, 23–35 (1975).

2. M. Inouye, "The computer as a wind tunnel," *Second USA-Japan computer conference proceedings*, Tokyo, sponsored by AFIPS and IPSJ (1975).

3. D. R. Hartree, *Numerical analysis*, Oxford U. P., Oxford, England (1958).

4. V. L. Streeter and E. B. Wylie, *Hydraulic transients*, McGraw-Hill, New York (1967).

5. N. E. Hoskin, "Solution by characteristics of the equations of one-dimensional unsteady flow," *Methods Comput. Phys.*, **3**, 265–293 (1964).

6. P. Fox, "The solution of hyperbolic partial differential equations," in *Mathematical methods for digital computers*, A. Ralston and S. Wilf, Eds., Wiley, New York (1960).

7. A. C. Spenser and S. Nakamura, "Implicit characteristic method for one-dimensional fluid flow," *Trans. Amer. Nucl. Soc.*, **17**, 247 (1973).

8. M. Lister, "The numerical solution of hyperbolic partial differential equations by the method of characteristics," in *Mathematical methods for digital computers*, A. Ralston and S. Wilf, Eds., Wiley, New York (1960).

9. S. Nakamura, M. A. Berger, and A. C. Spenser, "Implicit characteristic method for one-dimensional fluid flow," *Proceedings of the conference on computational method in nuclear engineering*, Conf-750413, National Technical Information Service, Springfield, Va. (1975).

10. C. A. Kot, "An improved constant time technique for the method of characteristics," *Proceedings of the third international conference on numerical methods in fluid mechanics*, Springer-Verlag, New York (1972).

11. K. Takeuchi and A. W. Gurcak, "Partial implicit characteristic method for one-dimensional fluid flow," *Trans. Amer. Nucl. Soc.*, **21**, 199 (1975).

12. A. C. Spencer, "Method of characteristics for solving two-dimensional core flows," *Proceedings of topical meeeting on water-reactor safety*, Conf-730304, National Technical Information Center, Springfield, Va. (1973).

13. S. G. Margolis and J. A. Redfield, *Flash, a program for digital simulation of the loss of coolant accident*, USAEC Rpt. WAPD-TM-534 (1966).

14. K. V. Moore and W. H. Rettig, *RELAP 2—A digital program for reactor blowdown and power excursion analysis*, IDO-17263, Idaho Falls National Laboratory, Idaho (1968).

15. T. A. Porshing, J. H. Murphy, J. A. Redfield, and V. C. Davis, *FLASH-4: A fully implicit FORTRAN-1 V program for the digital simulation of transients in a reactor plant*, WAPD-TM-840, Bettice Atomic Power Laboratory (1969).

16. T. A. Porsching, J. H. Murphy, and J. A. Redfield, "Stable numerical integration of conservation equations for hydraulic networks," *Nucl. Sci. Eng.*, **43**, 218–225 (1971).

17. W. H. Rettig, G. A. Jayne, K. V. Moore, C. E. Slater, and M. L. Uptmor, *RELAP 3—A computer program for reactor blowdown analysis*, IN-1321, Argonne Code Center Abstract No. 369, Argonne National Laboratory, Argonne, Ill. (1970).

18. P. L. Doan, D. D. Lanning, and N. C. Rasmussen, "Pressurized water reactor loss-of-coolant accidents by hypothetical vessel rupture," *Water-reactor safety*, CONF-730304, National Technical Information Service, Springfield, Va. (1973).

19. M. P. Khan, "Comparisons between results of the Westinghouse loss-of-coolant analyses and LOFT semiscale (ECC) test data, Part I: SATAN-5 code results," *Water-reactor safety*, CONF-730304, National Technical Information Service, Springfield, Va. (1973).

20. B. S. Massey, *Mechanics of fluid*, p. 231, Van Nostrand, Princeton, N.J. (1970).

21. K. N. Ghia and R. T. Davis, "Corner layer flow: optimization of numerical method of solution," *Computers Fluids*, **2**, 17–34 (1974).

22. C. Taylor and P. Hood, "A numerical solution of the Navier–Stokes equation using the finite element technique," *Computers Fluid*, **1**, 73–100 (1973).

23. F. H. Harlow and A. A. Amsden, "A numerical fluid dynamics calculation method for all flow speeds," *J. Comp. Phys.*, **8**, 197–213 (1971).

24. J. E. Welch, F. H. Harlow, J. P. Shannon, and B. J. Paly, "The MAC method," USAEC Rep. LA-3425, Los Alamos Scientific Laboratory, New Mexico (1966).

25. A. A. Amsden and F. H. Harlow, "The SMAC method," USAEC Rep. LA-4370, Los Alamos Scientific Laboratory, New Mexico (1970).

26. P. J. Roache, *Computational fluid dynamics*, Hermosa Albuquerque, N.M. (1972).

27. J. Veicelli, "A method for including arbitrary external boundaries in the MAC incompressible fluid computing technique," *J. Comput. Phys.* **4**, 543–551 (1969).

28. S. Nakamura and V. J. Esposito, "Coarse mesh rebalancing applied to hydraulic analysis," *Proceedings of the conference on mathematical models and computational techniques for analysis of nuclear systems*, CONF-7030414-P1, National Technical Information Service, Springfield, Va. (1973).

29. C. W. Hirt, "Heuristic stability for finite-difference equations," *J. Comput. Phys.*, **2**, 339–355 (1968).

Chapter 6 Weighted Residual Methods and Variational Principles

6.1 VARIATIONAL PRINCIPLE AS A MODERN COMPUTATIONAL TECHNIQUE

In the variational principle, the extrema of functionals are of interest just as the extrema of functions are of interest in differential calculus. A functional is a function of single or multiple functions. There are many kinds of quantities that appear as functionals in natural science. For example, the length of a curve from a point A to another point B is a functional of the curve chosen; the time ellapsed by a moving mass is a functional of the path; the energy of an electron bound by a proton is a functional of the wave function of the electron; the neutron absorption rate in a reactor is a functional of the flux distribution function. We might be interested in finding "the shortest distance from point A to B", "the shortest time for a mass moving from height A to B", or "the wave function that minimizes the energy of the electron". All of those problems are reduced to finding the argument function that makes the functional extremum. The differential or integral equation to describe a physical system are many times derived as the necessary condition for a functional to become extremum.

The importance of the variational principle in this book, however, is as a tool to derive numerical methods for a given equation. If we have an equation to solve and we are able to find the functional to which the given equation is the necessary condition for the extremum, then the solution of the given equation is found by looking for the function that makes the functional extremum. This approach is called the *direct method* and usually is applied in an approximating manner. As a primitive example, one may consider a number of "trial functions" and calculate the value of the functional for each trial function. The trial function that gives the maximum or minimum, depending upon the situation, is thought to be the best approximation among the tested trial functions. A more systematic approach is that a linear (or nonlinear) combination of known functions with undetermined coefficients is considered and the coefficients are determined so that the functional becomes the extremum within the limit of the postulated trial function.

Before electronic computers became available, the direct method of the variational principle was a powerful tool for finding an approximate solution for boundary value problem. It is well known that the first success in theoretical estimate for the chemical binding energy of a molecule was by means of the variational principle. In those days, the polynomials and step functions and orthogonal series were most favorably used as trial functions. As electronic computers became popular, the difference method replaced most roles once played by the variational principle. Nevertheless, the importance of the variational principle has not been diminished. This is because the variational principle is useful in deriving computer numerical methods. During the past 15 years, a tremendous number of papers were published on the variational principle applied to derive new numerical methods in conjunction with finite difference schemes, finite element method, piecewise polynomials, and spline functions. The beauty of the variational principle is that a numerical method can be derived on the physical principle whenever the variational principle exists.

The weighted residual method is mathematically straight-forward method to derive computational schemes and in most cases yields the same result as the variational principle. However, there are different limitations to each; thus one supplements the other. The weighted residual method has no particular physical meaning other than that the weighted average of the error is set to zero. It may be applied to any mathematical system including nonlinear equations. The Navier–Stokes equation is an example. On the other hand, the variational functional is more or less based on physical law, but may not be found for a given system of equations. In fact, no variational principle has been found for the Navier–Stokes equation.

The variational principle for linear systems cannot be discussed without the adjoint equations and the adjoint solutions, that are introduced in detail in Appendices I and II. The finite element method and the coarse-mesh rebalancing method in the later chapters are based on the variational principle, as well as on the weighted residual method. The variational principle is also essential to Monte Carlo calculations in Chapter 9.

6.2 WEIGHTED RESIDUAL METHOD

Consider a linear equation

$$L\psi(\mathbf{x}) = S(\mathbf{x}) \qquad (6.2.1)$$

where L is a linear operator including differential or integral operators, ψ is the field function, and S is the source or a prescribed function. Although $\mathbf{x}$ means simply a one-dimensional coordinate in the following sections, the definition of $\mathbf{x}$ is extended and interpreted as a generic coordinate in a multidimensional phase space. Some examples of $\mathbf{x}$ and the explicit form of L are shown in Table 6.1. We try to obtain an approximate solution in the form of a linear combination of given functions:

$$\phi(\mathbf{x}) = \sum_{i=1}^{K} a_i \eta_i(\mathbf{x}) \tag{6.2.2}$$

where a_i are undetermined coefficients and η_i are the prescribed trial functions.† We assume in this section that η_i and $L\eta_i$ are both non-singular and that η_i satisfy the boundary and initial conditions, if any. A part of these restrictions are waived in Section 6.6. Since ϕ cannot become an exact solution in general, the residual defined by

$$r(\mathbf{x}) = L\phi - S \tag{6.2.3}$$

Table 6.1   Illustration of Linear Operators

Operator	Interpretation of variable $\mathbf{x}$
$L = \dfrac{d}{dt}$	t
$L = -p(x)\dfrac{d^2}{dx^2} + q(x)$	x
$L = \dfrac{d^4}{dx^4}$	x
$L = \displaystyle\int dx' G(x' \to x)$	x
$L = \dfrac{\partial^2}{\partial x^2} + \dfrac{\partial^2}{\partial y^2} + \dfrac{\partial^2}{\partial z^2}$	(x, y, z)
$L = \Omega\nabla + \Sigma - \displaystyle\int d\Omega' \Sigma_s(\Omega' \to \Omega)$	(x, y, z, Ω)
$L = \dfrac{\partial}{\partial t} + \dfrac{\partial^2}{\partial x^2}$	(t, x)

† Prescribed trial functions may be called as *shape functions*.

cannot become zero in the entire domain. However, a weighted integral of r may be set to zero:

$$\int w(\mathbf{x})r(\mathbf{x})\,dV = \int w(\mathbf{x})[L\phi - S]\,dV = 0 \qquad (6.2.4)$$

where $w(\mathbf{x})$ is the weighting function. By doing this for K independent ws, we obtain K linear equations,

$$\sum_{i=1}^{K} \langle w_j, L\eta_i \rangle a_i = \langle w_j, S \rangle, \qquad j = 1, \ldots, K \qquad (6.2.5)$$

where $\langle\,,\,\rangle$ denotes volume integral of the product of two functions in the entire domain. Thus the undetermined coefficients are determined by solving Eq. (6.2.5).

This procedure can be equivalently stated by defining the total weighting function as

$$w(\mathbf{x}) = \sum_{j=1}^{K} b_j w_j(\mathbf{x}) \qquad (6.2.6)$$

and making $\langle w(x), r \rangle$ extremum with respect to all b_js as

$$\frac{\partial}{\partial b_j} \langle w(\mathbf{x}), r(\mathbf{x}) \rangle = 0, \qquad j = 1, 2, \ldots, K \qquad (6.2.7)$$

The number of weighting functions must equal the number of undetermined coefficients. Use of more trial functions and the matching number of weighting functions means that $r(\mathbf{x})$ is made orthogonal to more weighting functions. This leads to an increase of accuracy. The choice of the weighting functions $w_j(\mathbf{x})$ is quite arbitrary and provides various criteria in the weighted residual method.[1,2]

COLLOCATION METHOD

The weighting functions are the Dirac delta functions:

$$w_j(\mathbf{x}) = \delta(\mathbf{x} - \mathbf{x}_j) \qquad (6.2.8)$$

The differential equation is then satisfied exactly at K collocation points $\mathbf{x}_j$. How to choose the collocation points is rather arbitrary. As K increases, the residual vanishes at more and more points and presumably approaches zero throughout the entire region of interest.

REGION BALANCING METHOD

The entire region is divided into K disjunctive subdomains, V_j, $j = 1, 2, \ldots, K$, and the weighting functions are set as

$$w_j(\mathbf{x}) = 1, \quad \mathbf{x} \in V_j$$
$$= 0, \quad \mathbf{x} \notin V_j \quad (6.2.9)$$

then the differential equation is satisfied on the average in each of the K subdomains. As K increases, the size of the subdomains decreases. As a result, the differential equation is satisfied on the average in smaller and smaller subregions, and presumably the residual approaches zero everywhere.

LEAST SQUARE METHOD

This method uses $\partial r / \partial a_i = L\eta_i$ as weighting functions. This may be interpreted as the integral of the square residual $\langle r(\mathbf{x}), r(\mathbf{x}) \rangle$ being minimized with respect to the constants a_i.

GALERKIN METHOD

In this method the shape functions η_i are used as weighting functions. The Galerkin method than can be interpreted as making the residual orthogonal to members of the trial functions.

Example 6.1

Consider the equation

$$-\frac{d^2 \phi(x)}{dx^2} + 0.1\phi(x) = 1, \quad 0 \le x \le 10$$

and the boundary conditions $\phi'(0) = \phi(10) = 0$. Determine the coefficients of the trial function

$$\psi(x) = a_1 \cos \frac{\pi x}{20} + a_2 \cos \frac{3\pi x}{20} + a_3 \cos \frac{5\pi x}{20}$$

by using various types of weighting functions.

Four types of weighting functions are considered next and the coefficients determined are summarized as follows.

1. Collocation method. We need three collocation points corresponding to the number of unknowns. We choose

$$w(x) = \sum_{j=1}^{3} b_j \delta(x - x_j)$$

where

$$x_j = \frac{10}{6} + \frac{10}{3}(j-1)$$

2. Region balancing method. The three subdomains may be chosen as

$$V_1: \quad 0 \leq x < \frac{10}{3}$$

$$V_2: \frac{10}{3} \leq x < \frac{20}{3}$$

$$V_3: \frac{20}{3} \leq x \leq 10$$

3. Least-square method. The weighting function is

$$w(x) = L\left(b_1 \cos \frac{\pi x}{20} + b_2 \cos \frac{3\pi x}{20} + b_3 \cos \frac{5\pi x}{20}\right) - 1$$

4. Galerkin's method. The weighting function is

$$w(x) = b_1 \cos \frac{\pi x}{20} + b_2 \cos \frac{3\pi x}{20} + b_3 \cos \frac{5\pi x}{20}$$

```
                 Result  of  the  four  methods
```

Method	a_1	a_2	a_3
Collocation	10.33	-1.46	0.48
Region balancing	10.44	-1.61	0.67
Galerkin	10.21	-1.32	0.35
Least square	10.21	-1.32	0.35

Example 6.2

We solve the following problem approximately by using the weighted residual method.

$$\frac{d}{dt} y(t) - y(t) = 0, \qquad 0 < t < 1$$

$$y(0) = 1$$

Let us consider the approximate solution in the form

$$y(t) = 1 + a_1 t + a_2 t^2$$

where a_1 and a_2 are undetermined coefficients. The trial function given satisfies the initial condition. The coefficients are determined by two methods as follows:

1. Collocation method.

$$w(x) = b_1 \delta(t - \tfrac{1}{2}) + b_2 \delta(t - 1)$$

$$\left. \begin{array}{l} a_1 = 0.5 \\ a_2 = 1.0 \end{array} \right\}$$

2. Region balancing method.

$$V_1 : 0 \le t \le \tfrac{1}{2}, \qquad V_2 : \tfrac{1}{2} \le t \le 1$$

$$a_1 = a_2 = 0.8571$$

The weighted residual method also applies to a linear algebraic equation. We call this case the *discrete weighted residual method*. Most of the equations already developed in this section are valid for the discrete weighted residual method with minor changes in the definition of notations. The weighted residual method provides approximating methods for linear algebraic equations with a large order, especially for the linear equation that is derived as a discrete (finite difference or finite element) approximation for a boundary value problem in the form of Eq. (6.2.1).

In the remainder of this section we develop the weighted residual method for the linear equation

$$\mathbf{L}\boldsymbol{\psi} = \mathbf{S} \tag{6.2.11}$$

where $\mathbf{L}$ is a matrix of order N, $\boldsymbol{\psi}$ is the unknown vector, and $\mathbf{S}$ is a prescribed vector ($\mathbf{S} \ne \mathbf{0}$). We assume that $\mathbf{L}$ is not singular, so Eq. (6.2.11) has a unique solution. In accordance with Eq. (6.2.2) we try an approximate solution for Eq. (6.2.11) in the form

$$\boldsymbol{\phi} = \sum_{i=1}^{K} a_i \boldsymbol{\eta}_i \tag{6.2.12}$$

where a_i are the undetermined coefficients, $\boldsymbol{\eta}_i$ are the prescribed vectors of order N, and $K \ll N$. The residual vector is defined as

$$\mathbf{r} = \mathbf{L}\boldsymbol{\phi} - \mathbf{S} \tag{6.2.13}$$

Premultiplying Eq. '(6.2.13) by K independent weighting vectors, $\mathbf{w}_j$, $j = 1, 2, \ldots, K$, and setting the weighted residual to zero yield

$$\sum_{i=1}^{K} \langle \mathbf{w}_j, \mathbf{L}\boldsymbol{\eta}_i \rangle a_i = \langle \mathbf{w}_j, \mathbf{S} \rangle, \qquad j = 1, 2, \ldots, K \qquad (6.2.14)$$

where $\langle \mathbf{g}, \mathbf{h} \rangle$ means the scalar product of $\mathbf{g}$ and $\mathbf{h}$, or equivalently

$$\langle \mathbf{g}, \mathbf{h} \rangle = \mathbf{g} \cdot \mathbf{h} \qquad (6.2.15)$$

Equation (6.2.14) is the discrete version of Eq. (6.2.5).

The criteria for weighting vectors can be developed as follows:

COLLOCATION METHOD

K different numbers m are selected from $n = 1, 2, \ldots, N \gg K$ and numbered by i. The weighting vectors are then set to

$$\mathbf{w}_i = \mathbf{e}_{m(i)}, \qquad i = 1, 2, \ldots, K \qquad (6.2.16)$$

where $\mathbf{e}_m$ is a unit vector defined in Section 1.6.

REGION BALANCING METHOD

The N-dimensional vector space is partitioned into nonoverlapping subspaces, $\mathcal{D}_i$, $i = 1, 2, \ldots, K$. The element of the vector $\mathbf{w}_i$ is 1 if the element belongs to the subspace $\mathcal{D}_i$ and 0 otherwise. For example,

$$\begin{aligned} \mathbf{w}_1 &= \mathrm{col}\,[1, 1, 1, 0, 0, 0, \ldots \quad 0] \\ \mathbf{w}_2 &= \mathrm{col}\,[0, 0, 0, 1, 1, 1, 0, \ldots \quad] \\ \mathbf{w}_3 &= \mathrm{col}\,[\underbrace{0, 0, 0,}_{\mathcal{D}_1} \underbrace{0, 0, 0,}_{\mathcal{D}_2} \underbrace{1, 1,}_{\mathcal{D}_3} \ldots] \end{aligned} \qquad (6.2.17)$$

LEAST SQUARE METHOD

$$\mathbf{w}_i = \mathbf{L}\boldsymbol{\eta}_i, \qquad i = 1, 2, \ldots, K \qquad (6.2.18)$$

GALERKIN METHOD

$$\mathbf{w}_i = \boldsymbol{\eta}_i, \qquad i = 1, 2, \ldots, K \qquad (6.2.19)$$

6.3 VARIATIONAL PRINCIPLE AND EULER–LAGRANGE EQUATION

The weighted residual method is a mathematical procedure for deriving numerical schemes, but there is very little physical meaning behind the mathematical manipulation. For linear differential or integral equations describing a physical system, similar numerical schemes may be derived via the variational principle[3-9] based on the physical nature of the system.

For simplicity, we consider a specific problem: calculate the equilibrium shape of the uniform string of length l fixed at both ends.

There are an infinite number of admissible functions $y(x)$ that express possible shapes of the string, where x is the horizontal coordinate and $y(x)$ is the vertical position of the string at x. It is meant by admissible function that the string is continuous and passes the fixed points at both ends. We further require that $y(x)$ has the first and second derivatives. According to the principle of minimum potential energy,[3,8] the shape of the string is such that the total potential energy is minimized among all the admissible shapes. The total potential energy consists of the strain (internal potential) energy and the load (external potential) energy, which are given by

$$U = \frac{T}{2} \int_0^l \left(\frac{dy}{dx}\right)^2 dx \tag{6.3.1}$$

and

$$V = -\int_0^l fy\, dx \tag{6.3.2}$$

respectively, where l is the length of the string and T is the tension. The total potential energy is a function of the shape of the string and given by

$$I(y) = \int_0^l \left[\frac{T}{2}\left(\frac{dy}{dx}\right)^2 - fy\right] dx \tag{6.3.3}$$

where I is the total energy. The function of a function (or functions), such as $I(y)$, is called *functional*. We look for $y = y_0(x)$ that minimizes $I(y)$ among all admissible shapes. All the admissible y can be expressed in the form

$$y(x) = y_0(x) + \alpha \zeta(x) \tag{6.3.4}$$

where α is a parameter and ζ is an arbitrary continuous function with $\zeta(0) = \zeta(l) = 0$.

Since y_0 minimizes I by assumption, I must satisfy

$$\frac{\partial I}{\partial \alpha} = 0$$

or more explicitly,

$$0 = \int_0^l \left[T\frac{d\zeta}{dx}\frac{dy_0}{dx} - f\zeta\right] dx = -\int_0^l \left[T\frac{d^2y_0}{dx^2} + f\right]\zeta\, dx \tag{6.3.5}$$

where an integration by part and the boundary conditions for ζ are used. Since ζ in Eq. (6.3.5) is arbitrary, y_0 must satisfy

$$T\frac{d^2y_0}{dx^2}+f=0 \qquad (6.3.6)$$

Equation (6.3.6) is the equation that the shape of the string must satisfy. The equation that is derived as the necessary condition for a functional to be stationary is called Euler–Lagrange equation. Equation (6.3.6) is Euler–Lagrange equation for the functional, Eq. (6.3.3).

This derivation of Euler–Lagrange equation indicates that the solution for a given differential equation may be found by looking for the function that minimizes the corresponding functional. This approach to finding the solution of a differential equation is called *the direct method of variational principle*. In applying the direct method it is important to notice that, as a trial function y approaches the exact solution of Euler–Lagrange equation y_0, the functional approaches the true minimum (or maximum). This principle can be used to derive an approximating solution method or a numerical method for a given differential equation if the corresponding functional is found.

A primitive approach is to consider a number of trial functions for y_0 and select the one that makes I minimum as the best approximation. Another approach is to consider a linear combination of prescribed functions y_i as

$$y = \sum_{i=1}^{N} a_i y_i(x) \qquad (6.3.7)$$

and determine a_i so that I is minimized.

The minimum value of I that is found by using Eq. (6.3.7) is not the true minimum, in general. However, $\{a_i\}$ thus determined provides the best linear combination of the prescribed function, $\{y_i\}$.

Let us assume that $y_i(x)$s have the first and second derivatives and satisfy the boundary conditions $y_i(0) = y_i(l) = 0$. By substituting Eq. (6.3.6) into Eq. (6.3.3) and performing an integration by part, we have

$$I(y) = \int_0^l \left[\frac{T}{2}\left\{\sum_i a_i y_i'(x)\right\}^2 - f \sum_i a_i y_i(x) \right] dx$$

$$= -\int_0^l \sum_j a_j y_j \left[\frac{T}{2} \sum_i a_i y_i''(x) + f \right] dx \qquad (6.3.8)$$

$$= -\sum_j a_j \left[\sum_i \frac{T}{2}\langle y_i, y_i''\rangle a_i + \langle y_j, f\rangle \right]$$

where $\langle A \rangle$ is the integral of A in $[0, l]$. Since $\{y_i\}$ are prescribed functions, I is now a function of undetermined coefficients $\{a_i\}$. In order that I be minimized with respect to the coefficients, I must satisfy

$$\frac{dI}{da_i} = 0, \qquad i = 1, 2, \ldots, K \qquad (6.3.9)$$

Introducing Eq. (6.3.8) into Eq. (6.3.9) yields

$$\sum_{i=1}^{N} \langle y_j, y_i'' \rangle a_i + \langle y_j, f \rangle = 0, \qquad j = 1, 2, \ldots, K \qquad (6.3.10)$$

The undetermined coefficients a_is are determined by solving the linear algebraic equations, Eq. (6.3.10).

It should be noticed that Eq. (6.3.10) is identical to Eq. (6.2.5) of the weighted residual method using Galerkin's criterion. The approximating method for a differential equation based on the variational principle is a special case of the weighted residual method. However, the variational principle is based on physical law and has no arbitrariness in choosing the weighting function.

To be more general than the example considered, a few additional aspects must be pointed out. First, the variational principle may not be found for a given system of equations. For example, no variational principle has been found for the Navier–Stokes equations. For those systems, the weighted residual method will be useful. Second, even if the variational principle is found, the functional may not have any physical meaning. Nevertheless, the variational principle may provide useful informations about the nature of the equations for the system. Third, the minimum (or maximum) of a functional exists only if the operator of Euler–Lagrange equation is self-adjoint. When the operator is not self-adjoint, the functional becomes stationary without any extremum.

6.4 THE BASIC FUNCTIONALS AND DIRECT METHODS

In Section 6.3, Euler–Lagrange equation is obtained as the stationary condition for a functional. One should ask if it is possible to find the functional when Euler–Lagrange equation is given but the functional is not known. If this is possible, then a problem of solving a differential equation may be reduced to finding the stationary function of the functional. Historically, functionals were derived rather intuitively. The basic types of functionals in mathematical physics were established by Raussopolos,[10] Selengut,[11] Schwinger,[12] and Francis,[13] among others. All these functionals were postulated, and then the correctness was proved. By knowing the basic types of functionals and the corresponding Euler–Lagrange equation, it became possible to guess the functional for a given Euler–Lagrange equation. Since then, extension and generalization of functionals have been studied to widen the applicability of variational principles. In the meantime, systematic methods of finding functionals were investigated.[14-18] The basic types of functionals are introduced in this section.

We consider a linear equation

$$L\psi = S \tag{6.4.1}$$

where L is a linear operator including differential and integral operators, ψ is the solution, and S is the source term. The corresponding adjoint equation can be written

$$L^*\psi^* = S^* \tag{6.4.2}$$

where L^* is the adjoint operator, ψ^* is the adjoint solution, and S^* is the adjoint source. The derivation and physical meanings of adjoint equation are discussed in the Appendix I.

Roussopolos' variational functional is given by

$$J(\phi^*, \phi) = \langle S^*, \phi \rangle + \langle \phi^*, S \rangle - \langle \phi^*, L\phi \rangle \tag{6.4.3}$$

where $\langle \; \rangle$ denotes the volume integral in the entire domain. We consider a set of trial functions, ϕ_0 and ϕ_0^*, and another set of trial functions, ϕ and ϕ^*. Let us assume that the second set is slightly different from the first set and written as

$$\phi = \phi_o + \delta\phi \tag{6.4.4}$$

$$\phi^* = \phi_0^* + \delta\phi^* \tag{6.4.5}$$

where $\delta\phi$ and $\delta\phi^*$ are small compared with ϕ_0 and ϕ_0^*, respectively. We also assume that ϕ_0, ϕ_0^*, ϕ, and ϕ^* are nonsingular and satisfy the inner and outer boundary conditions and that $L\phi_0$, $L^*\phi_0^*$, $L\phi$, and $L^*\phi^*$ are nonsingular. Then the first variation† of $J(\phi^*, \phi)$ about $J(\phi_0^*, \phi_0)$ becomes

$$\delta J(\phi^*, \phi) = \langle \delta\phi^*, S - L\phi_0 \rangle + \langle \delta\phi, S^* - L^*\phi_0^* \rangle \tag{6.4.6}$$

---

† The difference between $J(\phi^*, \phi)$ and $J(\phi_0^*, \phi_0)$ may be written as

$$J(\phi^*, \phi) - J(\phi_0^*, \phi_0) = \delta J(\phi^*, \phi) + \delta^2 J(\phi^*, \phi)$$

where δJ is defined by Eq. (6.4.6) and $\delta^2 J$ is defined by

$$\delta^2 J = -\langle \delta\phi^*, L\delta\phi \rangle$$

The term δJ, which is first order in $\delta\phi^*$ and $\delta\phi$, is called the first variation, while $\delta^2\phi$, which is second order in $\delta\phi^*$ and $\delta\phi$, is called the second variation.

Therefore, in order that $\delta J = 0$ for any arbitrary $\delta\phi^*$ and $\delta\phi$, ϕ_0 and ϕ_0^* must satisfy Eqs. (6.4.1) and (6.4.2), respectively. This proves that Eq. (6.4.3) is the functional for which Eqs. (6.4.1) and (6.4.2) are the Euler–Lagrange equations.

The necessary conditions for $\delta J(\phi^*, \phi) = 0$ for any arbitrary change in ϕ^* and ϕ are called *stationary conditions*, and they are written, respectively, as

$$\frac{\delta J}{\delta\phi^*} = 0 \tag{6.4.7}$$

and

$$\frac{\delta J}{\delta\phi} = 0 \tag{6.4.8}$$

The functions that satisfy the stationary conditions are called the *stationary functions*.

The direct method that provides an approximate solution of Euler–Lagrange equation is illustrated as follows. Consider an approximate solution for Eq. (6.4.1) in the form,

$$\phi = \sum_{i=1}^{K} a_i \eta_i \tag{6.4.9}$$

where η_i are linearly independent prescribed functions that are continuous and satisfy the inner and outer boundary conditions, a_i are undetermined coefficients, and $L\eta_i$ are assumed to be nonsingular. We set the approximate solution for Eq. (6.4.2) in the similar form,

$$\phi^* = \sum_{i=1}^{K} a_i^* \eta_i^* \tag{6.4.10}$$

where we impose that the prescribed functions η_i^* satisfy the similar conditions as for η_i. By introducing the trial functions, Eqs. (6.4.9) and (6.4.10), into Eq. (6.4.3), the functional is reduced to

$$J(\phi^*, \phi) = J(a_1^*, a_2^*, \ldots, a_K^*; a_1, a_2, \ldots, a_K) \tag{6.4.11}$$

The right side of Eq. (6.4.11) is called a *reduced functional*.

Since J is now the function of unknown coefficients, the stationary point of J is sought with respect to changes of a_j^* and a_j. The stationary conditions for the reduced functional are

$$\frac{\partial J}{\partial a_j^*} = 0, \quad j = 1, 2, \ldots, K \tag{6.4.12}$$

$$\frac{\partial J}{\partial a_j} = 0, \quad j = 1, 2, \ldots, K \tag{6.4.13}$$

By using Eq. (6.4.3), Eqs. (6.4.12) and (6.4.13) become, respectively,

$$\sum_{i=1}^{K} \langle \eta_j, L\eta_i \rangle a_i = \langle \eta_j^*, S \rangle, \qquad j = 1, 2, \ldots, K \qquad (6.4.14)$$

$$\sum_{i=1}^{K} \langle \eta_j, L^*\eta_i^* \rangle a_i^* = \langle \eta_j, S^* \rangle \qquad j = 1, 2, \ldots, K \qquad (6.4.15)$$

The unknown coefficients, a_i, $i = 1 \sim K$, are determined by solving Eq. (6.4.14) and a_i^*, $i = 1 \sim K$ are determined by Eq. (6.4.15).

The stationary functions thus obtained in the forms of Eq. (6.4.9) and Eq. (6.4.10) are not generally the exact stationary functions for Eqs. (6.4.7) and (6.4.8). However, as the number of the prescribed functions in Eqs. (6.4.9) and (6.4.10) increase, those functions are presumed to approach the exact stationary functions. The direct method of the variational principle in finding an approximate solution for Eq. (6.4.1) is very similar to the weighted residual method. However, the important difference in the direct method from the weighted residual method is that the adjoint equation is considered. In the weighted residual method, there is no restriction in the choice of weighting functions. In the variational principle the adjoint trial function plays the role of the weighting function. Since the adjoint trial functions η_i^* are chosen on the basis of physical knowledge for the adjoint equation, their choice is not as arbitrary as in the weighted residual method.

Another important aspect of the variational principle is that Eq. (6.4.3) gives an estimate for $\langle S^*, \psi \rangle$ or $\langle \psi^*, S \rangle$ where ϕ^* and ϕ are trial functions, while ψ and ψ^* are the exact solutions. Furthermore, the error in this estimation is second order, while a direct estimation in the form of $\langle S^*, \phi \rangle$ or $\langle \phi^*, S \rangle$ involves a first order error. To prove this, we express trial functions in the forms

$$\phi = \psi + \delta\phi$$
$$\phi^* = \psi^* + \delta\phi^* \qquad (6.4.16)$$

where $\delta\phi$ and $\delta\phi^*$ are the errors in the trial functions. Introducing Eq. (6.4.16) into Eq. (6.4.3), we have

$$J(\phi^*, \phi) = \langle S^*, \psi \rangle - \langle \delta\phi^*, L\delta\phi \rangle = \langle \psi^*, S \rangle - \langle \delta\phi^*, L\delta\phi \rangle \quad (6.4.17)$$

The second term on the right side is the error that is second order with respect to the errors in ϕ^* and ϕ.

Example 6.3

Consider the equation,

$$-\frac{d^2}{dx^2}\phi(x) + \phi(x) = 1 \qquad (6.4.18)$$

$$\phi'(0) = \phi(1) = 0 \qquad (6.4.19)$$

and suppose we estimate

$$\int_0^{0.5} \phi(x)\, dx \tag{6.4.20}$$

without solving the equation exactly. We write the adjoint problem as

$$-\frac{d^2}{dx^2}\phi^*(x) + \phi^*(x) = S^*(x) \tag{6.4.21}$$

$$S^*(x) = \begin{cases} 1, & 0 < x < 0.5 \\ 0, & 0.5 < x < 1 \end{cases} \tag{6.4.22}$$

$$\phi^{*\prime}(0) = \phi^*(1) = 0$$

where S^* is chosen so that $J(\phi^*, \phi)$ becomes equal to Eq. (6.4.20) when ϕ^* and ϕ are exact.

If we select the trial functions as

$$\left. \begin{array}{l} \phi(x) = a \sin \pi x \\ \phi^*(x) = b \sin \pi x \end{array} \right\} \tag{6.4.23}$$

where a and b are unknown coefficients, Eq. (6.4.3) becomes

$$J(a, b) = \frac{a}{\pi} + \frac{2b}{\pi} - \frac{(\pi^2 + 1)ab}{2} \tag{6.4.24}$$

Using the stationary conditions,

$$\frac{\partial J}{\partial a} = 0, \qquad \frac{\partial J}{\partial b} = 0 \tag{6.4.25}$$

we obtain $a = 0.1171$ and $b = 0.05855$. Introducing those values into Eq. (6.4.24), an estimate for Eq. (6.4.20) is obtained as

$$\int_0^{0.5} \phi(x)\, dx \approx J(a, b) = 0.03727 \tag{6.4.26}$$

The exact value for Eq. (6.4.20) is 0.03789, so that the error of the estimation is -1.6%.

When $L^* = L$ and $S^* = S$ and $\phi^* = \phi$, Eq. (6.4.3) reduces to

$$J = 2\langle S, \phi \rangle - \langle \phi, L\phi \rangle \tag{6.4.27}$$

The functional used by Schwinger and Francis may be written[12,13]

$$J(\phi^*, \phi) = \frac{\langle \phi^*, S \rangle \langle S^*, \phi \rangle}{\langle \phi^*, L \rangle} \tag{6.4.28}$$

The equivalence of this functional to Eq. (6.4.3) can be seen as follows. If the trial functions ϕ^* and ϕ in Eq. (6.4.3) are replaced by $c^*\phi^*$ and $c\phi$, respectively, where c^* and c are undetermined constants, Eq. (6.4.3) becomes

$$J(c^*\phi^*, c\phi) = c\langle S^*, \phi\rangle + c^*\langle \phi^*, S\rangle - c^*c\langle \phi^*, L\phi\rangle \qquad (6.4.29)$$

The constants c^* and c are determined by the stationary conditions with respect to variations of c^* and c:

$$\frac{\partial J}{\partial c^*} = 0 \Rightarrow c = \frac{\langle \phi^*, S\rangle}{\langle \phi^*, L\phi\rangle} \qquad (6.4.30)$$

$$\frac{\partial J}{\partial c} = 0 \Rightarrow c^* = \frac{\langle S^*, \phi\rangle}{\langle \phi^*, L\phi\rangle} \qquad (6.4.31)$$

Eliminating c^* and c from Eq. (6.4.29) yields Eq. (6.4.28), which is free from normalization of ϕ^* and ϕ. Applications of the functionals to various physics problems are described in reference 18.

6.5 VARIATIONAL PRINCIPLE FOR EIGENVALUE PROBLEMS

The variational principle for an eigenvalue problem may be considered as a special case for the inhomogeneous problems. In order to apply the variational principle discussed in the previous section, we first set the source terms in Eqs. (6.4.1) and (6.4.2) to zero, and secondly include an undetermined parameter λ in L to assure the existence of solutions. An eigenvalue problem and its adjoint equation are written

$$L\psi \equiv (A - \lambda F)\psi = 0 \qquad (6.5.1)$$

$$L^*\psi^* \equiv (A^* - \lambda F^*)\psi^* = 0 \qquad (6.5.2)$$

where A and F are linear operators including differential or integral operators, λ is the undetermined parameter called the eigenvalue, and ψ and ψ^* are normal and adjoint solutions, respectively. Even though F and F^* can include differential or integral operators, they are usually functions.

If we apply Eq. (6.4.3) to Eqs. (6.5.1) and (6.5.2), the functional is reduced to

$$J(\phi^*, \phi) = \langle \phi^*, (A - \lambda F)\phi\rangle \qquad (6.5.3)$$

where the sign of J is altered for convenience. The first variation of J is

$$\delta J = \langle \delta\phi^*, (A - \lambda F)\phi\rangle + \langle (A^* - \lambda F^*)\phi^*, \delta\phi\rangle \qquad (6.5.4)$$

where we assumed that ϕ and ϕ^* both satisfy inner and outer boundary conditions if any and that there is no singularity. In order for the first variation to become zero for arbitrary variations of $\delta\phi^*$ and $\delta\phi$, Eq. (6.5.1) and (6.5.2) must be satisfied by ϕ and ϕ^*, respectively. Therefore, finding the stationary functions is equivalent to solving Eqs. (6.5.1) and (6.5.2).

The functional J becomes zero if ϕ^*, ϕ, and λ are exact. By forcing Eq. (6.5.3) to be zero with a given set of trial functions and solving the resulting equation for λ, we obtain an estimate for λ as

$$\lambda = \frac{\langle \phi^*, A\phi \rangle}{\langle \phi^*, F\phi \rangle} \qquad \text{Rayleigh quotient} \qquad (6.5.5)$$

It can be easily seen that if $\phi^* = \psi^*$ or $\phi = \psi$ then λ becomes the exact eigenvalue. The error in this estimation with a given set of trial functions may be evaluated as follows. We first express the trial function and the estimated eigenvalue as

$$\phi = \psi + \delta\phi$$

$$\phi^* = \psi^* + \delta\phi^* \qquad (6.5.6)$$

$$\lambda = \lambda_0 + \delta\lambda$$

where λ_0 is the exact eigenvalue of Eqs. (6.5.1) and (6.5.2). Introducing Eq. (6.5.6) into (6.5.5) yields

$$
\begin{aligned}
\lambda &= \frac{\langle \psi^*, A\psi \rangle + \langle \delta\phi^*, A\psi \rangle + \langle \delta\phi, A^*\psi^* \rangle + \langle \delta\phi^*, A\delta\phi \rangle}{\langle \psi^*, F\psi \rangle + \langle \delta\phi, F\psi \rangle + \langle \delta\phi, F^*\psi^* \rangle + \langle \delta\phi^*, F\delta\phi \rangle} \\[2mm]
&= \frac{\lambda_0[\langle \psi^*, F\psi \rangle + \langle \delta\phi^*, F\psi \rangle + \langle \delta\phi, F^*\psi^* \rangle] + \langle \delta\phi^*, A\delta\phi \rangle}{[\langle \psi^*, F\psi \rangle + \langle \delta\phi^*, F\psi \rangle + \langle \delta\phi, F^*\psi^* \rangle] + \langle \delta\phi^*, F\delta\phi \rangle} \qquad (6.5.8) \\[2mm]
&\approx \lambda_0 + \frac{\langle \delta\phi^*, (A - \lambda_0 F)\delta\phi \rangle}{[\langle \psi^*, F\psi \rangle + \langle \delta\phi^*, F\psi \rangle + \langle \delta\phi, F^*\psi^* \rangle]}
\end{aligned}
$$

In the above equation, the second term is the error, and it is second order with respect to the errors in the trial functions.

If Eq. (6.5.1) is self-adjoint ($L = L^*$ and $\psi = \psi^*$), the fundamental eigenvalue is equal to the minimum value of Eq. (6.5.5) among all admissible trial functions. Therefore, the minimum value of $J(\phi, \phi)$ among the actually tested trial functions gives the estimate for the upper bound of the fundamental eigenvalue. This principle was extensively used in estimating the upper bound of the energy level of an electron of an atom or molecule in the fundamental state. If the operator L is not self-adjoint, the eigenvalue estimated by Eq. (6.5.5) is not upper bound nor lower bound.

Example 6.4

Find the fundamental eigenvalue approximately for the self-adjoint eigenvalue problem given by

$$-\frac{d^2}{dx^2}\phi(x)+0.1\phi(x)=\lambda\phi(x) \tag{6.5.9}$$

where

$$0\le x\le 10$$

$$\phi(0)=\phi(10)=0$$

A trial function satisfying the boundary condition may be written

$$\phi^*(x)=\phi(x)=x(x-10) \tag{6.5.10}$$

There is no parameter in the trial function, so the eigenvalue estimation is given directly by introducing Eq. (6.5.10) into Eq. (6.5.5):

$$\lambda=J=\frac{\int_0^{10} x(x-10)(-2+0.1x(x-10))\,dx}{\int_0^{10} x^2(x-10)^2\,dx}=0.20000 \tag{6.5.11}$$

The exact solution for the fundamental eigenfunction and the corresponding eigenvalue for Eq. (6.5.9) are, respectively,

$$\psi(x)=\sin\left(\frac{\pi x}{10}\right) \tag{6.5.12}$$

$$\lambda_0=0.1+\left(\frac{\pi}{10}\right)^2=0.19869 \tag{6.5.13}$$

so the error of Eq. (6.5.11) is 0.66%.

Example 6.5

Estimate the eigenvalue of Eq. (6.5.9) by using the trial function,

$$\phi(x)=\phi^*(x)=a_1\eta_1(x)+a_2\eta_2(x) \tag{6.5.14}$$

where

$$\eta_1(x)=x(x-10)$$

$$\eta_2(x)=x(x-3)(x-7)(x-10)$$

We introduce Eq. (6.5.14) into Eq. (6.5.5). The stationary conditions are

$$\frac{\partial J}{\partial a_1}=\frac{\partial J}{\partial a_2}=0 \tag{6.5.15}$$

which yield a homogeneous set of equations for a_1 and a_2:

$$666.625a_1 + 189.671a_2 = \lambda(3333.34a_1 - 1428.58a_2)$$
$$189.671a_1 + 63122.2a_2 = \lambda(-1428.58a_1 + 57301.6a_2)$$
(6.5.16)

The eigenvalues of Eq. (6.5.16) are obtained as

$$\lambda = 0.19868, \qquad \lambda = 1.11981 \tag{6.5.17}$$

The first λ is the fundamental eigenvalue corresponding to Eq. (6.5.13). Notice that the error is zero with five significant figures. The second λ represents an approximation to the next symmetric eigenfunction of Eq. (6.5.9), namely,

$$\phi = \sin\frac{3\pi x}{10} \tag{6.5.18}$$

The exact eigenvalue for Eq. (6.5.18) is 0.98826, so the error of the second λ in Eq. (6.5.17) is 13%.

6.6 BOUNDARY CONDITIONS

In the previous sections we impose rather strict restrictions on the trial functions: the admissible trial functions satisfy the inner and outer boundary conditions and are nonsingular when operated by L. The purpose of this section is to relax those restrictions and include the trial functions that do not satisfy the previous restrictions in the admissible trial functions. We first consider the variational principle and then the weighted residual method.

Let us consider the functional,

$$J(\phi) = \int_0^H [p(x)\phi'^2 + q(x)\phi^2 - 2S(x)\phi(x)] \, dx \tag{6.6.1}$$

where continuities of $\phi(x)$, $p\phi'(x)$ are assumed, but $\phi(0)$ and $\phi(L)$ can take any value. Such problems with no specifications for boundary conditions are called *free-boundary problems*.[3] For the stationary of Eq. (6.6.1), the first variation of J is set to zero:

$$\delta J = -[2p\phi'\delta\phi]_{x=0} + [2p\phi'\delta\phi]_{x=H}$$
$$+ 2\int_0^H \left[-\frac{d}{dx}p\frac{d}{dx}\phi + q\phi - S\right]\delta\phi \, dx = 0 \tag{6.6.2}$$

Since $\delta\phi$ is arbitrary at $x = 0$, $x = H$, and $0 < x < H$, the stationary of J requires

$$\phi'(0) = 0$$
$$\phi'(H) = 0 \qquad (6.6.3)$$

and

$$-\frac{d}{dx}p\frac{d}{dx}\phi + q\phi - S = 0 \qquad (6.6.4)$$

Reversing this procedure, the solution to Eq. (6.6.4) with Eq. (6.6.3) can be obtained by finding the function that makes Eq. (6.6.1) stationary among all the admissible trial functions unconstrained at $x = 0$ and $x = H$. The direct method is developed by introducing into Eq. (6.6.1) a trial function in the form

$$\phi(x) = \sum_{i=1}^{N} a_i \eta_i(x) \qquad (6.6.5)$$

where η_i is a prescribed function unconstrained at $x = 0$ and $x = H$: $\eta_i(0)$ and $\eta_i(H)$ can take any value. The rest of the procedure to determine a_i is the same as described in Section 6.4. In general, the stationary function in the form of Eq. (6.6.5) does not satisfy Eq. (6.6.3) exactly. However, this is similar to the situation where Eq. (6.6.5) will not exactly satisfy Eq. (6.6.4). As the number of prescribed trial functions in Eq. (6.6.5) increases, the stationary function will become closer to the exact solution of Eqs. (6.6.4) and (6.6.3).

Let us next examine the functional:

$$J(\phi) = \int_0^H [p\phi'^2 + q\phi^2 - 2S\phi]\, dx + \gamma_L \phi^2(0) + \gamma_R \phi^2(H) \qquad (6.6.6)$$

where γ_L and γ_R are nonnegative constants. The necessary conditions for $\delta J = 0$ of this functional are given by Eq. (6.6.4) and

$$p(0)\phi'(0) - \gamma_L \phi(0) = 0$$
$$p(H)\phi'(H) + \gamma_R \phi(H) = 0 \qquad (6.6.7)$$

Therefore, the trial function unconstrained at $x = 0$ and $x = H$ that makes Eq. (6.6.6) stationary is the solution of Eqs. (6.6.4) and (6.6.7). The direct method can be developed by introducing Eq. (6.6.5) into Eq. (6.6.6). If γ_L and γ_R are set to ∞, then Eq. (6.6.7) become the Dirichlet boundary conditions, $\phi(0) = \phi(H) = 0$. If $\gamma_L = \gamma_R = 0$, Eq. (6.6.6) is reduced to Eq. (6.6.1).

Example 6.6

Consider an eigenvalue problem

$$-\frac{d^2}{dx^2}\psi(x) = \lambda\psi(x), \qquad 0 \le x \le 1 \tag{6.6.8}$$

where the boundary conditions are

$$\psi'(0) = \gamma\psi(0) \quad \text{Left boundary condition} \tag{6.6.9}$$

$$\psi(1) = 0 \qquad \text{Right boundary condition} \tag{6.6.10}$$

By using the direct method, find an approximate fundamental eigenvalue with the trial function in the form:

$$\phi = \sum_{n=1}^{I} a_n \cos\frac{(2n-1)\pi x}{2} \tag{6.6.11}$$

This trial function satisfies the right boundary condition. The functional that admits Eq. (6.6.11) as a trial function is written as

$$J(\phi) = \int_0^1 [\phi'^2 - \lambda\phi^2]\,dx + \gamma\phi^2(0) \tag{6.6.12}$$

Introducing Eq. (6.6.11) into Eq. (6.6.12) and using the stationary conditions, we obtain

$$\left\{\left(\frac{(2n-1)\pi}{2}\right)^2 - \lambda\right\}A_n a_n + \gamma\sum_{m=1}^{I} a_m = 0 \tag{6.6.13}$$

where $n = 1, 2, \ldots, I$ and

$$A_n = \int_0^1 \left\{\cos\frac{(2n-1)\pi x}{2}\right\}^2 dx = 0.5 \tag{6.6.14}$$

The fundamental eigenvalues of Eq. (6.6.13) for $\gamma = 0.1$ and 0.5 calculated with various I are listed and compared with the exact values below:

	$\gamma = 0.1$		$\gamma = 0.5$	
I	λ	error (%)	λ	error (%)
1	2.6674	0.15	3.4674	2.8
2	2.6654	0.07	3.4169	1.3
3	2.6647	0.05	3.4017	0.9
5	2.6642	0.03	3.3900	0.6
10	2.6638	0.02	3.3814	0.3
exact	2.6634	0	3.3710	0

We have assumed that the trial functions satisfy continuity of ϕ and $p\phi'$. It can be shown, however, that the continuity of $p\phi'$ is one of the necessary conditions for Eq. (6.6.1) to become stationary, and accordingly the requirement of continuty of $p\phi'$ can be waived. Suppose $p\phi'$ is not continuous at $x = x_1$ $(0 < x_1 < H)$, then the first variation of Eq. (6.6.1) becomes

$$\delta J = -[2p\phi'\delta\phi]_{x=0} + [2p\phi'\delta\phi]_{x=x_1-0} - [2p\phi'\delta\phi]_{x=x_1+0}$$
$$+ [2p\phi'\delta\phi]_{x=H} + 2\int_0^{x_1}\left[-\frac{d}{dx}p\frac{d}{dx}\phi + q\phi - S\right]\delta\phi\,dx$$
$$+ 2\int_{x_1}^H\left[-\frac{d}{dx}p\frac{d}{dx}\phi + q\phi - S\right]\delta\phi\,dx \tag{6.6.15}$$

where x_1-0 and x_1+0 denote the left and right sides of x_1, respectively. We obtain Eq. (6.6.4) and Eq. (6.6.3), as before. In addition, in order that $\delta J = 0$ for an arbitrary change of $\delta\phi$ at $x = x_1$ (ϕ and $\delta\phi$ are continuous at x_1), ϕ must satisfy

$$[2p\phi']_{x_1-0} = [2p\phi']_{x_1+0} \tag{6.6.16}$$

Thus it is proven that the continuity of $p\phi'$ is one of the necessary conditions for stationary of Eq. (6.6.1).

Example 6.7

Consider Eq. (6.6.4) and the boundary condition, Eq. (6.6.3). We assume the whole region is divided into two regions, $[0, x_1]$ where $p(x) = p_1$, and $[x_1, H]$ where $p(x) = p_2$; $p_1 \neq p_2$. The trial function in the form

$$\phi(x) = \sum_{n=1}^N a_n \cos\frac{n\pi x}{H} \tag{6.6.17}$$

does not satisfy the continuity of $p\phi'$ across the boundary between the two regions. Nevertheless, Eq. (6.6.17) is an admissible trial function for Eq. (6.6.1).

Example 6.8

Consider Eq. (6.6.4) and the boundary conditions given by Eq. (6.6.7). We assume $p(x) = $ const in the entire region. The following piecewise linear function is an admissible trial function to Eq. (6.6.6):

$$\phi(x) = \sum_{i=1}^N a_i\eta_i(x) \tag{6.6.18}$$

where

$$\eta_i(x) = \frac{x_{i+1} - x}{x_{i+1} - x_i}, \quad \text{if } x_i \leq x \leq x_{i+1}$$

$$= \frac{x - x_{i-1}}{x_i - x_{i-1}}, \quad \text{if } x_{i-1} \leq x \leq x_i \qquad (6.6.19)$$

$$= 0, \qquad \text{if } x < x_{i-1} \quad \text{or} \quad x_{i+1} < x$$

and where the x_is are grid points numbered sequentially from left to right ($x_1 = 0$, $x_N = H$). The first derivative of Eq. (6.6.18) is constant in each interval of $x_{i-1} < x < x_i$, but discontinuous across grid points x_i.

We now discuss the treatment of the boundary conditions in the weighted residual method. Suppose we seek an approximate solution of Eq. (6.6.4) that is subject to the boundary conditions, Eq. (6.6.7), by using the weighted residual method. We assume $p = \text{const}$ and ϕ has the second derivative. The weighted residual for Eq. (6.6.4) may be written

$$WR = \int_0^H w \left[-\frac{d}{dx} p \frac{d}{dx} \phi + q\phi - S \right] dx = 0 \qquad (6.6.20)$$

where w is assumed to be continuous and to have the first derivative. Integrating the first term of Eq. (6.6.20) by part yields

$$WR = \int_0^H [w'p\phi' + wq\phi - wS] + w(0)p\phi'(0) - w(H)p\phi'(H) = 0$$

$$(6.6.21)$$

Since the boundary conditions are given by Eq. (6.6.7), we introduce them into Eq. (6.6.21) and obtain

$$WR = \int_0^H [w'p\phi' + wq\phi - wS] \, dx + \gamma_L w(0)\phi(0) + \gamma_R w(H)\phi(H) = 0$$

$$(6.6.22)$$

Thus the outer boundary conditions are incorporated into the weighted residual. Compare Eq. (6.6.22) with Eq. (6.6.6).

Finally, we study the inner boundary conditions in the weighted residual method. Let us consider the same problem as Eq. (6.6.20) except that the trial function potentially has a discontinuity in $p\phi'$ at $x = x_1$. So, avoiding the integration across x_1, the weighted residual may be set as

$$WR = \int_0^{x_1} w \left[-\frac{d}{dx}(p\phi') + q\phi - S \right] dx$$

$$+ \int_{x_1}^H w \left[-\frac{d}{dx}(p\phi') + q\phi - S \right] dx = 0 \qquad (6.6.23)$$

Upon integrating the first terms in the brackets in Eq. (6.6.23) by part, we obtain

$$WR = \int_0^H [w'p\phi' + wq\phi - wS]\, dx + \gamma_1 w(0)\phi(0) + \gamma_2 w(H)p(H)$$

$$+ w(x_1)[-p(x_1-0)\phi'(x_1-0) + p(x_1+0)\phi'(x_1+0)] = 0 \qquad (6.6.24)$$

where the outer boundary conditions are used. Since the continuity of $p\phi'$ across x_1 requires the third term of Eq. (6.6.24) to vanish, we set the third term to zero. Thus Eq. (6.6.24) is reduced to Eq. (6.6.22). Because x_1 can be set to any value within the entire region, we conclude that the trial functions that do not satisfy the outer boundary conditions and continuity of $p\phi'$ are admissible to Eq. (6.6.21). It should be emphasized that, in order to admit discontinuity of $p\phi'$, w must be continuous and have the first derivative. In this sense, the weighting functions for Eq. (6.6.24) are no more so arbitrary as originally introduced in Section 6.1.

Let us reexamine Eq. (6.6.22), keeping in mind that $p\phi'$ can be discontinuous at x_1. Integrating the first term in the bracket of Eq. (6.6.22) by part, we obtain

$$WR = \int_0^{x_1} w\left[-\frac{d}{dx}(p\phi') + q\phi - S\right] dx$$

$$+ \int_{x_1}^H w\left[-\frac{d}{dx}(p\phi') + q\phi - S\right] dx + w(0)[-p(0)\phi'(0) + \gamma_L\phi(0)]$$

$$+ w(H)[p(H)\phi'(H) + \gamma_R\phi(H)] + w(x_1)[p(x_1-0)\phi'(x_1-0) \qquad (6.6.25)$$

$$- p(x_1+0)\phi'(x_1+0)]$$

We now observe that the third term is the product of the weighting function and the residual of the left boundary condition, the fourth term is the same for the right boundary, and the last term is the product of the weighting function and the residual of the continuity equation for $p\phi'$.

In summary, we have been able in the direct method of variational principle and the weighted residual method to wave the requirement for the trial function to satisfy the outer boundary conditions and continuity of $p\phi'$. Waving the requirement for continuity of ϕ is an interesting subject. In fact, many papers have been written on this subject.[17,18,22] Unfortunately, there are two major drawbacks in this approach: (a) while the number of admissible types of trial functions is increased, a significant error is introduced by the discontinuity of ϕ; (b) various mathematical difficulties and anomalies are encountered because of discontinuity of ϕ.[23]

6.7 CHOICE OF TRIAL FUNCTIONS AND USE OF PIECEWISE POLYNOMIALS

Selecting appropriate trial functions is a major key in success with the direct method or the weighted residual method. Although the trial function can include nonlinear parameters, this approach is not desirable unless the number of parameters is small. So we consider the linear combination of prescribed functions in the form of Eq. (6.2.2).

Some ideas for choosing prescribed functions are listed as follows:

a. Solutions of similar problems, which are already available or easier to solve than the given problem, may be used.
b. Orthogonal function series such as Fourier series, Bessel functions, Chebyshev polynomials, and so forth. (These are complicated and costly for two-dimensional problems.)
c. Eigenfunctions of a related eigenvalue problem. (Calculations of eigenfunctions can be costly unless the geometry is very simple.)
d. Polynomial interpolation formula. (A high-order polynomial has an undesirable nature as discussed in Section 1.1.)
e. Piecewise polynomial functions. (See Eq. (1.2.6) or Eq. (6.6.19) for a one-dimensional example; these are useful for multidimensional problems with a complicated geometry.)

Approaches (a) through (d) work reasonably well if the problem is one-dimensional and the number of linear combinations (or undetermined coefficients) is not large. For a multidimensional problem, (a) and (e) provide realistic approximation methods. The perturbation theory is closely related to (a). The synthesis method described in Section 6.9 is also based on (a). Choice (e) is the basis for the finite element method of Chapter 7 and provides the most versatile and systematic approach for multidimensional problems with complicated geometries. One of the major advantages with the piecewise polynomial is that each of the prescribed functions has nonzero values only in a small subregion, so that integrations to obtain the weighted residual becomes simple and fast, while integration with other choices should cover the whole domain in general.

In the remainder of this section, we apply the piecewise linear function to Eq. (6.6.4) with the boundary conditions given by Eq. (6.6.7). The weighted residual equation, Eq. (6.6.22), admits the piecewise linear function defined by Eq. (6.6.18) without any constraints to $\phi(0)$ and $\phi(H)$. Here we assume $p(x)$, $q(x)$, and $S(x)$ are constant in each mesh interval:

$$p(x) = p_i, \qquad q(x) = q_i, \quad S(x) = S_i \quad \text{for } x_{i-1} < x < x_i \qquad (6.7.1)$$

In accordance with the Galerkin's criterion, the weighting functions are set to

$$w(x) = \eta_j(x), \, j = 1, 2, \ldots, N \qquad (6.7.2)$$

By introducing Eqs. (6.6.18) and (6.7.2) into Eq. (6.6.22), we obtain

$$\int_0^H \left[\left(\frac{d}{dx} \eta_j \right) p(x) \left(\frac{d}{dx} \sum_{i=1}^N a_i \eta_i \right) \right.$$

$$\left. + q(x) \eta_j \sum_{i=1}^N a_i \eta_i - \eta_j S \right] dx + a_1 \gamma_L p(0) \eta_1^2(0) \qquad (6.7.3)$$

$$- a_N \gamma_R p(H) \eta_N^2(H) = 0$$

Notice the following relations:

$$\int_0^H p \eta_j' \eta_i' \, dx = \frac{p_j}{x_j - x_{j-1}} + \frac{p_{j+1}}{x_{j+1} - x_j}, \quad \text{if } i = j$$

$$= -\frac{p_{j+1}}{x_{j+1} - x_j}, \qquad \text{if } i = j+1$$

$$= -\frac{p_j}{x_j - x_{j-1}}, \qquad \text{if } i = j-1 \qquad (6.7.4)$$

$$= 0 \qquad \qquad \text{for all other cases}$$

$$\int_0^H q \eta_j \eta_i \, dx = \tfrac{1}{3}[q_j(x_j - x_{j-1}) + q_{j+1}(x_{j+1} - x_j)], \quad \text{if } i = j$$

$$= \tfrac{1}{6} q_{j+1}(x_{j+1} - x_j), \qquad \text{if } i = j+1$$

$$= \tfrac{1}{6} q_j(x_j - x_{j-1}), \qquad \text{if } i = j-1 \qquad (6.7.5)$$

$$\int_0^H \eta_j S \, dx = \tfrac{1}{2}[S_j(x_j - x_{j-1}) + S_{j+1}(x_{j+1} - x_j)] \qquad (6.7.6)$$

where p_i, q_i, and S_i are set to zero if $i \leq 1$ or $i > N$. By applying the above equations to Eq. (6.7.3), we have

$$
\frac{p_2}{x_2-x_1}a_1-\frac{p_2}{x_2-x_1}a_2+\left[\frac{q_2}{3}(x_2-x_1)+\gamma_L p_2\right]a_1
$$

$$
+\frac{q_2}{6}(x_2-x_1)a_2=\frac{1}{2}S_2(x_2-x_1) \tag{6.7.7}
$$

$$
-\frac{p_j}{x_j+x_{j+1}}a_{j-1}+\left[\frac{p_j}{x_j-x_{j-1}}+\frac{p_{j+1}}{x_{j+1}-x_j}\right]a_j-\frac{p_{j+1}}{x_{j+1}-x_j}a_{j+1}
$$

$$
+\frac{q_i}{6}(x_j-x_{j-1})a_{j-1}+\frac{1}{3}[q_i(x_j-x_{j-1})+q_{i+1}(x_{j+1}-x_j)]a_j \tag{6.7.8}
$$

$$
+\frac{q_{j+1}}{6}(x_{j+1}-x_j)a_{j+1}=\frac{1}{2}[s_j(x_j-x_{j-1})+S_{j+1}(x_{j+1}-x_j)], \qquad 2\le j\le N-1
$$

$$
-\frac{p_N}{x_N-x_{N-1}}a_{N-1}+\frac{p_N}{x_N-x_{N-1}}a_N+\frac{q_N}{6}(x_N-x_{N-1})a_{N-1}
$$

$$
+\left[\frac{q_N}{3}(x_N-x_{N-1})+\gamma_R p_N\right]a_N=\frac{1}{2}S_N(x_N-x_{N-1}) \tag{6.7.9}
$$

The set of Eqs. (6.7.7) through (6.7.9) consists of N linear equations to determine N undetermined coefficients, a_i, $i=1, 2,\ldots,N$. It may be expressed in the matrix form as Eq. (1.7.7).

It is interesting to compare Eq. (6.7.8) with the finite difference equations developed in Section 2.2. The point difference scheme applied to Eq. (6.6.4) becomes

$$
-\frac{p_j}{x_j-x_{j-1}}\phi_{j-1}+\left[\frac{p_j}{x_j-x_{j-1}}+\frac{p_{j+1}}{x_{j+1}-x_j}\right]\phi_j-\frac{p_{j+1}}{x_{j+1}-x_j}\phi_{j+1}
$$

$$
+\tfrac{1}{2}[q_j(x_j-x_{j-1})+q_{i+1}(x_{j+1}-x_j)]\phi_j \tag{6.7.10}
$$

$$
=\tfrac{1}{2}[s_j(x_j-x_{j-1})+s_{j+1}(x_{j+1}-x_j)]
$$

The difference between Eqs. (6.7.10) and (6.7.8) is seen in the treatment of $q\phi$. The similar observation can be made for Eqs. (6.7.7) and (6.7.9). As mentioned in Section 6.6, the boundary condition in this problem can become a Dirichlet boundary by setting $\gamma=\infty$, or a Neumann boundary by setting $\gamma=0$. Practically a sufficiently large but finite number for γ should be used in case of Dirichlet boundary because computers cannot deal with an infinity. The application of pricewise polynomials for two-dimensional problems are discussed in detail in Chapter 7 as the finite element method.

6.8 SEMIDIRECT METHOD AND FLUX SYNTHESIS

Any numerical calculations for a multidimensional partial differential (or integral) equation is expensive because of a large core memory required and long computing time. As a method of cutting down computing time, the weighted residual method may be used to reduce the multidimensional equation to a set of coupled ordinary differential equations. Then, the numerical schemes to solve ordinary differential (one-dimensional) equations may be used. This is much less expensive than applying a standard numerical scheme directly to the given multidimensional equation. In this section the general idea of the semidirect method is introduced and then applied to a three-dimensional neutron diffusion problem.

Suppose a three-dimensional equation written as

$$L\psi(x, y, z) = S(x, y, z) \tag{6.8.1}$$

The trial function and the total weighting function are written as

$$\phi = \sum_{l=1}^{K} Z_l(z)R_l(x, y) \tag{6.8.2}$$

$$w = \sum_{k=1}^{K} Z_k^*(z)R_k^*(x, y) \tag{6.8.3}$$

where R_l and R_k^* are prescribed functions of x and y; and Z_l and Z_k^* are undetermined functions of z. The weighted residual for Eq. (6.8.1) is written and set to zero as

$$WR = \int Z_k^*(z) \left[\sum_{l=1}^{K} \langle L \rangle_{kl} Z_l(z) - \langle S \rangle_k \right] dz = 0, \qquad k = 1, 2, \ldots, K \tag{6.8.4}$$

where $\langle L \rangle_{kl}$ is a linear operator on the z coordinate and $\langle S \rangle_k$ is a function of z defined by

$$\langle L \rangle_{kl} \equiv \int \int R_k^*(x, y) L R_l(x, y) \, dx \, dy \tag{6.8.5}$$

and

$$\langle S \rangle_k \equiv \int \int R_k^*(x, y) S(x, y) \, dx \, dy \tag{6.8.6}$$

respectively. Therefore, $WR = 0$ for any arbitrary change in $Z_k^*(z)$ requires that the following set of equations be satisfied:

$$\sum_{l=1}^{K} \langle L \rangle_{kl} Z_l(z) = \langle S \rangle_k, \qquad k = 1, 2, \ldots, K \tag{6.8.7}$$

If L is a partial differential operator, Eq. (6.8.7) appears to be coupled ordinary differential equations. The present method is designated as the *semidirect method*.

The synthesis method based on the semidirect method was originally proposed by Kaplan.[24] Since then, many alternative approaches were tried and published competitively by various authors working for neutron reactor physics and design.[22,23] For simplicity, we consider the one-group three-dimensional neutron diffusion equation:

$$-\nabla D \nabla \psi(x, y, z) + \Sigma_a \psi(x, y, z) = \lambda \nu \Sigma_f \psi(x, y, z) \qquad (6.8.8)$$

where D, Σ_a, $\nu\Sigma_f$ are the space-dependent coefficients and λ is the eigenvalue. Suppose a reactor that consists of the upper region $\mathscr{D}_1$ and the lower region $\mathscr{D}_2$ in each of which the material or structural properties do not change vertically. For example, a reactor whose control rods are all at the same position has such a configuration if we neglect the upper and lower reflectors. Let us consider the two-dimensional flux distribution at some height $z = z_0$ that is not close to the boundary between the two regions. One can imagine that, if z_0 is in the upper region, then $\psi(x, y, z_0)$ is similar to $H_1(x, y)$ which is the flux distribution for the infinite reactor having the same cross section as the upper region, and if z_0 is in the lower region, then $\psi(x, y, z_0)$ is similar to $H_2(x, y)$ for the infinite reactor having the same cross section as the lower region. The calculations for $H_1(x, y)$ and $H_2(x, y)$ are much easier and cheaper than for $\psi(x, y, z)$ because they are two-dimensional problems.

This observation leads us to a trial function given by

$$\phi(x, y, z) = H_1(x, y)\phi_1(z) + H_2(x, y)\phi_2(z) \qquad (6.8.9)$$

where H_1 and H_2 are the prescribed functions as discussed already, and $\phi_1(z)$ and $\phi_2(z)$ are undetermined functions of z. The equations to determine $\phi_1(z)$ and $\phi_2(z)$ are derived by using the weighted residual method with Galerkin's criterion. The weighting function is written as

$$w(x, y, z) = H_1(x, y)w_1(z) + H_2(x, y)w_2(z) \qquad (6.8.10)$$

In Eqs. (6.8.9) and (6.8.10), H_1 and H_2 are both continuous functions, but H_1 may not satisfy the inner boundary conditions in $\mathscr{D}_2$; the same is true for H_2 in $\mathscr{D}_1$ (we assume H_1 and H_2 both satisfy the outer boundary conditions). So, we have to write the weighted residual in the form (see Section 6.6 for the inner boundary conditions in the weighted residual):

$$WR = \int\int\int \left[\left(\frac{\partial}{\partial x} w\right) D \left(\frac{\partial}{\partial x} \phi\right) + \left(\frac{\partial}{\partial y} w\right) D \left(\frac{\partial}{\partial y} \phi\right) + \left(\frac{\partial}{\partial z} w\right) D \left(\frac{\partial}{\partial z} \phi\right) \right.$$
$$\left. + w(\Sigma_a - \lambda \nu \Sigma_f)\phi \right] dx \, dy \, dz \qquad (6.8.11)$$

Introducing Eqs. (6.8.9) and (6.8.10) into Eq. (6.8.11) and performing the integrations for x and y, Eq. (6.8.11) becomes

$$WR = \sum_{k=1}^{2} \sum_{l=1}^{2} \int \left\{ \left[\frac{d}{dz} w_k(z) \right] \langle D \rangle_{kl} \left[\frac{d}{dz} \phi_l(z) \right] \right. \tag{6.8.12}$$

$$\left. + w_k(z) [\langle DB_{xy}^2 \rangle_{kl} + \langle \Sigma_a \rangle_{kl} - \lambda \langle \nu \Sigma_f \rangle_{kl}] \phi_l(z) \right\} dz$$

where $\langle A \rangle_{kl}$ means

$$\langle A \rangle_{kl} = \int \int H_k(x, y) A(x, y, z) H_l(x, y) \, dx \, dy \tag{6.8.13}$$

and

$$\langle DB_{xy}^2 \rangle_{kl} = \int \int \left[\left(\frac{\partial}{\partial x} H_k \right) D \left(\frac{\partial}{\partial x} H_l \right) + \left(\frac{\partial}{\partial y} H_k \right) D \left(\frac{\partial}{\partial y} H_l \right) \right] dx \, dy \tag{6.8.14}$$

We assume the continuity of $\langle D \rangle_{kl} \phi_l'(z)$ along the z coordinate and that $\phi_l(z)$ satisfy the outer boundary conditions. Integrating the first term of Eq. (6.8.12) by parts yields

$$WR = \sum_{k=1}^{2} \int w_k(z) \sum_{l=1}^{2} \left[-\frac{d}{dz} \langle D_{kl} \rangle \frac{d}{dz} \right. \tag{6.8.15}$$

$$\left. + \langle DB_{xy}^2 \rangle_{kl} + \langle \Sigma_a \rangle_{kl} - \lambda \langle \nu \Sigma_f \rangle_{kl} \right] \phi_l(z) \, dz$$

By requiring $WR = 0$ for any arbitrary change in w_k, $k = 1, 2$, we obtain

$$\sum_{l=1}^{2} \left[-\frac{d}{dz} \langle D \rangle_{kl} \frac{d}{dz} \phi_l(z) + (\langle DB_{xy}^2 \rangle_{kl} + \langle \Sigma_a \rangle_{kl}) \phi_l(z) \right] \tag{6.8.16}$$

$$= \lambda \sum_{l=1}^{2} \langle \nu \Sigma_f \rangle_{kl} \phi_l(z), \qquad k = 1, 2$$

Equation (6.8.16) may be called as an eigenvalue problem of coupled ordinary differential equations and is subject to the conditions:

1. $\phi_l(z)$ satisfy the outer boundary conditions, and
2. continuity of $\phi_l(z)$ and $\langle D \rangle_{kl} \phi_l'(z)$.

Equation (6.8.16) is transformed to the finite difference equation by using the point scheme or box scheme as described in Section 2.2. The finite difference equation may be written in the matrix form as

$$\mathbf{A}\boldsymbol{\phi} = \lambda \mathbf{F}\boldsymbol{\phi} \tag{6.8.17}$$

where $\mathbf{A}$ is a block-tridiagonal matrix, $\mathbf{F}$ is a block-diagonal matrix, and $\boldsymbol{\phi}$ is a block vector, each block being of order 2. Each block of $\boldsymbol{\phi}$ consists of $\phi_1(z_i)$ and $\phi_2(z_i)$ where z_i is a grid point on the z coordinate. Equation (9.6.17) is similar to Eq. (2.7.11) for a one-dimensional two-group equation. The same iterative methods for Eq. (2.7.11) apply to Eq. (6.8.17).

Generally, the synthesis method works well to predict the eigenvalue of a reactor whose flux distribution is relatively smooth. For more complicated problems of multigroup diffusion equations than illustrated in this section, the success of the synthesis method depends on an appropriate selection of trial functions.[25-30] An efficient use of the method requires some experience. The synthesis method is not suitable for predicting the flux distribution because the synthesized flux can include significant local errors, for example, negative fluxes. The synthesis methods have been applied also to space–time and space–energy problems.[18]

PROBLEMS

1. An integral equation is given by

$$\phi(x) = \lambda \int_{-1}^{+1} K(x', x)\phi(x')\, dx'$$

By using the weighted residual method, reduce this equation to a simultaneous set of linear albegraic equations. Use the following trial function and the total weighting function:

$$\phi(x) = \sum_{i=1}^{N} a_i \eta_i(x)$$

$$w(x) = \sum_{j=1}^{N} w_j \delta(x - x_j)$$

where a_i is an undetermined coefficient, w_j is an arbitrary constant, and

$$\eta_i(x) = \begin{cases} 1, & 2\dfrac{i-1}{N} - 1 < x < \dfrac{2i}{N} - 1 \\ 0, & \text{otherwise} \end{cases}$$

$$x_j = \frac{1}{N} + \frac{2}{N}(j-1) - 1$$

2. Apply the weighted residual method to

$$\mu \frac{\partial}{\partial x} \psi(x, \mu) + \psi(x, \mu) = \frac{c}{2} \int_{-1}^{+1} \psi(x, \mu)\, d\mu, \quad 0 < c < 1$$

with the trial and the total weighting functions:

$$\phi(x, \mu) = \sum_{i=1}^{N} a_i(x)\eta_i(\mu)$$

$$w(x, \mu) = \sum_{j=1}^{N} w_j(x)\delta(\mu - \mu_j)$$

where $a_i(x)$ is an undetermined function, $w_j(x)$ is an arbitrary function, $\eta_i(\mu)$ is the step function defined in the previous problem.

3. Find the adjoint operators for

a. $L = \dfrac{d^4}{dx^4} + \dfrac{d^3}{dx^3} + \dfrac{d^2}{dx^2} + \dfrac{d}{dx}$

b. $L = \dfrac{1}{r}\dfrac{d}{dr}\left(r\dfrac{d}{dr}\right)$ (The volume element is $dv = 2\pi r\, dr$)

c. $L = \displaystyle\int_{-\infty}^{+\infty} dx'\, \exp\left(-|x - x'|\right)$

(Hint: see Appendix I for adjoint operators.)

4. Write the variational functional for the one-dimensional two-group neutron diffusion equation given by Eq. (2.7.1).

5. A fourth order ordinary differential equation is given by

$$\frac{d^4}{dx^4} y(x) = S(x), \qquad 0 \le x \le H$$

where the boundary conditions are assumed to be

$$y'(0) - \gamma_L y(0) = 0$$
$$y'(H) + \gamma_R y(H) = 0$$
$$y'''(0) - \omega_L y''(0) = 0$$
$$y'''(H) + \omega_R y''(H) = 0$$

Write the weighted residual method that admits the trial functions that do not satisfy the above boundary conditions.

6. The eigenvalue problem of an integral equation is given

$$\phi(x) = \lambda \int_0^1 K(x', x)\phi(x')\, dx', \qquad -\infty < x < +\infty$$

where

$$K(x', x) = \exp(-|x' - x|)$$

Write the variational principle for this equation and estimate the fundamental eigenvalue by using the trial function:

$$\phi(x) = \begin{cases} 1, & 0 \le x \le 1 \\ 0, & x < 0, 1 < x \end{cases}$$

7. Find the necessary condition for the following functional to become stationary

$$J(\phi) = \frac{\displaystyle\int_0^H (\phi')^2 \, dx}{\displaystyle\int_0^H \phi^2 \, dx}$$

Assume that the trial functions satisfy $\phi(0) = \phi(H) = 0$.

8. Find the variational functional for the following problem:

$$-\frac{d}{dx}(p(x)\phi'(x)) + q(x)\phi = \lambda\phi$$

$$-p(0)\phi'(0) + \gamma_L\phi(0) = 0$$

$$p(H)\phi'(H) + \gamma_R\phi(H) = 0$$

9. An inhomogeneous equation is given:

$$-\phi''(x) + \phi(x) = \begin{cases} 1, & 1 \le x \le 2 \\ 0, & 0 \le x < 1 \end{cases}$$

B.C.: $\phi'(0) = \phi(2) = 0$

a. Find an appropriate functional that gives an estimate for the following integral with a second order error:

$$\int_0^1 \phi(x) \, dx$$

b. Estimate the integral by using the trial function:

$$\phi(x) = A(1 - \tfrac{1}{4}x^2)$$

where A is an undetermined coefficient. (Hint: set the adjoint trial function as $\phi^*(x) = A^*(1 - \tfrac{1}{4}x^2)$.)

10. An eigenvalue problem for the one-dimensional cylindrical coordinate is given by

$$-\frac{10}{r}\frac{d}{dr}r\frac{d}{dr}\phi(r)+\phi(r)=\lambda\phi(r), \qquad 0.5\leq r\leq 20$$

with the boundary conditions

$$\phi'(0.5)=\phi(0.5)$$
$$\phi(20)=0$$

Calculate the lowest eigenvalue approximately by the direct method with the trial function:

$$\phi(r)=\sum_{i=0}^{4} a_iJ_0\left(\frac{j_ir}{20}\right)$$

where J_0 is the Bessel function of the first kind and j_i, $i=0, 1, 2, 3, 4$ are the first five roots of $J_0(x)=0$.

11. a. Make a FORTRAN program to solve the following one-dimensional problem by using the piecewise polynomial method in Section 6.7:

$$-p\frac{d^2}{dx^2}\phi(x)+q\phi(x)=S, \qquad 0\leq x\leq a$$

b. Calculate the solution with intervals of 3, 6, and 10 for the following two sets of conditions:

	case 1	case 2
p	1.0	1.0
q	0.0	0.01
a	3.0	3.0
$\phi(0)$	$\phi(0)$ = 2.0	$\phi'(0)$ = 0.0
$\phi(a)$	$\phi(a)$ = 0.0	$\phi(a)$ = 0.0

c. Solve the problem analytically, and check the validity of the numerical calculations.

12. a. Find the necessary conditions for stationary of the functional

$$J(j^*, j, \phi^*, \phi) = \int_0^H \left[q(x)\phi^*(x)\phi(x) + \phi^*(x)j'(x) + j^*(x)\phi'(x) \right.$$

$$+ \frac{1}{p(x)} j^*(x)j(x) - S^*(x)\phi(x) - \phi^*(x)S(x) \right] dx$$

$$+ j^*(0)[\phi(0) + j(0)\gamma_L/p(0)]$$

$$- j^*(H)[\phi(H) - j(H)\gamma_R/p(H)]$$

b. Show that, if j and j^* are eliminated from the Euler–Lagrange equation, the following equations are obtained:

$$-p\phi'' + q\phi = S$$

$$-p\phi^{*\prime\prime} + q\phi^* = S^*$$

13. The functional in Problem 12 admits trial functions that are discontinuous. So, set the trial functions as

$$\begin{aligned} \phi(x) &= \phi_i \text{ (const)} \\ \phi^*(x) &= \phi_i^* \end{aligned} \right\} x_i < x < x_{i+1}$$

$$\begin{aligned} j(x) &= j_i \text{ (const)} \\ j^*(x) &= j_i^* \end{aligned} \right\} \frac{x_{i-1} + x_i}{2} < x < \frac{x_i + x_{i+1}}{2}$$

where x_i is a grid point and ϕ_i, ϕ_i^*, j_i, and j_i^* are all constants. Introducing the above trial functions into the functional, find the necessary conditions for stationary. Show also that, by eliminating j_is and j_i^*'s, the finite difference formulas for the two equations in Problem 12.b are obtained.[31]

14. Consider the two-dimensional eigenvalue problem

$$-\left(\frac{\partial^2}{\partial x^2} + \frac{\partial^2}{\partial y^2} \right) \phi(x, y) + q(x, y)\phi(x, y) = 0.1\lambda\phi(x, y) \qquad (1)$$

The geometry and the value of $q(x, y)$ are shown in the figure. The boundary conditions are $\phi = 0$ for the top, bottom, and right boundaries and $\phi' = 0$ at the left boundary.

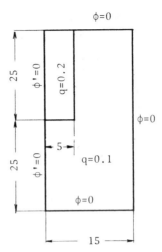

Calculate the eigenvalue by the method with the trial function in the form

$$\phi(x, y) = Y_1(y) \cos \frac{\pi x}{30} + Y_2(y) \cos \frac{3\pi x}{30} + Y_3(y) \cos \frac{5\pi x}{30}$$

where $Y_n(y)$ are undetermined functions of y. Hint:

a. Use the weighted residual method with Galerkin's criterion.

b. Reduce Eq. (1) to a simultaneous set of three ordinary differential equations for Y_ns.

c. Write the finite difference equations for the ordinary differential equations.

d. Express the difference equation in the matrix form with block matrices (3×3 square submatrices).

REFERENCES

1. B. A. Finlayson and L. E. Scriven, "The method of weighted residuals—a review," *Appl. Mech. Rev.*, **19**, 735–748 (1966).

2. B. A. Finlayson, *The method of weighted residuals and variational principles*, Academic, New York (1972).

3. R. Courant and D. Hilbert, *Methods of mathematical physics*, Vol. 1, Wiley-Interscience, New York (1953).

4. O. Bolza, *Lectures on the calculus of variations*, 3rd edit., Chelsea, New York (1973).

5. L. E. Elsgolc, *Calculus of variations*, Addison-Wesley, Reading, Mass. (1962).

6. I. M. Gelfand and S. V. Fomin, *Calculus of variations*, Prentice-Hall, Englewood Cliffs, N.J. (1963).

7. S. G. Mikhlin, *Variational methods in mathematical physics*, Macmillan, London (1964).

8. R. H. Gallagher, *Finite element analysis*, Prentice-Hall, Englewood Cliffs, N.J. (1975).

9. C. Lanczos, *The variational principles of mechanics*, University of Toronto, Toronto (1949).

10. P. Roussopolos, "Methodes variationueles en théories des collisions," *C.R. Acad. Sci.*, **236**, 1858 (1953).

11. D. S. Selengut, "Variational analysis of a multidimensional system," HW-59126, Hanford Laboratory, Richland, Wash. (1959).

12. H. Levine and J. Schwinger, "On the theory of diffraction by an aperture in an infinite plane screen," *Phys. Rev.*, **75**, 1423 (1943).

13. N. C. Francis, J. C. Stewart, L. S. Bohl, and T. J. Krieger, "Variational solutions of the transport equations," *Progr. Nucl. Energy Ser.* **3**, 360 (1959).

14. G. C. Pomraning, "Generalized variational principle for reactor analysis," *Proc. Int. Conf. Utilization Res. Reactors Math. Comput.*, Mexico D.F., Centro Nucl. de Mexico (1966).

15. G. C. Pomraning, "A derivation of variational principles for inhomogeneous equations," *Nucl. Sci. Eng.*, **29**, 220 (1967).

16. S. Kaplan and J. A. Davis, "Canonical and involuntary transformation of the variational problems of transport theory," *Nucl. Sci. Eng.*, **28**, 166–176 (1967).

17. S. Kaplan, "Variational methods in nuclear engineering," *Advances in nuclear science and technology*, **5**, 185 (1969).

18. W. M. Stacey, Jr., *Variational methods in nuclear reactor physics*, Academic, New York (1974).

19. D. S. Selengut, "On the derivation of a variational principle for linear systems," *Nucl. Sci. Eng.*, **17**, 310 (1963).

20. J. Lewins, *Importance: the adjoint function*, Pergamon, New York (1965).

21. M. Natelson and E. M. Gelbard, "A two overlapping-group transport computational method for thermal neutron problems," *Nucl. Sci. Eng.*, **49**, 202 (1972).

22. E. L. Wachspress, "On the use of different radial trial functions in different axial zones of a neutron flux synthesis computation," *Nucl. Sci. Eng.*, **34**, 342–343 (1968).

23. W. M. Stacey, "Variational flux synthesis methods for multigroup neutron diffusion theory," *Nucl. Sci. Eng.*, **47**, 449–469 (1972).

24. S. Kaplan, "Some new methods of flux synthesis," *Nucl. Sci. Eng.*, **13**, 22–31 (1962).

25. J. B. Yasinsky and S. Kaplan, "Anomalies arising from the use of adjoint weighting in a collapsed group-space synthesis model, *Nucl. Sci. Eng.*, **31**, 80–90 (1968).

26. C. H. Adams and W. J. Stacey, Jr., "An anomaly arising in the collapsed group flux synthesis approximation," *Nucl. Sci. Eng.*, **36**, 444–447 (1969).

27. R. Froehlich, "Anomalies in variational flux synthesis methods," *Trans. Amer. Nucl. Soc.*, **12**, 150 (1969).

28. V. Luco, "On the eigenvalue of the flux synthesis equations," *Trans. Amer. Nucl. Soc.*, **14**, 203 (1969).

29. R. Froehlich, "Flux synthesis methods versus difference approximation methods for the efficient determination of neutron flux distributions in fast and thermal reactors," *Numerical reactor calculations*, IAEA, Vienna (1972).

30. R. Froehlich, "Current problems in multi-dimensional reactor calculations," *Proc. Conf. Mathematical Models and Computational Techniques for Analysis of Nuclear Systems*, CONF-730414-P2, National Technical Information Service, Springfield, Va. (1973).

31. E. L. Wachspress, *Iterative solution of elliptic systems*, Prentice-Hall, Englewood Cliffs, N.J. (1966).

Chapter 7   Finite Element Methods

7.1   FINITE ELEMENT VERSUS FINITE DIFFERENCE

The finite element method was originally developed as a method for structural analyses.[1,2] The finite element method is more versatile than the finite difference method because of the freedom of selecting an arbitrary distribution of mesh points. The difficulty with the finite difference method occurs when the shape of the domain is irregular, because the mesh points based on the Cartesian coordinates cannot fit boundaries of such irregular shapes. The objects of the structural analyses are usually irregular, and this is the reason why the finite element method is more powerful than the finite difference method in structural analyses.

In other areas of boundary value problems than structural analyses, the finite difference method is still more widely and extensively used. However, the recent development in applying the finite element method in such areas as heat transfer, neutron transport and diffusion,[3-8] and fluid flow[14-20] is remarkable. Several reasons can be pointed out for this rapid growth in use of the finite element methods: (1) the finite element method is versatile in dealing with irregular geometries, (2) generally, the accuracy of the finite element method is higher than the finite difference method, (3) higher order approximations can be relatively easily obtained, (4) treatment of boundary conditions is easier than with the finite difference methods, and (5) in some cases, deriving finite element approximation is even easier and more straightforward than the finite difference approximations.

There are drawbacks to the finite element method, however. First, the calculations of the coefficients for the approximating equations are more lengthy than for the finite difference equations. Second, the matrix of the discretized system is irregular. Because of this, the data transfer of matrix elements of the finite element method becomes slow and inefficient. It is for the second reason that the direct inversion method is almost exclusively used for the finite element equations, while iterative techniques are widely used for the finite difference equations. Mathematically, there are some distinctions between the two approaches. The finite element methods are based on the variational principle as a physical law, or the weighted residual method if no variational principle exists for the system. The finite difference approach can be considered as a direct discretization of the differential operator, although this is not inherent in the finite difference approach since the finite difference approach can be derived through the weighted residual method. The finite element approach can also be characterized by the field definitions of the field variables, while the finite difference approach is based on the point definitions of the field variables.

In the rest of this chapter, we study the application of the finite element method to three areas: (a) the diffusion equation including the heat conduction problems, (b) the plane stress problem, and (c) the incompressible fluid flow problem. In all cases, two-dimensional problems are discussed.

7.2 FINITE ELEMENTS AND PIECEWISE POLYNOMIALS

The whole domain of interest, assuming it is on the two-dimensional cartesian coordinate system, is divided into nonoverlapping subdomains called finite elements. Any geometric shape may be used as finite elements. However, mathematical and practical constraints limit the possible varieties of finite elements.

Triangular elements have been most widely used because of their ability to fit curved or polygonal boundaries with relative ease. Expressions related to triangular finite elements are relatively simple. One of the usual requirements to the approximating functions is the continuity of the function across the element boundaries. In each triangular element, a field function may be approximated by the first order polynomial,

$$\phi(x, y) = a + bx + cy \qquad (7.2.1)$$

where a, b, and c are coefficients defined for each element. Another popular geometry is the rectangular element, in which the field function is approximated by

$$\phi(x, y) = a + bx + cy + dxy \qquad (7.2.2)$$

It is shown later that this polynomial satisfies the continuity requirement across element boundaries.

The coefficients of Eq. (7.2.1) or (7.2.2) are uniquely determined when the values of $\phi(x, y)$ are specified at the apices. In the case of a triangular element as shown in Fig. 7.1a, we require

$$\phi(x_i, y_i) = a + bx_i + cy_i = \phi_i \qquad (7.2.3)$$

$$\phi(x_j, y_j) = a + bx_j + cy_j = \phi_j \qquad (7.2.4)$$

$$\phi(x_k, y_k) = a + bx_k + cy_k = \phi_k \qquad (7.2.5)$$

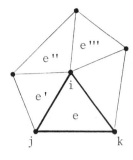
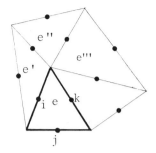

a.   Conforming   b.   Nonconforming
     element   element

Figure 7.1   Triangular element for piecewise linear function.

where ϕ_i, ϕ_j, and ϕ_k are prescribed values or unknowns to be determined. Solving Eqs. (7.2.3), (7.2.4), and (7.2.5) for a, b, and c yields

$$a = \frac{1}{2\Delta^{(e)}}\{(x_j y_k - x_k y_j)\phi_i + (x_k y_i - x_i y_k)\phi_j + (x_i y_j - x_j y_i)\phi_k\} \quad (7.2.6)$$

$$b = \frac{1}{2\Delta^{(e)}}\{(y_j - y_k)\phi_i + (y_k - y_i)\phi_j + (y_i - y_j)\phi_k\} \quad (7.2.7)$$

$$c = \frac{1}{2\Delta^{(e)}}\{(x_k - x_j)\phi_i + (x_i - x_k)\phi_j + (x_j - x_i)\phi_k\} \quad (7.2.8)$$

where

$$\Delta^{(e)} = \frac{1}{2}\det \begin{bmatrix} 1 & x_i & y_i \\ 1 & x_j & y_j \\ 1 & x_k & y_k \end{bmatrix} = \text{area of the triangular element } e$$

$$(7.2.9)$$

Substituting Eqs. (7.2.6) through (7.2.8) back into Eq. (7.2.1), we obtain

$$\phi(x, y) = \phi_i \eta_i^{(e)}(x, y) + \phi_j \eta_j^{(e)}(x, y) + \phi_k \eta_k^{(e)}(x, y) \quad (7.2.10)$$

where

$$\eta_i^{(e)}(x, y) = A_i^{(e)} + B_i^{(e)}x + C_i^{(e)}y \quad (7.2.11)$$

$$A_i^{(e)} = \frac{1}{2\Delta^{(e)}}(x_j y_k - x_k y_j) \quad (7.2.12)$$

$$B_i^{(e)} = \frac{1}{2\Delta^{(e)}}(y_j - y_k) \qquad (7.2.13a)$$

$$C_i^{(e)} = \frac{1}{2\Delta^{(e)}}(x_k - x_j) \qquad (7.2.13b)$$

The expressions for $\eta_j^{(e)}$ and $\eta_k^{(e)}$ are obtained by rotating i, j, and k. It should be noted that $\eta_i^{(e)}$, $\eta_j^{(e)}$, and $\eta_k^{(e)}$ are polynomials of the same order as Eq. (7.2.1), and furthermore, that they become unity at i, j, and k, respectively, and zero at other apices.

So far, we have considered only one finite element. Therefore, a, b, c or $\eta_i^{(e)}$, $\eta_j^{(e)}$, and $\eta_k^{(e)}$ are specific to the finite element e. The same expressions are used for the adjacent and all other elements. With these polynomials for all the finite elements, the approximating field function $\phi(x, y)$ in the entire domain is uniquely determined if ϕ_i for all apices are given. Consider a grid point i as shown in Fig. 7.1a. This grid point i is an apex of finite elements $e, e', e'', \ldots$. Therefore, $\eta_i^e, \eta_i^{e'}, \eta_i^{e''}, \ldots$ are defined in $e, e', e'', \ldots$, respectively. All of these $\eta_i^{(e)}$'s become unity at i, have common values at the boundary of two adjacent finite elements, and are zero along the polygon surrounding i. Including all of $\eta_i^{(e)}$'s we define a new function in the entire space:

$$\eta_i(x, y) = \eta_i^{(e)}(x, y), \quad \text{if } (x, y) \text{ is in } e \text{ that has } i \text{ as an apex}$$
$$= 0, \qquad \text{otherwise} \qquad (7.2.14)$$

In terms of η_i thus defined, all the piecewise linear functions in the entire region are expressed by the single formula,

$$\phi(x, y) = \sum_{i=1}^{J} \phi_i \eta_i(x, y) \qquad (7.2.15)$$

where J is the total number of grid points.

The piecewise polynomials in the rectangular finite elements also may be expressed in the form of Eq. (7.2.15). For a rectangular element illustrated in Fig. 7.2, $\eta_i^{(e)}(x, y)$ is defined by

$$\eta_i^{(e)}(x, y) = \frac{(x - x_j)(y - y_m)}{(x_i - x_j)(y_i - y_m)} \qquad (7.2.16)$$

which is unity at i and zero at j, k, and m. $\eta_i^{(e)}(x, y)$ is linear along the sides, i–j and i–m, zero along the sides, j–k and k–m. Similarly to Eq. (7.2.14), $\eta_i(x, y)$ for the entire domain divided into rectangular finite elements may be defined. η_i is unity at i and nonzero in the four rectangular elements around i. η_i is linear at the boundaries of two adjacent finite elements and accordingly continuous across the boundaries. Equation (7.2.14) may be

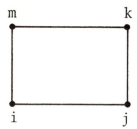

m   k

i   j   **Figure 7.2**   Rectangular element.

used to define η_i for rectangular finite elements. Thus the trial function for the entire region is expressed by Eq. (7.2.15), also.

Polynomials of a higher order are used (1) when higher accuracy of the field functions is required with the same order of finite element configurations, and (2) when use of higher order polynomials is a mathematical requirement.

Example 7.1

The second order polynomial that is used in a triangular element with six nodal points as shown in Fig. 7.3 is given by

$$\phi(x, y) = a_1 + a_2 x + a_3 y + a_4 x^2 + a_5 xy + a_6 y^2 \qquad (x, y) \in e \qquad (7.2.17)$$

The additional three grid points are selected to be located at the midpoints of the sides. $\phi(x, y)$ for the entire region may be expressed in the form of Eq. (7.2.15) again, if η_i are appropriately redefined. The shape function η_i will become unity at the grid point i and zero at all other grid points, and a quadratic function along any boundary of a triangular finite element. A quadratic function is uniquely determined by three grid points on a boundary. Therefore, $\eta_i(x, y)$ and accordingly $\phi(x, y)$ will became continuous across element boundaries.

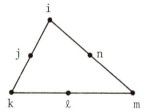

i

j   n

k ℓ m   **Figure 7.3**   Triangular element with six grids.

Example 7.2

When ϕ, $\partial\phi/\partial x$, and $\partial\phi/\partial y$ are specified at the apices of a triangular element, the polynomial must include nine coefficients as follows:

$$\phi(x, y) = a_1 + a_2 x + a_3 y + a_4 x^2 + a_5 xy + a_6 y^2$$
$$+ a_7 x^3 + a_8 (x^2 y + xy^2) + a_9 y^3 \qquad (7.2.18)$$

Example 7.3

When ϕ, ϕ_x, and ϕ_y are specified at the apices of a rectangular element, the polynomial must include 12 coefficients:

$$\phi(x, y) = a_1 + a_2 x + a_3 y + a_4 x^2 + a_5 xy + a_6 y^2$$
$$+ a_7 x^3 + a_8 x^2 y + a_9 xy^2 + a_{10} y^3 \qquad (7.2.19)$$
$$+ a_{11} x^3 y + a_{12} xy^3$$

Suppose the order of the highest derivative in the weighted residual equation (or functional) is m. A general criterion in choosing the finite element and the shape function is that the field function has the continuity of all the derivative up to the order $m - 1$ across the element boundaries. The finite element satisfying this criterion is called a conforming element. The weighted residual equation for a second order elliptic equation will have only first derivatives ($m = 1$) after applying Green's theorem of integration as shown in the Section 7.3. In this case, only the continuity of the field function is necessary for the piecewise polynomial in triangular elements to meet the criterion. Triangular elements using apices as grid points as shown in Fig. 7.1*a* is a conforming element, while the triangular element using midpoints on the corners as grid points as shown in Fig. 7.1*b* is a noncomforming element. A conforming element is not a necessary condition. In order that the finite element approximation converges to the exact solution as the size of finite element approaches zero, the set of the shape functions must be able to represent any constant field function as well as the derivatives up to order m within each element.

As the order of a polynomial increases, programming becomes more complicated and computational time increases. If a higher accuracy is desired, using the lowest order polynomial that is mathematically acceptable with finer finite elements rather than increasing the order of the polynomial is generally recommended. The isoparametric elements[21] consisting of curvilinear boundaries have been developed. They are useful when curvilinear external and internal boundaries must be well fitted.

7.3 APPLICATION OF THE FINITE ELEMENT METHOD TO AN ELLIPTIC PARTIAL DIFFERENTIAL EQUATION

In this section we apply the finite element method to the two-dimensional partial differential equation of elliptic type:

$$-\nabla p(x, y)\nabla \psi(x, y) + q(x, y)\psi(x, y) - r(x, y) = 0 \qquad (7.3.1)$$

Equation (7.3.1) is encountered in various fields. For example, it may be considered as the heat conduction equation with $q = 0$, or a neutron diffusion equation with absorption coefficient, $q \neq 0$.

Boundary conditions for Eq. (7.3.1) can be classified into the following three types:

a. $\psi = \omega(s)$ Dirichlet type   (7.3.2a)

b. $\dfrac{\partial}{\partial n}\psi = \pi(s)$ Neumann type   (7.3.2b)

c. $\dfrac{\partial}{\partial n}\psi + \gamma(s)\psi = \pi(s)$ Mixed type   (7.3.2c)

where $\partial/\partial n$ is the derivative outward normal, s is the coordinate on the boundary, and ω, π, and γ are prescribed functions of s. Equations (7.3.2a) and (7.3.2b) can be considered as special cases of Eq. (7.3.2c) as discussed in Section 3.1. In this section, however, the Neumann type will be included in the mixed type while the Dilichlet type will be treated independently.

We divide the whole region of interest into finite elements and assume p and q are constant in each finite element. We use Eq. (7.2.15) with Eq. (7.2.14) as an approximating function for ψ. To derive the finite element equation, we use the weighted residual method, although the same results can be obtained via a variational principle in the case of Eq. (7.3.1) (see Problem 3). Since Eq. (7.2.15) is an approximation, it does not exactly satisfy Eq. (7.3.1). However, in accordance with the weighted residual method, we require Eq. (7.2.15) to satisfy Eq. (7.3.1) on the weighted average basis:

$$\int w(x, y)[-\nabla p \, \nabla \phi + q\phi - r] \, dV = 0 \qquad (7.3.3)$$

where the integral is extended over the whole domain of interest.

One must be careful in introducing Eq. (7.2.15) into Eq. (7.3.3) because the first term becomes zero inside a finite element and $p\nabla\phi$ is not continuous across the element boundaries. To avoid the trouble due to the discontinuity of $p\nabla\phi$, we transform the first term by using Green's theorem as

$$\int [p(\nabla w)(\nabla \phi) + qw\phi - wr] \, dV - \int_\Gamma pw\frac{\partial}{\partial n}\phi \, ds = 0 \qquad (7.3.4)$$

where the second term is a surface integral along the external boundary and w is assumed to be continuous over the whole domain (this is a valid assumption since only the continuous shape functions are used as weighting functions). The surface integral along internal boundaries is eliminated by using the internal boundary conditions as discussed in Section 6.6.

The surface integral in Eq. (7.3.4) is divided into two parts:

$$\int_\Gamma pw\frac{\partial}{\partial n}\phi\,ds = \int_I pw\frac{\partial}{\partial n}\phi\,ds + \int_{II} pw\frac{\partial}{\partial n}\phi\,ds \qquad (7.3.5)$$

where I indicates the integral along the Dirichlet boundary and II the Neumann or the mixed boundary. The weighting function w is set to zero along the Dirichlet boundary, so the first surface integral becomes zero. Introducing Eq. (7.3.2c) into the second part of the above equation and introducing the result into Eq. (7.3.4) yield

$$WR \equiv \int [p(\nabla w)(\nabla\phi)+qw\phi - wr]\,dV + \int_{II} pw(\gamma\phi - \pi(s))\,ds = 0 \qquad (7.3.6)$$

We first assume the boundary conditions are all the mixed type. When the integrations in Eq. (7.3.6) are performed, a linear equation that includes at most J undetermined variables ϕ_i is obtained. Since there are J unknown variables to be determined, we need J linearly independent equations. Those equations are obtained by using J linearly independent weighting functions, $w_i\ i = 1, 2, \ldots, J$. Although there are several criteria in choosing weighting functions (see Section 6.2), the Galerkin criterion is used in the finite element method. This means to use J shape functions η_i as weighting functions.

Introducing Eq. (7.2.15) and $w = \eta_j$ into Eq. (7.3.6) yields

$$\sum_{i=1}^{J}\left[\langle p(\nabla\eta_j)(\nabla\eta_i)\rangle + \langle q\eta_j\eta_i\rangle + \int_{II} p\gamma(s)\eta_j\eta_i\,ds\right]\phi_i$$
$$= \langle\eta_j r\rangle + \int_{II}\eta_j p\pi(s)\,ds, \qquad j = 1, 2, \ldots, J \qquad (7.3.7)$$

where $\langle\ \rangle$ means the volume integral over the whole domain and the surface integral is performed along all the external boundary. By using matrix notations, Eq. (7.3.7) may be expressed as

$$\mathbf{H}\phi = \mathbf{r} \qquad (7.3.8)$$

where ϕ is a column vector,

$$\phi = \text{col}\,[\phi_1, \phi_2, \ldots, \phi_J] \qquad (7.3.9)$$

$\mathbf{H} = [H_{ji}]$ is a square matrix representing the coefficients of the ϕ_is in Eq. (7.3.7), and $\mathbf{r}$ is a column vector representing the right side of Eq. (7.3.7).

The matrix element H_{ji} which is equal to the inside of [ ] in Eq. (7.3.7) can be written as

$$H_{ji} = \sum_e \left\{ \int_e [p(\nabla \eta_j)(\nabla \eta_i) + q\eta_j\eta_i] \, dV + \int_{IIe} p\gamma\eta_j\eta_i \, ds \right\} \quad (7.3.10)$$

where the summation is over all the finite elements, the volume integrals are performed in each element, and the surface integrals are performed for the finite element e if the external boundary is along the side of the element e.

It is worthwhile to note the relation (see Appendix III):

$$\int_e \eta_j\eta_i \, dV = 0, \qquad \text{if } i \text{ or } j \text{ is not an apex of } e$$

$$\qquad\qquad (7.3.11a)$$

$$\doteq \frac{\Delta^{(e)}}{12}(1 + \delta_{ji}), \quad \text{if } i \text{ and } j \text{ are both apices of } e$$

$$\int_e \left[\left(\frac{\partial}{\partial x}\eta_j\right)\left(\frac{\partial}{\partial x}\eta_i\right) + \left(\frac{\partial}{\partial y}\eta_j\right)\left(\frac{\partial}{\partial y}\eta_i\right) \right] dV$$

$$= 0, \qquad\qquad \text{if } i \text{ or } j \text{ is not an apex of } e \quad (7.3.11b)$$

$$= \Delta^{(e)}[B_j^{(e)}B_i^{(e)} + C_j^{(e)}C_i^{(e)}], \quad \text{if } i \text{ and } j \text{ are both apices of } e$$

where $\Delta^{(e)}$ is the area of e and δ_{ji} is Kronecker's delta. As easily seen, H_{ji} becomes zero if j and i are not apices of a triangle. When the grids are adequately numbered, the matrix $\mathbf{H}$ becomes a band matrix even though the band is rather irregular. The maximum width of the band is affected by the way the grids are numbered. The computational time with direct inversion of $\mathbf{H}$ is rapidly increased as the bandwidth is increased. Therefore, an efficient numbering is important.

Example 7.4

Calculate H_{ji} in Eq. (7.3.10) by using the first order polynomial Eq. (7.2.1) for the system consisting of two triangular finite elements as shown in Fig. 7.4. Assume p, q, and r are constant in each element. Boundary conditions are (1) the Neumann boundary condition with $\pi = 0$ along 4–1–2–3, and (2) the mixed boundary condition with $\gamma = \gamma_a$ and $\pi = 0$ along 3–4.

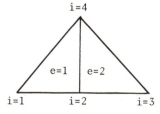

Figure 7.4   A geometry divided into two triangular finite elements.

By introducing Eq. (7.2.14) into Eq. (7.3.10) and using Eqs. (7.3.11a) and (7.3.11b), H_{ji} are obtained as (since $\mathbf{H}$ is symmetric, only H_{ji} with $j \leq i$ are shown):

$$H_{11} = \Delta^{(1)}[p^{(1)}(B_1^{(1)}B_1^{(1)} + C_1^{(1)}C_1^{(1)}) + \tfrac{1}{6}q^{(1)}]$$

$$H_{12} = \Delta^{(1)}[p^{(1)}(B_1^{(1)}B_2^{(1)} + C_1^{(1)}C_2^{(1)}) + \tfrac{1}{12}q^{(1)}]$$

$$H_{13} = 0$$

$$H_{14} = \Delta^{(1)}[p^{(1)}(B_1^{(1)}B_4^{(1)} + C_1^{(1)}C_4^{(1)}) + \tfrac{1}{12}q^{(1)}]$$

$$H_{22} = \sum_{e=1}^{2} \Delta^{(e)}[p^{(e)}(B_2^{(e)}B_2^{(e)} + C_2^{(e)}C_2^{(e)}) + \tfrac{1}{6}q^{(e)}]$$

$$H_{23} = \Delta^{(2)}[p^{(2)}(B_2^{(2)}B_3^{(2)} + C_2^{(2)}C_3^{(2)}) + \tfrac{1}{12}q^{(2)}]$$

$$H_{24} = \sum_{e=1}^{2} \Delta^{(e)}[p^{(e)}(B_2^{(e)}B_4^{(e)} + C_2^{(e)}C_4^{(e)}) + \tfrac{1}{12}q^{(e)}]$$

$$H_{33} = \Delta^{(2)}[p^{(2)}(B_3^{(2)}B_3^{(2)} + C_3^{(2)}C_3^{(2)}) + \tfrac{1}{6}q^{(2)}] + \tfrac{1}{3}p^{(2)}\gamma_a L$$

$$H_{34} = \Delta^{(2)}[p^{(2)}(B_3^{(2)}B_4^{(2)} + C_3^{(2)}C_4^{(2)}) + \tfrac{1}{12}q^{(2)}] + \tfrac{1}{6}p^{(2)}\gamma_a L$$

$$H_{44} = \sum_{e=1}^{2} \Delta^{(e)}[p^{(e)}(B_4^{(e)}B_4^{(e)} + C_4^{(e)}C_4^{(e)}) + \tfrac{1}{6}q^{(e)}] + \tfrac{1}{3}p^{(2)}\gamma_a L$$

where L is the length of boundary 3–4.

Equation (7.3.7) was obtained by introducing $w(x, y) = \eta_j(x, y)$ for $j = 1, 2, \ldots, J$. This derivation may be equivalently stated as follows. Set the weighting function as

$$w = \sum_{j=1}^{J} \phi_j^* \eta_j(x, y) \tag{7.3.12}$$

where ϕ_j^* are arbitrary constants and η_j are the shape functions. The degree of freedom in Eq. (7.3.12) is J in accordance with that of Eq. (7.2.15). We introduce Eqs. (7.3.12) and (7.2.15) into Eq. (7.3.6) and require $WR = 0$ for any arbitrary change in w. Since the degree of freedom in Eq. (7.3.12) is J, this requirement is equivalent to

$$\frac{\partial WR}{\partial \phi_i^*} = 0, \qquad i = 1, 2, \ldots, J \tag{7.3.13}$$

Performing the partial differentiation we obtain Eq. (7.3.7) again.

Let us now consider the case when the Dirichlet boundary is included. Suppose m grid points on the boundary are subject to the Dirichlet boundary and specified by the prescribed values. Then the degree of freedom in Eq. (7.2.15) is reduced to $N = J - m$. To obtain the same degree of freedom in Eq. (7.3.10), we prescribe the values of ϕ_i^* for the corresponding points. It is most convenient to set $\phi_i^* = 0$ whenever ϕ_i is specified by the Dirichlet boundary conditions. Thus the weighting function, Eq. (7.3.12), becomes zero along the Dirichlet boundary. Applying Eq. (7.3.13) for N variables ϕ_is, N linear equations are obtained. The set of equations thus obtained can be expressed in the matrix notations as

$$\mathbf{H}'\boldsymbol{\phi}' = \mathbf{r}' \qquad (7.3.14)$$

where $\boldsymbol{\phi}'$ is a vector representing N undetermined ϕ_is, $\mathbf{H}'$ is a $N \times N$ matrix, and $\mathbf{r}'$ is a known vector.

Generally, however, it is preferred to use Eq. (7.3.8) with order J with a slight modification to $\mathbf{H}$ rather than reducing the order of the matrix to N. This can be done by considering ϕ_i for all the grid points as unknowns, including those on the Dirichlet boundary, and by regarding the Dirichlet boundary conditions as members of simultaneous equations. If ϕ_k is prescribed by the Dirichlet boundary condition as $\phi_k = \bar{\phi}_k$, then the kth row of $\mathbf{H}$ is $H_{ki} = 0$ for all i except $H_{kk} = 1$ and $r_k = \bar{\phi}_k$. The advantage of including all ϕ_i in $\boldsymbol{\phi}$ is twofold. First, $\boldsymbol{\phi}$ include all the grid points so that the size of $\mathbf{H}$, $\boldsymbol{\phi}_k$, and $\mathbf{r}$ and the correspondence between elements and the grid points do not change if the position of the Dirichlet boundary is changed in a sequence of computations with different boundary configurations. Second, the algorithm of calculating elements of $\mathbf{H}$ is as simple as follows:

1. Calculate $\mathbf{H}$ as if all the Dirichlet boundaries were temporarily altered to the Neumann boundary with $\pi(s) = 0$.
2. For those grids subject to the Dirichlet boundary condition ($\phi_k = \bar{\phi}_k$), set H_{ki} as $H_{ki} = \delta_{ki}$ and r_k as $r_k = \bar{\phi}_k$, where δ_{ki} is Kronecker's delta.

Example 7.5

Calculate H_{ji} in Eq. (7.3.10) for the same problem as Example 7.4 except that the boundary condition for side 3–4 is altered to the Dirichlet boundary condition $\phi = \phi_B$ (const). The boundary conditions for the discrete variables are $\phi_3 = \phi_4 = \phi_B$.

H_{ji} and r_j are all the same as for Example 7.4 except

$$H_{3i} = \delta_{3i}$$

$$H_{4i} = \delta_{4i}$$

$$r_3 = r_4 = \phi_B$$

Equation (7.3.8) may be solved by the Gaussian elimination scheme or an iterative method such as SOR. Usually the former is preferred to the latter because the computational time is shorter unless the number of unknowns is very large. **H** is a sparse matrix, and its configuration is greatly affected by the order of numbering the grids. By appropriate numbering, **H** becomes a band diagonal matrix with a certain bandwidth. The computational time for

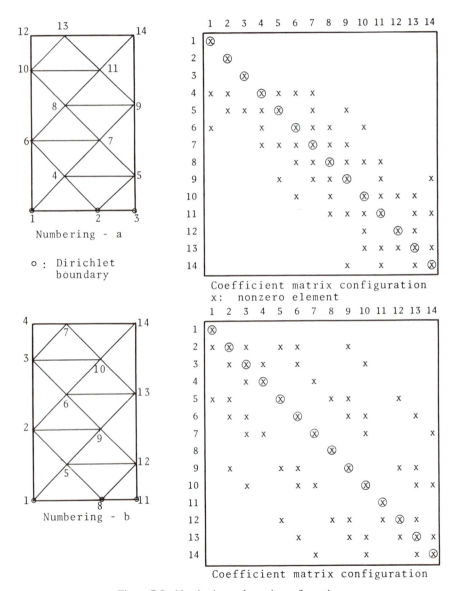

Figure 7.5   Numbering and matrix configuration.

a Gaussian elimination is approximately proportional to $Im^2/2$, where I is the order of $\mathbf{H}$ and m is the half-bandwidth, so the grids must be numbered so as to minimize the bandwidth. See Reference 1 for an efficient solution of symmetric band-diagonal matrices.

Example 7.6

The configuration of $\mathbf{H}$ for two different ways of numbering for a rectangular geometry is shown in Fig. 7.5. The grid points marked by a circle are specified by the Dirichlet boundary conditions. Nonzero elements are shown by $\times$.

The iterative approach[1,8,11,12,29] will become more attractive as the number of grid points becomes very large. Several studies have been done to investigate the feasibility of iterative schemes to Eq. (7.3.8) and to improve the efficiency of iterative approaches. The matrix $\mathbf{H}$ must have certain properties for an iterative solution to be convergent. The successive-over-relaxation method is convergent if $\mathbf{H}$ is a positive definite matrix.[11] The matrix $\mathbf{H}$ is positive definite if

$$\langle \boldsymbol{\phi}, \mathbf{H}\boldsymbol{\phi} \rangle > 0 \qquad (7.3.15)$$

for any nonzero vector $\boldsymbol{\phi}$, where the brackets denote an scalar product. Suppose $\boldsymbol{\phi}$ represents the coefficients of the trial function given by Eq. (7.2.15). We assume for simplicity $\omega(s) = 0$ for any Dirichlet boundary and that the trial function satisfies this boundary condition. Then it is easily seen that the left side of Eq. (7.3.15) is equal to

$$\int [p(\nabla\phi)^2 + q\phi^2]\, dV + \int_{II} p\gamma\phi^2\, ds \qquad (7.3.16)$$

which is positive for any ϕ.

Example 7.7

Consider a plane geometry that is divided into triangular finite elements as shown in Fig. 7.6. The equation $-\nabla^2 T(x, y) = 2.0$ was solved by the finite element method. The boundary conditions are specified in Fig. 7.6. This is a simple model for a unit cell of the heated material with infinite array of cooling pipes. The solution at the grids are shown in Fig. 7.6, where only the values on the lower triangular part are given because of the 45-degree symmetry.

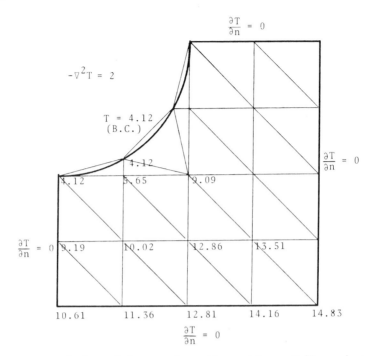

Figure 7.6 A sample heat-transfer problem and the result (the numbers show the temperature at the grid points).

7.4 FINITE ELEMENT STRESS ANALYSIS

This section introduces a finite element stress analysis method. We consider the plane-stress problem. Unlike most textbooks on this subject, we start with the partial differential equations for stress distribution functions and apply the weighted residual method in a straightforward manner.

Consider a plane with a finite thickness unstrained to the thickness direction. Referring to Fig. 7.7, a differential element on the x–y plane is subject to normal stresses, the shear stress, and external body forces (forces per unit volume). The stresses are internal forces that act on the surface of any arbitary volume element. The balance of force equation for the x direction is

$$\left(\sigma_x + \frac{\partial \sigma_x}{\partial x}\,dx\right)dy - \sigma_x\,dy + F_x\,dx\,dy$$

$$+\left(\tau_{yx} + \frac{\partial \tau_{yx}}{\partial y}\,dy\right)dx - \tau_{yx}\,dx = 0 \qquad (7.4.1)$$

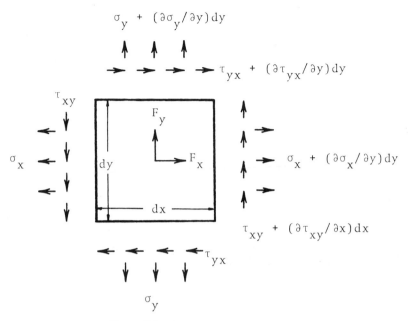

Figure 7.7 Forces to a volume element.

where σ_x is the normal stress in the x direction, τ_{xy} is the shear stress, and F_x is the body force in the x direction. After rearranging and dividing by $dx\,dy$, Eq. (7.4.1) becomes

$$\frac{\partial \sigma_x}{\partial x} + \frac{\partial \tau_{yx}}{\partial y} + F_x = 0 \qquad (7.4.2)$$

Similarly, for the y direction we have

$$\frac{\partial \sigma_y}{\partial y} + \frac{\partial \tau_{xy}}{\partial x} + F_y = 0 \qquad (7.4.3)$$

where σ_y and F_y are the normal stress and the body force, respectively, in the y direction.

We define the displacement functions by $u(x, y)$ and $v(x, y)$ for the x and y directions, respectively. The displacement function $u(x, y)$ represents the displacement of an original point (x, y) in the x direction.

The external boundary conditions with respect to $u(x, y)$ may be divided into two kinds. We denote the whole external boundary by Γ. The first kind is given by

$$u(s) = 0, \qquad s \in \Gamma_x \qquad (7.4.4)$$

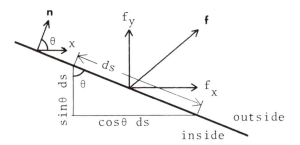

Figure 7.8 Forces to a surface element.

where s denotes the coordinate along the external boundary and Γ_x denotes that portion of the external boundary where the displacement in the x direction is zero. The second kind includes the rest of the external boundary where surface forces in the x direction apply in general. Referring to Fig. 7.8, the balance of forces for a differential element at an external surface is

$$f_x(s)\,ds = \tau_{xy}\sin\theta\,ds + \sigma_x\cos\theta\,ds, \qquad s \in \Gamma - \Gamma_x$$

or, dividing by ds,

$$f_x(s) = \tau_{xy}\sin\theta + \sigma_x\cos\theta, \qquad s \in \Gamma - \Gamma_x \tag{7.4.5}$$

where θ is the angle between the unit vector outward normal to the surface and the x coordinate, $f_x(s)$ denotes the external force in the x direction applied at s along the external boundary, and Γ represents the entire boundary.

The boundary conditions with respect to the displacement in the y direction are given by

$$v(s) = 0, \quad s \in \Gamma_y \tag{7.4.6}$$

and

$$f_y(s) = \sigma_y\sin\theta + \tau_{xy}\cos\theta, \quad s \in \Gamma - \Gamma_y \tag{7.4.7}$$

where Γ_y is that portion of the external boundary where the displacement in the y direction is zero, and $f_y(s)$ is the external force in the y direction at s.

Normal and shear strains are related to displacements by

$$\epsilon_x = \frac{\partial u}{\partial x}$$

$$\epsilon_y = \frac{\partial v}{\partial y} \tag{7.4.8}$$

$$\gamma_{xy} = \frac{\partial u}{\partial y} + \frac{\partial v}{\partial x}$$

where ϵ_x and ϵ_y are the normal strains and γ_{xy} is the shear strain ($\gamma_{xy} = \gamma_{yx}$). In the plane stress analysis, stress and strains are related by

$$\epsilon_x = \frac{\sigma_x}{E} - \nu\frac{\sigma_y}{E}$$

$$\epsilon_y = \frac{\sigma_y}{E} - \nu\frac{\sigma_x}{E} \qquad (7.4.9)$$

$$\gamma_{xy} = 2(1+\nu)\frac{\tau_{xy}}{E}$$

where E is Young's modulus and ν is the Poisson ratio. Equation (7.4.9) may be reversed and expressed as

$$\sigma_x = \alpha(\epsilon_x + \nu\epsilon_y)$$

$$\sigma_y = \alpha(\epsilon_y + \nu\epsilon_x) \qquad (7.4.10)$$

$$\tau_{xy} = \beta\gamma_{xy}$$

where $\tau_{xy} = \tau_{yx}$ and

$$\alpha = \frac{E}{1-\nu^2}$$

$$\beta = \frac{E}{2(1+\nu)}$$

In developing the finite element method, the displacement functions are approximated by the trial functions in the form of Eq. (7.2.15) (or see Eqs. (7.4.20) and (7.4.21)). The strain and stress functions are then given by Eqs. (7.4.8) and (7.4.10), respectively. This means that σ_x and τ_{xy} will be expressed in terms of the undetermined coefficients in the trial functions of u and v. Those undetermined coefficients are determined by requiring the weighted residuals for Eqs. (7.4.2) and (7.4.3) to become zero. The weighted residual for Eq. (7.4.2) is written and set to zero as

$$\int w\left[\frac{\partial\sigma_x}{\partial x} + \frac{\partial\tau_{yx}}{\partial y} + F_x\right]dV = 0 \qquad (7.4.11)$$

where w is a weighting function and the integral is extended over the entire domain, or equivalently,

$$\sum_e \int_e w\left[\frac{\partial\sigma_x}{\partial x} + \frac{\partial\tau_{yx}}{\partial y} + F_x\right]dV = 0 \qquad (7.4.12)$$

where e denotes a finite element and the integrals are extended over each element.

Using the Gauss theorem of integration, each integral in Eq. (7.4.12) becomes

$$\int_e w \left[\frac{\partial \sigma_x}{\partial x} + \frac{\partial \tau_{yx}}{\partial y} + F_x \right] dV$$

$$= \int_e \left[-\frac{\partial w}{\partial x} \sigma_x - \frac{\partial w}{\partial y} \tau_{xy} + w F_x \right] dV \qquad (7.4.13)$$

$$+ \int_{\Gamma_e} w [\sigma_x \cos \theta + \tau_{xy} \sin \theta] \, ds$$

where Γ_e denotes the surface integral along the element boundary. Introducing Eq. (7.4.13) into Eq. (7.4.12), we find that the surface integral along an internal boundary between two finite elements is counted twice, once for the element left to the boundary and another for the element right to the boundary. Across any nonslipping internal boundary, σ_x, σ_y, and τ_{xy} must be continuous. The angle θ at an internal boundary with respect to an element is different by π from that θ at the same position with respect to another element. When w is continuous across internal boundaries, the two surface integrals cancel each other. Thus all the surface integrals along internal boundaries disappear, and Eq. (7.4.11) becomes

$$\int \left[-\frac{\partial w}{\partial x} \sigma_x - \frac{\partial w}{\partial y} \tau_{xy} + w F_x \right] dV + \int_{\Gamma} w [\sigma_x \cos \theta + \tau_{xy} \sin \theta] \, ds = 0$$

$$(7.4.14)$$

where the first integral is extended over the whole domain and Γ in the second term means the whole external boundary.

The second term of Eq. (7.4.14) is now divided into two parts in accordance with the two types of boundary conditions:

$$\int_{\Gamma} = \int_{\Gamma_x} + \int_{\Gamma - \Gamma_x} \qquad (7.4.15)$$

The first part is where the displacement $u(s)$ is prescribed by Eq. (7.4.4). We can prescribe the weighting function as $w = 0$ along Γ_x. In fact, this is a natural consequence of using the Galerkin criterion in selecting weighting functions. Therefore, the first term of Eq. (7.4.15) vanishes. By using Eq. (7.4.5), the second term of Eq. (7.4.15) becomes

$$\int_{\Gamma - \Gamma_x} w f_x(s) \, ds \qquad (7.4.16)$$

Equation (7.4.14) now becomes

$$\int \left[-\frac{\partial w}{\partial x} \sigma_x - \frac{\partial w}{\partial y} \tau_{xy} + wF_x \right] dV + \int_{\Gamma - \Gamma_x} wf_x(s) \, ds = 0 \qquad (7.4.17)$$

Expressing σ_x and τ_{xy} in terms of u and v by using Eqs. (7.4.8) and (7.4.10), and introducing them into Eq. (7.4.17) yields

$$\int \left[-\frac{\partial w}{\partial x} \alpha \left(\frac{\partial u}{\partial x} + v \frac{\partial v}{\partial y} \right) - \frac{\partial w}{\partial y} \beta \left(\frac{\partial u}{\partial y} + \frac{\partial v}{\partial x} \right) + wF_x \right] dV$$

$$+ \int_{\Gamma - \Gamma_x} wf_x(s) \, ds = 0 \qquad (7.4.18)$$

Deriving the weighted residual for Eq. (7.4.3) and transforming it in a similar way, we obtain

$$\int \left[-\frac{\partial z}{\partial y} \alpha \left(v \frac{\partial u}{\partial x} + \frac{\partial v}{\partial y} \right) - \frac{\partial z}{\partial x} \beta \left(\frac{\partial u}{\partial y} + \frac{\partial v}{\partial x} \right) + zF_y \right] dV$$

$$+ \int_{\Gamma - \Gamma_y} zf_y(s) \, ds = 0 \qquad (7.4.19)$$

where $z = z(x, y)$ is a weighting function for Eq. (7.4.3).

In accordance with the finite element method, the trial function for u is given by

$$u(x, y) = \sum_{i=1}^{J} u_i \eta_i(x, y) \qquad (7.4.20)$$

where u_i is the displacement at the grid i and η_i is a prescribed shape function. In this summation, u_i are zero for the points on Γ_x. With the Galerkin criterion w is set to $\eta_i(x, y)$, $i = 1, 2, \ldots, J$, except for i on Γ_x. Similarly, v is written as

$$v(x, y) = \sum_{i=1}^{J} v_i \eta_i(x, y) \qquad (7.4.21)$$

where $v_i = 0$ for i on Γ_y. The weighting functions are set $z = \eta_i$, $i = 1, 2, \ldots, J$, except for i on Γ_y.

Introducing those trial functions and the weighting functions into Eq. (7.4.18) yields

$$\sum_i \left\{ \left\langle \alpha \frac{\partial \eta_j}{\partial x} \frac{\partial \eta_i}{\partial x} \right\rangle u_i + \left\langle \alpha \nu \frac{\partial \eta_j}{\partial x} \frac{\partial \eta_i}{\partial y} \right\rangle v_i + \left\langle \beta \frac{\partial \eta_j}{\partial y} \frac{\partial \eta_i}{\partial y} \right\rangle u_i + \left\langle \beta \frac{\partial \eta_j}{\partial y} \frac{\partial \eta_i}{\partial x} \right\rangle v_i \right\}$$

$$= \langle \eta_j F_x \rangle + \int_{\Gamma - \Gamma_x} \eta_j f_x(s)\, ds \qquad (7.4.22)$$

where the brackets denote the volume integral over the whole domain and $j = 1, 2, \ldots, J$ except for the grids where u_i are prescribed. Similarly, Eq. (7.4.19) becomes

$$\sum_i \left\{ \left\langle \alpha \nu \frac{\partial \eta_j}{\partial y} \frac{\partial \eta_i}{\partial x} \right\rangle u_i + \left\langle \alpha \frac{\partial \eta_j}{\partial y} \frac{\partial \eta_i}{\partial y} \right\rangle v_i + \left\langle \beta \frac{\partial \eta_j}{\partial x} \frac{\partial \eta_i}{\partial y} \right\rangle u_i + \left\langle \beta \frac{\partial \eta_j}{\partial x} \frac{\partial \eta_i}{\partial x} \right\rangle v_i \right\}$$

$$= \langle \eta_j F_y \rangle + \int_{\Gamma - \Gamma_y} \eta_j f_y(s)\, ds \qquad (7.4.23)$$

where $j = 1, 2, \ldots, J$ except for the grids on Γ_y.

If we use the linear shape functions defined by Eq. (7.2.14) for triangular finite elements, each term on the left side of Eq. (7.4.22) becomes

$$\left\langle \alpha \frac{\partial \eta_j}{\partial x} \frac{\partial \eta_i}{\partial x} \right\rangle = \sum_e (\alpha B_j B_i \Delta)^{(e)}$$

$$\left\langle \alpha \nu \frac{\partial \eta_j}{\partial x} \frac{\partial \eta_i}{\partial y} \right\rangle = \sum_e (\alpha \nu B_j C_i \Delta)^{(e)}$$

$$\left\langle \beta \frac{\partial \eta_j}{\partial y} \frac{\partial \eta_i}{\partial y} \right\rangle = \sum_e (\beta C_j C_i \Delta)^{(e)}$$

$$\left\langle \beta \frac{\partial \eta_j}{\partial y} \frac{\partial \eta_i}{\partial x} \right\rangle = \sum_e (\beta C_j B_i \Delta)^{(e)}$$

where $(ab)^{(e)}$ is the abbreviation of $a^{(e)} b^{(e)}$, and $\Delta^{(e)}$ is the area of the triangle e, and $B_i^{(e)} = C_i^{(e)} = 0$ if i is not an apex of e.

Example 7.8

Derive the finite element equation for the plane stress of the triangular plane material shown in Fig. 7.9 under the following conditions: (a) Apices 1 and 2 are fixed, (b) a force with the components $\bar{f}_x$ and $\bar{f}_y$ are applied to the apex 3, (c) we assume there is no body force, and (d) we use only one triangular element.

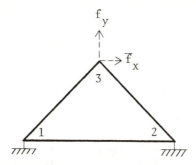

Figure 7.9 A triangular plane material fixed at two apices.

The forces applied to the apex 3 may be interpreted as $f_x = \bar{f}_x \delta(s - s_3)$ and $f_y = \bar{f}_y \delta(s - s_3)$. The boundary conditions for apices 1 and 2 are $v_1 = u_1 = v_2 = u_2 = 0$. Therefore, Eqs. (7.4.22) and (7.4.23) are reduced to two equations:

$$\int \left[\alpha \left(\frac{\partial \eta_3}{\partial x} \right)^2 u_3 + \beta \left(\frac{\partial \eta_3}{\partial y} \right)^2 u_3 + (\alpha \nu + \beta) \frac{\partial \eta_3}{\partial x} \frac{\partial \eta_3}{\partial y} v_3 \right] dV = \bar{f}_x$$

$$\int \left[(\alpha \nu + \beta) \frac{\partial \eta_3}{\partial x} \frac{\partial \eta_3}{\partial y} u_3 + \alpha \left(\frac{\partial \eta_3}{\partial y} \right)^2 v_3 + \beta \left(\frac{\partial \eta_3}{\partial x} \right)^2 v_3 \right] dV = \bar{f}_y$$

Introducing (7.2.14) and upon integration, the above equations become

$$(\alpha B_3^2 + \beta C_3^2) u_3 + (\alpha \nu + \beta)(B_3 C_3) v_3 = \frac{\bar{f}_x}{\Delta}$$

$$(\alpha \nu + \beta)(B_3 C_3) u_3 + (\alpha C_3^2 + \beta B_3^2) v_3 = \frac{\bar{f}_y}{\Delta}$$

where both equations are divided by Δ and the superscript (e) has been omitted.

Example 7.9

The boundary condition for apex 2 of Example 7.9 is now altered. Apex 2 is free in the x direction as shown in Fig. 7.10 but cannot move in the y direction. Derive the finite element equation, assuming other conditions are the same as Example 7.9.

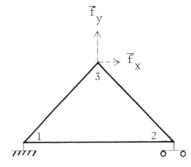

Figure 7.10 A triangular plane material with a sliding apex.

The boundary conditions for this problem are $u_1 = v_1 = v_2 = 0$. The three equations for unknowns v_2, u_3, and v_3 are as follows:

$$(\alpha B_2^2 + \beta C_2^2)u_2 + (\alpha B_2 B_3 + \beta C_2 C_3)u_3$$
$$+ (\alpha \nu B_2 C_3 + \beta C_2 B_3)v_3 = 0$$
$$(\alpha B_3 B_2 + \beta C_3 C_2)u_2 + (\alpha \beta_3^2 + \beta C_3^2)u_3$$
$$+ (\alpha \nu + \beta)(B_3 C_3)v_3 = \frac{\bar{f}_x}{\Delta}$$
$$(\alpha \nu C_3 B_2 + \beta B_3 C_2)u_2 + (\alpha \nu + \beta)(B_3 C_3)u_3$$
$$+ (\alpha C_3^2 + \beta B_3^2)v_3 = \frac{\bar{f}_y}{\Delta}$$

Displacements of some grid points are always prescribed, so the number of unknowns becomes less than $2J$ where J is the total number of grid points. It is, however, more convenient to deal with $2J$ variables as if all of them are unknowns, because when the finite element equations are expressed in matrix notations the order of unknowns (or rows) is not affected by the presence of prescribed displacements. As discussed in Section 7.3, this can be done by considering the boundary conditions as members of simultaneous equations.

The finite element equations for the whole system are derived through the following steps:

1. Derive the finite element equations by using Eqs. (7.4.22) and (7.4.23) as if none of the displacements were prescribed.
2. Express the finite element equations in the matrix and vector notations as

$$\mathbf{K}\boldsymbol{\phi} = \mathbf{F} \qquad\qquad (7.4.24)$$

where

$$\mathbf{\phi} = \text{col}\,[u_1, v_1, u_2, v_2, \ldots] \tag{7.4.25}$$

$\mathbf{K}$ is a square matrix of order $2J$ (it is called the *stiffness matrix*), and $\mathbf{F}$ is a vector representing the external forces.

3. a. Set all the elements in the row of $\mathbf{K}$, which corresponds to a prescribed displacement, to zero.

   b. Set the diagonal element of the row to unity.

   c. Set the corresponding element in $\mathbf{F}$ to the prescribed value of the displacement.

4. Repeat Step 3 to all the prescribed displacements.

As the number of the triangular elements increases, the derivation of the finite element equation for each grid become more complicated. There is another approach that is simpler than applying Eqs. (7.4.22) and (7.4.23) in a straightforward manner. We consider a single triangular element e and derive the finite element equations for it, assuming there are no constraints to the three apices, i, j, and k (Fig. 7.11). The two components of forces at an apex i of e are denoted by $f_{xi}^{(e)}$ and $f_{yi}^{(e)}$. The equation for e that involves $f_{xi}^{(e)}$ may be written in the form,

$$g_{xii}^{(e)}u_i + h_{xii}^{(e)}v_i + g_{xij}^{(e)}u_j + h_{xij}^{(e)}v_j + g_{xik}^{(e)}u_k + h_{xik}^{(e)}v_k = f_{xi}^{(e)} \tag{7.4.26}$$

where g and h are coefficients, and the body force is neglected for simplicity. A similar equation for another element e'' that shares i as an apex may be written

$$g_{xii}^{(e'')}u_i + h_{xii}^{(e'')}v_i + g_{xin}^{(e'')}u_n + h_{xin}^{(e'')}v_n + g_{xim}^{(e'')}u_m + h_{xim}^{(e'')}v_m = f_{xi}^{(e'')} \tag{7.4.27}$$

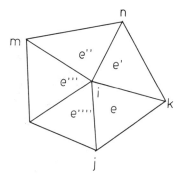

Figure 7.11 Triangular finite elements around the grid point i.

Since an external force at i, that is, f_{xi} (usually zero if i is an internal point), must balance with the total of the forces at each elements, we have

$$f_{xi}^{(e)} + f_{xi}^{(e')} + f_{xi}^{(e'')} + \cdots = f_{xi} \qquad (7.4.28)$$

Introducing Eq. (7.4.26), Eq. (7.4.27), and the equivalents for other elements to the left side of Eq. (7.4.28), we obtain a finite element equation for the whole system. Repeating this procedure to other components and all other grid points, $2J$ equations are obtained. The remainder of the procedures is the same as Steps 2, 3, and 4 of the other approach. This approach will make the programming easier.

The properties of the stiffness matrix **K** in Eq. (7.4.24) are somewhat similar to **H** of Eq. (7.3.8), although the order of **K** is twice as large for the same number of grid points. **K** is a sparce matrix and positive definite. Numbering the grid points is very important to minimize the bandwidth of **K**, similarly to the case of **H**. Equation (7.4.24) can be solved by the Gaussian elimination or an iterative scheme.[1]

7.5 DERIVATION OF THE FINITE ELEMENT EQUATIONS FOR VISCOUS INCOMPRESSIBLE FLUID FLOW

Finite element fluid flow analysis is relatively new. However, because of its simple algorithm and versatile applicability, the use of the finite element method in fluid flow analysis has rapidly increased during the past few years.†[14–28] Finite element analysis for viscous incompressible fluid flow is more difficult than for stress or heat transfer analyses for the following reasons. First, Navier–Stokes equations are highly nonlinear. There is no variational principle found for viscous incompressible fluid flow except for special cases. The weighted residual method is, therefore, used to derive the finite element equations. Second, the boundary conditions for a flow system cannot be uniquely defined in many cases. For example, as discussed in Section 5.6, the boundary conditions are affected by upstream and downstream conditions. This problem is not, however, particular to the finite element approach but rather inherent in the fluid flow problems. Third, the solution involves a wide range of numbers that are significantly different in orders of magnitude. Since the number of digits in a computer memory is limited, the round-off error could be a serious problem.

In this section, we derive the discrete approximation to the Navier–Stokes equation for two-dimensional incompressible fluid flow. Several methods to solve nonlinear simultaneous equations are introduced in Section 7.6. Finite element methods can be applied either to the vorticity and stream function

† See also References 30–32.

approach, or to the velocity and pressure approach.[30] Several investigators, however, have pointed out that the velocity and pressure system of the variables is much more preferred to the vorticity and stream function approach. This is first because little difficulty is encountered with the continuity condition in the finite element method and second because boundary conditions can be incorporated with the velocity and pressure system in an easy and straightforward manner. We consider the two-dimensional Navier–Stokes equation and the continuity equation represented by Eqs. (5.6.1) through (5.6.3). The Navier–Stokes equation may be also expressed in terms of a stress tensor. This expression is more convenient if the boundary condition is given in the form of surface stress. We first study the finite element method with the boundary conditions specified individually for u, v, and p, and later with the surface stress boundary conditions.

7.5.1 With Boundary Conditions for u, v, and p

The boundary conditions for u may be divided into two kinds:

1. $$u = U_0, \quad \text{prescribed value of } u \text{ (Dirichlet type)} \tag{7.5.1}$$

2. $$\frac{\partial u}{\partial n} = \gamma_x, \quad \text{prescribed gradient of } u \text{ (Neumann type)} \tag{7.5.2}$$

where $\partial/\partial n$ is the derivative outward normal to the boundary. For simplicity, let us assume the boundaries are parallel to the x or y coordinate. Boundary condition 1 applies to the inlet boundary if the inlet velocity is specified, or to the horizontal and vertical walls if the walls are nonslip boundary ($U_0 = 0$). Boundary condition 2 applies to the horizontal free-slip wall in the form $\partial u/\partial y = 0$, or to the vertical inlet or outlet boundary if only the pressure is specified (if the parallel flow is assumed behind the outlet boundary, we set $\partial u/\partial y = 0$). The boundary conditions for v are similarly written

3. $$v = V_0 \tag{7.5.3}$$

4. $$\frac{\partial v}{\partial n} = \gamma_y \tag{7.5.4}$$

The boundary condition for p must be considered only where p is prescribed. If the absolute value of p is not specified anywhere, the pressure distribution function is arbitrary by a constant. In order to assure the numerical solution for p, the absolute value of one grid point must be specified. The Nuemann boundary condition for the pressure is required in the MAC method in Section 5.7 wherever p is not prescribed. This is unnecessary in the finite element method because the pressure and velocities are simultaneously solved.

The weighting functions for the three equations are denoted, respectively, by u^*, v^*, and p^*. The superscript $*$ indicates that these functions are just weighting functions in accordance with the Galerkin criterion, but there is no particular physical meaning in those notations such as adjoint functions. The boundary conditions for u^*, v^*, and p^* are similar to those for u, v, and p, respectively. The weighting function u^* is set to a constant wherever u is specified by the Dirichlet boundary condition. We set $u^* = 0$ for simplicity. The weighting functions v^* and p^* are also set to zero if v and p are specified, respectively, by the Dirichlet boundary conditions.

Before introducing any particular forms for the field variables and weighting functions, we premultiply Eqs. (5.6.1), (5.6.2), and (5.6.3) by u^*, v^*, and p^*, respectively, integrate over the whole domain, and set each integral to zero. Thus we have

$$\int u^*\left[\frac{\partial u}{\partial t}+u\frac{\partial u}{\partial x}+v\frac{\partial u}{\partial y}+\frac{1}{\rho}\frac{\partial p}{\partial x}-\nu\nabla^2 u-F_x\right]dV=0 \qquad (7.5.5)$$

$$\int v^*\left[\frac{\partial v}{\partial t}+u\frac{\partial v}{\partial x}+v\frac{\partial v}{\partial y}+\frac{1}{\rho}\frac{\partial p}{\partial y}-\nu\nabla^2 v-F_y\right]dV=0 \qquad (7.5.6)$$

$$\int p^*\left[\frac{\partial u}{\partial x}+\frac{\partial v}{\partial y}\right]dV=0 \qquad (7.5.7)$$

where body forces are represented by F_x and F_y. By using the Green's theorem, the fifth term in Eq. (7.5.5) is transformed to

$$-\nu\int u^*\nabla^2 u\,dV=-\nu\int_\Gamma u^*\frac{\partial}{\partial n}u\,ds+\nu\int(\nabla u^*)(\nabla u)\,dV \qquad (7.5.8)$$

where Γ denotes that the surface integral is performed along the whole external boundary. The surface integral in Eq. (7.5.8) is further divided into two parts

$$\int_\Gamma u^*\frac{\partial}{\partial n}u\,ds=\int_I u^*\frac{\partial}{\partial n}u\,ds+\int_{II} u^*\frac{\partial}{\partial n}u\,ds \qquad (7.5.9)$$

where the first part is for the portion of the Dirichlet boundary and becomes zero because $u^* = 0$, and the second part is for the Neumann boundary. Introducing Eq. (7.5.2) into Eq. (7.5.9) yields

$$\int_\Gamma u^*\frac{\partial}{\partial n}u\,ds=\int_{II} u^*\gamma_x(s)\,ds \qquad (7.5.10)$$

Thus, we rewrite Eqs. (7.5.5) and (7.5.6) as

$$\int \left[u^* \left(\frac{\partial u}{\partial t} + u \frac{\partial u}{\partial x} + v \frac{\partial u}{\partial y} + \frac{1}{\rho} \frac{\partial p}{\partial x} - F_x \right) \right.$$

$$\left. + \nu (\nabla u^*)(\nabla u) \right] dV - \nu \int_{II} u^* \gamma_x(s) \, ds = 0 \qquad (7.5.5a)$$

$$\int \left[v^* \left(\frac{\partial v}{\partial t} + u \frac{\partial v}{\partial x} + v \frac{\partial v}{\partial y} + \frac{1}{\rho} \frac{\partial p}{\partial y} - F_y \right) \right.$$

$$\left. + \nu (\nabla v^*)(\nabla v) \right] dV - \nu \int_{II} v^* \gamma_y(s) \, ds = 0 \qquad (7.5.6a)$$

In applying the finite element method, the time-dependent field variables, $u(x, y, t)$, $v(x, y, t)$, and $p(x, y, t)$ are approximated by linear combination of prescribed shape functions as

$$u(x, y, t) = \sum_i u_i(t) \eta_i(x, y) \qquad (7.5.11)$$

$$v(x, y, t) = \sum_i v_i(t) \eta_i(x, y) \qquad (7.5.12)$$

$$p(x, y, t) = \sum_i p_i(t) \eta_i(x, y) \qquad (7.5.13)$$

where $u_i(t)$, $v_i(t)$, and $p_i(t)$ are the values of u, v, and p at grid i and time t; η are the shape functions. Different types of shape function may be applied to each of u, v, and p. The values u_i, v_i, and p_i are unknowns unless specified by the Dirichlet boundary conditions. If, for example, piecewise linear functions with triangular element are used, the total number of unknowns is $N = 3J - M$, where J is the total number of grid points and M is the total number of u_i, v_i, and p_i specified by the Dirichlet boundary conditions. Choosing the type of finite elements and the order of shape function is still a state of art, but no definite criteria is established other than the general criteria stated in Section 7.2. Oden[20] used linear functions for all pressures and velocities, and later second order polynomials[28] for all components. Temam[26] used linear functions for all pressures and velocities with nonconforming triangular elements. Kawahara,[23,24] Hood, and Yakawa[22] used second order polynomials for velocity components and linear functions for pressures with triangular elements.

Adopting Galerkin's criterion (see Section 6.2), the weighting functions
are written as

$$u^*(x, y, t) = \sum_i u_i^*(t)\eta_i(x, y) \tag{7.5.14}$$

$$v^*(x, y, t) = \sum_i v_i^*(t)\eta_i(x, y) \tag{7.5.15}$$

$$p^*(x, y, t) = \sum_i p_i^*(t)\eta_i(x, y) \tag{7.5.16}$$

where * denotes that the weighting functions are chosen in accordance with
the Galerkin criterion. In Eqs. (7.5.14) through (7.5.16) the values of u^*, v_i^*,
and p_i^* are set to zero if the corresponding values of u_i, v_i, and p_i are specified
by the Dirichlet boundary conditions. Thus the number of unknowns is
matched with the number of degrees of freedom in the weighting functions.

Equations (7.5.11) through (7.5.16) are introduced into Eqs. (7.5.5a),
(7.5.6a), and (7.5.7). In accordance with the weighted residual method, Eqs.
(7.5.5a), (7.5.6a), and (7.5.7) must be satisfied for any arbitrary changes of
u_j^*, v_j^*, and p_j^*, respectively. Therefore, by taking the partial derivatives of
the integrals with respect to u_j^*, v_j^*, and p_j^*, respectively, and setting each
result to zero, we obtain

$$\sum_i \left[\langle \eta_j\eta_i \rangle \frac{\partial}{\partial t} + \sum_k \left\langle \eta_j\eta_k \frac{\partial\eta_i}{\partial x} \right\rangle u_k + \sum_k \left\langle \eta_j\eta_k \frac{\partial\eta_i}{\partial y} \right\rangle v_k \right.$$

$$\left. + \nu \langle (\nabla\eta_j)(\nabla\eta_i) \rangle \right] u_i(t) + \frac{1}{\rho}\sum_i \left\langle \eta_j \frac{\partial\eta_k}{\partial x} \right\rangle p_i(t) \tag{7.5.17}$$

$$= \nu \int_{\text{II}} \gamma_x\eta_j \, ds + \langle \eta_j F_x \rangle \equiv R_{x_j}(t)$$

$$\sum_i \left[\langle \eta_j\eta_i \rangle \frac{d}{dt} + \sum_k \left\langle \eta_j\eta_k \frac{\partial\eta_i}{\partial x} \right\rangle u_k(t) + \sum_k \left\langle \eta_j\eta_k \frac{\partial\eta_i}{\partial y} \right\rangle v_k(t) \right.$$

$$\left. + \nu \langle (\nabla\eta_j)(\nabla\eta_i) \rangle \right] v_i(t) + \frac{1}{\rho}\sum_i \left\langle \eta_i \frac{\partial\eta_i}{\partial y} \right\rangle p_i(t) \tag{7.5.18}$$

$$= \nu \int_{\text{II}} \eta_j\gamma_y \, ds + \langle \eta_j F_y \rangle \equiv R_{yj}(t)$$

$$\sum_i \left\langle \eta_j \frac{\partial}{\partial x}\eta_i \right\rangle u_i(t) + \sum_i \left\langle \eta_j \frac{\partial}{\partial y}\eta_i \right\rangle v_i(t) = 0 \tag{7.5.19}$$

In Eqs. (7.5.17) through (7.5.19), $\langle\ \rangle$ means the volume integral over the
whole region. Equation (7.5.17) applies to $j = 1, 2, \ldots, J$ except for those
grids where u_j are specified by the Dirichlet boundary conditions. These
missing equations occur because u_j^* for those points were set to zero. For the

the same reason, Eq. .(7.5.18) is missing for j if v_j is specified, and Eq. (7.5.19) is missing for j if p_j is specified.

Equations (7.5.17) through (7.5.19) represent a simultaneous set of first order, nonlinear, ordinary differential equations and are called *finite element equations*. For a steady state problem, the time derivatives in Eqs. (7.5.17) and (7.5.18) are set to zero and the nonlinear simultaneous equations are solved. For a time dependent problem, both forward and backward schemes with respect to the time derivatives may be developed. With the backward difference scheme the equations for each time step become fully implicit so the same solution techniques as for steady state problems may be used. With the forward difference scheme the equations become explicit but are subject to instability unless very short time step is used. The solution for the finite element equations is discussed in Section 7.6.

7.5.2 With Surface Forces as Boundary Conditions[23,24,30,31]

We consider here the steady state case for simplicity. The normal and shear stress are given by

$$\tau_{xx} = -p + 2\mu \frac{\partial u}{\partial x} \tag{7.5.20}$$

$$\tau_{xy} = \mu \left(\frac{\partial v}{\partial x} + \frac{\partial u}{\partial y} \right) \tag{7.5.21}$$

$$\tau_{yy} = -p + 2\mu \frac{\partial v}{\partial y} \tag{7.5.22}$$

where τ_{xx} and τ_{yy} are normal stress, τ_{xy} is shear stress and $\mu = \nu\rho$. By using these relations, Eqs. (5.6.1) and (5.6.2) become

$$u \frac{\partial u}{\partial x} + v \frac{\partial u}{\partial y} - \frac{1}{\rho} \left(\frac{\partial \tau_{xx}}{\partial x} + \frac{\partial \tau_{xy}}{\partial y} \right) - F_x = 0 \tag{7.5.23}$$

$$u \frac{\partial v}{\partial x} + v \frac{\partial v}{\partial y} - \frac{1}{\rho} \left(\frac{\partial \tau_{xy}}{\partial x} + \frac{\partial \tau_{yy}}{\partial y} \right) - F_y = 0 \tag{7.5.24}$$

respectively, where Eq. (5.6.3) was also used and the time derivatives were set to zero. The boundary condition for the Navier–Stokes equation with the stress tensor may be divided into the two parts: (1) u and v are prescribed, and (2) the surface forces are specified as

$$\tau_{xx}l + \tau_{xy}m = f_x \tag{7.5.25}$$

$$\tau_{xy}l + \tau_{yy}m = f_y \tag{7.5.26}$$

where f_x and f_y are prescribed surface forces, and l and m are the directional cosine of the unit normal vector on the boundary surface.

In accordance with the weighted residual method, we multiply Eq. (7.5.23) by the weighting function $u^*(x, y)$ and integrate over the whole domain as

$$\int u^* \left[u \frac{\partial u}{\partial x} + v \frac{\partial u}{\partial y} - \frac{1}{\rho} \left(\frac{\partial \tau_{xx}}{\partial x} + \frac{\partial \tau_{xy}}{\partial y} \right) - F_x \right] dV = 0 \qquad (7.5.27)$$

Applying the Gauss theorem of integration to the derivatives of the stress terms in Eq. (7.5.27) yields

$$\int \left[u^* \left(u \frac{\partial u}{\partial x} + v \frac{\partial u}{\partial y} \right) + \frac{1}{\rho} \left(\tau_{xx} \frac{\partial u^*}{\partial x} + \tau_{xy} \frac{\partial u^*}{\partial y} \right) - F_x \right] dV$$
$$- \frac{1}{\rho} \int_{\Gamma} u^* (\tau_{xx} l + \tau_{xy} m) \, ds = 0 \qquad (7.5.28)$$

The surface integral in Eq. (7.5.28) is divided into two parts,

$$\int_{\Gamma} = \int_{I} + \int_{II} \qquad (7.5.29)$$

In Eq. (7.5.29), I is the boundary on which u and v are specified, and u^* and v^* are set to zero. Therefore, the first surface integral in Eq. (7.5.29) vanishes. By using Eqs. (7.5.25) the surface integral in Eq. (7.5.28) becomes

$$\int_{II} u^* f_x \, ds \qquad (7.5.30)$$

Thus Eq. (7.5.28) becomes

$$\int \left[u^* \left(u \frac{\partial u}{\partial x} + v \frac{\partial u}{\partial y} \right) + \frac{1}{\rho} \left(\tau_{xx} \frac{\partial u^*}{\partial x} + \tau_{xy} \frac{\partial u^*}{\partial y} \right) \right] dV$$
$$= \frac{1}{\rho} \int_{II} u^* f_x \, ds + \int u^* F_x \, dV \qquad (7.5.31)$$

Applying the similar procedure to Eq. (7.5.24), we obtain

$$\int \left[v^* \left(u \frac{\partial v}{\partial x} + v \frac{\partial v}{\partial y} \right) + \frac{1}{\rho} \left(\tau_{xy} \frac{\partial v^*}{\partial x} + \tau_{yy} \frac{\partial v^*}{\partial y} \right) \right] dV$$
$$= \frac{1}{\rho} \int_{II} v^* f_y \, ds + \int v^* F_y \, dV \qquad (7.5.32)$$

The derivation of the finite element equations is similar to the case in Section 7.5.1 and straightforward by introducing Eqs. (7.5.11) through (7.5.16) into Eqs. (7.5.20) through (7.5.22) and Eqs. (7.5.31) and (7.5.32).

Example 7.10[30]

The Navier–Stokes equation, Eqs. (5.6.1) through (5.6.3), was solved for a conduit as shown in Fig. 7.12. The boundary conditions were set as

1. $u = v = 0$, along the solid walls,
2. $u = U_0(y)$ and $v = 0$ at the left boundary, and
3. $f_x = f_y = 0$ along the right boundary.

The polynomial of sixth order in the form of Eq. (7.2.17) was used for u, u^*, v, and v^* for each triangular element with six grid points as shown in Fig. 7.3. On the other hand, piecewise linear functions were used for p with three apices of each triangular element as grid points. The result of a computation is illustrated in Fig. 7.12. The Newton–Raphson method was used to solve the nonlinear finite element equations. The arrows in the figure show the magnitude and direction of velocities, and the numbers in parentheses are the pressure.

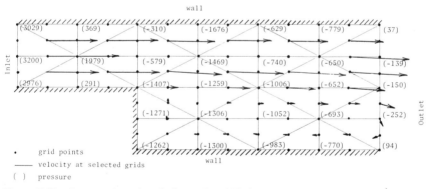

Figure 7.12 Computed steady velocity at $R_e = 240$. (Reproduced from Fig. 6 of Reference 30).

7.6 SOLUTION OF THE NONLINEAR FINITE ELEMENT EQUATIONS FOR FLUID FLOW

The total number of finite element equations derived in Section 7.5 is less than $3J$ if some of u_i, v_i, and p_i are specified by the Dirichlet boundary conditions. However, it is more convenient to deal with all u_i, v_i, and p_i as unknowns, so we consider all of the velocities and pressures as unknowns and the Dirichlet boundary conditions as members of simultaneous equations. Therefore, we have $3J$ equations including the Dirichlet boundary conditions for $3J$ unknowns.

We now define the vector representing all the unknowns:

$$\mathbf{z}(t) = \text{col}\,[u_1, v_1, p_1, u_2, v_2, p_2, \ldots, u_J, v_J, p_J] \qquad (7.6.1)$$

The system of nonlinear finite element equations and the Dirichlet boundary conditions can be compactly expressed in the matrix notations as

$$\mathbf{M}\frac{d}{dt}\mathbf{z}(t) + \mathbf{K}(\mathbf{z})\mathbf{z} + \mathbf{L}\mathbf{z} = \mathbf{R} \qquad (7.6.2)$$

where $\mathbf{M}$, $\mathbf{K}$, and $\mathbf{L}$ are square matrices and $\mathbf{R}$ is the vector representing the inhomogeneous terms. The elements of $\mathbf{R}$ consist of the external force and the boundary values of the velocity and pressure. The matrix $\mathbf{K}$ is a function of $\mathbf{z}$ because of the nonlinearity of the finite element equations. $\mathbf{K}$ represents the coefficients of the finite element equations that are linear homogeneous functions of $\mathbf{z}$. $\mathbf{L}$ includes all the constant coefficients.

When the first term of Eq. (7.6.2) is set to a null vector, Eq. (7.6.2) is a steady state problem. We first consider the three techniques to solve the steady state equation:

$$\mathbf{\Omega z} \equiv [\mathbf{K}(\mathbf{z}) + \mathbf{L}]\mathbf{z} = \mathbf{R} \qquad (7.6.3)$$

SUCCESSIVE SUBSTITUTION METHOD

Denoting the iterative number by n, this method is expressed by

$$[\mathbf{K}(\mathbf{z}_{n-1}) + \mathbf{L}]\mathbf{z}_n = \mathbf{R} \qquad (7.6.4)$$

The results of the previous iterations are used to calculate the coefficients of the nonlinear terms, namely, $\mathbf{K}$. The initial vector, $\mathbf{z}_0$, is set to any known approximation or simply to a null vector. Equation (7.6.4) is solved by the Gaussian elimination method.

The success with this method is reported to be less sensitive to the initial guess compared with the next method.

NEWTON–RAPHSON METHOD

The successive approximation is again denoted by $\mathbf{z}_n$ and the initial guess by $\mathbf{z}_0$. If the approximation $\mathbf{z}_{n-1}$ is known, the next approximation can be expressed by

$$\mathbf{z}_n = \mathbf{z}_{n-1} + \delta\mathbf{z}_n \qquad (7.6.5)$$

where δz_n is an incremental vector. Introducing Eq. (7.6.5) into Eq. (7.6.3) yields

$$[\mathbf{K}(z_{n-1}) + \mathbf{K}(\delta z_n) + \mathbf{L}](z_{n-1} + \delta z_n) = \mathbf{R} \qquad (7.6.6)$$

where the homogeneity of the elements of $\mathbf{K}$ in z is used. By using the same property of $\mathbf{K}$ we can find such a matrix $\mathbf{K}'$ as

$$\mathbf{K}(z_a)z_b = \mathbf{K}'(z_b)z_a \qquad (7.6.7)$$

where z_a and z_b are independent vectors. Using Eq. (7.6.7) and neglecting the second order term, Eq. (7.6.6) may be written as

$$[\mathbf{K}'(z_{n-1}) + \mathbf{K}(z_{n-1}) + \mathbf{L}]\delta z_n = \mathbf{R} - [\mathbf{K}(z_{n-1}) + \mathbf{L}]z_{n-1} \qquad (7.6.8)$$

In Eq. (7.6.7), the coefficient matrix for δz_n and the right side is known since z_{n-1} is given. Solving Eq. (7.6.8) and substituting the solution into Eq. (7.6.5) yields the new approximation.

It is expected that if z_{n-1} is close to the exact solution, δz_n is a small vector, so that neglecting the second order term in Eq. (7.6.6) has a very small effect, thus yielding a more accurate solution z_n than z_{n-1}. If a good initial guess z_0 is available, the convergence of this method is faster than the previous method. On the other hand, if the initial estimate z_0 is not close to the exact solution, the present scheme may not converge to the exact solution.

PERTURBATION METHOD

This method is useful when the solution to a reference problem that is similar to the given problem is known. The reference problem is assumed to be different from the given problem in boundary conditions. The given problem $\mathbf{\Omega}z = \mathbf{R}$ may be related to the reference problem $\mathbf{\Omega}z_0 = \mathbf{R}_0$ by

$$\mathbf{R} = \mathbf{R}_0 + \epsilon \mathbf{R}_1 \qquad (7.6.9)$$

$$z = z_0 + \epsilon z_1 + \epsilon^2 z_2 + \epsilon^3 z_3 + \cdots \qquad (7.6.10)$$

where ϵ is a parameter.

Introducing Eqs. (7.6.9) and (7.6.10) into Eq. (7.6.3) and equating the terms of the same order in ϵ, the following equations are obtained:

$$\epsilon^0. \quad [\mathbf{K}(z_0) + \mathbf{L}]z_0 = \mathbf{R}_0 \qquad (7.6.11a)$$

$$\epsilon^1. \quad \mathbf{K}(z_1)z_0 + \mathbf{K}(z_0)z_1 + \mathbf{L}z_1 = \mathbf{R}_1 \qquad (7.6.11b)$$

$$\epsilon^2. \quad \mathbf{K}(z_2)z_0 + \mathbf{K}(z_0)z_2 + \mathbf{L}z_2 = -\mathbf{K}(z_1)z_1 \qquad (7.6.11c)$$

$$\epsilon^3. \quad \mathbf{K}(z_3)z_0 + \mathbf{K}(z_0)z_3 + \mathbf{L}z_3 = -\mathbf{K}(z_1)z_2 - \mathbf{K}(z_2)z_1 \qquad (7.6.11d)$$

$$\epsilon^n. \quad \mathbf{K}(z_n)z_0 + \mathbf{K}(z_0)z_n + \mathbf{L}z_n = -\sum_{m=1}^{n-1} \mathbf{K}(z_m)z_{n-m} \qquad (7.6.11e)$$

By using Eq. (7.6.7), Eqs. (7.6.11b) through (7.6.11e) can be written in the form:

$$[\mathbf{K}'(\mathbf{z}_0) + \mathbf{K}(\mathbf{z}_0) + \mathbf{L}]\mathbf{z}_n = \mathbf{T}_n, \qquad n \geq 1 \qquad (7.6.12)$$

Equation (7.6.12) is successively solved until sufficient convergence of Eq. (7.6.10) is attained.

Even when the reference problem is not close to the given problem, one can apply the present method with several steps. Consider the sequence of the inhomogeneous terms, $\mathbf{R}_0$, $\mathbf{R}'$, $\mathbf{R}''$, ..., $\mathbf{R}$, where $\mathbf{R}_0$ is the reference vector, $\mathbf{R}$ is the given vector, and $\mathbf{R}'$, $\mathbf{R}''$, ... gradually approach $\mathbf{R}$. The perturbation method is applied to each step of $\mathbf{R}_0 \rightarrow \mathbf{R}' \rightarrow \mathbf{R}'' \rightarrow \cdots \rightarrow \mathbf{R}$.

Once the solutions for a steady state problem are established, a transient problem can be solved by applying them in each time step interval. By using the backward difference formula to Eq. (7.6.2), we have

$$\mathbf{K}(\mathbf{z}^{(m+1)})\mathbf{z}^{(m+1)} + \left(\frac{1}{\Delta t}\mathbf{M} + \mathbf{L}\right)\mathbf{z}^{(m+1)} = \mathbf{R} + \frac{1}{\Delta t}\mathbf{M}\mathbf{z}^{(m)} \qquad (7.6.13)$$

where Δt is the time interval and $\mathbf{z}^{(m)} = \mathbf{z}(m\,\Delta t)$. Equation (7.6.13) has the same form as Eq. (7.6.3), so the three solution methods for Eq. (7.6.3) also apply to Eq. (7.6.13) for each time step. The solution for the previous step is a good initial guess in the successive substitution and Newton–Raphson methods. It can be also used as the reference solution for the perturbation method.

Example 7.11

Let us define $\mathbf{K}$, $\mathbf{L}$, and $\mathbf{R}$ in Eq. (7.6.3) by

$$\mathbf{K}(\mathbf{z}) = \alpha \begin{bmatrix} (z)_1 & 0 & 0 \\ 0 & (z)_2 & 0 \\ 0 & 0 & (z)_3 \end{bmatrix}, \quad \mathbf{L} = \begin{bmatrix} 1 & -1 & 0 \\ -1 & 2 & -1 \\ 0 & -1 & 2 \end{bmatrix}, \quad \mathbf{R} = \begin{bmatrix} 1 \\ 0 \\ 0 \end{bmatrix}$$

where $(z)_i$ is the ith element of $\mathbf{z}$. If $\alpha = 0$, the solution to Eq. (7.6.3) is $\mathbf{z} = \text{col}[3, 2, 1]$. We use the following reference equation:

$$\mathbf{\Omega}\mathbf{z}_0 \equiv (\mathbf{K}(\mathbf{z}_0) + \mathbf{L})\mathbf{z}_0 = \mathbf{R}_0$$

where

$$\mathbf{z}_0 \equiv \begin{bmatrix} (z_0)_1 \\ (z_0)_2 \\ (z_0)_3 \end{bmatrix} = \begin{bmatrix} 3 \\ 2 \\ 1 \end{bmatrix}, \quad \mathbf{R}_0 \equiv \begin{bmatrix} 1 + 9\alpha \\ 4\alpha \\ \alpha \end{bmatrix}$$

and where $(z_0)_i$ is the ith element of z_0. Then, Eqs. (7.6.11b) and (7.6.11c) become,

$$\begin{bmatrix} 1+2\alpha(z_0)_1, & -1, & 0 \\ -1, & 2+2\alpha(z_0)_2, & -1 \\ 0, & -1, & 2+2\alpha(z_0)_3 \end{bmatrix} \begin{bmatrix} (z_1)_1 \\ (z_1)_2 \\ (z_1)_3 \end{bmatrix} = \begin{bmatrix} -9\alpha \\ -4\alpha \\ -\alpha \end{bmatrix}$$

and

$$\begin{bmatrix} 1+2\alpha(z_0)_1, & -1, & 0 \\ -1, & 2+2\alpha(z_0)_2, & -1 \\ 0, & -1, & 2+2\alpha(z_0)_3 \end{bmatrix} \begin{bmatrix} (z_2)_1 \\ (z_2)_2 \\ (z_2)_3 \end{bmatrix} = -\alpha \begin{bmatrix} (z_1)_1^2 \\ (z_1)_2^2 \\ (z_1)_3^2 \end{bmatrix}$$

respectively. The rest of Eq. (7.6.11) may be treated in a similar way. The numerical results for $\alpha = -0.01$ and 0.01 are shown next:

$\alpha = -0.01$

$$z_1 = \text{col}\,[0.4787, 0.3600, 0.1869]$$
$$z_2 = \text{col}\,[0.0131, 0.0100, 0.0052]$$
$$z \approx z_0 + z_1 + z_2 = \text{col}\,[3.4656, 2.3700, 1.1921]$$

$\alpha = 0.01$

$$z_1 = \text{col}\,[-0.8855, -0.2159, -0.1118]$$
$$z_2 = \text{col}\,[-0.0028, -0.0022, -0.0011]$$
$$z \approx z_0 + z_1 + z_2 = \text{col}\,[2.1117, 1.7819, 0.8871]$$

PROBLEMS

1. Show that Eq. (7.2.17) becomes a quadratic function along any boundary of a triangular element.
2. Can the piecewise polynomial defined by Eq. (7.2.18) be made continuous across any side of a triangular finite element with respect to ϕ, $\partial\phi/\partial x$, and $\partial\phi/\partial y$?
3. A functional is given by

$$J(\phi) = \int [p(\nabla\phi)^2 + q\phi - 2r\phi]\,dV + \int_\Gamma p\gamma\phi^2\,dS$$

where the integral is performed over the volume (or area) of interest and Γ is the surface (external boundary) of the volume. Use the trial function given by Eq. (7.2.15) with Eq. (7.2.14), and derive the necessary conditions to make $J(\phi)$ stationary.

4. Derive the finite element equation for Eq. (7.3.1) for the following configurations of triangular finite elements:

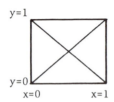
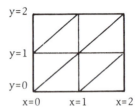

The boundary conditions are:

a. $\dfrac{\partial}{\partial n}\psi + \gamma\psi = 0$ along the top and right sides

b. $\dfrac{\partial}{\partial n}\psi = 0$ along the left and bottom sides

5. Derive the finite element equations for Eq. (7.3.1) for the following configurations of rectangular finite elements (apply the same boundary conditions as Problem 4).

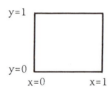
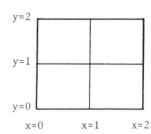

6. Consider the finite element equation for Eq. (7.3.1) for a rectangular geometry consisting of two triangular elements as shown below. If Eq. (7.2.17) is used and the boundary condition for all the sides is the mixed type, what is the configuration of matrix **H** in Eq. (7.3.8)? (Show **H** by using 0 for zero elements and x for nonzero elements.)

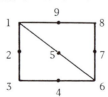

7. Derive the set of nine linear equations to determine the nine coefficients of Eq. (7.2.18) by using ϕ, ϕ_x, and ϕ_y at three apices of a triangle element.

8. The input for the finite element code for Eq. (7.3.1) with a general geometry consists of the following kinds:

   a. Information about each grid point: x and y coordinates.

   b. Information about each triangle: the grid numbers of apices of the triangle; p, q, r for that element.

   c. Information about the boundary conditions.

   Write an input routine that is sufficient for the finite element code.

9. Matrix **H** in Eq. (7.3.8) is generally a band diagonal matrix. Code a subprogram to solve Eq. (7.3.8) by using the Gaussian elimination method.

10. Based on the input given by Problem 8, write a subprogram to calculate the matrix elements of **H'** in Eq. (7.3.14).

11. Code a FORTRAN program to solve

$$-0.2\nabla^2\phi(x, y)+0.1\phi(x, y)=1.0$$

   by using the finite element method for the following geometry (assume mesh spacings are $\Delta x = \Delta y = 2.0$):

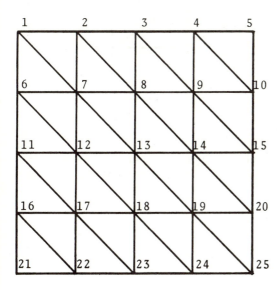

Boundary conditions:

$$-\nabla_n\phi = \frac{1}{0.4}\phi \quad \begin{cases} 1\text{-}3 \\ 1\text{-}11 \end{cases}$$

$$\nabla_n\phi = 0 \quad \begin{cases} 3\text{-}5\text{-}25 \\ -21\text{-}11 \end{cases}$$

12. Prove the following statements:

   a. **H** in Eq. (7.3.8) is a symmetric matrix.

b. K in Eq. (7.4.24) is a symmetric matrix.

c. H′ in Eq. (7.3.14) can be made symmetric by removing the matrix elements related to the grid points on the Dirichlet boundary.

13. Derive the finite element equation for plane stress by using a rectangular element and Eq. (7.2.16) for the following configurations:

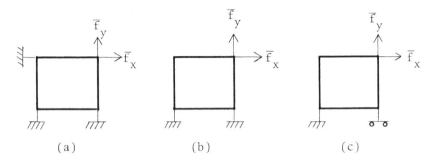

(a)   (b)   (c)

14. Write the matrix elements of **K**, **L**, and **R** in Eq. (7.6.2) explicitly for the following geometry:

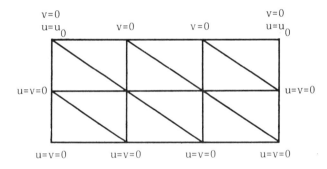

15. Derive Eqs. (7.6.11a) through (7.6.12).

16. Solve the equation in Example 7.11 by using the successive substitution and Newton–Raphson method for $\alpha = 0.01$, $\alpha = -0.01$, $\alpha = -0.1$, $\alpha = -0.5$.

REFERENCES

1. C. C. Zienkiewicz, *The finite element method in engineering sciences*, McGraw-Hill, New York (1971).

2. J. R. Whiteman, *A bibliography for finite elements*, Academic, New York (1975).

3. K. F. Hansen and C. M. Kang, "Finite element methods in reactor physics analysis," *Advances Nucl. Sci. Tech.*, **8**, 174–252 (1975).

4. G. Strang and G. J. Fix, *An analysis of the finite element method*, Prentice-Hall, Englewood Cliffs, N.J. (1973).

5. T. Ohnishi, "Application of finite element solution techniques to neutron diffusion and transport equations," ANS National Topical Meeting, Idaho Falls, CONF-710302, National Technical Information Service, Springfield, Va. (1971).

6. L. A. Semenza, E. E. Lewis, and E. C. Rossow, "The application of the finite element to the multigroup neutron diffusion," *Nucl. Sci. Eng.*, **47**, 302 (1972).

7. G. Kaper, K. Leaf, and A. J. Lindeman, "Applications of finite element methods in reactor mathematics," ANL-7925, Argonne National Laboratory, Argonne, Ill. (1972).

8. A. Nakamura and T. Ohnishi, "The iterative solutions for the finite-element method," *Numerical Reactor Calculations*, Publication of Int. Atomic Energy Agency, Vienna (1972).

9. D. J. Rose, *Symmetric elimination on sparse positive definite systems and the potential flow problem*, Ph.D. thesis, Harvard University (1970).

10. B. M. Irons and K. Y. Kan, "Equation solving algorithms for the finite-element method," *Numerical and computer methods in structural mechanics*, Academic, New York (1973).

11. R. S. Varga, "Extensions of the successive overrelaxation theory with applications to finite element approximations," Kent State University, Kent, Ohio (1972).

12. L. D. Eisenhart, *Application of higher order variational methods to the solution of the problems of reactor physics*, Ph. D. thesis, Nuclear Science and Engineering, Carnegie-Mellon University (1973).

13. I. Sokolnikoff and R. Redheffer, *Mathematics and modern engineering*, 2nd edit., pp. 370–375, McGraw-Hill, New York (1966).

14. R. H. Gallagher and S. T. K. Chang, "Higher-order finite element analysis of lake circulation," *Computers Fluids*, **1**, 119–132 (1973).

15. G. DeVries and D. H. Norrie, "the application of the finite element technique to potential flow problems," 71-APM-22, Trans. ASME Appl. Mech. Div., pp. 798–802 (1971).

16. O. Z. Zienkiewicz and Y. K. Cheung, "Finite elements in the solution of field problems," *The Engineer*, **220**, 507–510 (1965).

17. D. K. Gartling, "Finite element analysis of viscous incompressible fluid flow," *Proceeding of the Conference on Computational Methods in Nuclear Engineering*, Vol. 2, CONF-750413, National Technical Information Service, Springfield, Va. (1975).

18. D. K. Gartling, *Finite element analysis of viscous, incompressible fluid flow*, Ph.D. thesis, University of Texas, Austin (1975).

19. R. T. Cheng, "Numerical solution of the Navier–Stokes equation by the finite element method," *Physics Fluids*, **12**, 2098–2105 (1972).

20. J. T. Oden, "Finite element analogue of Navier–Stokes equations," *Proc. ASCE, J. Eng. Mech. Div.*, **96**, 529–534 (1970).

21. J. T. Oden, *Finite elements of nonlinear continua*, McGraw-Hill, New York (1972).

22. G. Yagawa, Y. Ishida, and Y. Ando, "An application of finite element method of nonlinear continuum mechanics to Navier–Stokes equation," *J. Atomic Energy Soc. Japan*, **16**, 591–594 (1974).

23. M. Kawahara, "On finite-element analysis in fluid flow," *Theor. Appl. Mech.*, **22**, 105–113 (1972).

24. M. Kawahara, N. Yoshimura, and K. Nakagawa, "Steady laminar and turbulent flow analysis of incompressible viscous fluid by the finite element method," *Proceedings of the International Symposium on Finite Element Methods in Flow Problems*, U. Wales, Swansea (1974).

25. C. Taylor and P. Hood, "A numerical solution of the Navier–Stokes equations using the finite element technique," *Computers Fluids*, **1**, 73–100 (1973).

26. R. Temam, *Finite element methods in fluid flow*, Report No. FM-75-2, College of Engineering, University of California, Berkeley (1975).

27. P. Hood and C. Taylor, "Navier–Stokes equation using mixed interpolation," *Finite element method in flow problems*, University of Alabama Press, Huntsville (1974).

28. J. T. Oden and L. C. Wellford, Jr., "Analysis of flow of viscous fluid by the finite element method," *AIAA J.*, **10**, 1590–1599 (1972).

29. I. K. Abu-Shumays and L. A. Hageman, "Development and comparison of practical discretization methods for the neutron diffusion equation over general quadrilateral partitions," *Proc. Conf. Computational Methods in Nuclear Engineering*, Vol. 1 CONF-750413, National Technical Information Service, Springfield, Va. (1975).

30. M. Kawahara, N. Yoshimura, K. Nakagawa, and H. Ohsaka, "Steady and unsteady finite element analysis of incompressible viscous fluid," *Int. J. Num. Method in Eng.*, **10**, 437–456 (1976).

31. M. Kawahara and T. Okamoto, "Finite element analysis of steady flow of viscous fluid using stream function," *Proc. JSCE*, No. 247, 123–135 (1976).

32. R. H. Gallagher, J. T. Oden, C. Taylor, and O. C. Zienkiewicz, Ed., *Finite Elements in Fluids*, Vol. 1 & 2, Wiley-Interscience, New York (1975).

Chapter 8   Coarse-Mesh
Rebalancing Method

8.1   RATIONALE OF COARSE-MESH REBALANCING

Iterative solutions are suited to modern electronic computers to solve large linear systems associated with finite difference approximations for partial differential equations of the elliptic type and eigenvalue problems of ordinary and partial differential equations. The fast direct solution method for a large linear system as described in Section 10.7 is an alternative, but it works only for limited configurations. A large fraction of the total computing cost spent in the areas of computational fluid dynamics, heat conduction analysis, and nuclear reactor analysis is due to lengthy computing time of iterative solutions for large linear systems. There is a remarkable fact that, as the capability of electronic computers increases, engineers want more detailed and more accurate calculations. In the case of partial differential equations of the elliptic type, this means an increase of mesh points, which not only increases the computational time per iteration but also slows down the convergence rate. Another fact is that the size or the domain of the system to be studied, such as aircraft and nuclear reactors, increases as the technology is advanced. The convergence rate of eigenvalue problems is considerably slowed down as the size of the interested domain is increased. For this reason, no matter how large and fast a computer becomes, it will soon be fully loaded with larger and longer iterative problems.

   To relieve this problem, the coarse-mesh rebalancing technique has been successfully used to accelerate iterative solution of large homogeneous and inhomogeneous linear systems such as multigroup neutron diffusion equations,[1-6] Poisson's equations in fluid dynamics,[7] stiffness matrices in finite element stress analysis,[8] and finite element diffusion calculations.[9] Coarse-mesh rebalancing is extremely useful for accelerating slowly convergent iterative schemes for very large linear systems.

   The convergence rate of iterative methods such as SOR or the Chebyshev polynomial method can be made faster by using a low estimate for spectral radius or dominance ratio, if the low frequency eigenvector components in the error vector are decayed efficiently by some other means than the iterative scheme.[1] On the other hand, the low-frequency eigenvectors in the iterative residue may be efficiently decayed by coarse-mesh rebalancing if the high-frequency eigenvector components are sufficiently small compared with the low-frequency components. Therefore, in the coarse-mesh-accelerated iterative schemes, a short iteration cycle (typically three to ten iterations) with a low estimate for the spectral radius or the dominance ratio and coarse-mesh rebalancing are alternatively applied until convergence.

There are two disadvantages with coarse-mesh rebalancing. First, an additional programming effort is necessary. This effort can be substantially greater than that for a simple iterative scheme such as the point SOR method. Second, the computational cost for each coarse-mesh rebalancing is relatively high compared with one cycle of iteration such as SOR. Therefore, the effort and cost for coarse-mesh rebalancing will not be paid off unless the saving of the total computing cost is substantial. However, coarse-mesh rebalancing is worth the additional effort and computing time if the convergence of the iteration scheme is very slow by itself or the calculation for each iteration cycle is time consuming. For example, coarse-mesh rebalancing has been found extremely useful for accelerating the outer iteration of the standard multigroup neutron diffusion code. This is because each outer iteration involves a number of inner iterations and the computing time required for coarse-mesh rebalancing adds only a very small fraction of the outer iteration, so any saving of the number of outer iterations saves the overall computing costs. As another example, coarse-mesh rebalancing saves the computing cost of the SOR scheme if the spectral radius of the SOR matrix is very close to unity or if each SOR iteration requires a heavy data transfer between fast and slow memory systems.

In spite of its potential advantage, the coarse-mesh rebalancing technique has not been extensively used. A few reasons may be pointed out: (1) this technique is not well known, (2) an additional programming effort is required, (3) the methods are not standardized, (4) successful utilization needs a thorough knowledge and experience with the standard iterative techniques, (5) successful utilization needs skill and experience, and (6) theoretical backup of coarse-mesh rebalancing has not been satisfactory. For these reasons, many attempts with coarse-mesh rebalancing have resulted in unsuccessful ends.

In Sections 8.2 through 8.5, the basic techniques of coarse-mesh rebalancing, mathematical properties of coarse mesh equations, iterative solution of coarse-mesh equations, a theoretical analysis of coarse mesh rebalancing effect, and numerical illustrations are described. The multiplicative type of coarse-mesh rebalancing is considered in those four sections. Section 8.6

introduces the additive type of coarse-mesh rebalancing. It is hoped that the methods and analyses summarized in this chapter will help the reader understand and utilize the coarse-mesh rebalancing technique. Successful results of applying coarse-mesh rebalancing are illustrated in Section 8.5.

8.2 PARTITIONING AND WEIGHTING

In a coarse-mesh-rebalanced iterative scheme, the iterative scheme is interrupted at every nth increment of iteration times, where n is a small integer, typically 3 to 10. We call the iterative vector to be rebalanced a *prerebalanced vector* and denote it by ψ_0. A prerebalanced vector is the result of a preceding short iterative cycle with a low estimate for the spectral radius or the dominance ratio. The result of coarse mesh rebalancing is called a *rebalanced vector* and denoted by ψ, which becomes the initial guess for the subsequent short iterative cycles.

A prerebalanced vector and the rebalanced vector are related by

$$\psi = \left(\sum_{k=1}^{K} \Phi_k \mathbf{P}_k \right) \psi_0 \tag{8.2.1}$$

where Φ_k are arbitrary coefficients to be determined by the discrete weighted residual method (see Section 6.1), K is the number of coarse mesh regions (or points) much smaller than the order of ψ, and $\mathbf{P}_k$ are prescribed diagonal matrices satisfying

$$\sum_{k=1}^{K} \mathbf{P}_k = \mathbf{I} \tag{8.2.2}$$

where $\mathbf{I}$ is an identity matrix. We designate $\mathbf{P}_k$ as *partitioning matrices.*

The iterative schemes considered here are either for an inhomogeneous equation

$$\mathbf{A}\boldsymbol{\phi} = \mathbf{S} \tag{8.2.3}$$

where $\mathbf{A}$ is a matrix of a large order N and $\mathbf{S}$ is a known vector or for a homogeneous equation

$$\mathbf{A}\boldsymbol{\phi} = \lambda \mathbf{F} \boldsymbol{\phi} \tag{8.2.4}$$

where $\mathbf{F}$ is another matrix. We assume that Eq. (8.2.3) represents the finite difference equation for a partial differential equation of elliptic type or similar equations. The iterative scheme adopted here for the inhomogeneous problems is typically the SOR or the Chebyshev polynomial method not based on the cyclic properties. In the case of Eq. (8.2.4), we seek the fundamental eigenvector and the corresponding eigenvalue, assuming there exists a unique positive eigenvector corresponding to the smallest and real eigenvalue. Such problems include one- to three-dimensional multigroup neutron diffusion equations and the equation of elastic vibration of plate, and so forth. The iterative scheme for Eq. (8.2.4) is typically a fixed-parameter extrapolation method or a Chebyshev polynomial method. Since the main role of the iterative schemes is to decay the high-frequency eigenvector components, the extrapolation parameters must be based on a low estimate for the spectral radius or the dominance ratio. If the extrapolation parameter optimized for standard iterative schemes is used, coarse-mesh rebalancing will not work. There are iterative schemes that are not compatible with coarse-mesh rebalancing, such as the Jacobi iterative method and the cyclic Chebyshev polynomial method. In addition, the iterative method based on initial value techniques rather than the relaxation technique, such as two-dimensional DS_n iterations, cannot be effectively accelerated by coarse-mesh rebalancing.[12]

The K arbitrary coefficients in Eq. (8.2.1) are determined by the weighted residual method with the same number of independent weighting vectors, $\mathbf{w}_k, k = 1, 2, \ldots, K$. In the case of an inhomogeneous problem, the residual is given by

$$\mathbf{r} = \mathbf{A}\left(\sum_k \Phi_k \mathbf{P}_k \right)\psi_0 - \mathbf{S} \qquad (8.2.5)$$

By making $\mathbf{r}$ orthogonal to $\mathbf{w}_k$ in accordance with the weighted residual method, the *coarse mesh equation* is obtained as

$$\sum_k \langle \mathbf{w}_l, \mathbf{AP}_k\psi_0 \rangle \Phi_k = \langle \mathbf{w}_l, \mathbf{S} \rangle, \qquad l = 1, 2, \ldots, K \qquad (8.2.6)$$

where $\langle \ \rangle$ denotes a scalar product of two vectors. Equation (8.2.6) determines the arbitrary coefficients, Φ_k. If K is not large, this equation may be solved by the Gaussian elimination method. An iterative solution of the coarse-mesh equation requires properties similar to those of the *fine-mesh* equation so that the iterative convergence is guaranteed. The properties of the coarse mesh equations are discussed in Section 8.3. In the case of a homogeneous problem, the coarse-mesh equation becomes

$$\sum_{k=1}^{K} \langle \mathbf{w}_l, \mathbf{AP}_k\psi_0 \rangle \Phi_k = \lambda \sum_{k=1}^{K} \langle w_l, \mathbf{FP}_k\psi_0 \rangle \Phi_k \qquad (8.2.7)$$

where λ is now the eigenvalue of the coarse-mesh equation. The fundamental eigenvector and the corresponding eigenvalue of Eq. (8.2.7) are to be found. As ψ approaches the exact solution, the fundamental eigenvalue of

Eq. (8.2.7) approaches the fundamental eigenvalue of Eq. (8.2.4). To solve Eq. (8.2.7), an iterative method similar to that for the *fine-mesh* equation is necessary. Equation (8.2.7) is required to have mathematical properties similar to those of Eq. (8.2.4) so that there is a unique positive solution corresponding to the fundamental eigenvalue. The properties of homogeneous coarse-mesh equations are also discussed in Section 8.3.

Although there are several possible types of partitioning matrices, partitioning matrices of all types must satisfy Eq. (8.2.2). This is because, if ψ_0 is the exact solution, ψ must also be the exact solution with $\Phi_k = 1$ for all k. The following two partitioning methods are important and useful among others.

DISJUNCTIVE PARTITIONING

K disjunctive spatial subdomains $\mathcal{D}_k$ are defined so that each element of a vector of order N belongs to one of the subdomains. The disjunctive partitioning is characterized by the diagonal matrix $\mathbf{P}_k$ such that all the elements are zero except that the diagonal elements belonging to the subdomain $\mathcal{D}_k$ are unity. For example, if $N = 6$ and $K = 3$, $\mathcal{D}_1$ include the first and second element, $\mathcal{D}_2$ the third and fourth, $\mathcal{D}_3$ the fifth and sixth. Then $\mathbf{P}_k$ becomes

$$\mathbf{P}_1 = \text{diag}\,(110000)$$

$$\mathbf{P}_2 = \text{diag}\,(001100) \qquad (8.2.8)$$

$$\mathbf{P}_3 = \text{diag}\,(000011)$$

In two dimensions, we use the two-dimensional index (k, k') for the partitioning matrices and coefficients respectively, for example, $\mathbf{P}_{k,k'}$ and $\Phi_{k,k'}$. It is rather cumbersome to write down a partitioning matrix explicitly. Instead, we show how an element of ψ is related to ψ_0. Considering a two-dimensional mesh space as shown in Fig. 8.1, an element of ψ_0 represents the field value at a grid, that is, (i, j). We assume that the fine-mesh system is divided into coarse subdomains, $\mathcal{D}_{k,k'}$, by dotted lines, then the elements of $\mathbf{P}_{k,k'}$, are all zero except that the diagonal elements of $\mathbf{P}_{k,k'}$ corresponding to the grids in $\mathcal{D}_{k,k'}$, are all unity. The rebalanced vector and the prerebalanced vectors are related, in terms of their elements, by

$$\psi_{i,j} = \Phi_{k,k'}\psi_{0,i,j} \qquad (8.2.9)$$

where $\psi_{i,j}$ is the element of ψ corresponding to grid (i, j), and (i, j) is assumed to belong to $\mathcal{D}_{k,k'}$.

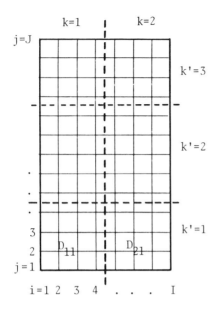

Figure 8.1 Coarse-mesh system (disjunctive partitioning).

PYRAMID PARTITIONING

Suppose that a boundary value problem of a difference equation is defined in $0 \leq x \leq H$ on the one-dimensional coordinate, and the grid points are at $x = x_i$, $i = 1, 2, \ldots, I$. We define a coarse grid system on $0 \leq x \leq H$ by X_k, $k = 1, 2, \ldots, K$. The diagonal elements of $\mathbf{P}_k$ are all zero except

$$(\mathbf{P}_k)_{i,i} = \frac{x_i - X_{k-1}}{X_k - X_{k-1}}, \quad \text{if } X_{k-1} \leq x_i \leq X_k$$

$$= \frac{X_{k+1} - x_i}{X_{k+1} - X_k}, \quad \text{if } X_k \leq x_i \leq X_{k+1} \tag{8.2.10}$$

where $(\mathbf{P}_k)_{i,i}$ is the ith diagonal element of $\mathbf{P}_k$.

In two dimensions,[5] the fine-mesh and coarse-mesh systems are shown in Fig. 8.2. The coarse-mesh system is defined by $(X_k, Y_{k'})$ such that $X_1 = x_1$, $X_K = x_I$, $Y_1 = y_1$, $Y_{K'} = y_J$ and every coarse-mesh line coincides with one of the fine-mesh lines. The elements in ψ_0 and ψ corresponding to the grid (i, j) are related by

$$\psi_{i,j} = \left[\sum_{k=1}^{K} \sum_{k'=1}^{K'} \Delta_{k,k'}(x_i, y_j) \Phi_{k,k'} \right] \psi_{0,i,j} \tag{8.2.11}$$

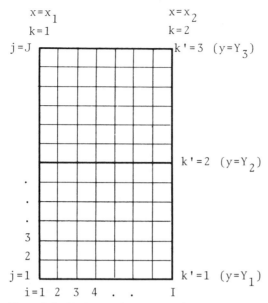

Figure 8.2 Coarse-mesh system (pyramid partitioning).

where $\Phi_{k,k'}$ are coefficients and $\Delta_{k,k'}(x, y) = 0$, except for

$$
\begin{aligned}
\Delta_{k,k'}(x, y) &= \frac{(X_{k+1}-x)(Y_{k'+1}-y)}{(X_{k+1}-X_k)(Y_{k'+1}-Y_{k'})}, && \text{if} \quad \begin{matrix} X_k \le x \le X_{k+1} \\ Y_{k'} \le y \le Y_{k'+1} \end{matrix} \\[2mm]
&= \frac{(X_{k-1}-x)(Y_{k'+1}-y)}{(X_{k-1}-X_k)(Y_{k'+1}-Y_{k'})}, && \text{if} \quad \begin{matrix} X_{k-1} \le x \le X_k \\ Y_{k'} \le y \le Y_{k'+1} \end{matrix} \\[2mm]
&= \frac{(X_{k+1}-x)(Y_{k'-1}-y)}{(X_{k+1}-X_k)(Y_{k'-1}-Y_{k'})}, && \text{if} \quad \begin{matrix} X_k \le x \le X_{k+1} \\ Y_{k'-1} \le y \le Y_{k'} \end{matrix} \quad (8.2.12) \\[2mm]
&= \frac{(X_{k-1}-x)(Y_{k'-1}-y)}{(X_{k-1}-X_k)(Y_{k'-1}-Y_{k'})}, && \text{if} \quad \begin{matrix} X_{k-1} \le x \le X_k \\ Y_{k'-1} \le y \le Y_{k'} \end{matrix}
\end{aligned}
$$

We refer to $\Delta_{k,k'}$ as the modified pyramid function.

The following choices for the weighting vectors are considered.

REGION BALANCING

$$\mathbf{w}_k = \mathbf{P}_k \mathbf{1} \tag{8.2.13}$$

where $\mathbf{P}_k$ is the disjunctive partitioning matrix and $\mathbf{1}$ is the sum vector defined in Section 1.6.

GALERKIN'S METHOD

$$\mathbf{w}_k = \mathbf{P}_k \mathbf{\psi}_0 \qquad (8.2.14)$$

LEAST SQUARE

$$\mathbf{w}_k = \mathbf{A} \mathbf{P}_k \mathbf{\psi}_0 \qquad (8.2.15)$$

PYRAMID WEIGHTING

$$\mathbf{w}_k = \mathbf{P}_k \mathbf{1} \qquad (8.2.16)$$

where $\mathbf{P}_k$ is the pyramid partitioning matrix.

The acceleration effect is affected by the combination of trial and weighting function. The effect of typical combinations are briefly discussed in the following.

1. Disjunctive partitioning and region balancing: Programming is easy. Coarse-mesh equations retain the mathematical properties of the fine-mesh equation and can be solved iteratively. The acceleration effect is moderate with sufficiently small coarse mesh subdomains. However, the acceleration effect is lost as the ratio of the fine-mesh size to the coarse-mesh size is decreased. This method is improved by introducing a *decoupling coefficient* as described later in this section.

2. Disjunctive partitioning and Galerkin's weighting vectors: The mathematical properties of the fine-mesh equation are retained under certain conditions. Wachspress' original method[1] of variational coarse-mesh rebalancing falls in this category. In his method for two dimensions, the coarse-mesh system is coarse only for one direction but uses the fine mesh for another direction.

3. Pyramid partitioning and pyramid weighting: The accelerating effect is the best among other combinations tested. Programming is complicated. The iterative solution mostly converges, but may diverge, especially if a diagonal element of the fine-mesh equation has a large positive value representing a strong sink that causes a sharp spatial variation of the solution. Iterative solutions of coarse-mesh equations for inhomogeneous equations converge if coarse-mesh rebalancing is applied before any fine-mesh iteration is initiated and all the elements of $\mathbf{\psi}_0$ are unity. The mathematical properties have not been well studied.

4. Pyramid partitioning and region balancing: The accelerating effect is nearly the same as case 3, while the programming effort and the computational time for coarse-mesh rebalancing are significantly reduced. This combination has been revised further to save more programming effort and computational time by using decoupling factors.

5. Disjunctive partitioning and least square weighting: This combination has not been numerically tested. However, a theoretical analysis predicts that the least square method is very sensitive to the high-frequency eigenvector components and not effective in accelerating iterative convergence.

Example 8.1

The difficulty with the disjunctive partitioning with Galerkin's weighting is illustrated and proven in this analysis. We consider a self-adjoint eigenvalue problem so that the choice of partitioning and weighting functions makes the weighted residual method equivalent to Rayleigh's principle.

Suppose we seek the minimum eigenvalue of the finite difference equation

$$2\phi_i - \phi_{i-1} - \phi_{i+1} = \lambda h^2 \phi_i, \qquad i = 1, 2, \ldots, (I-1) \qquad (8.2.17)$$

for a slab geometry with half thickness H, where $h = H/(I-1)$ and the boundary conditions are $\phi_0 = \phi_2$ and $\phi_I = 0$ (I is an even integer). Assuming that an approximate solution is given by $f_i = f(x_i)$, where $f(x)$ and its derivative are continuous, $f(x) = f(-x)$, $f(H) = 0$, and $x_{i+1} = ih$, we first derive an approximation for the eigenvalue by coarse-mesh rebalancing with two coarse-mesh regions. This may be done by setting the trial functions as

$$\phi_i = f_i, \quad \text{for } 1 \le i \le I/2.$$

$$\phi_i = af_i, \quad \text{for } I/2 < i \le I-1.$$

and by defining the functional

$$\lambda(a) = \frac{\sum_{i=2}^{I/2}(f_i - f_{i-1})^2 + (af_{1/2+1} - f_{1/2})^2 + a^2 \sum_{i=I/2+2}^{I}(f_i - f_{i-1})^2}{h^2(\sum_{i=1}^{I/2} f_i^2 + a^2 \sum_{i=I/2+1}^{I-1} f_i^2)} \qquad (8.2.18)$$

If we let $h \to 0$ (increase of fine-mesh points), the functional in Eq. (8.2.18) becomes

$$\lambda(a) = \frac{\int_0^{x_j} f'^2(x)\,dx + (af_{1/2+1} - f_{1/2})^2/h + a^2 \int_{x_{j+1}}^{H} f'^2(x)\,dx}{\int_0^{x_j} f^2(x)\,dx + a^2 \int_{x_{j+1}}^{H} f^2(x)\,dx} \qquad (8.2.19)$$

where $j = I/2$. Now, we observe that, if $a \ne 1$ the contribution of the second term in the numerator increases as H decreases. This indicates that with $h \to 0$, the value of a, about which λ is stationary, approaches unity, and thus the effectiveness of the coarse-mesh rebalancing is lost.

In order to compromise the good mathematical property of case 1 or 2 and the good accelerating effect of case 3 or 4, the method of decoupling has been used. The main reason why case 1 loses the acceleration effect is that, as the fine-mesh interval becomes smaller, the adjacent two coarse-mesh regions are coupled more strongly as if the average geometrical distance between the two $\mathcal{D}$s is equal to the fine-mesh interval. With the pyramid partitioning, this effect does not occur. One way to remedy this erratic coupling effect with disjunctive partitioning is to introduce a *decoupling coefficient* so that the average distance of the two coarse-mesh regions is mathematically reflected in the coarse-mesh equation.

With the decoupling coefficients, Eq. (8.2.6) is revised as

$$\left[\langle \mathbf{w}_l, \mathbf{AP}_l \psi_0 \rangle + \sum_{k \neq l} (1 - \alpha_{l,k}) \langle \mathbf{w}_l, \mathbf{AP}_k \psi_0 \rangle \right] \Phi_l$$
$$+ \sum_{k \neq l} \alpha_{l,k} \langle \mathbf{w}_l, \mathbf{AP}_k \psi_0 \rangle \Phi_k = \langle \mathbf{w}_l, \mathbf{S} \rangle, \qquad (8.2.20)$$

$$0 < \alpha_{l,k} < 1, \quad l = 1, 2, \ldots, K$$

where $\alpha_{l,k}$ is the decoupling coefficient. When $\alpha_{l,k} = 1$, Eq. (8.2.20) is reduced to Eq. (8.2.6). As $\alpha_{l,k}$ decreases, $\mathcal{D}_l$ and $\mathcal{D}_k$ are more loosely coupled, and the accelerating effect is enhanced. A measure to determine $\alpha_{l,k}$ is

$$\alpha_{l,k} = \frac{\Delta x}{\Delta X} \qquad (8.2.21)$$

where Δx is the fine-mesh interval between $\mathcal{D}_l$ and $\mathcal{D}_k$, while ΔX is the distance from the center of $\mathcal{D}_l$ to the center of $\mathcal{D}_k$. However, if the decoupling effect is overdosed, the accelerating effect is again lost. The effect of the decoupling coefficient is illustrated in Section 8.5.

8.3. PROPERTIES OF COARSE-MESH REBALANCING EQUATIONS

It is important that the coarse-mesh equation preserves some of the matrix properties that the fine-mesh equation associated with an elliptic partial differential equation has. If the fine-mesh equation has nonnegative inhomogeneous terms and, therefore, only positive solutions are expected, then the coarse-mesh equation must have a positive solution. It is also desirable that the same iterative scheme as for the fine-mesh equation is applicable to the coarse-mesh equation. In order that coarse-mesh rebalancing for a multigroup neutron diffusion equation works properly, it is necessary that the coarse-mesh equation has a unique positive eigenvector and the corresponding positive eigenvalue that is larger in absolute value than any other eigenvalues. If the coarse-mesh equations do not have those properties, their solutions can be irrelevant to the fine-mesh equation, and consequently coarse-mesh rebalancing may disturb the whole iterative procedure. The negative solution and any abnormal solution of the coarse-mesh rebalancing equation are called *anomalies*. We call the disturbances or disruptions of a coarse-mesh rebalanced iterative scheme caused by anomalies as *instability* of the coarse-mesh rebalancing. In this section we do not question the effectiveness of coarse-mesh rebalancing, but concentrate our attention on the stability of coarse-mesh rebalancing.

The theoretical foundation of coarse-mesh rebalancing methods has been limited to a few specific combinations of the partitioning matrix and the weighting vectors and may not be for general applications. Nevertheless, the theoretical foundation for special cases helps in understanding the properties of coarse-mesh equations and will serve as a guideline to fix anomalies if encountered.

8.3.1 Coarse-mesh Inhomogeneous Problems

Equation (8.2.6) may be written as

$$\mathscr{A}\Phi = d \geq 0 \tag{8.3.1}$$

where Φ is the vector with elements Φ_k; the elements of $\mathscr{A}$ and d are

$$\mathscr{A}_{l,k} = \langle \mathbf{w}_l, \mathbf{A}\mathbf{P}_k \boldsymbol{\phi} \rangle \tag{8.3.2}$$

and

$$d_l = \langle \mathbf{w}_l, \mathbf{S} \rangle \tag{8.3.3}$$

respectively. In Eq. (8.3.2), it is assumed that $\mathbf{P}_k$ satisfies Eq. (8.2.2).

We assume that $\mathbf{A} = [a_{m,n}]$ is the matrix representing the finite difference operator for an elliptic differential equation. It has the following properties:

1. The diagonal elements are positive and off-diagonal elements are zero or negative,
2. $\mathbf{A}$ is irreducibly diagonal dominant, and
3. $\mathbf{A}$ is symmetric.

The sufficient condition that the solution of Eq. (8.3.1) has a positive solution vector is that all the elements of $\mathscr{A}^{-1}$ are positive. The following theorem[3] is important to evaluate the property of $\mathscr{A}$.

Theorem 8.1. If $\mathbf{M} = (m_{i,j})$ is a real, irreducibly diagonally dominant matrix with $m_{i,i} > 0$ for all i, then $\mathbf{M}^{-1} > 0$. The diagonal dominance may be replaced by column diagonal dominance.

Proof. If **M** is symmetric, it is an S-matrix, and the theorem is reduced to Theorem 3.3. The proof for Theorem 3.3 is more general than just for an S-matrix and applies to nonsymmetric **M** also. Thus $\mathbf{M}^{-1} > \mathbf{0}$ (and the point Jacobi iterative scheme for **M** is convergent). If **M** is irreducibly column-diagonally dominant, then $\mathbf{M}' = \mathbf{M}^T$ has a positive inverse $\mathbf{M}'^{-1} = (\mathbf{M}^T)^{-1} > \mathbf{0}$ according to the first part. Therefore, $\mathbf{M}^{-1} > \mathbf{0}$.

Among various combinations of weighting vectors and partitioning matrices, the combination of the disjunctive partitioning matrices and the region balance weighting has the most favourable mathematical properties in the sense that, if $\boldsymbol{\phi} > \mathbf{0}$, the elements of $\mathscr{A}^{-1}$ are all positive, and accordingly the solution of the coarse-mesh equation is positive. The positivity of $\mathscr{A}^{-1}$ is easily shown by using Theorem 8.1 and the next theorem.[3]

Theorem 8.2. If $\mathbf{P}_k$ are the disjunctive partitioning matrices and $\mathbf{w}_l$ is given by Eq. (8.2.13) and $\boldsymbol{\phi} > \mathbf{0}$, then $\mathscr{A}$ has positive diagonal elements, nonpositive off-diagonal elements, and irreducibly column diagonal dominance.

Proof. Obviously $\mathscr{A}_{l,k} \leq 0$ for $l \neq k$ with strict inequality for $l = k \pm 1$, where $\mathscr{A}_{l,k}$ is defined by Eq. (8.3.2). Therefore, $\mathscr{A}$ is irreducible. To prove positive diagonal elements and irreducibly column diagonal dominance, it is sufficient to say the column summation of $\mathscr{A}_{l,k}$ for every k is nonnegative. The column summation for any k becomes

$$\sum_{l=1}^{K} \mathscr{A}_{l,k} = \sum_{l=1}^{K} (\mathbf{P}_l \mathbf{1})^T \mathbf{A} \mathbf{P}_k \boldsymbol{\phi} = \mathbf{1}^T \mathbf{A} \mathbf{P}_k \boldsymbol{\phi}$$

$$= \sum_{m=1}^{N} \sum_{n \in k}^{N} a_{m,n} \phi_n = \sum_{n \in k} \phi_n \sum_{m=1}^{N} a_{m,n} \tag{8.3.4}$$

where Eq. (8.2.2) is used after the first equality sign. Since **A** is irreducibly diagonally dominant, we have

$$\sum_{m=1}^{N} a_{m,n} \geq 0 \tag{8.3.5}$$

with strict inequality for at least one n. Therefore, we have

$$\sum_{l=1}^{K} \mathscr{A}_{l,k} \geq 0 \tag{8.3.6}$$

with strict inequality for at least one k.

The pyramid partitioning combined with the region balance weighting is effective for accelerating an iterative convergence as shown in Section 8.5.

However, this combination is sometimes encountered with instabilities, especially when one or more diagonal elements have very large positive values. The next example[3] shows that $\mathscr{A}$ may have a positive off-diagonal element, and thus $\mathscr{A}^{-1}$ may include negative elements.

Example 8.2

Consider the fine-mesh equation

$$\begin{bmatrix} 50 & -1 \\ -1 & 2 \end{bmatrix}\begin{bmatrix} \phi_1 \\ \phi_2 \end{bmatrix} = \begin{bmatrix} 1 \\ 0 \end{bmatrix}$$

in which the matrix on the left is an S-matrix. Assuming that the order of the coarse-mesh system is identical to that of the fine-mesh system, we set

$$\mathbf{P}_1 = \text{diag}\,(0.9, 0.1), \mathbf{P}_2 = \text{diag}\,(0.1, 0.9)$$

$$\mathbf{W}_1 = \text{col}\,(1, 0), \mathbf{W}_2 = \text{col}\,(0, 1), \boldsymbol{\phi} = \text{col}\,(1, 1)$$

Then, $\mathscr{A}$ and $\mathscr{A}^{-1}$ become

$$\mathscr{A} = \begin{bmatrix} 44.9, & 4.1 \\ -0.7, & 1.7 \end{bmatrix}, \quad \mathscr{A}^{-1} = \begin{bmatrix} 0.02146, & 0.00884 \\ -0.05176 & 0.56692 \end{bmatrix}$$

It is seen that a negative element is involved in $\mathscr{A}^{-1}$.

In the case of Galerkin's method combined with the disjunctive partitioning, a more restrictive condition for $\boldsymbol{\phi}$ is necessary[3] to guarantee $\mathscr{A}^{-1} > 0$.

Theorem 8.3   If $\boldsymbol{\phi}$ satisfies the following inequalities

$$\boldsymbol{\phi} > 0, \qquad \mathbf{A}\boldsymbol{\phi} = \mathbf{S} \geq 0, \qquad \mathbf{S} \neq 0 \qquad\qquad (8.3.7)$$

then the disjunctive Galerkin matrix $\mathscr{A} = (\mathscr{A}_{l,k})$ has $\mathscr{A}_{l,l} > 0$ for all l, $\mathscr{A}_{l,k} \leq 0$ for $l \neq k$ and irreducibly row-diagonal dominance.

Proof.   Obviously $\mathscr{A}_{l,k} \leq 0$ for $l \neq k$ with strict inequality for $l = k+1$. Therefore, $\mathscr{A}$ is irreducible. Since

$$\sum_k \mathscr{A}_{l,k} = \sum_k (\mathbf{P}_l \boldsymbol{\phi})^T \mathbf{A}(\mathbf{P}_k \boldsymbol{\phi}) = (\mathbf{P}_l \boldsymbol{\phi})^T \mathbf{A}\boldsymbol{\phi} = (\mathbf{P}_l \boldsymbol{\phi})^T \mathbf{S} \geq 0 \qquad (8.3.8)$$

$\mathscr{A}_{l,l}$ are positive for all l, and $\mathscr{A}$ is row diagonally dominant.

Equation (8.3.7) is satisfied in most cases when the coarse-mesh rebalancing is applied to discrete multigroup diffusion equations. If $\boldsymbol{\phi}$ does not satisfy Eq. (8.3.7), $\mathscr{A}^{-1} > 0$ is not guaranteed. However, the following theorem guarantees the convergence of the SOR method.

Theorem 8.4.   If $\mathbf{A}$ is a real positive definite matrix and Galerkin's method is used, then $\mathscr{A}$ becomes real, symmetric, and positive definite, regardless of the types of partitioning and weighting.

Proof.   Since $\mathbf{A}$ is real and symmetric, $\mathscr{A}$ is also real and symmetric if the Galerkin's method is used. A real matrix $\mathscr{A}$ is positive definite if $\mathbf{\Phi}^T\!\mathscr{A}\mathbf{\Phi}>0$ for any nonnull vector $\mathbf{\Phi}$. This is proven as follows

$$\mathbf{\Phi}^T\!\mathscr{A}\mathbf{\Phi}=\sum_l\sum_k \Phi_l\,\mathscr{A}_{l,k}\Phi_k =\left(\sum_k \Phi_k\mathbf{P}_k\boldsymbol{\phi}\right)^T\mathbf{A}\left(\sum_k \Phi_k\mathbf{P}_k\boldsymbol{\phi}\right) \qquad (8.3.9)$$

If $\mathbf{\Phi}\neq\mathbf{0}$ and $\boldsymbol{\phi}\neq\mathbf{0}$, then Eq. (8.3.9) is positive definite since $\mathbf{X}^T\mathbf{A}\mathbf{X}>0$ for any $\mathbf{X}\neq\mathbf{0}$.

8.3.2.   Coarse-Mesh Eigenvalue Problem

As an eigenvalue problem, we consider a two-group, two-dimensional diffusion equation given by Eq. (3.10.2). The coarse-mesh rebalancing equation for Eq. (3.10.2) may be written as

$$\mathscr{A}_1\mathbf{\Phi}_1 = \frac{1}{k}[\mathscr{F}_1\mathbf{\Phi}_1+\mathscr{F}_2\mathbf{\Phi}_2]$$

$$\mathscr{A}_2\mathbf{\Phi}_2 = \mathscr{Q}\,\mathbf{\Phi}_1 \qquad\qquad (8.3.10)$$

where

$$(\mathscr{A}_g)_{k,l} = \langle \mathbf{w}_{g,k}, \mathbf{A}_g\mathbf{P}_l\boldsymbol{\phi}_g\rangle, \qquad g=1,2 \qquad (8.3.11)$$

$$(\mathscr{F}_g)_{k,l} = \langle \mathbf{w}_{1,k}, \mathbf{F}_g\mathbf{P}_l\boldsymbol{\phi}_g\rangle, \qquad g=1,2 \qquad (8.3.12)$$

$$(\mathscr{Q})_{k,l} = \langle \mathbf{w}_{2,k}, \mathbf{Q}\mathbf{P}_l\boldsymbol{\phi}_1\rangle \qquad\qquad (8.3.13)$$

$$\mathbf{w}_{g,k} = \mathbf{P}_k\mathbf{1} \quad \text{or} \quad \mathbf{P}_k\boldsymbol{\phi}_g \qquad\qquad (8.3.14)$$

Since $\mathbf{A}_1$ and $\mathbf{A}_2$ in Eq. (3.3.2) are S-matrices, we have $\mathscr{A}^{-1}>0$ and $\mathscr{A}_2^{-1}>0$ for disjunctive partitioning and region balance weighting in accordance with Theorem 8.2.

In the case of Galerkin's weighting, we assume

$$\mathbf{A}_1\boldsymbol{\phi}_1\geq\mathbf{0} \quad \text{and} \quad \mathbf{A}_2\boldsymbol{\phi}_2>\mathbf{0} \qquad\qquad (8.3.15)$$

so that $\mathscr{A}_1^{-1}$ and $\mathscr{A}_2^{-1}>\mathbf{0}$. Equation (8.3.15) is a reasonable assumption for the following reasons: If the first coarse-mesh rebalancing is applied before any fine-mesh iteration is performed, when $\boldsymbol{\phi}_1$ and $\boldsymbol{\phi}_2$ are set to the sum vector (see Eq. (1.7.8)) as initial guess, Eq. (8.3.15) is satisfied because of the diagonal dominance of the neutron-diffusion equation. If the coarse-mesh rebalancing is applied after at least one outer iteration is performed, the inner iterations in the last outer iteration should have solved $\mathbf{A}_1\boldsymbol{\phi}_1 = \mathbf{S}_1 \geq \mathbf{0}$ and $\mathbf{A}_2\boldsymbol{\phi}_2 = \mathbf{S}_2 > \mathbf{0}$, where $\mathbf{S}_1$ and $\mathbf{S}_2$ are the fission source and the slowing-down source, respectively, with a reasonable accuracy.

We now ask if there is a unique positive eigenvector, $\boldsymbol{\Phi} = \mathrm{col}\,(\boldsymbol{\Phi}_1, \boldsymbol{\Phi}_2) > \mathbf{0}$, and the corresponding single positive eigenvalue k that is greater in absolute value than any other eigenvalue of Eq. (8.3.10). This proof can be obtained in a manner similar to that for Eq. (2.7.2) in Section 2.7.

It is easily seen that $\mathscr{F}_1$, $\mathscr{F}_2$, and $\mathscr{Q}$ are nonnegative matrices for both the region-balancing and Galerkin methods. Assuming Eq. (8.3.15) is satisfied, Eq. (8.3.10) may be written as

$$y = \frac{1}{k}\mathscr{R}y \qquad (8.3.16)$$

where

$$y = \mathscr{F}_1\boldsymbol{\Phi}_1 + \mathscr{F}_2\boldsymbol{\Phi}_2 \qquad (8.3.17)$$

$$\mathscr{R} = \mathscr{F}_1\mathscr{A}_1^{-1} + \mathscr{F}_2\mathscr{A}_2^{-1}\mathscr{Q}\mathscr{A}_1^{-1} \qquad (8.3.18)$$

The matrix $\mathscr{R}$ is a nonnegative matrix. If y includes any zero elements, an auxiliary vector y' of a reduced order may be obtained by removing the zero elements from y. We also obtain an auxiliary matrix $\mathscr{R}'$, which is a positive matrix by removing the corresponding rows and columns from $\mathscr{R}$. Thus we have

$$y' = \frac{1}{k}\mathscr{R}y' \qquad (8.3.19)$$

According to the Perron's theorem, which is introduced in Section 2.7, a positive matrix has a unique positive eigenvector and the corresponding eigenvalue that is larger in the absolute value than any other eigenvalues. This completes the proof for the positivity property of the coarse-mesh eigenvalue problem.

8.4 THEORETICAL ANALYSIS OF THE COARSE-MESH REBALANCING EFFECT

Why coarse-mesh rebalancing accelerates an iterative convergence is an interesting question, but no satisfactory answer has been given.[15] In fact, this is not a trivial question to answer. Coarse-mesh rebalancing is a nonlinear transformation of an iterative vector into another vector through a weighted residual method.[14] The major difficulty in the theoretical study on the effect of coarse-mesh rebalancing lies in its nonlinear nature. In this section we attempt to analyze the coarse-mesh rebalancing effect of homogeneous problems mathematically. Although the present analysis[16] is limited to a self-adjoint eigenvalue problem for simplicity of analysis, it is hoped that the theorems and discussions are useful for better understanding and utilization of coarse-mesh rebalancing.

8.4.1 Definitions and Assumptions

We consider a self-adjoint eigenvalue problem on a fine mesh system,

$$\mathbf{A}\boldsymbol{\phi} = \lambda \mathbf{F}\boldsymbol{\phi} \tag{8.4.1}$$

where λ is the eigenvalue, $\mathbf{A}$ and $\mathbf{F}$ are symmetric matrices of order N, and $\boldsymbol{\phi}$ is the solution vector. Equation (8.4.1) appears as the finite difference approximation to a homogeneous elliptic partial differential equation. We seek the fundamental eigenvector of Eq. (8.4.1) via an iterative scheme. The eigenvectors and eigenvalues are defined by

$$\mathbf{A}\mathbf{u}_n = \lambda_n \mathbf{F}\mathbf{u}_n, \qquad n = 1, 2, \ldots, N \tag{8.4.2}$$

$$\lambda_1 < \lambda_2 < \lambda_3 < \cdots < \lambda_N$$

where all the eigenvalues are assumed to be distinct and $\mathbf{u}_n$ are the orthonormal eigenvectors in the sense

$$\langle \mathbf{u}_m, \mathbf{F}\mathbf{u}_n \rangle = \delta_{m,n} \tag{8.4.3}$$

where $\langle , \rangle$ denotes the scalar product of two vectors and $\delta_{m,n}$ is Kronecker's delta.

We call the vector to be rebalanced the *prerebalanced vector* and denote it by $\boldsymbol{\psi}_0$. It is presumed that a prerebalanced vector is the result of a few iterations based on the Chebyshev polynomial method with a low dominance ratio. (Practically, the fixed-parameter extrapolation method, which is a special case of the Chebyshev polynomial method, is sufficient.) We also make the following assumptions:

1. The dominance ratio estimate is set up in such a way that the first p eigenvectors are out of the range of uniform decay of the Chebyshev polynomial method, where p is a small integer typically less than 10, although it may vary depending upon the size of the fine-mesh and coarse-mesh schemes.
2. Because of the low estimate for the dominance ratio, the last N-p eigenvector components among N in the prerebalanced vector have

decayed uniformly during the Chebyshev polynomial iteration, but the magnitude of the second through pth eigenvector components remains significantly high in the prerebalanced vector.

The objective of coarse-mesh rebalancing is that the components of $\mathbf{u}_2$ through $\mathbf{u}_p$ in $\boldsymbol{\psi}_0$ be completely eliminated or at least substantially suppressed via rebalancing.

The rebalanced vector denoted by $\boldsymbol{\psi}$ is expressed by

$$\boldsymbol{\psi} = \sum_{k=1}^{K} \Phi_k \mathbf{f}_k \qquad (8.4.4)$$

where $\mathbf{f}_k$ are the trial vectors for the weighted residual method defined as

$$\mathbf{f}_k = \mathbf{P}_k \boldsymbol{\psi}_0$$

In this equation, Φ_k are undetermined coefficients, K is the number of freedom of $\boldsymbol{\psi}$, $\boldsymbol{\psi}_0$ is the prerebalanced vector, and $\mathbf{P}_k$ is the partitioning matrix of order N satisfying

$$\sum_{k=1}^{K} \mathbf{P}_k = \mathbf{I} \qquad \text{Identity matrix} \qquad (8.4.5)$$

where $\mathbf{I}$ is the identity matrix of order N. If Φ_k for all k is unity, $\boldsymbol{\psi}$ is reduced to $\boldsymbol{\psi}_0$. Therefore, we have the relation

$$\boldsymbol{\psi}_0 = \sum_{k=1}^{K} \mathbf{f}_k \qquad (8.4.6)$$

In terms of eigenvectors, each of $\mathbf{f}_k$ may be expanded as

$$\mathbf{f}_k = \sum_{n=1}^{N} B_{n,k} \mathbf{u}_n, \qquad k = 1, 2, \ldots, K \qquad (8.4.7)$$

where $B_{n,k}$ are expansion coefficients. Introducing Eq. (8.4.7) into Eq. (8.4.4) yields

$$\boldsymbol{\psi} = \sum_{n=1}^{N} \left(\sum_{k=1}^{K} B_{n,k} \Phi_k \right) \mathbf{u}_n \qquad (8.4.8)$$

We now classify all the eigenvectors other than the fundamental eigenvectors into the low eigenvectors and high eigenvectors as follows.

Definition. Low eigenvectors are $\mathbf{u}_n$, $n = 2, 3, \ldots, K$, excluding the fundamental eigenvector, $\mathbf{u}_1$, where K is the number of trial vectors in $\boldsymbol{\psi}$. High eigenvectors are $\mathbf{u}_n$, $K < n \leq N$. In general, a higher eigenvector has more geometrical frequencies than a lower eigenvector.

8.4.2   Ideal Weighting Vectors

We first study the effect of rebalancing with ideal weighting vectors.

Theorem 8.5. If the low eigenvectors are known and used as weighting vectors to determine Φ_k of Eq. (8.4.4), the low eigenvectors are completely eliminated from ψ.

Proof. Define the residual vector by

$$\mathbf{r} = (\mathbf{A} - \lambda \mathbf{F})\psi \qquad (8.4.9)$$

In accordance with the weighted residual method, we make $\mathbf{r}$ orthogonal to the low eigenvectors:

$$\langle \mathbf{u}_l, \mathbf{r} \rangle = \langle \mathbf{u}_l, (\mathbf{A} - \lambda \mathbf{F})\psi \rangle = 0, \qquad l = 2, 3, \ldots, K \qquad (8.4.10)$$

where $\langle , \rangle$ denotes a scalar product of two vectors. In Eq. (8.4.10), we assume that λ is known exactly or approximately as $\tilde{\lambda}$ ($\tilde{\lambda}$ is different from any of λ_n, $n = 2, 3, \ldots, K$). In effect, λ may be obtained as the fundamental eigenvalue of Eq. (8.4.10) plus an additional homogeneous equation that is derived with another independent weighting vector.

By using Eqs. (8.4.8) and (8.4.9), Eq. (8.4.10) becomes

$$(\lambda_l - \tilde{\lambda}) \sum_{k=1}^{K} B_{l,k}\Phi_k = 0, \qquad l = 2, 3, \ldots, K \qquad (8.4.11)$$

Since we have assumed $\lambda_l - \tilde{\lambda} \neq 0$ for $2 \leq l \leq K$, Eq. (8.4.11) becomes

$$\sum_{k=1}^{K} B_{l,k}\Phi_k = 0, \qquad l = 2, 3, \ldots, K \qquad (8.4.12)$$

Because the left side of Eq. (8.4.12) is the coefficient of $\mathbf{u}_n$, $n = 2, 3, \ldots, K$, in ψ, Eq. (8.4.12) implies that the coefficients of those $K-1$ eigenvectors in ψ become zero, thus completing the proof.

The undetermined coefficients Φ_k are determined by solving Eq. (8.4.12) plus the additional equation. This solution may be normalized by

$$\sum_{k=1}^{K} B_{1,k}\Phi_k = 1 \qquad (8.4.13)$$

By using Eq. (8.4.12) and Eq. (8.4.13), ψ of Eq. (8.4.8) becomes

$$\psi = \mathbf{u}_1 + \sum_{n=K+1}^{N} \left(\sum_{k=1}^{K} B_{n,k}\Phi_k \right)\mathbf{u}_n \qquad (8.4.14)$$

Thus, ψ does not include the low eigenvector components $\mathbf{u}_n$, $n = 2, \ldots, K$.

The second term of Eq. (8.4.14) represents the eigenvector components that were not eliminated, or that were excited via coarse-mesh rebalancing.

Corollary. If Φ_k of Eq. (8.4.4) are determined by making $\mathbf{r}$ orthogonal to K independent vectors that consist of only the fundamental and low eigenvectors, then the rebalanced vector will not include the low eigenvector components.

Proof. Let $\mathbf{w}_l$, $l = 1, 2, \ldots, K$, be K independent weighting vectors, which are expanded in accordance with the assumption as

$$\mathbf{w}_l = \sum_{m=1}^{K} C_{m,l}\mathbf{u}_m \tag{8.4.15}$$

where $C_{m,l}$ are the coefficients. Since $\mathbf{w}_l$ are all independent, the coefficient matrix $\mathbf{C} \equiv [C_{m,l}]$ has its inverse $\mathbf{C}^{-1} \equiv [d_{m,l}]$:

$$\sum_{l=1}^{K} C_{m,l}d_{l,n} = \delta_{m,n} \tag{8.4.16}$$

On the other hand, orthogonalizing $\mathbf{r}$ to $\mathbf{w}_l$, $l = 1, 2, 3, \ldots, K$, yields

$$\langle \mathbf{w}_l, \mathbf{r} \rangle = \sum_{m=1}^{K} C_{m,l}\mathbf{u}_m \sum_{k=1}^{K} \Phi_k \sum_{n=1}^{N} B_{n,k}(\lambda_n - \lambda)\mathbf{F}\mathbf{u}_n$$

$$= \sum_{m=1}^{K} (\lambda_m - \lambda)C_{m,l}\left(\sum_{k=1}^{K} B_{m,k}\Phi_k \right) = 0, \tag{8.4.17}$$

$$l = 1, 2, \ldots, K$$

By multiplying Eq. (8.4.17) by $d_{l,n}$ and summing over l, we obtain

$$(\lambda_n - \lambda) \sum_{k=1}^{K} B_{n,k}\Phi_k = 0 \qquad n = 1, 2, \ldots, K \tag{8.4.18}$$

In this case, $\lambda = \lambda_1$ satisfies Eq. (8.4.18) as the fundamental eigenvalue. With $\lambda = \lambda_1$, Eq. (8.4.18) for $n = 2, 3, \ldots, K$ is equivalent to Eq. (8.4.12).

The weighting vectors considered in this theorem and corollary are impractical because it is impossible to avoid high eigenvectors in the weighting vectors. Therefore, it is important to study the effect of high eigenvectors in the weighting vectors.

8.4.3 Parasitic Effects of High Eigenvector Components in the Weighting Vectors

We assume here that prerebalanced vector does not include the high eigenvector components; thus

$$\sum_{k=1}^{K} B_{n,k} = 0, \qquad n > K \tag{8.4.19}$$

and

$$\boldsymbol{\psi}_0 = \sum_{n=1}^{K} \left(\sum_{k=1}^{K} B_{n,k} \right) \mathbf{u}_n \tag{8.4.20}$$

Each weighting vector is assumed to include all the eigenvector components:

$$\mathbf{w}_l = \sum_{n=1}^{N} C_{n,l} \mathbf{u}_n, \qquad l = 1, 2, \ldots, K \tag{8.4.21}$$

Instead of making $\mathbf{r}$ orthogonal to each $\mathbf{w}_l$ in a straightforward manner, we first apply a linear transformation to provide a new set of weighting vectors as

$$\mathbf{v}_l = \sum_{k=1}^{K} \alpha_{k,l} \mathbf{w}_k \tag{8.4.22}$$

where $\alpha_{k,l}$ are the transformation coefficients. If $\mathbf{v}_l$ are all independent, orthogonalizing $\mathbf{r}$ to $\mathbf{w}_l$s is equivalent to orthogonalizing $\mathbf{r}$ to $\mathbf{v}_l$s. The transformation is chosen in such a way that $\mathbf{v}_l$ becomes

$$\mathbf{v}_l = \mathbf{u}_l + \sum_{n=K+1}^{N} \gamma_{n,l} \mathbf{u}_n, \qquad l = 1, 2, \ldots, K \tag{8.4.23}$$

We now show that this transformation is possible. Equating Eq. (8.4.23) with Eq. (8.4.22), we have

$$\mathbf{u}_l + \sum_{n=K+1}^{N} \gamma_{n,l} \mathbf{u}_n = \sum_{k=1}^{K} \alpha_{k,l} \mathbf{w}_k \tag{8.4.24}$$

Introducing Eq. (8.4.21) into Eq. (8.4.24) yields

$$\mathbf{u}_l + \sum_{n=K+1}^{N} \gamma_{n,l} \mathbf{u}_n = \sum_{k=1}^{K} \alpha_{k,l} \sum_{n=1}^{N} C_{n,k} \mathbf{u}_n \tag{8.4.25}$$

By equating the coefficients of $\mathbf{u}_n$, the following relations are obtained:

$$\sum_{k=1}^{K} \alpha_{k,l} C_{n,k} = \delta_{n,l}, \qquad n \leq K \tag{8.4.26}$$

$$\gamma_{l,n} = \sum_{k=1}^{K} \alpha_{k,l} C_{n,k}, \qquad n > K, l = 1, 2, \ldots, K \tag{8.4.27}$$

where $\delta_{n,l}$ is Kronecker's delta. The transformation coefficients $\alpha_{k,l}$ are uniquely determined by Eq. (8.4.26). The coefficients $\gamma_{l,n}$ are given by Eq. (8.4.27) after $\alpha_{k,l}$ are determined.

The magnitude of the low eigenvector components in ψ are dependent on the higher eigenvectors in ψ as shown in the following. By making $\mathbf{r}$ orthogonal to $\mathbf{v}_l$, we obtain the linear equations to determine Φ_k:

$$\langle \mathbf{v}_l, \mathbf{r} \rangle = \left\langle \left(\mathbf{u}_l + \sum_{m=K+1}^{N} \gamma_{m,l} \mathbf{u}_m \right), \sum_{k=1}^{K} \Phi_k \sum_{n=1}^{N} B_{n,k}(\lambda_n - \lambda)\mathbf{F}\mathbf{u}_n \right\rangle$$

$$= \sum_{k=1}^{K} \left[(\lambda_l - \lambda)B_{l,k} + \sum_{n=K+1}^{N} (\lambda_n - \lambda)\gamma_{n,l}B_{n,k} \right] \Phi_k = 0 \tag{8.4.28}$$

In Eq. (8.4.28), λ is an eigenvalue of Eq. (8.4.28) itself. The lowest eigenvalue of Eq. (8.4.28) is denoted by $\tilde{\lambda}_1$, which is generally not identical to λ_1. By introducing $\lambda = \tilde{\lambda}_1$ and rearranging, Eq. (8.4.28) becomes

$$\sum_{k=1}^{K} B_{l,k}\Phi_k = - \sum_{n=K+1}^{N} \frac{\lambda_n - \tilde{\lambda}_1}{\lambda_l - \tilde{\lambda}_1} \gamma_{n,l} g_n, \qquad l = 1, 2, \ldots, K \tag{8.4.29}$$

$$g_n = \sum_{k=1}^{K} B_{n,k}\Phi_k, \qquad n > K \tag{8.4.30}$$

where g_n is the magnitude of the high eigenvector $\mathbf{u}_n$ in ψ. Introducing Eq. (8.4.29) into Eq. (8.4.8), we have

$$\psi = \sum_{n=1}^{K} \left[- \sum_{m=K+1}^{N} \frac{\lambda_m - \tilde{\lambda}_1}{\lambda_n - \tilde{\lambda}_1} \gamma_{m,n} g_m \right] \mathbf{u}_n + \sum_{n=K+1}^{N} g_n \mathbf{u}_n \tag{8.4.31}$$

In Eq. (8.4.31), the first term on the right side includes the first K eigenvector components, while the second term includes the rest of the eigenvector components. Since ψ_0 does not include the higher eigenvector components by assumption, the second term in Eq. (8.4.31) represents the high eigenvectors excited by coarse-mesh rebalancing. The absolute value of the coefficient of low eigenvectors in Eq. (8.4.31) satisfies

$$\left| \sum_{m=K+1}^{N} \frac{\lambda_m - \tilde{\lambda}_1}{\lambda_n - \tilde{\lambda}_1} \gamma_{m,n} g_m \right| \leq \sum_{m=K+1}^{N} \frac{\lambda_m - \tilde{\lambda}_1}{\lambda_n - \tilde{\lambda}_1} |\gamma_{m,n}||g_m|, \qquad n = 2, 3, \ldots, K \tag{8.4.32}$$

The right side of Eq. (8.4.32) includes $\gamma_{m,n}$, which is the magnitude of the high eigenvector component $\mathbf{u}_m$ involved in $\mathbf{v}_n$ as seen in Eq. (8.4.23) and uniquely determined when the original set of the weighting vectors are chosen. The other factor g_m is the magnitude of the high eigenvector $\mathbf{u}_m$ excited via coarse-mesh rebalancing. It is clear that to make the left side of Eq. (8.4.32) smaller, either $\gamma_{m,n}$, $m > K$ and $2 \leq n \leq K$, or g_m or both must be minimal.

8.4.4 Accelerating Effect

Definition. Let the prerebalanced and rebalanced vectors be expressed by

$$\psi_0 = \sum_{n=1}^{N} \left(\sum_{k=1}^{K} B_{n,k} \right) \mathbf{u}_n \tag{8.4.33}$$

and

$$\psi = \sum_{n=1}^{N} \left(\sum_{k=1}^{K} B_{n,k} \Phi_k \right) \mathbf{u}_n \tag{8.4.34}$$

respectively. The accelerating effect is positive if the coefficients of $\mathbf{u}_n$, $n = 2$, $3, \ldots, p$, in ψ_0 and ψ satisfy the inequality

$$R_n \equiv \frac{\left| \sum_{k=1}^{K} B_{n,k} \Phi_k \right|}{\left| \sum_{k=1}^{K} B_{1,k} \Phi_k \right|} < \frac{\left| \sum_{k=1}^{K} B_{n,k} \right|}{\left| \sum_{k=1}^{K} B_{1,k} \right|} \tag{8.4.35}$$

for each of $n = 2, 3, \ldots, p$.

Example 8.3

This example shows how the intensity of eigenvector components in a prerebalanced vector change via coarse-mesh rebalancing. Suppose an iterative scheme to find the fundamental eigenvector of Eq. (8.2.17). We set $I = 7$ and the boundary conditions as $\phi_0 = \phi_7 = 0$. With disjunctive partitioning and region balance weighting, Eq. (8.2.7) becomes

$$(\phi_1 + \phi_2)\Phi_1 - \phi_3\Phi_2 = \lambda(\phi_1 + \phi_2)\Phi_1$$

$$-\phi_2\Phi_1 + (\phi_3 + \phi_4)\Phi_2 - \phi_5\Phi_3 = \lambda(\phi_3 + \phi_4)\Phi_2$$

$$-\phi_4\Phi_2 + (\phi_5 + \phi_6)\Phi_3 = \lambda(\phi_5 + \phi_6)\Phi_3$$

where ϕ_i denote the prerebalanced iterative solution. This equation is an eigenvalue problem of three unknowns: Φ_1, Φ_2, and Φ_3.

In order to see the effect of the coarse-mesh rebalancing scheme, we express the prerebalanced vector and the rebalanced vector, respectively, in terms of eigenvectors of Eq. (8.2.17) as

$$\phi_i = \sin\frac{\pi i}{7} + \sum_{k=2}^{6} a_k \sin\frac{k\pi i}{7}, \qquad i = 1, 2, \ldots, 6$$

$$\psi_i = \sin\frac{\pi i}{7} + \sum_{k=2}^{6} b_k \sin\frac{k\pi i}{7}, \qquad i = 1, 2, \ldots, 6$$

where both ϕ_i and ψ_i are normalized so that the coefficient of the first term is unity, and $\sin(k\pi i/7)$, $k = 1, 2, \ldots, 6$, are eigenfunctions of Eq. (8.2.17).

The response of $\{b_k\}$ to various sets of $\{a_k\}$ via the coarse-mesh rebalancing is shown in Table 8.1. The first two cases include only one low eigenvector and show positive acceleration effects. The third and fourth cases show the effect of the last eigenvector in the prerebalanced vector. It is seen that as a_6 increases, the acceleration effect represented by a_2/b_2 becomes poorer.

As shown in Section 8.4.3, the positive acceleration effect depends on the high eigenvector components excited in ψ and the high eigenvector components in the weighting vectors. It also depends on the high eigenvector components involved in the prerebalanced vector. In fact, we can show that, if the magnitude of a high eigenvector in ψ_0 is much larger than the magnitude of a low eigenvector in ψ_0, then the acceleration effect becomes negative. In this section we assume $p = 2$. In order to avoid the complexity introduced by the high eigenvector components in ψ_0, we further assume that the prerebalanced vector does not include high eigenvectors. This assumption is equivalent to having

$$\psi_0 = \sum_{n=1}^{2} \left(\sum_{k=1}^{K} B_{n,k} \right) \mathbf{u}_n \qquad (8.4.36)$$

$$\sum_{k=1}^{K} B_{n,k} = 0 \quad \text{for } n = 3, 4, \ldots, N \qquad (8.4.37)$$

With Eq. (8.4.29), the left side of Eq. (8.4.35) for $n = 2$ becomes

$$R_2 = \frac{\lambda_1 - \tilde{\lambda}_1}{\lambda_2 - \tilde{\lambda}_1} \frac{\sum_{m=K+1}^{N} (\lambda_m - \tilde{\lambda}_1) \gamma_{m,2} g_m}{\sum_{m=K+1}^{N} (\lambda_m - \tilde{\lambda}_1) \gamma_{m,1} g_m} \qquad (8.4.38)$$

If $\tilde{\lambda}_1$ approaches λ_1 in Eq. (8.4.38), R_2 obviously approaches zero, satisfying Eq. (8.4.35). However, how $\tilde{\lambda}_1$ approaches λ_1 as a result of rebalancing is to be clarified. To dissolve this problem, we first use Galerkin's method and then the general weighting vector.

GALERKIN'S WEIGHTING VECTORS

In Galerkin's method, the weighting vectors are

$$\mathbf{w}_k = \mathbf{f}_k = \sum_{n=1}^{N} B_{n,k} \mathbf{u}_n \qquad (8.4.39)$$

Therefore, if we apply the transformation defined by Eq. (8.4.23) to $\mathbf{w}_k$, it

Table 8.1   Numerical Study of the Accelerating Effect[a]

Case		k=1	2	3	4	5	6	$R_2=b_2/a_2$
1	a_k	1.0	0.1	0	0	0	0	
	b_k	1.0	0.0073	0.0024	-0.0057	-0.016	0.034	0.073
2	a_k	1.0	0.4	0	0	0	0	
	b_k	1.0	0.022	0.0036	0.0009	-0.050	0.1191	0.055
3	a_k	1.0	0.4	0	0	0	0.05	
	b_k	1.0	0.060	-0.0045	0.0074	-0.059	0.153	0.150
4	a_k	1.0	0.4	0	0	0	0.1	
	b_k	1.0	0.141	-0.0123	-0.0147	-0.067	0.23	0.35

[a] The acceleration effect is positive if R_2 is less than 1. The smaller the R_2, the greater the acceleration effect. See. Eq. (8.4.35) for definition of R_2.

automatically applies to $\mathbf{f}_k$. After this transformation, we use $\mathbf{v}_k$ for both trial and weighting vectors. The rebalanced vector is now written as

$$\psi = \sum_{k=1}^{K} h_k \mathbf{v}_k = \sum_{k=1}^{K} h_k \left(\mathbf{u}_k + \sum_{n=K+1}^{N} \gamma_{n,k} \mathbf{u}_n \right) \qquad (8.4.40)$$

where h_k are undetermined coefficients. The prerebalanced vector is also expressed by

$$\psi_0 = \sum_{k=1}^{K} h_k^0 \left(\mathbf{u}_k + \sum_{n=K+1}^{N} \gamma_{n,k} \mathbf{u}_n \right) \qquad (8.4.41)$$

where h_k^0 is determined when ψ_0 is given. Because we have assumed that ψ_0 does not include eigenvector components higher than p, we have

$$h_3^0 = h_4^0 = \cdots = h_K^0 = 0 \qquad (8.4.42)$$

and

$$h_1^0 \gamma_{n,1} + h_2^0 \gamma_{n,2} = 0 \quad \text{for } n > K \qquad (8.4.43)$$

Therefore, Eq. (8.4.41) is reduced to

$$\psi_0 = h_1^0 \mathbf{u}_1 + h_2^0 \mathbf{u}_2 \qquad (8.4.44)$$

Since Eq. (8.4.1) is self-adjoint and positive definite, Galerkin's method is equivalent to minimizing the Rayleigh quotient

$$J(\psi) = \frac{\langle \psi, \mathbf{A}\psi \rangle}{\langle \psi, \mathbf{F}\psi \rangle} \qquad (8.4.45)$$

If $J(\psi)$ is not stationary about $\psi = \psi_0$, then $J(\psi_0) > J_{\min}$, where $J_{\min}$ is the extremum of $J(\psi)$. As shown later, if $J(\psi_0) > J_{\min}$, the accelerating effect is positive. So we first prove that $J(\psi_0) > J_{\min}$ or, equivalently, that $h_k = h_k^0$ does not satisfy the stationary conditions.

Introducing Eq. (8.4.40) into Eq. (8.4.45), the stationary conditions are

$$\frac{\partial J}{\partial h_k} = 0, \qquad k = 1, 2, \ldots, K \qquad (8.4.46)$$

or equivalently

$$\sum_{k=1}^{K} \left[\lambda_k \delta_{k,l} + \sum_{n=K+1}^{N} \lambda_n (\gamma_{n,l} \gamma_{n,k}) \right] h_k$$

$$= \lambda \sum_{k=1}^{K} \left[\delta_{k,l} + \sum_{n=K+1}^{N} (\gamma_{n,l} \gamma_{n,k}) \right] h_k, \qquad l = 1, 2, \ldots, K \qquad (8.4.47)$$

where $\lambda = J(\mathbf{\psi})$ is introduced as an eigenvalue to be found. Equation (8.4.47) may be written compactly as

$$\mathscr{A}\mathbf{h} = \lambda \mathscr{F}\mathbf{h} \tag{8.4.48}$$

where $\mathscr{A}$ and $\mathscr{F}$ are matrices representing the coefficients of Eq. (8.4.47) and $\mathbf{h} = \mathrm{col}\,(h_1, h_2, \ldots, h_K)$. $\mathscr{A}$ and $\mathscr{F}$ are symmetric matrices, and Eq. (8.4.47) is evidently a positive-definite eigenvalue problem. We assume that none of the elements of $\mathscr{A}$ and $\mathscr{F}$ are zero. This is a natural assumption because a trial vector generally includes all the high eigenvectors independently of another trial vector, and accordingly $\gamma_{n,k} \neq 0$ for all n.

We now show that the eigenvector of Eq. (8.4.48) cannot have the following values of solution:

$$h_1 = h_1^0 \neq 0 \quad \text{and} \quad h_2 = h_2^0 \neq 0 \quad \text{with} \quad h_3 = h_4 = \cdots = h_K = 0 \tag{8.4.49}$$

If Eq. (8.4.49) is the solution, then h_1^0 and h_2^0 have to satisfy

$$\sum_{k=1}^{2} \left[\lambda_k \delta_{k,l} + \sum_{n=K+1}^{N} \lambda_n (\gamma_{n,l}\gamma_{n,k}) \right] h_k^0$$

$$= \lambda \sum_{k=1}^{2} \left[\delta_{k,l} + \sum_{n=K+1}^{N} (\gamma_{n,l}\gamma_{n,k}) \right] h_k^0, \qquad l = 1, 2 \tag{8.4.50}$$

which is the first two of Eq. (8.4.47). Because of Eq. (8.4.43), we have

$$\sum_{k=1}^{2} \left[\sum_{n=K+1}^{N} \lambda_n (\gamma_{n,l}\gamma_{n,k}) \right] h_k^0 = 0 \tag{8.4.51}$$

and

$$\sum_{k=1}^{2} \left[\sum_{n=K+1}^{N} (\gamma_{n,l}\gamma_{n,k}) \right] h_k^0 = 0 \tag{8.4.52}$$

Therefore, Eq. (8.4.50) becomes

$$\left. \begin{array}{l} \lambda_1 h_1^0 = \lambda h_1^0 \\ \lambda_2 h_2^0 = \lambda h_2^0 \end{array} \right\} \tag{8.4.53}$$

The solution of this equation is either $h_1^0 =$ arbitrary while $h_2^0 = 0$, or $h_1^0 = 0$ and $h_2^0 =$ arbitrary, which are inconsistent with the assumption that $h_1^0 \neq 0$ and $h_2^0 \neq 0$. Therefore, Eq. (8.4.49) is not the solution of Eq. (8.4.47), and accordingly $J(\mathbf{\psi}_0) > J_{\min}$.

In the next theorem, we prove that the acceleration effect is positive if $J(\mathbf{\psi}_0) \neq J_{\min}$.

Theorem 8.6. If $J_{\min} < J(\psi_0)$, the inequality

$$\left|\frac{h_2}{h_1}\right| < \left|\frac{h_2^0}{h_1^0}\right| \tag{8.4.54}$$

is satisfied, which is equivalent to Eq. (8.4.35) for $p = 2$.

Proof. $J_{\min}$ and $J(\psi_0)$ are written as

$$
\begin{aligned}
J_{\min} &= \frac{\sum_{l=1}^{K} \sum_{k=1}^{K} [\lambda_k \, \delta_{l,k} + \sum_{n=K+1}^{N} \lambda_n \gamma_{n,l} \gamma_{n,k}] h_l h_k}{\sum_{l=1}^{K} \sum_{k=1}^{K} [\delta_{k,l} + \sum_{n=K+1}^{N} \gamma_{n,l} \gamma_{n,k}] h_l h_k} \\[2mm]
&= \frac{\lambda_1 + \sum_{k=2}^{K} \lambda_k (h_k/h_1)^2 + \sum_{n=K+1}^{N} \lambda_n [\sum_l \gamma_{n,l}(h_l/h_1)]^2}{1 + \sum_{k=2}^{K} (h_k/h_1)^2 + \sum_{n=K+1}^{N} [\sum_l \gamma_{n,l}(h_l/h_1)]^2}
\end{aligned}
\tag{8.4.55}
$$

and

$$J(\psi_0) = \frac{\lambda_1 + \lambda_2 (h_2^0/h_1^0)^2}{1 + (h_2^0/h_1^0)^2} \tag{8.4.56}$$

respectively. Because of $\lambda_1 < \lambda_2 < \lambda_3 \cdots < \lambda_N$, we have

$$
\begin{aligned}
&\sum_{k=3}^{K} \lambda_k \left(\frac{h_k}{h_1}\right)^2 + \sum_{n=K+1}^{N} \lambda_n \left(\sum_l \gamma_{n,l}\frac{h_l}{h_1}\right)^2 \\[2mm]
&\qquad \geq \lambda_3 \left[\sum_{k=3}^{K} \left(\frac{h_k}{h_1}\right)^2 + \sum_{n=K+1}^{N} \left(\sum_l \gamma_{n,l}\frac{h_l}{h_1}\right)^2 \right]
\end{aligned}
\tag{8.4.57}
$$

if at least one of the h_ks for $k \geq 3$ is nonzero. Equation (8.4.57) and $J_{\min} < J(\psi_0)$ imply

$$\left(\frac{h_2}{h_1}\right)^2 < \left(\frac{h_2^0}{h_1^0}\right)^2 \quad \text{or} \quad \left|\frac{h_2}{h_1}\right| < \left|\frac{h_2^0}{h_1^0}\right| \tag{8.4.58}$$

thus completing the proof.

NON-GALERKIN WEIGHTING VECTORS

The analysis of the coarse mesh accelerating effect with non-Galerkin weighting vectors is more difficult than with Galerkin weighting vectors. As a basic case, we consider only two partitioning matrices and the corresponding two weighting vectors: $K = p = 2$.

We assume the prerebalanced vector is given by Eq. (8.4.36) with $K = 2$. If we normalize ψ_0 by

$$B_{1,1} + B_{1,2} = 1 \tag{8.4.59}$$

and define

$$\alpha_0 = B_{2,1} + B_{2,2} \tag{8.4.60}$$

then ψ_0 becomes

$$\psi_0 = \mathbf{u}_0 + \alpha_0 \mathbf{u}_1 \tag{8.4.61}$$

With $K = 2$, Eq. (8.4.37) also becomes

$$B_{n,2} = -B_{n,1} \qquad \text{for } n > 2 \tag{8.4.62}$$

The rebalanced vector given by Eq. (8.4.34) with $K = 2$ may be normalized by

$$B_{1,1}\Phi_1 + B_{1,2}\Phi_2 = 1 \tag{8.4.63}$$

Equation (8.4.63) is reduced to Eq. (8.4.59) when $\Phi_1 = \Phi_2 = 1$. Introducing Eqs. (8.4.62) and (8.4.63) into Eq. (8.4.34) with $K = 2$ yields

$$\psi = \mathbf{u}_1 + \alpha\mathbf{u}_2 + (\Phi_1 - \Phi_2) \sum_{n=1}^{N} B_{n,1}\mathbf{u}_n \tag{8.4.64}$$

where

$$\alpha = B_{2,1}\Phi_1 + B_{2,2}\Phi_2 \tag{8.4.65}$$

Equation (8.4.65) is reduced to Eq. (8.4.60) if $\Phi_1 = \Phi_2 = 1$. We can eliminate $(\Phi_1 - \Phi_2)$ by using Eqs. (8.4.63) and (8.4.65) to obtain

$$\psi = \mathbf{u}_1 + \alpha\mathbf{u}_2 + \frac{\alpha_0 - \alpha}{\Delta} \sum_{n=3}^{N} B_{n,1}\mathbf{u}_n \tag{8.4.66}$$

where $\Delta = B_{1,1}B_{2,2} - B_{1,2}B_{2,1}$ and Eqs. (8.4.60) and (8.4.61) were also used. Equation (8.4.66) has only one arbitrary parameter α, although two weighting vectors are necessary to determine it. Notice that ψ is reduced to ψ_0 if $\alpha = \alpha_0$.

Comparing Eq. (8.4.66) with Eq. (8.4.34), we notice that the acceleration effect is positive if

$$|\alpha| < |\alpha_0| \quad \text{or} \quad \left|\frac{\alpha}{\alpha_0}\right| < 1 \tag{8.4.68}$$

is satisfied. In order to find whether this condition is satisfied with non-Galerkin weighting vectors, we first derive the conditions for Eq. (8.4.68) to be satisfied.

As we have pointed out previously, the residual vector is orthogonalized to two independent weighting vectors:

$$\langle \mathbf{w}_l, (\mathbf{A} - \lambda\mathbf{F})\psi \rangle = 0, \qquad l = 1, 2 \tag{8.4.69}$$

Expanding $\mathbf{w}_l$ in the form of Eq. (8.4.21) with $K=2$ and using Eq. (8.4.66), Eq. (8.4.69) becomes

$$(\lambda_1 - \lambda) + \alpha(\lambda_2 - \lambda)C_{2,l} + \frac{\alpha_0 - \alpha}{\Delta}\sum_{n=3}^{N}(\lambda_n - \lambda)C_{n,l}B_{n,1} = 0, \qquad l = 1, 2$$

(8.4.70)

where $\mathbf{w}_l$ are assumed to be normalized by

$$C_{1,1} = C_{1,2} = 1 \tag{8.4.71}$$

In Eq. (8.4.70), λ is an eigenvalue of itself. Dividing Eq. (8.4.70) by λ_1 and solving for λ/λ_1 yields

$$\frac{\lambda}{\lambda_1} = \frac{1 + \alpha(\lambda_2/\lambda_1)C_{2,l} + [(\alpha_0 - \alpha)/\Delta]\sum_{n=3}^{N}(\lambda_n/\lambda_1)C_{n,l}B_{n,1}}{1 + \alpha C_{2,l} + [(\alpha_0 - \alpha)/\Delta]\sum_{n=3}^{N}C_{n,l}B_{n,1}}$$

(8.4.72)

We now assume that the second and third terms in both numerator and denominator of Eq. (8.4.72) are much smaller in magnitude than unity. Except for an early stage of an iterative scheme, this is a valid assumption. We expand the denominator into a power series. After truncating the second order terms and rearranging terms, Eq. (8.4.72) becomes

$$\frac{\lambda}{\lambda_1} \approx 1 + \alpha\left(\frac{\lambda_2}{\lambda_1} - 1\right)C_{2,l} + \frac{\alpha_0 - \alpha}{\Delta}\sum_{n=3}^{N}\left(\frac{\lambda_n}{\lambda_1} - 1\right)C_{n,l}B_{n,1}, \qquad l = 1, 2$$

(8.4.73)

By equating the right side of Eq. (8.4.73) for $l=1$ with that for $l=2$ and rearranging, we obtain

$$\frac{\alpha}{\alpha_0} = \frac{1}{1 - \xi} \tag{8.4.74}$$

where

$$\xi = \frac{(\lambda_2/\lambda_1 - 1)(C_{2,1} - C_{2,2})\Delta}{\sum_{n=3}^{N}(\lambda_n/\lambda_1 - 1)(C_{n,1} - C_{n,2})B_{n,1}} \tag{8.4.75}$$

In order that Eq. (8.4.68) be satisfied, ξ must be

$$\xi < 0 \quad \text{or} \quad \xi > 2 \tag{8.4.76}$$

Equation (8.4.76) is the necessary condition for the acceleration effect to be positive. For $\xi < 0$, α will have the same sign as α_0, which means a monotonic approach of α from α_0 to zero via coarse-mesh rebalancing. If $\xi > 2$, on the

Table 8.2

Range of ξ	Accelerating effect	Overshooting
$\xi < 0$	positive	no
$0 < \xi \ 1$	negative	no
$1 < \xi < 2$	negative	yes
$2 < \xi$	positive	yes

other hand, α will have a different sign from α_0, which means an overshooting effect. Table 8.2 summarizes the accelerating and overshooting effects for four ranges of ξ.

The investigation of the value of ξ involves the signs and magnitude of $C_{n,1}, C_{n,2}, B_{n,1}$, and Δ, which depend on the particular partitioning matrices and weighting vectors selected. The study for various combinations of partitioning and weighting is left open for future contributions. We, however, show the results of an analysis on a typical case.

Consider a symmetric one-dimensional geometry and its mesh system consisting of $2I$ grids. The grid points are indexed from left to right in the sequence, $-I, -I+1, \ldots, -1, 1, \ldots I-1, I$. A vector and the field value on the grids are related by

$$\mathbf{z} = \text{col}\,(z_{-I}, z_{-I+1}, \ldots, z_{I-1}, z_I) \tag{8.4.77}$$

where $\mathbf{z}$ is any vector in the present analysis. As a consequence of considering a symmetric geometry, the eigenvectors have odd and even parities as

$$
\begin{aligned}
u_{n,-i} = u_{n,i} \quad &\text{for } n \text{ odd} \\
u_{n,-i} = -u_{n,i} \quad &\text{for } n \text{ even}
\end{aligned}
\tag{8.4.78}
$$

where $u_{n,i}$ is an element of $\mathbf{u}_n$. We normalize $\mathbf{u}_n$ in such a way that $u_{n,I}$ will be positive for all n, as schematically illustrated in Fig. 8.3.

The trial vectors $\mathbf{f}_k \geq 0$ based on pyramid partitioning and disjunctive partitioning generally satisfy the property:

$$
\left.
\begin{aligned}
f_{1,-i} \leq f_{1,i} \\
f_{2,-i} \geq f_{2,i}
\end{aligned}
\right\}
\tag{8.4.79}
$$

where $f_{k,i}$ is an element of $\mathbf{f}_k$ and $K = p = 2$ is assumed. The weighting

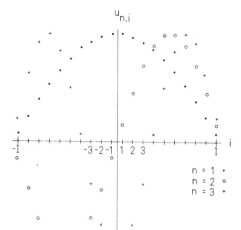

Figure 8.3 Schematic illustration of components of the first three eigenvectors for a slab geometry.

vectors $\mathbf{w}_k \geq 0$ based on region balancing or pyramid partitioning strictly satisfy

$$w_{1,-i} \leq w_{1,i}$$
$$w_{2,-i} \geq w_{2,i}$$

(8.4.80)

We also assume a symmetry relation between $\mathbf{w}_1$ and $\mathbf{w}_2$ as

$$w_{2,-i} = w_{1,i}$$

(8.4.81)

With those assumptions, the following relations are easily proven:

$$\left. \begin{array}{ll} B_{1,1}>0, & B_{1,2}>0 \\ B_{2,1}>0, & B_{2,2}<0 \end{array} \right\} \Rightarrow \Delta < 0$$

(8.4.82)

$$C_{2,1}>0, \quad C_{2,2}<0 \Rightarrow C_{2,1}-C_{2,2}>0$$

(8.4.83)

$$C_{n,1} = C_{n,2} \quad \text{for all } n \text{ odd}$$

(8.4.84)

$$C_{n,1} = -C_{n,2} \quad \text{for } n \text{ even}$$

(8.4.85)

If the weighting vectors are selected to be similar to the trial vectors, namely

$$\mathbf{w}_k \approx \mathbf{f}_k$$

(8.4.86)

then $C_{n,1}$ and $B_{n,1}$ will have the same sign. This leads to

$$\sum_{n=3}^{K} (C_{n,1} - C_{n,2})B_{n,1} = 2 \sum_{\substack{n=4 \\ n:\text{even}}}^{K} C_{n,1}B_{n,1} > 0 \qquad (8.4.87)$$

Because of Eqs. (8.4.82), (8.4.83), and (8.4.87), Eq. (8.4.75) becomes negative. Therefore, the acceleration effect is positive without any overshooting. If the weighting vectors are not similar to the trial vectors, the signs of $C_{n,1}B_{n,1}$ for $n > 3$ are unpredictable for general cases without further study. Overshooting occurs if $C_{n,1}B_{n,1}$ for even ns are negative. It is interesting that, with a symmetric choice of the weighting vectors, the value of ξ, namely, the accelerating effect, is not affected by $\mathbf{u}_n$ with n even.

8.4.5 Conclusion

The coarse-mesh rebalancing effect on an iterative scheme to obtain the fundamental eigenvector of a self-adjoint eigenvalue problem has been mathematically studied. It has been presumed that every coarse-mesh rebalancing is preceded by a short Chebyshev polynomial iteration with a low dominance ratio estimate that uniformly decays the last $N-p$ high eigenvector components. The acceleration effect has been defined as positive if the low eigenvector components except the fundamental eigenvectors are decayed via coarse-mesh rebalancing.

As a result of the study, we have shown:

1. As an ideal case, if the first p eigenvectors are expressed exactly by linear combination of K weighting vectors, the $p-1$ low eigenvectors in iterative residue are completely eliminated via coarse-mesh rebalancing.
2. The accelerating effect with non-Galerkin weighting vectors is a function of interacting effect between the high eigenvector components in the weighting vectors and the high eigenvector components excited via coarse-mesh rebalancing in the rebalanced vector.
3. High eigenvector components in the weighting vectors are parasitic to the accelerating effect.
4. The Galerkin weighting vectors have a positive accelerating effect, provided that the magnitude of high eigenvector components is sufficiently low compared with low eigenvector components.
5. With non-Galerkin weighting vectors, the accelerating effect is positive and there is no overshooting if the weighting vectors are similar to the trial vectors. If the weighting vectors are dissimilar to the trial vectors, the overshooting effect can occur and the positive acceleration effect cannot be guaranteed.

8.5 NUMERICAL ILLUSTRATIONS

This section describes four examples of application of coarse-mesh rebalancing.

Case 1.   One-Dimensional Eigenvalue Problem

We apply the coarse-mesh rebalancing method to the one-dimensional difference equation,

$$-\frac{\tau}{h^2}\phi_{i-1}+\left(1+2\frac{\tau}{h^2}\right)\phi_i-\frac{\tau}{h^2}\phi_{i+1}=\frac{1}{\lambda}\phi_i, \qquad i=1,2,\ldots,(I-1) \quad (8.5.1)$$

where $I=50$, $\phi_0=\phi_2$, $\phi_I=0$, $\tau=50$, and $h=150/49=3.06$. The fixed-parameter extrapolation method described in Section 2.8 is combined with

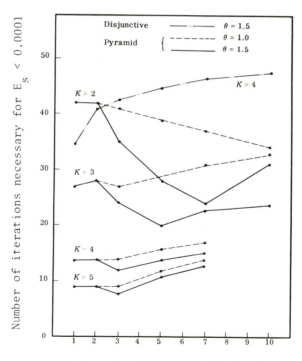

Figure 8.4 Number of iterations for convergence (one-dimensional case).

various coarse-mesh rebalancing parameters. We define the iterative error as

$$E_s^{(t)} = \max_i \left(\frac{\phi_i^{(t)}}{\phi_i^{(t-1)}} \right) - \min_i \left(\frac{\phi_i^{(t)}}{\phi_i^{(t-1)}} \right) \tag{8.5.2}$$

where t is the iteration number.

The following two combinations of partitioning and weightings were tested: (a) disjunctive partitioning and region balancing, and (b) pyramid partitioning and pyramid weighting. The former combination is designated as *disjunctive*, while the latter is designated as *pyramid*. As the parameters for a sensitivity study, we varied (1) the number of iterations between two consecutive coarse mesh rebalancings, (2) the number of coarse-mesh regions (or points) K, and (3) the extrapolation parameter θ. With the one-parameter Chebyshev polynomial of fifth order using the exact dominance ratio, the number of iterations to satisfy $E_s^{(t)} < 0.0001$ was 47. The number of iterations of the tested schemes for the same convergence criterion is shown in Fig. 8.4.

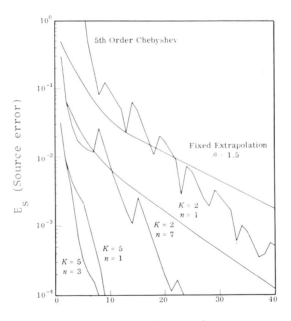

Figure 8.5 Decay of source error by several methods (one-dimensional case).

It is found at first that the convergence with the coarse-mesh rebalancing, even in the poorest case with $K = 2$, is faster than with the Chebyshev extrapolation, and also than the disjunctive partitioning with a smaller number of coarse-mesh points. The use of 1.5 for θ is better than 1.0. The accelerating effect increases with increasing K. When K is as small as 2 or 3, the convergence is fastest at some number n about 6. The optimum value of n decreases with an increasing K, whereas an optimum number greater than unity has not been observed in connection with the disjunctive partitioning. The decay of E_s, when accelerated by the present method, is compared in Fig. 8.5, with the Chebyshev parameter extrapolation method and also with the fixed parameter of $\theta = 1.5$.

In discussing the mechanism of coarse-mesh rebalancing, the decay of E_s with a rather long period of $n = 10$ is compared in Fig. 8.6 with the fixed-parameter extrapolation method. From this figure, we see that the decay of E_s is very fast during the first few iterations after coarse-mesh

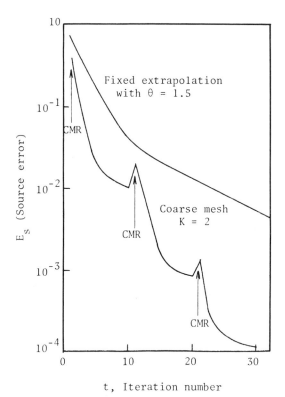

Figure 8.6 Decay of source error (one-dimensional case).

rebalancing. Then the speed is gradually decreased until it becomes almost the same as that of the fixed extrapolation method. This phenomenon demonstrates that low error modes are substantially eliminated by coarse-mesh rebalancing, while high error modes are retained. During a few iterations after a coarse-mesh rebalancing, those high modes decay rapidly and low modes that are still alive become dominant. Each time the coarse-mesh rebalancing is applied, the same procedure is repeated. It is also noticed that E_s increases slightly each time just after a coarse-mesh rebalancing. This means that high modes are slightly excited by coarse-mesh rebalancing in compensation for the substantial elimination of low modes.

From this discussion it can be inferred that there exists an optimum number n for each K. If coarse-mesh rebalancing is used too frequently, high modes have little chance to decay. With an increase in K, the effectiveness of coarse-mesh rebalancing for higher modes is also increased. This is the reason for the decrease of the optimum value of n with an increase in K.

Case 2. Iterative Finite Element Solution[9]

This example shows the results of the coarse-mesh-accelerated SOR iteration applied to a finite element problem for the geometry shown in Fig. 8.7.

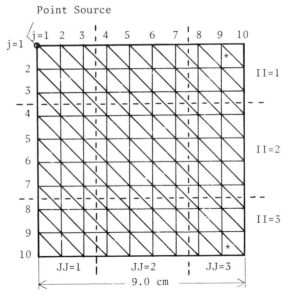

Figure 8.7 Finite element problem. (For all triangles, $D = 1.0$, $\Sigma_a = 0.0$ except for two triangles with asterisks, where $\Sigma_a = 0.02$. Boundary conditions: zero current).

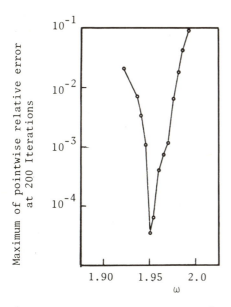

Figure 8.8 Effect of ω for Case 2.

A one-energy group neutron equation,

$$-D\nabla^2\phi(x, y) + \Sigma_a\phi(x, y) = \delta(x)\delta(y) \qquad (8.5.3)$$

was solved where δ is the Dirac delta function $\Sigma_a = 0$, except in the two triangular elements with * shown in Fig. 8.7, where $\Sigma_a = 0.02$. The boundary condition is zero-current at all the external surfaces.

This problem was first solved iteratively without coarse-mesh rebalancing. Figure 8.8 shows the maximum pointwise relative error after 200 iterations. The optimum extrapolation parameter is found from Fig. 8.8 to be $\omega_{opt} = 1.95$. With $\omega_{opt} = 1.95$, 174 iterations are required to satisfy the convergence criterion for this problem: the maximum pointwise relative error is less than 10^{-4}. Next, ω was fixed at 1.5 and the SOR scheme was accelerated by coarse-mesh rebalancing with various values of the decoupling coefficient. The coarse-mesh rebalancing was applied after every one, two, or three fine-mesh iterations. Figure 8.9 shows the number of fine-mesh iterations to satisfy the convergence criterion. It is seen that the acceleration effect of coarse-mesh rebalancing is optimized with $\alpha = 0.4$ for this case. Overall computing speed with the best combination of n and α was 7.5 times faster than the SOR with ω_{opt} but without coarse-mesh rebalancing.

Case 3. Three-Group Two-Dimensional Neutron-Diffusion Equation[5]

The three-group two-dimensional neutron-diffusion equation for a light water power reactor was solved by the standard inner and outer iteration

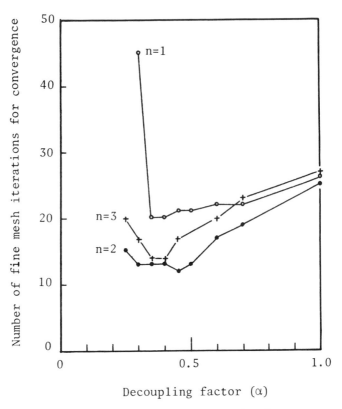

Figure 8.9 Number of iterations for Case 2.

schemes, which is basically the same as shown in Fig. 3.13. The coarse-mesh rebalancing schemes with the following combinations were applied after every five outer iterations with the fixed extrapolation parameter, $\theta = 1.5$: (a) pyramid partitioning and pyramid weighting, and (b) pyramid partitioning and region balancing. Since coarse mesh was applied to each group, the coarse-mesh equation was also a three-group equation. As shown in Fig. 8.10, the numbers of fine-mesh and coarse-mesh points were 50×50 and 8×8, respectively. The convergence of eigenvalues is compared in Table 8.3 to the one-parameter Chebyshev polynomial method of 6th order with an accurate estimate for the dominance ratio. The decay of the error similar to Eq. (8.5.2) was defined as

$$E_s^{(t)} = \max_{i,j} \left(\frac{S_{i,j}^{(t)}}{S_{i,j}^{(t-1)}} \right) - \min_{i,j} \left(\frac{S_{i,j}^{(t)}}{S_{i,j}^{(t-1)}} \right) \tag{8.5.4}$$

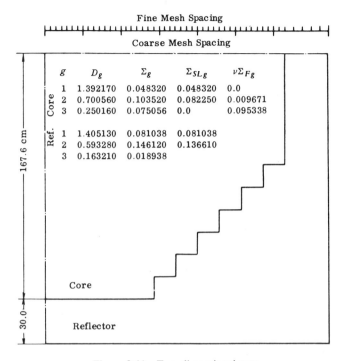

	g	D_g	Σ_g	Σ_{SLg}	$\nu\Sigma_{Fg}$
Core	1	1.392170	0.048320	0.048320	0.0
	2	0.700560	0.103520	0.082250	0.009671
	3	0.250160	0.075056	0.0	0.095338
Ref.	1	1.405130	0.081038	0.081038	
	2	0.593280	0.146120	0.136610	
	3	0.163210	0.018938		

Figure 8.10 Two-dimensional core.

where $S_{i,j}^{(t)}$ is the tth iterative for the fission source at (i, j). It is interesting that even the first coarse-mesh eigenvalue has a sufficient accuracy for most purposes, and also that the eigenvalue after five outer iterations with coarse mesh rebalancing is as accurate as the eigenvalue after 50 outer iterations with the Chebyshev polynomial extrapolation.

In Table 8.4, the number of iterations and elapsed computing time of coarse-mesh-accelerated scheme are compared with the solution of the 6th order Chebyshev polynomial. The effects of differences between pyramid weighting and region balance weighting are also compared in Table 8.4. No significant difference in the acceleration effect between these two weighting schemes was observed, but the computer time with region balance weighting is much smaller than with the other.

Case 4 Equipoise Iterative Scheme

The iterative solution method for few-group diffusion equations known as the "Equipoise method"[12] has been extensively used[13–16] as an alternative to the standard inner–outer iteration schemes. There are several advantages

Table 8.3 Eigenvalue Convergence (Two-Dimensional Case)

Iteration Number t	6th Order Chebyshev		Coarse-mesh-accelerated with $\tilde{\Delta}$[b]	
	$\lambda^{(t)}$	$E^{(t)}$	$\lambda^{(t)}$	$E_e^{(t)}$
			(1.094442)[a]	(0.000516)
1	1.069634	0.025324	1.093357	0.001601
2	1.073994	0.020964	1.093619	0.001339
3	1.077359	0.017599	1.094055	0.000903
4	1.079809	0.015149	1.094291	0.000667
5	1.081640	0.013318	1.094428	0.000530
			(1.094938)	(0.000020)
6	1.090862	0.004096	1.094963	0.000005
7	1.089255	0.005703	1.094964	0.000006
8	1.090353	0.004605	1.094966	0.000010
9	1.090616	0.004342	1.094961	0.000003
10	1.090818	0.004140	1.094962	0.000004
			(1.094958)	(0.000000)
11	1.090988	0.003970	1.094955	0.000004
12	1.091144	0.003814	1.094961	0.000003
13	1.093263	0.001695	1.094960	0.000002
14	1.093447	0.001511	1.094959	0.000001
15	1.093551	0.001407	1.094959	0.000001
			(1.094958)	(0.000000)
20	1.094278	0.000680	1.094958	0.000000
25	1.094490	0.000468	1.094958	0.000000
30	1.094782	0.000176		
40	1.094904	0.000054		
50	1.094951	0.000007		
70	1.094957	0.000001		

[a]The values in parentheses are eigenvalues from coarse mesh equation.

[b]$\tilde{\Delta}$ denotes the combination of pyramid partitioning and region balancing.

in the Equipoise method over the inner–outer scheme. However, when applied to large reactor systems which require very fine mesh points, frequently the convergence speed becomes so slow that no satisfactory convergence can be reached even after hundreds of iterations.

In order to improve the convergence speed, the coarse-mesh rebalancing method using decoupling coefficients was incorporated into a two-group two-dimensional diffusion code that is basically the same as the Exterminator code.[19] In applying the coarse mesh rebalancing method, the

Table 8.4 Comparison of Iteration Number and Computing Time for Two-Dimensional Calculations

Acceleration Method	Chebyshev	Coarse mesh with Δ[c]	Coarse mesh with $\tilde{\Delta}$[d]
Number of fine mesh outer iterations[a] (25 sec/iteration)[b]	66	13	13
Number of coarse mesh rebalancing	– –	3 (73 sec/acceleration)	3 (45 sec/acceleration)
Total elapsed time (sec)[b]	1650	544	460

[a]Convergence Criterion: $E_s < 0.00001$.

[b]HITAC 5020F.

[c]Pyramid partitioning and pyramid weighting.

[d]Pyramid partitioning and region balancing.

two energy-group equations were collapsed into one group. If the coarse-mesh rebalancing equation is written for a one-dimensional problem for simplicity it becomes

$$-\alpha_J \left(\sum_{g=1}^{2} \frac{D_g \psi_{g,m-1}}{h_m} \right) \Phi_{J-1} + \left[(\alpha_J - 1) \left(\sum_{g=1}^{2} \frac{D_g \psi_{g,m-1}}{h_m} \right) \right.$$

$$+ \sum_{g=1}^{2} \left(\frac{D_g \psi_{g,m}}{h_m} + \frac{D_g \psi_{g,n}}{h_{n+1}} \right) + (\alpha_{J+1} - 1)$$

$$\left. \times \sum_{g=1}^{2} \frac{D_g \psi_{g,n+1}}{h_{n+1}} + \sum_{i \in J} \frac{h_i + h_{i+1}}{2} \left(\sum_{g=1}^{2} \Sigma_{ag} \psi_{g,i} - \Sigma_{sl} \psi_{1,i} \right) \right] \Phi_J$$

$$-\alpha_{J+1} \left(\sum_{g=1}^{2} \frac{D_g \psi_{g,n+1}}{h_{n+1}} \right) \Phi_{J+1} = \frac{1}{k} \left(\sum_{i \in J} \frac{h_i + h_{i+1}}{2} \sum_{g=1}^{2} \nu \Sigma_{fg} \psi_{g,i} \right) \Phi_J$$

where $\psi_{g,i}$ is the prerebalanced flux of group g at mesh point i, h_i is the fine mesh space, Φ_J are unknown coarse mesh rebalancing values, α_J are decoupling factors, m and n are the left-most and right-most grids in coarse-mesh region J, respectively.

Table 8.5 illustrates the performance of the code with and without coarse-mesh rebalancing. Problems 1 and 2 are 1/8 PWR core, and Problem 3 is 1/2 core with a one off-centred control rod. Convergence

Table 8.5 Comparison of Computational Time (CDC 6600)

	Without C. M.	With C. M.	Ratio
Problem #1 (total fine mesh = 1606, total coarse mesh = 99)			
No. of iterations	313	34	
Time/iteration	1.70 sec.	1.70 sec.	
No. of c. m. rebalancing	0	6	
Time/c. m. rebalancing	--	1.06 sec.	
Total time	532.1 sec.	64.1 sec.	1/8.3
Problem #2 (total fine mesh = 6825, total coarse mesh = 99)			
No. of iterations	234	45	
Time/iterations	7.33 sec.	7.33 sec.	
No. of c. m. rebalancing	0	8	
Time/c. m. rebalancing	--	1.13 sec.	
Total time	1715 sec.	339 sec.	1/5.0
Problem #3 (total fine mesh = 5000, total coarse mesh = 162)			
No. of iterations	745*	51	
Time/iterations	6.13 sec.	6.13 sec.	
No. of c. m. rebalancing	0	10	
Time/c. m. rebalancing	--	1.01 sec.	
Total time	4567 sec.	323 sec.	<1/14.1

*Computation was terminated because of a time limit

criteria used were flux error < 0.0001 and eigenvalue error < 0.00001. All the problems tested were found to converge within 30–50 iterations starting with flat initial flux. The computing time was reduced to $1/5$–$1/14$ of the original computing time. More gain is obtained when the number of mesh points is increased. Considering that the present code does not separately perform inner and outer iterations and the code includes feedbacks, the total number of iterations such as 30–50 is much smaller than the total number of inner iterations in the standard inner–outer scheme.

8.6 COARSE-MESH REBLANCING USING ADDITIVE CORRECTIONS

Another form of coarse-mesh rebalancing may be derived by using additive corrections rather than multiplicative corrections.[22,24] In this approach a prerebalanced vector $\mathbf{\psi}_0$ and the rebalanced vector $\mathbf{\psi}$ are related by

$$\mathbf{\psi} = \mathbf{\psi}_0 + \sum_{k=1}^{K} \Phi_k \mathbf{u}_k \qquad (8.6.1)$$

where the second term is the correction, $\mathbf{u}_k$ is a prescribed vector, and Φ_k is an arbitrary coefficient to be determined by the weighted residual method. The residual vector (see Section 6.2) for the inhomogeneous equation Eq. (8.2.3) becomes

$$\mathbf{r} = \mathbf{A}\left(\mathbf{\psi}_0 + \sum_{k=1}^{K} \Phi_k \mathbf{u}_k\right) - \mathbf{S} \qquad (8.6.2)$$

Premultiplying Eq. (8.6.2) by K independent weighting vectors $\mathbf{w}_l$, $l = 1, 2, \ldots, K$, and requiring the weighted residual to be zero, we obtain

$$\sum_{k=1}^{K} \langle \mathbf{w}_l, \mathbf{A}\mathbf{u}_k \rangle \Phi_k = \langle \mathbf{w}_l, \mathbf{S} - \mathbf{A}\mathbf{\psi}_0 \rangle, \qquad l = 1, 2, \ldots, K \qquad (8.6.3)$$

The undetermined coefficients are determined by solving Eq. (8.6.3). The selection of $\mathbf{u}_k$ and $\mathbf{w}_k$ is important for short computing time and good acceleration effect. The various types of partitioning matrices defined in Section 8.2 may be used to provide $\mathbf{u}_k$ and $\mathbf{w}_k$, for example,

$$\mathbf{u}_k = \mathbf{w}_k = \mathbf{P}_k \mathbf{1} \qquad (8.6.4)$$

where $\mathbf{1}$ is the sumvector (all the elements are unity).

The present additive correction was first applied for two-dimensional neutron diffusion codes by using the disjunctive partitioning matrices. It was soon found[22] that the acceleration effect is significantly improved by the bilinear functions. When $\mathbf{u}_k$ are chosen as piecewise polynomials, the present additive coarse-mesh rebalancing turns out to be a direct application of the finite element method to accelerating iterative convergence. The merit of using the finite element method for the correction term is due to the smoothness of the piecewise polynomial functions. As pointed out in Section 8.1, slow convergence of an iterative solution is usually due to slowly varying function in space so that they can be well approximated by smooth piecewise polynomials. An advantage of the additive form of coarse-mesh rebalancing over the multiplicative form lies in the fact that the coefficients of Φ_k in Eq. (8.6.3) are not dependent on the prerebalanced vector, ψ_0. Therefore, once the coefficients of Φ_k are calculated at the first application of coarse-mesh rebalancing, they can be stored in computer memories to be repeatedly used in later coarse-mesh rebalancing, thus saving computing time for calculating the coefficients.

If the whole domain of interest is homogeneous, a more straightforward derivation of the coarse-mesh equation is possible without the weighted residual method.[25] The additive coarse-mesh rebalancing has a great potential in accelerating relaxation schemes to solve non-linear equations that appear in fluid dynamic problems.[25,26]

In case additive coarse-mesh rebalancing is applied to an eigenvalue problem, the residual vector is written as

$$\mathbf{r} = \mathbf{A}\psi - \lambda \mathbf{F}\psi \qquad (8.6.5)$$

where ψ is given by Eq. (8.6.1). In order to obtain K equations for the K unknowns, Eq. (8.6.5) is premultiplied by K independent weighting vectors, $\mathbf{w}_k$:

$$\sum_k \langle \mathbf{w}_l, \mathbf{A}\mathbf{u}_k \rangle \Phi_k + \langle \mathbf{w}_l, \mathbf{A}\psi_0 \rangle$$
$$= \lambda \left[\sum_k \langle \mathbf{w}_l, \mathbf{F}\mathbf{u}_k \rangle \Phi_k + \langle \mathbf{w}_l, \mathbf{F}\psi_0 \rangle \right], \qquad l = 1, 2, \ldots, K \qquad (8.6.6)$$

Equation (8.6.6) is a coarse-mesh eigenvalue problem for Φ_k. One may be curious, however, about the second term on the left side, as well as the second term of the right side, which may look to be an inhomogeneous term. In answering this question, it should be noticed that Eq. (8.6.3) could have been written as

$$\psi = \Phi_0 \psi_0 + \sum_{k=1}^{K} \Phi_k \mathbf{P}_k \mathbf{1} \qquad (8.6.7)$$

where Φ_0 is another arbitrary coefficient. By using $K+1$ weighting vectors, we could have obtained a set of the corresponding number of homogeneous linear equations. Since the solution of an eigenvalue problem is arbitrary by a constant, we could have normalized the eigenvector so that $\Phi_0 = 1$, thus obtaining Eq. (8.6.1). The fundamental eigenvalue and the corresponding eigenvector can be obtained by the iterative procedure in the form

$$\sum_k A_{l,k}\Phi_k^{(n+1)} = \lambda^{(n)}\left[\sum_{k=1}^{K} C_{l,k}\Phi_k^{(n)} + D_l\right] - B_l \qquad (8.6.8)$$

$$\lambda^{(n)} = \frac{\sum_l V_l\left[\sum_{k=1}^{K} A_{l,k}\Phi_k^{(n)} + B_l\right]}{\sum_l V_l\left[\sum_{k=1}^{K} C_{l,k}\Phi_k^{(n)} + D_l\right]} \qquad (8.6.9)$$

where n is the iteration number; V_l, $l = 1, 2, \ldots, K$, are weights; and

$$A_{l,k} = \langle w_l, Au_k \rangle \qquad (8.6.10)$$

$$B_l = \langle w_l, A\psi_0 \rangle \qquad (8.6.11)$$

$$C_{l,k} = \langle w_l, Fu_k \rangle \qquad (8.6.12)$$

$$D_l = \langle w_l, F\psi_0 \rangle \qquad (8.6.13)$$

PROBLEMS

1. Derive the coarse-mesh rebalancing equation for Eq. (8.2.17) with $I = 50$ and $K = 5$ by using disjunctive partitioning and Galerkin's weighting.
2. Repeat Problem 1 using pyramid partitioning and pyramid weighting.
3. A matrix is called M-matrix[23] if it has column diagonal dominance with positive diagonal elements and nonpositive offdiagonal elements (irreducibility and symmetry are not required). Show that the matrix on the left side of Eq. (8.3.1) is a M-matrix.
4. Equation (8.3.10) may be written as

$$\begin{bmatrix} \mathscr{A}_1 & 0 \\ -\mathscr{Q} & \mathscr{A}_2 \end{bmatrix}\begin{bmatrix} \phi_1 \\ \phi_2 \end{bmatrix} = \frac{1}{k}\begin{bmatrix} \mathscr{F}_1 & \mathscr{F}_2 \\ 0 & 0 \end{bmatrix}\begin{bmatrix} \phi_1 \\ \phi_2 \end{bmatrix}$$

Show that if the region balance weighting is used, the matrix on the left side is a M-matrix.
5. Show that the inverse of the matrix on the left side of the equation in Problem 4 is

$$\begin{bmatrix} \mathscr{A}_1^{-1} & 0 \\ \mathscr{A}_2^{-1}\mathscr{Q}\mathscr{A}_1^{-1} & \mathscr{A}_2^{-1} \end{bmatrix}$$

and a nonegative matrix.

6. Consider the equation

$$\mathbf{A}\boldsymbol{\phi} = \mathbf{S}$$

where $\mathbf{A}$ is an S-matrix and $\mathbf{S}$ includes both positive and negative elements. Show that $\boldsymbol{\phi}$ may be given by

$$\boldsymbol{\phi} = \boldsymbol{\phi}_+ + \boldsymbol{\phi}_-$$
$$\boldsymbol{\phi}_+ = \mathbf{A}^{-1}\mathbf{S}_+ > \mathbf{0}$$
$$\boldsymbol{\phi}_- = \mathbf{A}^{-1}\mathbf{S}_- < \mathbf{0}$$

where $\mathbf{S}_+$ and $\mathbf{S}_-$ are nonnegative and nonpositive vectors satisfying

$$\mathbf{S}_+ + \mathbf{S}_- = \mathbf{S}$$

(Use this formulation in applying coarse-mesh rebalancing to an iterative solution involving negative solutions. See Reference 10.)

7. Discuss why convergence of the SOR scheme with ω_{opt} cannot be accelerated by coarse-mesh rebalancing.

8. Provided that $\mathbf{A}$ is an S-matrix, show that the matrix of the coefficients of Eq. (8.6.5) is positive definite regardless of the partitioning scheme selected if Galerkin's weighting is used.

9. If $\mathbf{A}$ is an S-matrix and the pyramid partitioning and the region balance is used, the coefficient matrix for Eq. (8.6.5) will have a positive inverse. Prove this.

10. Discuss the advantage of the additive coarse-mesh rebalancing with respect to stability.

11. The additive coarse-mesh rebalancing may result in a negative rebalanced vector. Explain the reason and prove this possibility mathematically.

REFERENCES

1. E. L. Wachspress, *Iterative solution of elliptic systems and applications to the neutron diffusion equations of reactor physics*, Prentice-Hall, Englewood Cliffs, N.J. (1966).

2. R. A. Froehlich, "Computer independence of large reactor physics codes with reference to well balanced computer configurations," *Proceedings of the Conference on the Effective Use of Computers in the Nuclear Industry*, USAEC CONF-690401, National Technical Information Service, Springfield, Va. (1969).

3. R. A. Froehlich, "A theoretical foundation for coarse mesh variational techniques," GA-7870, General Atomic, San Diego, Calif. (1967).

4. R. A. Froehlich, "Flux synthesis methods versus difference approximation methods for efficient determination of neutron flux distributions in fast and thermal reactors," *Numerical Reactor Calculations*, printed by International Atomic Energy Agency, Vienna, Austria (1973).

5. S. Nakamura, "A variational rebalancing method for linear iterative convergence scheme of neutron diffusion and transport equations," *Nucl. Sci. Eng.*, **39**, 278–283 (1970).

6. S. Nakamura, "Coarse mesh acceleration of iterative solution of neutron diffusion equation," *Nucl. Sci. Eng.*, **43**, 116–120 (1971).

7. S. Nakamura and V. E. Esposito, "Coarse mesh rebalancing applied to hydraulic analysis," CONF-730414, National Technical Information Center, Springfield, Va. (1973).

8. S. Nakamura and P. Su, unpublished work.

9. S. Nakamura and T. Ohnishi, "The iterative solutions for the finite-element method," *Numerical Reactor Calculations*, printed by International Atomic Energy Agency, Vienna, Austria (1973).

10. S. Nakamura, "New formulation and coarse mesh acceleration for two-dimensional DS_n and P_n methods," *Numerical Reactor Calculations*, printed by International Atomic Energy Agency, Vienna, Austria (1973).

11. B. A. Finlayson and L. E. Scriven, "The method of weighted residuals—A review," *Appl. Mech. Rev.*, **19**, 735–748 (1966).

12. B. A. Finlayson, *The method of weighted residual and variational principles*, Academic, New York (1972).

13. R. Bonalmi, "Integral transport theory and numerical analysis methods in the design of a heavy-water reactor," *Numerical Reactor Calculations*, printed by International Atomic Energy Agency, Vienna, Austria (1973).

14. S. Nakamura, "Effect of weighting functions on the coarse mesh rebalancing acceleration," *Mathematical models and computational techniques for analysis of reactor systems*, CONF-730414, National Technical Information Centre, Springfield, Va. (1973).

15. R. Froehlich, "Current problems in multidimensional reactor calculations," CONF-730414, National Technical Information Center, Springfield, Va. (1973).

16. S. Nakamura, "Analysis of coarse mesh rebalancing effect," *Nucl. Sci. Eng.*, **61**, 98–106 (1976).

17. T. B. Fowler and M. L. Tobias, *EQUIPOISE-3: A two-dimensional, two-group, neutron diffusion code*, ORNL-3199, Oak Ridge National Laboratory, Tenn. (1962).

18. M. L. Tobias and T. B. Fowler, *The TWENTY GRAND program for the numerical solution of few-group neutron diffusion equations in two dimensions*, ORNL-3200, Oak Ridge National Laboratory, Tenn. (1962).

19. T. B. Fowler, M. L. Tobias, and D. R. Vondy, *EXTERMINATOR-A multigroup code for solving neutron diffusion equation in one and two dimensions*, ORNL-TM-842, Oak Ridge National Laboratory, Tenn. (1966).

20. T. B. Fowler and D. R. Vondy, *Nuclear reactor core analysis code: CITATION*, ORNL-TM-2406, Rev. 1, Oak Ridge National Laboratory, Tenn. (1970).

21. D. L. Delp, D. L. Fisher, J. M. Harriman, and M. J. Stedwell, *FLARE, a three-dimensional boiling water reactor simulator*, GEAP-4598, General Electric Company, San Jose, Calif. (1964).

22. M. Melice and C. Hunin, "Acceleration of neutron diffusion calculation by a coarse mesh rebalancing method using additive bilinear corrections," paper presented and distributed at Belgian Section of annual meeting of Amer. Nucl. Soc. (1973).

23. R. S. Varga, *Matrix iterative analysis*, p. 85, Prentice-Hall, Englewood Cliffs, N.J. (1962).

24. F. De La Vallee Poussin, "An accelerated relaxation algorithm for iterative solutions of elliptic equations," *SIAM J. Num. Anal.*, **5**, 2 (1968).

25. J. C. South, Jr. and A. Brandt, "The multigrid method: fast relaxation for transonic flows," *Advances in Eng. Sci.*, **4**, NASA CP-2001, National Technical Information Service, Springfield, Va. (1976).

26. S. Nakamura and P. P. Su, "Coarse-mesh accelerated relaxation scheme for mesh generation equations," to be published.

Chapter 9   Monte Carlo Methods for Particle Transport and Heat Transfer

9.1 RANDOM WALK OF PARTICLES

The Monte Carlo method is a special numerical method using random numbers. It can be applied to a variety of problems. In this chapter we are interested in Monte Carlo methods as numerical methods to solve particle-tranport[1-12] and heat-transfer problems. In Sections 9.1 through 9.8, we mainly consider neutrons, although the methods apply to photon-transport problems by appropriately dealing with cross sections. In later sections, applications to heat transfer are discussed. Fundamentals for those later sections are laid in the early sections.

The particle (neutron or photon) transport is stochastic in nature. The migration of a particle is governed by probability distributions such as scattering-angle probability distribution, flight-length probability distribution, energy transfer probability distribution, and so forth. The behavior of multiple particles is also probabilistic and fluctuates about the mean. However, if we take the average for a very long time or the average of a very large number of particles, the distribution of particles becomes the expected value that is the solution of the Boltzmann equation.[13-15] For example, there are so many neutrons in the neutron chain reactor at a high power level that the total behavior of neutrons in steady state is equal to the average behavior.

By solving the Boltzmann equation, one obtains the expected distribution of collision density (or particle flux) and the reaction rate in various regions. While derivation of a rigorous Boltzmann equation is not difficult, solving the Boltzmann equation either analytically or numerically (finite difference methods) is impossible unless a drastic simplification of the equation is made. The Monte Carlo calculation is a simulation of random walk of particles. Each process of random walk is governed by simple probability distribution and can be simulated by using random numbers. The simulation is by far simpler than trying to solve the Boltzmann equation by a standard numerical method. The importance of the Monte Carlo method as a solution for the linear-transport equation is due to this reason. Even though the answer of the random walk simulation is of a stochastic nature, it is expected that, by simulating the history of a large number of particles and by taking the average of the results, the solution of the Boltzmann equation is estimated.

The Monte Carlo method is not restricted by complexity of the geometry or number of independent variables or combination of probability distributions. The drawback, however, is that the computational time to follow random walks is lengthy and expensive. The Monte Carlo calculation is subject to variance, and the answer always fluctuates about the exact one because only a finite number of particles is simulated. The accuracy of a Monte Carlo calculation is proportional to the square root of the number of histories simulated rather than to the number of histories.

Example 9.1

Consider the monoenergetic neutron Boltzmann equation,

$$[\mathbf{\Omega}\nabla + \Sigma(\mathbf{r})]\psi(\mathbf{r}, \mathbf{\Omega}) = \int \psi(\mathbf{r}, \mathbf{\Omega}')\Sigma_s(\mathbf{r}, \mathbf{\Omega}' \to \mathbf{\Omega})\, d\mathbf{\Omega}'$$
$$+ \delta(r - r_0)\delta(\omega - \omega_0)\delta(z - z_0)$$

for the uniform cylindrical geometry shown in Fig. 9.1, and estimate the probability that a neutron is absorbed in the cylinder. In this equation $\mathbf{r} \equiv (r, \omega, z)$, $\mathbf{\Omega} \equiv (\theta, \phi)$ is the angular coordinate; Σ and Σ_s are the total and differential scattering cross sections, respectively; and ψ is the angular flux distribution. With the solution of the above equation, $\Sigma\, dV \int \psi\, d\mathbf{\Omega}$ is the probability that a neutron emitted from the source S at (r_0, ω_0, z_0) will have a collision in the volume element dV throughout its life.

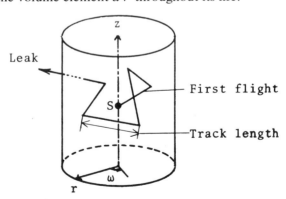

Figure 9.1 Random walk in a cylindrical medium (S=point source).

The Monte Carlo simulation of a particle starts with selection of the direction of its first flight, θ and ϕ, by using random numbers. Then the track length of the flight in that direction is decided. If the particle exits the surface, the life of the particle is terminated. Otherwise, the type of collision at the end of the track is determined by a random number. In this example, it is either scattering or absorption. If absorption occurs, the absorption is scored and the life of the neutron is terminated. If scattering occurs, then the new direction of flight is determined. This procedure is repeated until the particle life is terminated either by leakage or absorption. Figure 9.2 shows

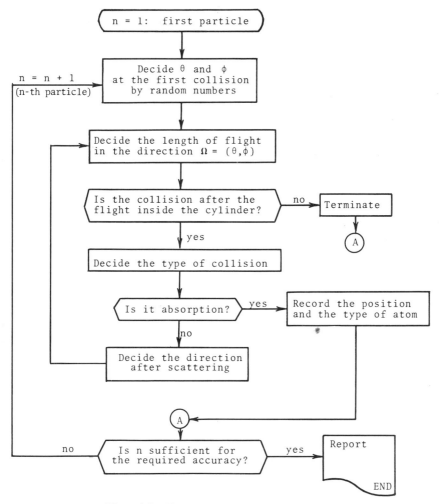

Figure 9.2 Flow of calculation for Example 9.1.

the flow diagram of the Monte Carlo simulation of a monoenergetic neutron transport.

9.2 DECISION OF EVENTS BY RANDOM NUMBERS

Random numbers ξ are numbers between 0 and 1 and are uniformly distributed in this interval. In other words, the probability that a random number ξ falls in any interval Δ in $[0, 1]$ is equal to Δ. They can be generated by using various natural phenomena. The most successful random numbers, which are free from any lack of randomness, are generated by using the radio isotopic decay.[17] However, this kind of method is not necessarily appropriate for Monte Carlo studies except for special cases. There are several numerical routines for generating random numbers.[18] The formula most frequently used to generate a sequence of random numbers is the multiplicative congruential method:[19-21]

$$\left.\begin{array}{l} x_i = [ax_{i-1}+b](\mathrm{mod}\ m) \\[2mm] \xi_i = \dfrac{x_i}{m} \end{array}\right\} i = 1, 2, 3, \ldots$$

where x_i, a, and b are positive integers; and m is a positive and large integer. The notation $x = y(\mathrm{mod}\ m)$ means x is the remainder of y upon division by m, $0 \leq x < m$. In FORTRAN language, these equations read:

$$IX = IX - ((IA*IX+IB)/M)*M$$
$$XI = FLOAT(IX)/FLOAT(M)$$

In the first statement, IX on the right side is the random number in $[0, m]$ generated for the last use. It should be noticed that ξs thus generated are subject to more or less periodicities[17] and called pseudorandom numbers. The period depends on the choice of a, b, m, and x_0, (initial value of x_i). Therefore, the set of parameters must be tested by examining ξs statistically. However, an appropriate choice of the parameters will provide sufficiently long periods so that practically most Monte Carlo calculations can be done successfully with this method. For example, with $a = 2^7 + 1$, $b = 1$, $m = 2^{35}$, the period is reported to become 2^{35}. See References 19 and 20 for more detail. Hereafter, we assume random numbers are generated by a random-number generator whenever necessary.

In simulating random walk of particles, all the events are determined by the corresponding probability distribution functions and random numbers. To start with a simple problem, suppose a particle having three possible targets, C_1, C_2, and C_3, for which the probabilities of collision are 0.2, 0.3, and 0.5, respectively. It is intuitively clear that, if N random numbers uniformly distributed in the range $[0, 1]$ are generated, then approximately $0.2N$ will fall on the interval $0 \le \xi < 0.2$, $0.3N$ will fall on the interval $0.2 \le \xi < 0.5$, and $0.5N$ will fall on the interval $0.5 \le \xi < 1$. This approximation will improve as N increases. This nature of the random numbers may be used to decide the target of each collision if a sequence of collisions is to be numerically simulated. More generally, if $A_1 \ldots A_J$ are independent and mutually exclusive events with probabilities $p_1 \ldots p_J$, respectively, and the total of p_j is unity, then we agree that the event A_i occurs when a random number ξ falls in the interval

$$\sum_{j=1}^{i-1} p_j \le \xi < \sum_{j=1}^{i} p_j \qquad (9.2.1)$$

where $p_0 = 0$. This problem is called a *discrete problem* because there is only a finite number of eventualities. *Discrete Monte Carlo* refers to Monte Carlo games that involve only discrete problems.

We may extend the present procedure to a *continuous problem*, where a random variable x that can take any value in a given interval is to be determined (*continuous Monte Carlo*). Let us assume that x is defined in $-\infty < x < \infty$ and $p(x)\, dx$ is the probability that x falls in dx about x. The function $p(x)$ is called the *probability distribution* and satisfies

$$\int_{-\infty}^{\infty} p(x)\, dx = 1 \qquad (9.2.2)$$

The value of x is determined with $p(x)$ and a random number ξ, $0 \le \xi \le 1$, by solving

$$\xi = \Gamma(x) \equiv \int_{-\infty}^{x} p(s)\, ds \qquad (9.2.3)$$

where $\Gamma(x)$ is called the *cumulative probability distribution*. If x and $p(x)$ are defined in a finite interval, $a < x < b$, the range of the random variable x may be altered to $-\infty < x < +\infty$ by defining $p(x)$ in the widened range as

$$p(x) = 0 \quad \text{for } x < a \quad \text{and} \quad x > b \qquad (9.2.4)$$

Figure 9.3 schematically illustrates the relation between $p(x)$ and the decision of x with a random number.

The random walk of a particle in particle-transport problems consists of many eventualities, each of which is either of discrete or continuous type and governed by a relatively simple probability distribution. The answer of the whole history of the particle can be either discrete or continuous, depending on the problem. Even though it is difficult to analytically express the probability distribution for the individual answer of a particle, we can discuss

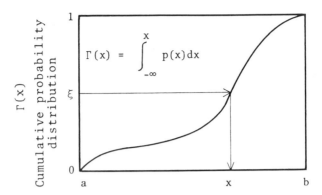

Figure 9.3   Decision of x by using a random number.

tne nature of the probability distribution for the average answer without the knowledge of the probability distribution for individual events.

Let us consider a discrete problem first. Assume that the count of particle "n" is scored as $f_n = 1$ if the particle is absorbed in a specific region and $f_n = 0$ otherwise. The average answer of N particles is defined as

$$\langle f \rangle_N \equiv \frac{1}{N} \sum_{n=1}^{N} f_n \tag{9.2.5}$$

This average gives an estimate for the probability of having $f = 1$ with a particle. The probability of having k answers of "yes" ($f = 1$) out of N trials is given by the binomial distribution[16]

$$B_k^N = \frac{N!}{k!(N-k)!} p^k (1-p)^{N-k} \tag{9.2.6}$$

where p is the probability for a particle answering "yes." The expected value for k is given in terms of B_k^N by

$$\langle k \rangle = \sum_{k=0}^{N} k B_k^N = pN \tag{9.2.7}$$

The variance of k is given by

$$\sigma_k^2 = \langle (k - pN)^2 \rangle = \langle k^2 \rangle - \langle k \rangle^2 = p(1-p)N \tag{9.2.8}$$

where $\langle \ \rangle$ is the expected value (or equivalently the average of an infinite number of trials).† The standard deviation is the square of the variance:

$$\sigma_k = \sqrt{p(1-p)N} \tag{9.2.9}$$

---

† The notation $\langle \ \rangle_N$ means here the average of N trials, while $\langle \ \rangle$ without any subscript means the expected value or equivalently $\langle \ \rangle_\infty$.

It is known that when N becomes very large, the binomial distribution approaches the normal distribution in the sense

$$P(\langle k \rangle - m < k < \langle k \rangle + m) = \frac{1}{\sqrt{2\pi}} \int_{-\delta}^{+\delta} \exp\left(\frac{-z^2}{2}\right) dz \qquad (9.2.10)$$

where

$$\delta = \frac{m + 0.5}{\sigma_k} \qquad (9.2.11)$$

$$z = \frac{k - \langle k \rangle}{\sigma_k} \qquad (9.2.12)$$

In Eqs. (9.2.10) and (9.2.11) m is an integer and $P(\alpha < k < \beta)$ is the probability that k falls in the interval $\alpha < k < \beta$. It may seem peculiar that the lower and upper bounds of the integral in Eq. (9.2.10) are, respectively, $\pm(m + 0.5)/\sigma_k$ rather than $\pm m/\sigma_k$. This is the result of approximating the discrete distribution by a continuous function. In effect, $0.5/\sigma_k$ is a correction term that approaches zero as N increases ($\sigma_k \to \infty$ as $N \to \infty$).

Since $\langle f \rangle_N = k/N$, it is easily seen that the standard deviation for $\langle f \rangle_N$ is given by

$$\sigma_{fN} = \sqrt{\frac{p(1-p)}{N}} \qquad (9.2.13)$$

where σ_{fN} is the standard deviation of $\langle f \rangle_N$. By using Eqs. (9.2.10) and (9.2.13) and by assuming N is sufficiently large [neglect 0.5 in Eq. (9.2.11)], the probability distribution is given as

$$P(\langle f \rangle - \epsilon\sigma_{fN} < \langle f \rangle_N < \langle f \rangle + \epsilon\sigma_{fN}) = \frac{1}{\sqrt{2\pi}} \int_{-\epsilon}^{+\epsilon} \exp\left(\frac{-z^2}{2}\right) dz \quad (9.2.14)$$

where $\langle f \rangle$ is the true mean and

$$z = \frac{\langle f \rangle_N - \langle f \rangle}{\sigma_{fN}} \qquad (9.2.15)$$

and ϵ is a small parameter. Equation (9.2.14) gives the probability that $\langle f \rangle_N$ is not deviating from the true average more than $\epsilon\sigma_{fN}$.

Next, we consider the probability distribution of $\langle g \rangle_N$, where g refers to the answer to a continuous problem and $\langle g \rangle_N$ is the average of N trials. If the probability distribution of g is known, the probability distribution for $\langle g \rangle_N$ can be analytically or numerically calculated. This calculation, however, becomes difficult when N is large. Fortunately, it is well known as the central limit theorem[16] that as N becomes large the distribution of $\langle g \rangle_N$ approaches the normal distribution regardless of the distribution of g. The only condition required is that there exists the average and the variance of g. Therefore, the probability that $\langle g \rangle_N$ is not deviating from the true average more than $\epsilon \sigma_{gN}$ is given by

$$P(\langle g \rangle - \epsilon \sigma_{gN} < \langle g \rangle_N < \langle g \rangle + \epsilon \sigma_{gN}) = \frac{1}{\sqrt{2\pi}} \int_{-\epsilon}^{+\epsilon} \exp\left(\frac{-z^2}{2}\right) dz \quad (9.2.16)$$

where $\langle g \rangle$ is the true mean and σ_{gN} is the standard deviation of $\langle g \rangle_N$ given by

$$\sigma_{gN} = \sqrt{\frac{\sigma_g^2}{N}} \quad (9.2.17)$$

in which σ_g^2 is the variance of g and

$$z = \frac{\langle g \rangle_N - \langle g \rangle}{\sigma_{gN}} \quad (9.2.18)$$

Thus we find that the average answers of discrete and continuous problems both belong to the normal distribution regardless of the probability distributions for the individual trials.

The interval defined by

$$\langle h \rangle - \epsilon \sigma_{hN} < \langle h \rangle_N < \langle h \rangle + \epsilon \sigma_{hN} \quad (9.2.19)$$

is called the confidence interval where h is either f or g, and the probability given by Eqs. (9.2.14) or (9.2.16) is called confidence level. The confidence levels for several confidence intervals are listed in Table 9.1. Equation

Table 9.1 Confidence Interval and Confidence Levels for Normal Distribution[a]

ϵ	Confidence interval $\langle h \rangle \pm \epsilon \sigma_h$	Confidence levels $100(1-\alpha)\%$
0.5	$\langle h \rangle \pm 0.5\sigma_h$	38.3
1.0	$\langle h \rangle \pm \sigma_h$	68.3
1.5	$\langle h \rangle \pm 1.5\sigma_h$	86.6
2.0	$\langle h \rangle \pm 2.0\sigma_h$	95.4
3.0	$\langle h \rangle \pm 3.0\sigma_h$	99.7

[a]Confidence interval for different ϵ may be found easily by using the normal distribution table.

(9.2.19) and Table 9.1 indicate that if the variance of the distribution h is large then a large N is required in order to suppress the variance in the average answer. There are two types of variance reduction methods: scoring methods and nonanalog Monte Carlo as discussed in Sections 9.4 and 9.7, respectively. The success of large-scale Monte Carlo calculations depends on how the variance reduction methods are effectively used.

9.3 RANDOM WALK IN THE PHASE SPACE

In this section we consider the direct simulation of a particle history, which is called *analog Monte Carlo*. For general problems of particle transport in the steady state, we postulate a phase space that is a six-dimensional continuum: three independent variables representing the position, one the energy, and two the angle of the particle flight. If the symbols x, x_k, and so forth are used to denote generic coordinate in such a phase space, a particle random walk is characterized by the first collision at some state x_1, the subsequent transmissions through a sequence of states x_2, x_3, . . . , and finally, the termination at some state x_K.

A continuous random walk in the nonmultiplying medium, therefore, is completely specified by the following probability distribution functions:

$S(x)\,dx$ the probability that the first collision is made in the volume dx of the phase space about x

$K(x' \to x)\,dx$ the probability that if a collision occurs at x' the next collision is in dx about x.

The kernel $K(x' \to x)$ involves several probabilities as follows: the probability that the particle makes the collision with one type of atom among others, the probability that the collision is scattering or absorbing, the probability that the emergent particle is in the solid angle in which the small volume element dx about x is embodied, the probability that the particle makes the next collision in the small volume dx about x, and so on. The probability distribution function for each process in most cases is simple. The decision at each process can be made by using a random number or a set of random numbers as described in the following subsections.

The history of a particle is terminated whenever the particle is absorbed or escapes from the system. The answer of a random walk is obtained by scoring the events that occur in the specified region $\mathcal{D}$ of the phase space. If the absorption in $\mathcal{D}$ is to be estimated, the answer can be obtained by scoring only those particles that are actually absorbed in $\mathcal{D}$ (absorption estimator). If the collision density in $\mathcal{D}$ is to be estimated, the number of collisions made in

$\mathscr{D}$ is counted (collision estimator). Provided $\mathscr{D}$ is a homogeneous region, the absorption rate is the product of Σ_a/Σ_t and the collision rate, where Σ_t is the total cross section and Σ_a is the absorption cross section, so the absorption rate can be estimated by using the collision rate and vice versa. However, since the number of collisions in a region is generally greater than the absorptions, the collision estimator has less variance than the absorption estimator. It is wiser to use the collision estimator than to use the absorption estimator. Such devices for scoring are called the *scoring methods*.

9.3.1 Decision of the Collision Process

Suppose a particle entering a collision in a medium. For simplicity, we assume there is no inelastic scattering nor multiplication of particles such as fission. The total cross section of the medium may be written as

$$\Sigma_t(E') = \Sigma_s(E') + \Sigma_a(E') \tag{9.3.1}$$

where E' is the energy of the incident particle, Σ_t is the total cross section, Σ_s is the scattering cross section, and Σ_a is the absorption cross section. Using a random number $\xi (0 \leq \xi \leq 1)$, the collision is decided to be

1. scattering if $0 \leq \xi < \Sigma_s/\Sigma_t$, or
2. absorption if $\Sigma_s/\Sigma_t \leq \xi \leq 1$.

Once the type of reaction is determined, which element the particle collides with is then determined by using another random number. The particle is thought to collide with the jth element if ξ satisfies the inequality

$$\frac{1}{\Sigma_\alpha} \sum_{i=1}^{j-1} N_i \sigma_{\alpha i} \leq \xi < \frac{1}{\Sigma_\alpha} \sum_{i=1}^{j} N_i \sigma_{\alpha i} \tag{9.3.2}$$

where N_i is the number density of element i, $\sigma_{\alpha i}$ is the micro cross section of element i for the reaction of type α (s or a in this case), and

$$\Sigma_\alpha = \sum_{i=1}^{I} N_i \sigma_{\alpha i} \tag{9.3.3}$$

where the summation is over all the elements in the medium.

In case an elastic scattering collision occurs, the energy and the direction of flight after the collision are determined by using random numbers and the transfer probability function

$$f_i(E', \mathbf{\Omega}' \rightarrow E, \mathbf{\Omega}) = \frac{\sigma_{si}(E', \mathbf{\Omega}' \rightarrow E, \mathbf{\Omega})}{\sigma_{si}(E')} \tag{9.3.4}$$

where the numerator is the differential scattering cross section and the denominator is the total scattering cross section of element i. Since $\sigma_{si}(E')$ and the differential cross section are related by

$$\sigma_{si}(E') = \int_{4\pi} \int \sigma_{si}(E', \mathbf{\Omega}' \to E, \mathbf{\Omega}) \, dE \, d\mathbf{\Omega} \tag{9.3.5}$$

the integral of f_i over E and $\mathbf{\Omega}$ becomes 1.

If the elastic scattering is spherically symmetric in the center-of-mass system, the differential cross section of scattering is given[13] by

$$\sigma_{si}(E', \mathbf{\Omega}' \to E, \mathbf{\Omega}) = \frac{\sigma_{si}(E')}{4\pi A} \left\{ \frac{A^2 - 1 + 2\tilde{\mu}^2}{(A^2 - 1 + \tilde{\mu}^2)^{1/2}} + 2\tilde{\mu} \right\}$$

$$\times \delta\left(E - \frac{E'}{(A+1)^2} [(A^2 - 1 + \tilde{\mu}^2)^{1/2} + \tilde{\mu}]^2 \right) \tag{9.3.6}$$

where $\tilde{\mu}$ is cosine of the angle between $\mathbf{\Omega}'$ and $\mathbf{\Omega}$ in the laboratory system, A is the mass of the atom, δ is the delta function, and the suffix i for σ_s and A is omitted for simplicity. Since E and $\tilde{\mu}$ are correlated through the delta function, $\tilde{\mu}$ is automatically determined if E is decided. By setting the argument of the delta function to zero and solving it for $\tilde{\mu}$, we obtain the relation between $\tilde{\mu}$ and E as

$$\tilde{\mu} = \frac{1}{2}\left[\sqrt{\frac{E}{E'}}(A+1) - \sqrt{\frac{E'}{E}}(A-1) \right] \tag{9.3.7}$$

The azimuthal angle $\tilde{\phi}$ around the incident direction $\mathbf{\Omega}'$ is not included in the right side of Eq. (9.3.6).

In order to determine the emerging energy E, we integrate Eq. (9.3.4) over $\mathbf{\Omega}$ and obtain first the energy transfer probability function:

$$f_{Ei}(E' \to E) = \frac{1}{\sigma_{si}(E')} \int \sigma_{si}(E', \mathbf{\Omega}' \to E, \mathbf{\Omega}) \, d\mathbf{\Omega}$$

$$= \begin{cases} \dfrac{1}{(1-\alpha_i)E'} & \text{if } \alpha_i E' < E < E' \\ 0 & \text{if } \alpha_i E' > E \quad \text{or} \quad E' < E \end{cases} \tag{9.3.8}$$

where Eq. (9.3.6) was used and

$$\alpha_i = \frac{(A_i - 1)^2}{(A_i + 1)^2} \tag{9.3.9}$$

It is clear that $f_{Ei}(E' \to E) \, dE$ is the probability that if a neutron makes an elastic collision at E' the emergent neutron will have the energy in dE about E. The emerging energy can be determined by a random number and f_{Ei} as

$$\xi = \int_{E}^{E'} f_{Ei}(E' \rightarrow E'') \, dE'' \tag{9.3.10}$$

where E is the emergent energy to be determined. Since the angular distribution of scattering is axisymmetric about the incident direction, the probability that the azimuthal angle of emergent direction is in $d\tilde{\phi}$ about $\tilde{\phi}$ about the axis of the incident direction is equal to $d\tilde{\phi}/2\pi$. Therefore, $\tilde{\phi}$ is determined by

$$\xi = \frac{\tilde{\phi}}{2\pi} \tag{9.3.11}$$

In the monoenergetic approximation, the differential cross section may be written as

$$\sigma_{si}(\boldsymbol{\Omega}' \rightarrow \boldsymbol{\Omega}) = \sigma_{si} f_{\Omega i}(\boldsymbol{\Omega}' \rightarrow \boldsymbol{\Omega}) \tag{9.3.12}$$

where

$$\sigma_{si} = \int_{4\pi} \sigma_{si}(\boldsymbol{\Omega}' \rightarrow \boldsymbol{\Omega}) \, d\boldsymbol{\Omega}$$

Since the azimuthal angular distribution of scattering is constant, $f_{\Omega i}$ is the function of only the angle between $\boldsymbol{\Omega}'$ and $\boldsymbol{\Omega}$, and may be written in the form

$$f_{\Omega i}(\boldsymbol{\Omega}' \rightarrow \boldsymbol{\Omega}) = \frac{1}{2\pi} g_{\Omega i}(\tilde{\mu}) \tag{9.3.13}$$

where

$$\tilde{\mu} = \langle \boldsymbol{\Omega}', \boldsymbol{\Omega} \rangle$$

Notice that $f_{\Omega i}(\boldsymbol{\Omega}' \rightarrow \boldsymbol{\Omega})$ and $g_{\Omega i}(\tilde{\mu})$ are normalized as

$$\int_{4\pi} f_{\Omega i}(\boldsymbol{\Omega}' \rightarrow \boldsymbol{\Omega}) \, d\boldsymbol{\Omega} = 1$$

$$\int_{-1}^{1} g_{\Omega i}(\tilde{\mu}) \, d\tilde{\mu} = 1$$

The value of $\tilde{\mu}$ is determined by solving

$$\xi = \int_{-1}^{\tilde{\mu}} g_{\Omega i}(\mu) \, d\mu \tag{9.3.14}$$

The azimuthal angle $\tilde{\phi}$ is determined by Eq. (9.3.11).

Example 9.2

A particle flying in the direction $\Omega_0 = (\theta_0, \phi_0)$ made a collision in the medium consisting of two kinds of atoms where the cross sections are given by

$$\Sigma_t = \Sigma_a + \Sigma_s$$

$$\Sigma_a = \Sigma_{aA} + \Sigma_{aB}, \qquad \Sigma_s = \Sigma_{sA} + \Sigma_{sB}$$

$$\Sigma_{aA} = 0.05, \quad \Sigma_{aB} = 0.07, \quad \Sigma_{sA} = 0.5, \quad \Sigma_{sB} = 0.8$$

Assuming the scattering is isotropic, determine the type of collision and the new direction of flight if the collision is scattering.

Suppose we use the following random numbers sequentially: $\xi_1 = 0.7811$, $\xi_2 = 0.3404$, $\xi_3 = 0.1874$, $\xi_4 = 0.2436$. By using the first rule after Eq. (9.3.1) and ξ_1, namely, $\xi_1 = 0.7811 < \Sigma_s/\Sigma_t = 0.9155$, the collision is found to be scattering. The scattering collision is with A since $\Sigma_{sA}/\Sigma_s = 0.5/(0.5 + 0.8) = 0.3846$. The new direction of flight is determined by Eqs. (9.3.11) and (3.4.14), respectively, with ξ_3 and ξ_4. In the case of isotropic scattering, the directional angles are selected in reference to the laboratory coordinate independently of the incident angle. The azimuthal angle ϕ about z-coordinate is determined as $\phi = 2\pi\xi_3 = 1.1774$ (radian). The longitudinal angle to the z-coordinate is determined by $\mu = \cos\theta = 2\xi_4 - 1 = -0.5128$, namely $\theta = 2.1092$ (radian).

9.3.2 Decision of Flight Length

Suppose that a particle starts a flight from the origin in an infinite homogeneous medium. The probability that this particle travels a straight path of distance s and collides within the next small distance ds is

$$\Sigma \exp(-\Sigma s)\, ds \tag{9.3.15}$$

where Σ is the total cross section of the medium. Therefore, the collision probability distribution is defined as

$$p(s) = \Sigma \exp(-\Sigma s) \tag{9.3.16}$$

where the variable lies between 0 and ∞. It is easily seen that

$$\int_0^\infty p(s)\, ds = 1 \tag{9.3.17}$$

The distance of the straight travel s is determined by equating a random number with a cumulative probability function as

$$\xi' = \Gamma(s) \equiv \int_0^s p(s') \, ds' = 1 - \exp(-\Sigma s) \qquad (9.3.18)$$

where Γ is the cumulative probability distribution. If ξ' is uniformly distributed in $[0, 1]$, so is $\xi = 1 - \xi'$ in the same range. Therefore, Eq. (9.3.18) may be solved as

$$s = -\frac{1}{\Sigma} \ln \xi \qquad (9.3.19)$$

In the case of more general geometry, a particle traverses various regions with different total cross sections. Generally, $p(s)$ can be written as

$$p(s) = \Sigma(s) \exp[-\lambda(s)] \qquad (9.3.20)$$

where

$$\lambda(s) = \int_0^s \Sigma(s') \, ds' \qquad (9.3.21)$$

Integrating Eq. (9.3.20), the cumulative probability distribution becomes

$$\Gamma(s) \equiv \int_0^s p(s') \, ds' = 1 - \exp[-\lambda(s)] \qquad (9.3.22)$$

Equating $\Gamma(s)$ with a random number $\xi' = 1 - \xi$ yields

$$\xi = \exp[-\lambda(s)]$$

or equivalently

$$\lambda(s) = -\ln \xi \qquad (9.3.23)$$

In the case of piecewise homogeneous medium, s of Eq. (9.3.23) is solved as follows: Referring to Fig. 9.4, the length of the segments of the flight path is sequentially denoted by $s_1, s_2, \ldots, s_N$. If $\lambda(s)$ satisfies

$$\sum_{j=1}^{n-1} \Sigma_j s_j < \lambda(s) < \sum_{j=1}^{n} \Sigma_j s_j \qquad (9.3.24)$$

s also satisfies

$$\sum_{j=1}^{n-1} s_j < s < \sum_{j=1}^{n} s_j \qquad (9.3.25)$$

Therefore, s is given by

$$s = \frac{1}{\Sigma_n} \left(\lambda(s) - \sum_{j=1}^{n-1} \Sigma_j s_j \right) + \sum_{j=1}^{n-1} s_j \qquad (9.3.26)$$

where the first term is the distance of travel in the nth region.

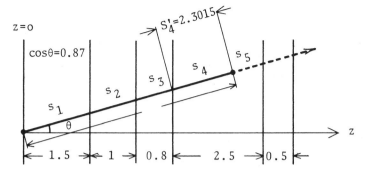

z = 0

cosθ=0.87

$S'_4 = 2.3015$

S_5

S_4

S_3

S_2

S_1

θ

z

← 1.5 → ← 1 → 0.8 ← 2.5 → 0.5 ←

Figure 9.4 A path in a heterogeneous slab geometry.

Example 9.3

Suppose a slab consisting of multiple layers of different materials as shown in Fig. 9.4. A particle is departing $z = 0$ for the right direction with angle θ to the z coordinate. Determine the flight length.

If a random number is $\xi = 0.25924$, the $\lambda = 1.3500$ from Eq. (9.3.23). By using Table 9.2 and Eq. (9.3.25), we find that the collision occurs in the 4th slab. The length of flight in the 4th slab is $s'_n = (1.3500 - 0.7285)/0.27 = 2.3015$.

Table 9.2

j	Σ_j	s_j	$\Sigma_j s_j$	$\sum\limits_{i=1}^{j} \Sigma_j s_j$
1	0.30	1.724	0.5172	0.5172
2	0.12	1.149	0.1379	0.6551
3	0.08	0.919	0.0735	0.7286
4	0.27	2.874	0.7760	1.5046
5	0.79	0.574	0.4535	1.9581

Example 9.4

We simulate numerically the random walk of a particle that travels on the x axis to the left and right in the range $[0, 10]$. If the particle passes $x = 0$ from the right or $x = 10$ from the left, we assume that the particle escapes from the system. The particle collides with the medium in dx with the probability $\Sigma\, dx$

as it passes dx. We assume that if a particle enters a collision, the probability of termination of travel (absorption) is $p_a = 0.1$, the probability of going to the left is $p_l = 0.45$, and the probability of going to the right is $p_r = 0.45$. The original departure of the particle is at $x = 0$ in the positive direction.

The length of a single flight is determined by $\xi = \exp(-0.5s)$ where s is the path length. The result of a collision is decided by using p_a, p_l, and p_r with ξ. The average number of collisions that a batch of 5000 particles make in each of 1-cm regions, $(i-1) < x < i$, was calculated ten times, and the standard deviations of the answers of the batches were calculated. Table 9.3 shows the average numbers of collisions and absorption per particle in a few selected regions based on 50,000 particle histories. The standard deviations of the answer of 5000 particles are also shown.

Table 9.3   Monte Carlo Results with 50,000 Particles

Region number, i	1	2	4	7	10
Average number[+] of collisions	1.2996	0.9478	0.5056	0.1902	0.0485
Standard deviation[++] for 5000 particles, %	1.3	1.7	2.0	4.0	6.0
Average number[+] of absorption	0.1273	0.0958	0.0501	0.0184	0.0051
Standard deviation[++] for 5000 particles, %	3.5	4.1	4.2	7.5	21.7

[+]per particle
[++]calculated with 10 batches of 5000 particles

9.4 SCORING METHODS

Suppose one desires to estimate the total collisions of N particles in a specified region, $\mathscr{D}$. The most straightforward way is to count the actual collisions in $\mathscr{D}$ throughout the histories of N particles. The number of absorptions or other reactions can be counted in the same way. These approaches are called *direct analog estimators*.

It is easily seen that the error in the absorption rate directly estimated is larger than that of the collision rate, especially if the absorption rate is much smaller than the total collision. This difference is amplified as the size of $\mathscr{D}$ becomes smaller since, as the chance of absorption in $\mathscr{D}$ decreases, the variance of the estimated absorption rate increases. The direct estimate of absorption rate is also called a last-event estimator. Alternatively, the absorption rate can be estimated by using the collision rate. If $\mathscr{D}$ is a homogeneous region, the absorption rate is equal to Σ_a/Σ_t times the collision rate. If $\mathcal{D}$ consists of heterogeneous materials, the absorption rate may be expressed in terms of the collision density ψ as

$$R_a = \int_{\mathcal{D}} \frac{\Sigma_a(\mathbf{x})}{\Sigma_t(\mathbf{x})} \psi(\mathbf{x})\,d\mathbf{x} = \int \frac{\Sigma_a(\mathbf{x})}{\Sigma_t(\mathbf{x})} \chi(\mathbf{x})\psi(\mathbf{x})\,d\mathbf{x} \qquad (9.4.1)$$

where the integral after the second equality extends the entire domain and

$$\chi(\mathbf{x}) = 1 \quad \text{for } \mathbf{x} \in \mathcal{D}$$
$$= 0 \quad \text{for } \mathbf{x} \notin \mathcal{D}$$

Since the variance of the collision rate is smaller than that of the absorption rate, the variance of the absorption rate thus estimated has the same variance as the collision rate and is smaller than the variance of the absorption estimator.

The advantage of using the collision estimator can be seen from Table 9.3, where the standard deviation (square root of variance × 100/mean, percentage) of the absorption estimators are three to four times larger than those of the collision estimators.

From the random walk of the nth particle, a set of positions of collisions in the phase space $[\mathbf{x}_{n,1}, \mathbf{x}_{n,2}, \dots, \mathbf{x}_{n,k}, \dots, \mathbf{x}_{n,K(n)}]$ is obtained, where K is the total number of collisions of the nth particle. By using N sets of random walks, the collision rate in $\mathcal{D}$ is approximated by

$$\int \psi(\mathbf{x})\chi(\mathbf{x})\,d\mathbf{x} \approx \frac{1}{N} \sum_{n=1}^{N} \sum_{k=1}^{K(n)} \chi(\mathbf{x}_{n,k}) \qquad (9.4.2)$$

where the collision rate is normalized for a single particle. Equation (9.4.2) may be written

$$\int \psi(\mathbf{x})\chi(\mathbf{x})\,d\mathbf{x} \approx \frac{1}{N} \sum_{n=1}^{N} \sum_{k=1}^{K(n)} \int \chi(\mathbf{x})\delta(\mathbf{x}-\mathbf{x}_{n,k})\,d\mathbf{x} \qquad (9.4.3)$$

where δ is Dirac's delta function. Since $\mathcal{D}$ can be defined quite arbitrarily, the collision density distribution must satisfy

$$\psi(\mathbf{x}) \approx \frac{1}{N} \sum_{n=1}^{N} \sum_{k=1}^{K(n)} \delta(\mathbf{x}-\mathbf{x}_{n,k}) \qquad (9.4.4)$$

Therefore, by introducing Eq. (9.4.4) into Eq. (9.4.1), the absorption rate is estimated by

$$R_a = \frac{1}{N} \sum_{n=1}^{N} \sum_{k=1}^{K(n)} \frac{\Sigma_a(\mathbf{x}_{n,k})}{\Sigma_t(\mathbf{x}_{n,k})} \chi(\mathbf{x}_{n,k}) \qquad (9.4.5)$$

Equation (9.4.1) may be generalized for any kind of reaction. For reaction type α, the Monte Carlo estimate is given by

$$R_\alpha = \frac{1}{N} \sum_{n=1}^{N} \eta_n \qquad (9.4.6)$$

where η_n is the answer of the nth particle and is given by

$$\eta_n = \sum_{n=1}^{K(n)} C(\mathbf{x}_{n,k}) \qquad (9.4.7)$$

$$C(\mathbf{x}) = \frac{\Sigma_\alpha(\mathbf{x})}{\Sigma_t(\mathbf{x})} \chi(\mathbf{x}) \qquad (9.4.8)$$

where $C(\mathbf{x})$ is the contribution of a collision at $\mathbf{x}$ to the answer of the particle and Σ_α is the cross section for the reaction of interest.

In the shielding calculations, it frequently happens that a reaction rate of the region very remote from the neutron source or a reaction rate of a very thin region has to be estimated. In those regions the rate of collisions is so small that the collision estimator is still inefficient. In order to increase the accuracy of estimation, use can be made of nondirect estimators with which a particle can contribute to the answer without actual collision in $\mathcal{D}$. In the remainder of this section, the nondirect estimators are discussed.

Since we are interested in the mean value of R_α, we may use "estimates" for η_n rather than the direct results of random walks. When the position of the kth collision is known as $\mathbf{x}_{n,k}$, the analytical estimate for the contribution of the next collision to η_n is given by

$$\int K(\mathbf{x}_{n,k} \to \mathbf{x}) C(\mathbf{x})\, d\mathbf{x} \qquad (9.4.9)$$

Therefore, the estimate to the total contribution of a particle may be given

$$\eta_n = \sum_{k=0}^{K(n)-1} \int K(\mathbf{x}_{n,k} \to \mathbf{x}) C(\mathbf{x})\, d\mathbf{x} \qquad (9.4.10)$$

where $\mathbf{x}_{n,0}$ is the position in the source where the nth particle was emitted. The estimate for R is obtained by introducing Eq. (9.4.10) into Eq. (9.4.6). This method is called the next-event estimator. Since $K(\mathbf{x}_k \to \mathbf{x}) \neq 0$ except for $\mathbf{x}$ in the vacuum, η_n becomes nonzero even if there is no direct collision in $\mathcal{D}$.

In order to illustrate how the next-event estimator works, let us consider the kernel,

$$K(\mathbf{x}' \to \mathbf{x}) = \gamma(\mathbf{x}') K_\Omega(\mathbf{\Omega}' \to \mathbf{\Omega}) K_r(\mathbf{r}' \to \mathbf{r}) \qquad (9.4.11)$$

where γ is the probability given by $\gamma(\mathbf{x}') = \Sigma_s / \Sigma_t$ that the particle survives the collision at $\mathbf{x}'$; $K_\Omega(\mathbf{\Omega}' \to \mathbf{\Omega})\, d\mathbf{\Omega}$ is the probability that, if the particle survives, the emergent angle is in $d\mathbf{\Omega}$ about $\mathbf{\Omega}$; and $K_r(\mathbf{r}' \to \mathbf{r})\, ds$ is the probability that the particle starting from $\mathbf{r}'$ terminates the flight in ds about s where $\mathbf{r} = s\mathbf{\Omega} + \mathbf{r}'$.

As the simplest case, suppose $\mathcal{D}$ is a very small homogeneous region and accordingly the solid angle $\Delta\mathbf{\Omega}$ in which $\mathcal{D}$ is embodied is also small (Fig. 9.5). Then by denoting $\mathbf{x}' = \mathbf{x}_k$, the estimate may be approximated by

$$E_{n,k+1} = \int K(\mathbf{x}' \to \mathbf{x}) C(\mathbf{x})\, d\mathbf{x} = \frac{\Sigma_\alpha(\mathbf{x}_0)}{\Sigma_t(\mathbf{x}_0)} \int_{\mathcal{D}} K(\mathbf{x}' \to \mathbf{x})\, d\mathbf{x}$$

$$= \gamma(\mathbf{x}') \frac{\Sigma_\alpha(\mathbf{x}_0)}{\Sigma_t(\mathbf{x}_0)} \int_{\mathcal{D}} \int K_\Omega(\mathbf{\Omega}' \to \mathbf{\Omega}) K_r(\mathbf{r}' \to \mathbf{r})\, d\mathbf{\Omega}\, ds \quad (9.4.12)$$

$$= \gamma(\mathbf{x}') \frac{\Sigma_\alpha(\mathbf{x}_0)}{\Sigma_t(\mathbf{x}_0)} K_\Omega(\mathbf{\Omega}' \to \mathbf{\Omega}_0) K_r(\mathbf{r}' \to \mathbf{r}_0) \Delta v$$

where $\mathbf{\Omega}_0$ is the angle of the center of $\mathcal{D}$, $\mathbf{r}_0$ is the spatial coordinate of the center of $\mathcal{D}$, and Δv is the volume of $\mathcal{D}$. As the volume of $\mathcal{D}$ increases, a more accurate integration scheme becomes necessary for the third line of Eq. (9.4.12). One approach is to divide $\mathcal{D}$ into sufficiently small subdomains and apply Eq. (9.4.12) to each of the subdomains. The same technique is used for the numerical integrations of the integral transport equation[22,23] for non-scattering mediums.

The next-event estimation can be done partially with random numbers, especially when $\mathcal{D}$ is not sufficiently small. Draw a couple of random numbers and decide the emergent angle $\mathbf{\Omega}$. If $\mathbf{\Omega}$ is not pointing any part of $\mathcal{D}$, then the estimate is set as

$$E_{n,k+1} = 0 \quad (9.4.13)$$

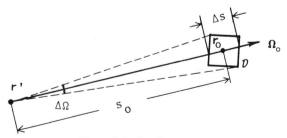

Figure 9.5 Small counter.

and if $\boldsymbol{\Omega}$ is within the solid angle $\Delta\Omega$ then

$$
\begin{aligned}
E_{n,k+1} &= \gamma(\mathbf{x}') \int_{s_1}^{s_2} \frac{\Sigma_\alpha(\mathbf{r})}{\Sigma_t(\mathbf{r})} K_r(\mathbf{r}' \to \mathbf{r}) \, ds \\
&= \gamma(\mathbf{x}') \int_{s_1}^{s_2} \Sigma_\alpha(\mathbf{r}) \exp\left[-\int_0^s \Sigma_t(s'') \, ds''\right] ds
\end{aligned}
\tag{9.4.14}
$$

where s_1 is the distance from the collision site to the entering surface of $\mathscr{D}$, s_2 is the distance to the exiting surface, and γ is the probability for the particle to survive the kth collision. This calculation can be done independently of the determination of the actual $(k+1)$th collision site. However, computing time can be saved, if this procedure is confined with the actual random walk as follows. Namely, we first determine the actual direction of the next flight for the $(k+1)$th collision and use this angle as $\boldsymbol{\Omega}$. If the flight direction is not in the direction of $\Delta\Omega$ we use Eq. (9.4.13), and if the flight is in the direction of $\Delta\Omega$ we use Eq. (9.4.14) whether $\mathbf{x}_{k+1}$ is in $\mathscr{D}$ or not.

We can further extend the use of random variables in the next-event estimator and still simplify the calculations. For example, the integration in Eq. (9.4.14) may be partially done by using a random number. The integral on the right side of Eq. (9.4.14) may be written as

$$
\begin{aligned}
&\int_{s_1}^{s_2} \exp\left[-\int_0^s \Sigma_t(s'') \, ds''\right] ds \\
&= \exp\left[-\int_0^{s_1} \Sigma_t(s'') \, ds''\right] \int_{s_1}^{s_2} \Sigma_\alpha(s) \exp\left[-\int_{s_1}^s \Sigma_t(s'') \, ds''\right] ds
\end{aligned}
\tag{9.4.15}
$$

In Eq. (9.4.15), the first exponential factor is the probability that the particle reaches the surface of $\mathscr{D}$ and the second exponential factor is the probability that the particle entering $\mathscr{D}$ is collided in $\mathscr{D}$. By using random numbers we can determine if the particle reaches $\mathscr{D}$. If the particle does not reach $\mathscr{D}$, then we set

$$
E_{n,k+1} = 0 \tag{9.4.16}
$$

and if the particle reaches $\mathscr{D}$ we set

$$
E_{n,k+1} = \int_{s_1}^{s_2} \Sigma_\alpha(\mathbf{r}) \exp\left[-\int_{s_1}^s \Sigma_t(s'') \, ds''\right] ds \tag{9.4.17}
$$

It is intuitively clear that the variance in estimating the $(k+1)$th event is decreased as the more previous events are traced back analytically. In fact, if all the previous events are traced back analytically, the estimation is reduced to the analytical solution of the integral transport equation, so the estimation becomes exact.

9.5   INTEGRAL TRANSPORT EQUATION AND RANDOM WALKS

It is shown here that the expected (or average) answer of a Monte Carlo calculation satisfies the integral transport equation and that, if the number of simulated particles is sufficiently large, then the estimated answer approaches the exact solution of the integral equation.

The estimate for the collision density distribution given by Eq. (9.4.4) may be written, by changing the order of summations, as

$$\psi(\mathbf{x}) = \sum_{k=1}^{K} \psi_k(\mathbf{x}) \tag{9.5.1}$$

where $\psi_k(\mathbf{x})$ is the Monte Carlo estimate for the kth collision density given by

$$\psi_k(\mathbf{x}) = \frac{1}{N} \sum_{n=1}^{N} \delta(\mathbf{x} - \mathbf{x}_{n,k}) \beta_{n,k} \tag{9.5.2}$$

where

$$\beta_{n,k} = 1 \quad \text{if } k \leq K(n)$$
$$= 0 \quad \text{if } k > K(n)$$

The site of the kth collision of particles may be equivalently sampled as follows. Assuming the source distribution is normalized, the first collision site of a particle belongs to the probability distribution function defined by

$$\phi_1(\mathbf{x}) = \int K(\mathbf{x}' \to \mathbf{x}) S(\mathbf{x}') \, d\mathbf{x}' \tag{9.5.3}$$

where $S(\mathbf{x})$ is the source distribution. Then the kth collision probability distribution is

$$\phi_k(\mathbf{x}) = \int K(\mathbf{x}' \to \mathbf{x}) \phi_{k-1}(\mathbf{x}') \, d\mathbf{x}' = \int\int K(\mathbf{x}' \to \mathbf{x}) K(\mathbf{x}'' \to \mathbf{x}') \phi_{k-2}(\mathbf{x}'') \, d\mathbf{x}'' \, d\mathbf{x}'$$

$$= \int\int \cdots \int K(\mathbf{x}' \to \mathbf{x}) K(\mathbf{x}'' \to \mathbf{x}') \cdots S(\mathbf{x}'''') \, d\mathbf{x}'''' \cdots d\mathbf{x}' \tag{9.5.4}$$

An equivalent sampling for the kth collision sites of N particles may be obtained by applying N sets of random numbers directly to $\phi_k(\mathbf{x})$. Since the integral of $K(\mathbf{x}' \to \mathbf{x})$ over $\mathbf{x}$ is generally less than 1, we have

$$\int \phi_k(\mathbf{x}) \, d\mathbf{x} < 1 \tag{9.5.5}$$

If a random number satisfies the inequality

$$\int \phi_k(\mathbf{x}) \, d\mathbf{x} > \xi \tag{9.5.6}$$

then there is no collision site found for the trial. The number of kth collisions out of N trials decreases as k increases. The only difference in the direct sampling of the kth collision sites from the sampling with random walks is that the sequence of collision sites for each particle is lost.

The expected distribution of the total collisions is the summation of all the collision probability distribution:

$$\phi(\mathbf{x}) = \sum_{k=1}^{\infty} \phi_k(\mathbf{x}) \qquad (9.5.7)$$

Introducing Eq. (9.5.4) into Eq. (9.5.7), we have

$$\phi(\mathbf{x}) = \sum_{k=2}^{\infty} \int K(\mathbf{x}' \to \mathbf{x})\phi_{k-1}(\mathbf{x}')\,d\mathbf{x}' + \phi_1(\mathbf{x})$$

$$= \sum_{k=1}^{\infty} \int K(\mathbf{x}' \to \mathbf{x})\phi_k(\mathbf{x}')\,d\mathbf{x}' + \phi_1(\mathbf{x}) \qquad (9.5.8)$$

Using Eq. (9.5.7) in the first term of Eq. (9.5.8), we have the integral transport equation,

$$\phi(\mathbf{x}) = \int K(\mathbf{x}' \to \mathbf{x})\phi(\mathbf{x}')\,d\mathbf{x}' + \phi_1(\mathbf{x}) \qquad (9.5.9)$$

Thus we obtain the integral transport equation for the expected distribution of the total collisions. Equation (9.5.9) is a Fredholm equation of the second kind.

Since the kth collision sites are sampled from $\phi_k(\mathbf{x})$, Eq. (9.5.2) approaches ϕ_k as N increases, and accordingly, $\psi(\mathbf{x})$ also approaches ϕ as N increases. Thus it is proven that the result of a Monte Carlo calculation approaches the solution of Eq. (9.5.9).

9.6 ADJOINT MONTE CARLO

In this section, two subjects related to the adjoint transport equations are discussed. The first subject is to show that the reaction rate can be estimated by playing the adjoint Monte Carlo game, and the second subject is to derive an analytical expression for the variance of the analog Monte Carlo method. For the definition and physical meaning of adjoint equations, see Appendix I.

The answer of an individual particle is a random η. The mean answer of an infinite number of particles becomes equal to

$$\langle \eta \rangle = \int \phi(\mathbf{x})C(\mathbf{x})\,d\mathbf{x} \qquad (9.6.1)$$

where $C(\mathbf{x})$ is defined by Eq. (9.4.8), $\phi(\mathbf{x})$ is the solution of Eq. (9.5.9), and $\langle \eta \rangle$ is the expected value.

It is interesting to see that $\langle \eta \rangle$ can be estimated by considering the adjoint problem. Suppose a particle entering a collision at $\mathbf{x}$. The total contribution to the answer of the present collision and the subsequent collisions may be written[30] as

$$\zeta(\mathbf{x}) = C(\mathbf{x}) + \zeta(\mathbf{x}') \qquad (9.6.2)$$

where $\mathbf{x}'$ is the site of the next collision. In Eq. (9.6.2), $C(\mathbf{x})$ on the right side is the contribution of the present collision and $\zeta(\mathbf{x}')$ is the contribution of the next and later collisions. If the collision at $\mathbf{x}$ is the terminal, ζ becomes

$$\zeta(\mathbf{x}) = C(\mathbf{x}) \qquad (9.6.3)$$

The expected value of $\zeta(\mathbf{x})$ is the sum of the contributions from Eqs. (9.6.2) and (9.6.3) and satisfies

$$\langle \zeta(\mathbf{x}) \rangle = \alpha C(\mathbf{x}) + \int K(\mathbf{x} \to \mathbf{x}')[C(\mathbf{x}) + \langle \zeta(\mathbf{x}') \rangle]\, d\mathbf{x}'$$

or equivalently,

$$\langle \zeta(\mathbf{x}) \rangle = C(\mathbf{x}) + \int K(\mathbf{x} \to \mathbf{x}')\langle \zeta(\mathbf{x}') \rangle\, d\mathbf{x}' \qquad (9.6.4)$$

where $K(\mathbf{x} \to \mathbf{x}')$ is defined early in Section 9.3, and α is the probability that the present collision is the terminal.

By comparing Eq. (9.6.4) with Eq. (9.5.9), we find that Eq. (9.6.4) is an adjoint equation for Eq. (9.5.9), because the roles of $\mathbf{x}$ and $\mathbf{x}'$ in K are interchanged. $C(\mathbf{x})$ in Eq. (9.6.4) plays the role of an adjoint source.

Once $\langle \zeta(\mathbf{x}) \rangle$ is obtained, the expected value of η is given by

$$\langle \eta \rangle = \int S(\mathbf{x})\langle \zeta(\mathbf{x}) \rangle\, d\mathbf{x} \qquad (9.6.5)$$

The distribution $\zeta(\mathbf{x})$ can be estimated by the adjoint Monte Carlo as follows. We consider the random walks of "adjoint" particles. The first collision site, $\mathbf{x}_1$, is determined by $C(\mathbf{x})$ and a random number. The subsequent flight direction and the second collision site are determined by $K(\mathbf{x} \to \mathbf{x}')$. Since the kernel is reversed, the order of determining the type of reactions, the angle of the flight, and the flight distance is also reversed with respect to the normal Monte Carlo.

We rewrite the kernel given by Eq. (9.4.11) in the form

$$K(\mathbf{x} \to \mathbf{x}') = \gamma K_{\Omega}(-\mathbf{\Omega} \to -\mathbf{\Omega}')K_r(\mathbf{r} \to \mathbf{r}') \qquad (9.6.6)$$

The meaning of each term in the adjoint Monte Carlo is as follows. If an adjoint particle is emitted at $\mathbf{r}'$ in the direction $-\mathbf{\Omega}'$, then $K_r(\mathbf{r} \to \mathbf{r}')$ is the probability that the adjoint particle reaches $\mathbf{r} = -s\mathbf{\Omega} + \mathbf{r}'$, where s is the flight distance. So s is decided by $\xi = K_r(-s\mathbf{\Omega} + \mathbf{r}' \to \mathbf{r}')$. Once the particle stops at $\mathbf{r}$, then the new direction $-\mathbf{\Omega}$ is determined by random numbers and $K_\Omega(-\mathbf{\Omega} \to -\mathbf{\Omega}')$. The adjoint particle leaving for $-\mathbf{\Omega}$ is tested with a random number ξ and whether the collision at $\mathbf{r}$ was really scattering is determined. The reaction is determined to be absorption if $\xi \le \gamma$ and scattering if $\xi > \gamma$. If $\xi > \gamma$ the adjoint particle starts the next flight for the $-\mathbf{\Omega}$ direction. The test of reaction type can be done before determining $-\mathbf{\Omega}$ in the case of the monoenergetic transport. (If the problem is energy dependent, γ cannot be calculated until the new $-\mathbf{\Omega}$ and the new energy are determined.)

By using the result of the random walks of N adjoint particles, the estimate for $\zeta(x)$ is given by

$$\zeta(\mathbf{x}) = \frac{1}{N} \sum_{n=1}^{N} \sum_{k=1}^{K(n)} \delta(\mathbf{x} - \mathbf{x}_{k,n}) \tag{9.6.7}$$

where $\mathbf{x}_{k,n}$ is the kth collision site of the nth particle. An estimate for $\langle \eta \rangle$ is obtained by introducing Eq. (9.6.7) into $\langle \zeta \rangle$ in Eq. (9.6.5).

Another interesting application of the adjoint equation is to develop the analytical expression for the variance of η, which is defined by

$$\sigma^2(\eta) = \langle \eta^2 \rangle - \langle \eta \rangle^2 \tag{9.6.8}$$

If such an analytical expression for the variance can be given even approximately, one can estimate the number of games required to satisfy a given confidence level.

Suppose we have $\langle \eta \rangle$ already. Then $\langle \eta^2 \rangle$ must be calculated. Let us first consider ζ^2. The mean value of ζ^2 consists of the contributions of the square of Eq. (9.6.2) as well as Eq. (9.6.3) and is given by

$$\langle \zeta^2(\mathbf{x}) \rangle = \alpha(\mathbf{x})C^2(\mathbf{x}) + \int K(\mathbf{x} \to \mathbf{x}')\langle C(\mathbf{x}) + \zeta(\mathbf{x}') \rangle^2 \, d\mathbf{x}'$$
$$= C^2(\mathbf{x}) + 2C(\mathbf{x}) \int K(\mathbf{x} \to \mathbf{x}')\langle \zeta(\mathbf{x}') \rangle \, d\mathbf{x}' + \int K(\mathbf{x} \to \mathbf{x}')\langle \zeta^2(\mathbf{x}') \rangle d\mathbf{x}' \tag{9.6.9}$$

Rearranging and using Eq. (9.6.4), the above equation is reduced to

$$\langle \zeta^2(\mathbf{x}) \rangle = C(\mathbf{x})[2\langle \zeta(\mathbf{x}) \rangle - C(\mathbf{x})] + \int K(\mathbf{x} \to \mathbf{x}')\langle \zeta^2(\mathbf{x}') \rangle \, d\mathbf{x}' \tag{9.6.10}$$

Equation (9.6.10) is the integral equation for $\langle \zeta^2 \rangle$ and is an adjoint equation of Eq. (9.5.9) also because $\mathbf{x}$ and $\mathbf{x}'$ in the kernel are interchanged relative to

Eq. (9.5.9). In terms of $\langle \zeta^2 \rangle$, $\langle \eta^2 \rangle$ is given by

$$\langle \eta^2 \rangle = \int S(\mathbf{x}) \langle \zeta^2(\mathbf{x}) \rangle \, d\mathbf{x} \qquad (9.6.11)$$

Thus we find that the variance of the random walks can be calculated by solving two adjoint equations, analytically or numerically, or even by using the Monte Carlo method.

9.7 NONANALOG MONTE CARLO

When the size of the counter is very small or remote from the source, the probability for a particle to reach the counter becomes small, so the variance of the counting becomes large. For such problems, *altered games* of random walk are effective to improve accuracy. An altered game is played by a new source distribution $\tilde{S}(\mathbf{x})$ and a new collision kernel $\tilde{K}(\mathbf{x}' \to \mathbf{x})$, instead of using the original probability distributions $S(\mathbf{x})$ and $K(\mathbf{x}' \to \mathbf{x})$. We can easily choose such $\tilde{S}(\mathbf{x})$ and $\tilde{K}(\mathbf{x}' \to \mathbf{x})$ that result in more particles reaching the counter region. In order for the altered game to be mathematically equivalent to the original game or to yield the same average answer, we introduce the concept of *weight* for the particle. The weight of a particle is changed whenever the altered probability distributions are used, so the expected result of the altered game is unchanged. The altered games are also called *nonanalog* Monte Carlo method.[5,7]

The only requirements to $\tilde{S}(\mathbf{x})$ and $\tilde{K}(\mathbf{x}' \to \mathbf{x})$ are

1. $\tilde{S}(\mathbf{x}) \neq 0$ \qquad if $S(\mathbf{x}) \neq 0$, \hfill (9.7.1)

2. $\tilde{K}(\mathbf{x}' \to \mathbf{x}) \neq 0$ \quad if $K(\mathbf{x}' \to \mathbf{x}) \neq 0$ \hfill (9.7.2)

We assume for simplicity that $\tilde{S}$ and $\tilde{K}$ satisfy the conditions

$$\int \tilde{S}(\mathbf{x}) \, d\mathbf{x} = 1 \qquad (9.7.3)$$

$$\int \tilde{K}(\mathbf{x}' \to \mathbf{x}) \, d\mathbf{x} \leq 1 \qquad (9.7.4)$$

These two conditions, Eqs. (9.7.3) and (9.7.4), are waived later.

In the altered game, the first collision site, $\mathbf{x}_1$, is selected from $\tilde{S}(\mathbf{x})$. The weight of the particle is then given by

$$w_1 = \frac{S(\mathbf{x}_1)}{\tilde{S}(\mathbf{x}_1)} \qquad (9.7.5)$$

where w_1 is the weight of the particle at the first collision. If $\tilde{S}(\mathbf{x}_1) > S(\mathbf{x}_1)$, then the probability of having the first collision in $d\mathbf{x}$ about $\mathbf{x}_1$ in the altered game is greater than in the original game by the factor $\tilde{S}(\mathbf{x}_1)/S(\mathbf{x}_1)$. However, this effect is exactly offset by considering the weight just as a reciprocal of the factor. The second site of collision $\mathbf{x}_2$ is determined by $\tilde{K}(\mathbf{x}_1 \to \mathbf{x})$. The weight of the particle at the second collision is set to

$$w_2 = w_1 \frac{K(\mathbf{x}_1 \to \mathbf{x}_2)}{\tilde{K}(\mathbf{x}_1 \to \mathbf{x}_2)} \tag{9.7.6}$$

In general, the kth collision site is selected by $\tilde{K}(\mathbf{x}_{k-1} \to \mathbf{x})$, and the weight at the kth collision is

$$w_k = w_{k-1} \frac{K(\mathbf{x}_{k-1} \to \mathbf{x}_k)}{\tilde{K}(\mathbf{x}_{k-1} \to \mathbf{x}_k)} \tag{9.7.7}$$

The probability that the particle terminates its random walk at the kth collision is $1 - \tilde{\alpha}(\mathbf{x}_k)$, where $\tilde{\alpha}$ is the probability that the random walk in the altered game is not terminated is given by

$$\tilde{\alpha}(\mathbf{x}_k) = \int \tilde{K}(\mathbf{x}_k \to \mathbf{x}') \, d\mathbf{x}' \tag{9.7.8}$$

Once the set of collision sites $[\mathbf{x}_1, \mathbf{x}_2, \ldots, \mathbf{x}_K]$ and the weights at the collision sites $[w_1, w_2, \ldots, w_K]$ are obtained, the answer of the altered game with a particle is given by

$$\eta = \sum_{k=1}^{K} w_k C(\mathbf{x}_k) \tag{9.7.9}$$

where $C(\mathbf{x})$ is the contribution of the collision at $\mathbf{x}$ to the answer as defined by Eq. (9.4.8).

The conditions given by Eqs. (9.7.3) and (9.7.4) are not always necessary. If $\tilde{S}(\mathbf{x})$ does not satisfy Eq. (9.7.3) the first collision site is decided by using the normalized altered source distribution

$$\tilde{S}'(\mathbf{x}) = \frac{\tilde{S}(\mathbf{x})}{A} \tag{9.7.10}$$

where

$$A = \int \tilde{S}(\mathbf{x}) \, d\mathbf{x} \tag{9.7.11}$$

In order that the selection of $\mathbf{x}_1$ from $\tilde{S}'(\mathbf{x})$ is equivalent to the selection from the original source $S(\mathbf{x})$, the weight of the particle is set to

$$w_1 = \frac{S(\mathbf{x}_1)}{\tilde{S}'(\mathbf{x}_1)} = \frac{S(\mathbf{x}_1)}{\tilde{S}(\mathbf{x}_1)} \frac{\tilde{S}(\mathbf{x}_1)}{\tilde{S}'(\mathbf{x}_1)} = A \frac{S(\mathbf{x}_1)}{\tilde{S}(\mathbf{x}_1)} \tag{9.7.12}$$

The treatment of $\tilde{K}$ not satisfying Eq. (9.7.4) is more difficult because $\tilde{K}$ is generally a complicated function including space, angle, and energy variables. In case the integration of $\tilde{K}(\mathbf{x} \to \mathbf{x}')$ over $\mathbf{x}'$ is easy, then the kth collision site can be determined by using the normalized kernel,

$$\tilde{K}'(\mathbf{x}_{k-1} \to \mathbf{x}) = \frac{\tilde{K}(\mathbf{x}_{k-1} \to \mathbf{x})}{B} \qquad (9.7.13)$$

where $\mathbf{x}_{k-1}$ is the location of the $(k-1)$th collision, $\mathbf{x}$ is the location of the next collision, and

$$B = \int \tilde{K}(\mathbf{x}_{k-1} \to \mathbf{x}') \, d\mathbf{x}' \qquad (9.7.14)$$

The weight of the particle at the kth collision is then

$$w_k = w_{k-1} \frac{K(\mathbf{x}_{k-1} \to \mathbf{x}_k)}{\tilde{K}'(\mathbf{x}_{k-1} \to \mathbf{x}_k)} = w_{k-1} \frac{BK(\mathbf{x}_{k-1} \to \mathbf{x}_k)}{\tilde{K}(\mathbf{x}_{k-1} \to \mathbf{x}_k)} \qquad (9.7.15)$$

In most cases, the integration in Eq. (9.7.14) is impossible. However, there are still a few ways of dealing with such $\tilde{K}$ without actually integrating it. The step biasing introduced later is one such example.

There are a number of methods in choosing the altered probability distributions.[5,7,28,29] Some popular methods are outlined next:

SOURCE BIASING

The source distribution $S(\mathbf{x})$ is replaced by the altered source $\tilde{S}(\mathbf{x})$, which has a greater probability of generating the particles that reach the counter region than $S(\mathbf{x})$ does. For example, $\tilde{S}(\mathbf{x})$ is greater than $S(\mathbf{x})$ as $\mathbf{x}$ approaches the counter. As another example, $\tilde{S}(\mathbf{x})$ may generate more particles in the direction of the counter than $S(\mathbf{x})$.

SURVIVAL BIASING

This method suppresses the absorption in the altered kernel $\tilde{K}(\mathbf{x}' \to \mathbf{x})$. While the probability of survival at a collision is always made unity in this method, the weight of the particle is decreased at the collision by the factor Σ_s/Σ_t. The present alteration prohibits the termination of a random walk except by leakage from the system, thus increasing the chance for a particle reaching the counter region.

The drawback of this method, however, is that the weight of a particle will become smaller and smaller unless the particle leaks from the system. The cost of computation is the same no matter how small the weight becomes and accordingly no matter how the contribution to the answer is small. Large variety of weights introduce another variance to the answer. In order to avoid those problems, the particles with unnecessarily small weight are suppressed by setting the lower limit of weights w_{min}. If w falls below w_{min}, the particle is tested for its survival with a random number ξ. If $\xi \leq w/w_s$, where w_s is a prescribed standard, the particle survives with the new weight w_s ($w_s \gg w_{min}$). If $\xi > w/w_s$, the particle history is terminated.

Example 9.4

Consider a homogeneous medium in which $\Sigma_s/\Sigma_t = 0.9$. We set $w_{min} = 0.3$ and $w_s = 0.5$. The weight of the particle at the kth collision is $w = (0.9)^{k-1}$. With survival biasing, a particle experiences 12 collisions without any chance of termination, because $(0.9)^{11} = 0.3138 > w_{min} = 0.3 > (0.9)^{12} = 0.2824$. After the 12th collision, the particle is terminated if $\xi > 0.2824/0.5 = 0.5648$ and survives with the weight 0.5 if $\xi < 0.5648$.

SCATTERING-ANGLE BIASING

With the altered kernel $\tilde{K}$, the probability for a scattered particle to emerge in the direction of the counter region is made larger than with the original kernel. Scattering-angle biasing is particularly useful in combination with source biasing. In fact, unless the scattering angle is accurately biased at every collision, the benefit from source biasing can be largely wiped out by the fluctuations created in the collisions. Scattering-angle biasing itself is not, however, always easy if a variety of collisions, for example, $(n, 2n)$, elastic and inelastic, isotropic and anisotropic, and so forth, are involved.

DENSITY BIASING

In this method, an arbitrary positive function $I(\mathbf{x})$ is introduced, by which the altered probability distributions are calculated as

$$\tilde{S}(\mathbf{x}) = \frac{S(\mathbf{x})I(\mathbf{x})}{A} \tag{9.7.16}$$

$$\tilde{K}(\mathbf{x}' \to \mathbf{x}) = K(\mathbf{x}' \to \mathbf{x})\frac{I(\mathbf{x})}{I(\mathbf{x}')} \tag{9.7.17}$$

where A is the normalization constant given by

$$A = \int S(\mathbf{x})I(\mathbf{x})\,d\mathbf{x} \tag{9.7.18}$$

In accordance with Eqs. (9.7.5) and (9.7.7), the weight at each collision becomes

$$w_1 = \frac{S(\mathbf{x}_1)}{\tilde{S}(\mathbf{x}_1)} = \frac{A}{I(\mathbf{x}_1)} \tag{9.7.19}$$

$$w_k = w_1 \frac{I(\mathbf{x}_1)}{I(\mathbf{x}_2)} \cdots \frac{I(\mathbf{x}_{k-1})}{I(\mathbf{x}_k)} = \frac{A}{I(\mathbf{x}_k)} \qquad (9.7.20)$$

where $\mathbf{x}_k$ is the kth collision site. The answer of a particle is given by

$$\eta = \sum_{k=1}^{K} \frac{AC(\mathbf{x}_k)}{I(\mathbf{x}_k)} \qquad (9.7.21)$$

This method provides a systematic way of determining the altered probability distributions. If $I(\mathbf{x})$ is increased as $\mathbf{x}$ approaches the counter region, then the altered game will yield more collisions in the counter region than the original game.

STEP BIASING (RUSSIAN ROULETTE)

The whole spatial domain is divided into nonoverlapping subregions. In each subregion, $I(\mathbf{x})$ is set to a constant. For simplicity, suppose there are only two subregions: $I(\mathbf{x}) = I_1$ in $\mathscr{D}_1$ (region 1) and $I(\mathbf{x}) = I_2$ in $\mathscr{D}_2$ (region 2) including the counter. We set $I_2 > I_1$, assuming the counter is in $\mathscr{D}_2$. For a particle collided in region 1, the altered kernel is then

$$\tilde{K}(\mathbf{x}' \to \mathbf{x}) = K(\mathbf{x}' \to \mathbf{x}) \qquad \text{for } \mathbf{x} \text{ in } \mathscr{D}_1$$

$$= \frac{I_2}{I_1} K(\mathbf{x}' \to \mathbf{x}) \qquad \text{for } \mathbf{x} \text{ in } \mathscr{D}_2 \qquad (9.7.22)$$

where $\mathbf{x}'$ is the location of the present collision in $\mathscr{D}_i$. The paths of the particle after $\mathbf{x}'$ are determined as follows. At collision site $\mathbf{x}'$, the original kernel is used to decide the location of the next collision. If the particle passes the border from $\mathscr{D}_1$ to $\mathscr{D}_2$, then the number of particle is increased by the factor I_2/I_1, but the weight of each particle is divided by I_2/I_1. If I_2/I_1 is not an integer, the number of the split particles is adjusted to become an integer by using the formula

$$M = [I_1/I_1]_{\text{int}} + 1 \quad \text{if } \xi \le \frac{I_2}{I_1} - \left[\frac{I_2}{I_1}\right]_{\text{int}}$$

$$= [I_2/I_1]_{\text{int}} + 1 \quad \text{if } \xi > \frac{I_2}{I_1} - \left[\frac{I_2}{I_1}\right]_{\text{int}} \qquad (9.7.23)$$

where M is the number of the split particles, ξ is a random number, and $[\ ]_{\text{int}}$ means an integer. The above process of increasing the number of particles is called *splitting*.

Since the number of particles increases when passing the border from $\mathscr{D}_1$ to $\mathscr{D}_2$, the reverse is necessary when passing the border from $\mathscr{D}_2$ to $\mathscr{D}_1$. This is called *Russian roulette* and is done by terminating the random walk with probability

$$1 - \frac{I_1}{I_2} \qquad (9.7.24)$$

Example 9.5

Consider the boundary x_0 between two regions, $\mathcal{D}_1$ and $\mathcal{D}_2$, and $I_1 = 1.1$, $I_2 = 3$.

a. A particle with weight w is passing x_0 from $\mathcal{D}_1$ to $\mathcal{D}_2$. Since $2 < I_2/I_1 = 2.7272 < 3$, the particle splits into 2 or 3 depending on ξ. If $\xi < I_2/I_1 - [I_2/I_1]_{int} = 0.7272$, the particle splits into 3 with weight $1/3 W = 0.3333w$. If $\xi > 0.7272$, the particle splits into 2 with $0.5000w$.

b. A particle with w is passing x_0 from $\mathcal{D}_2$ to $\mathcal{D}_1$. The particle is terminated if $\xi > I_1/I_2 = 0.3667$. The particle survives with $wI_2/I_1 = 2.7272w$ if $\xi < 0.3667$.

ADJOINT BIASING

It is shown here that the ultimate biasing function is the adjoint solution to the transport equation. The adjoint equation is given by

$$\varphi^*(\mathbf{x}) = \int K(\mathbf{x} \to \mathbf{x}')\varphi^*(\mathbf{x}')\, d\mathbf{x}' + C(\mathbf{x}) \qquad (9.7.25)$$

where $\varphi^*(\mathbf{x})$ is the adjoint collision density, and $C(\mathbf{x})$ is the contribution of a collision at $\mathbf{x}$ to the answer as defined by Eq. (9.4.8) and plays the role of the adjoint source. For simplicity, we assume the counter region $\mathcal{D}$ is a pure absorber: $C(\mathbf{x}) = 1$ for $\mathbf{x}$ in $\mathcal{D}$ and 0 otherwise.

Introducing $I(\mathbf{x}) = \varphi^*(\mathbf{x})$, the altered source and the altered collision kernel are, respectively,

$$\tilde{S}(\mathbf{x}) = \frac{S(\mathbf{x})\varphi^*(\mathbf{x})}{A} \qquad (9.7.26)$$

$$\tilde{K}(\mathbf{x}' \to \mathbf{x}) = K(\mathbf{x}' \to \mathbf{x})\frac{\varphi^*(\mathbf{x})}{\varphi^*(\mathbf{x}')} \qquad (9.7.27)$$

The probability for a particle to survive a collision at $\mathbf{x}$ is

$$\tilde{\alpha}(\mathbf{x}) = \int \tilde{K}(\mathbf{x} \to \mathbf{x}')\, d\mathbf{x}' = \frac{1}{\varphi^*(\mathbf{x})}\int K(\mathbf{x} \to \mathbf{x}')\varphi^*(\mathbf{x}')\, d\mathbf{x}'$$

$$\qquad (9.7.28)$$

$$= 1 - \frac{C(\mathbf{x})}{\varphi^*(\mathbf{x})}$$

where Eq. (9.7.27) was used to obtain the second equality and Eq. (9.7.25) to obtain the third equality. We have assumed that the counter is a pure absorber. Once a collision occurs in the counter region, the particle is always absorbed. This means $\tilde{K}(\mathbf{x} \to \mathbf{x}') = 0$ for $\mathbf{x}$ in the counter region, and accordingly Eq. (9.7.25) becomes $\varphi^*(\mathbf{x}) = C(\mathbf{x})$ for $\mathbf{x}$ in the counter region. We also have $C(\mathbf{x}) = 0$ for $\mathbf{x}$ outside the counter. Finally, we have

$$\tilde{\alpha}(\mathbf{x}) = 0 \quad \text{for } \mathbf{x} \text{ in the counter}$$
$$= 1 \quad \text{otherwise} \tag{9.7.29}$$

This result shows that the particle in the altered game never terminates its histroy until it is absorbed in the counter region. The count of the particle in the altered game is always unity.

According to Eq. (9.7.19), the weight of the particle at the first collision is

$$w_1 = \frac{S(\mathbf{x}_1)}{\tilde{S}(\mathbf{x}_1)} = \frac{A}{\varphi^*(\mathbf{x}_1)} \tag{9.7.30}$$

where

$$A = \int \tilde{S}(\mathbf{x}) \, d\mathbf{x} = \int S(\mathbf{x})\varphi^*(\mathbf{x}) \, d\mathbf{x} \tag{9.7.31}$$

Referring to Eq. (9.7.20), the weights at the subsequent collisions are

$$w_2 = w_1 \frac{\varphi^*(\mathbf{x}_1)}{\varphi^*(\mathbf{x}_2)} = \frac{A}{\varphi^*(\mathbf{x}_1)} \frac{\varphi^*(\mathbf{x}_1)}{\varphi^*(\mathbf{x}_2)} = \frac{A}{\varphi^*(\mathbf{x}_2)} \tag{9.7.32}$$
$$\vdots$$
$$w_k = \frac{A}{\varphi^*(\mathbf{x}_k)} \tag{9.7.33}$$

The weight at the last collision in the counter region is

$$w_K = \frac{A}{\varphi^*(\mathbf{x}_K)} = \frac{A}{C(\mathbf{x}_K)} = \frac{\int S(\mathbf{x})\varphi^*(\mathbf{x}) \, d\mathbf{x}}{C(\mathbf{x}_K)} \tag{9.7.34}$$

In Eq. (9.7.34), $\varphi^*(\mathbf{x}_K) = C(\mathbf{x}_K)$ for $\mathbf{x}_K$ in the counter region was used. Therefore, Eq. (9.7.9) becomes

$$\eta = \sum_{k=1}^{K} w_k C(\mathbf{x}_k) = w_K C(\mathbf{x}_K) = \int S(\mathbf{x})\varphi^*(\mathbf{x}) \, d\mathbf{x} \tag{9.7.35}$$

The right side of Eq. (9.7.35) is the exact solution as proven in the following. Denoting $\phi(\mathbf{x})$ as the expected collision density satisfying Eq. (9.5.9), the expected answer is obviously given by

$$\int C(\mathbf{x})\phi(\mathbf{x}) \, d\mathbf{x} \tag{9.7.36}$$

Multiplying Eq. (9.5.9) and Eq. (9.7.25), respectively, by $\varphi^*(\mathbf{x})$ and $\phi(\mathbf{x})$, integrating the products over the entire space, and subtracting one from the other, we find the right side of Eq. (9.7.35) becomes equal to Eq. (9.7.36) [ϕ_1 in Eq. (9.5.9) is identical with S by the definition given in Section 9.3].

Although this is a remarkable result, the use of the adjoint solution in the exact sense has little practical value. This is because the solution of the adjoint equation is as difficult as the normal problem. Furthermore, if the adjoint solution is available, the exact answer is obtained by performing the integration of the right side of Eq. (9.7.35) rather than performing any Monte Carlo calculations. Nevertheless, the importance of the adjoint biasing is in that, as the density biasing function $I(\mathbf{x})$ approaches the adjoint solution, the variance of the altered game approaches zero. Even if the solution of the exact adjoint solution is difficult, an approximate solution may be obtained by alternative approaches[28] with various degrees of accuracy.

DELTA SCATTERING

Delta scattering involves adding an artificial material to some region. The artificial material does nothing but scatter neutrons directly forward without changing the direction so the expected value of the Monte Carlo game is not affected by the presence of the artificial material. At the delta-scattering collision, the neutron simply acquires another flight in the same direction.

The advantage of delta scattering is that it sometimes simplifies the geometrical tracking of neutrons. For example, suppose a unit cell calculation has an internal void. In straightforward Monte Carlo calculation, the distance to the void and whether the neutron intersects the void must be determined. If so, it must be tracted through the void. By adding a delta scatterer with the same total cross section as the external material, one need only determine a collision point and then determine whether the collision was in the void or not. If it was, a new flight length is chosen, and the history continues with the same direction. The delta scatterer may also be used to make the total cross section in two adjacent regions (nonvoided) equal. A careful use of delta scattering can drastically reduce the time required to track neutrons, allowing more neutrons to be followed in the same overall time required for a conventional Monte Carlo calculation.

CORRELATED SAMPLING

When a small difference between two almost identical configurations is of interest, it would be totally masked by the variance if the Monte Carlo method is applied independently to each of the configurations. Under this circumstance, the small difference can be computed by correlating the two sets of histories. For the first configuration the Monte Carlo calculation is performed as usual. The paths and events in the second configuration are assumed to be exactly the same as the first configuration, but only the weights of the particles are independently calculated. The difference in the final answers are then caused by the differences in weights at the events. With this method, a meaningul difference can be computed, even when the total difference is only half as large as the standard deviation on either individual Monte Carlo answer. Correlated sampling is a limited technique, but it can be useful when it is applicable.

9.8 EIGENVALUE PROBLEM

The fundamental eigenvalue of the neutron transport equation for a steady-state reactor can be estimated by the Monte Carlo method. The homogeneous neutron transport equation may be written as

$$\psi(\mathbf{x}) = \int K(\mathbf{x}' \to \mathbf{x})\psi(\mathbf{x}')\,d\mathbf{x}' + \lambda \int K_\nu(\mathbf{x}' \to \mathbf{x})\psi(\mathbf{x}')\,d\mathbf{x}' \qquad (9.8.1)$$

where λ is the eigenvalue (reciprocal of the effective multiplication factor), $K_\nu(\mathbf{x}' \to \mathbf{x})\,d\mathbf{x}$ is the probability that a neutron entering a collision at $\mathbf{x}'$ causes a fission reaction, and subsequently one of the fission neutrons born will have its first collision in $d\mathbf{x}$ at $\mathbf{x}$.

The Monte Carlo eigenvalue calculation is essentially based on the power method discussed in Chapter 2. The power method for Eq. (9.8.1) is written as

$$\varphi^{(t)}(\mathbf{x}) = \int K(\mathbf{x}' \to \mathbf{x})\varphi^{(t)}(\mathbf{x}')\,d\mathbf{x}' + \lambda^{(t-1)}S^{(t-1)}(\mathbf{x}) \qquad (9.8.2)$$

$$S^{(t-1)}(\mathbf{x}) = \int K_\nu(\mathbf{x}' \to \mathbf{x})\varphi^{(t-1)}(\mathbf{x}')\,d\mathbf{x} \qquad (9.8.3)$$

$$\lambda^{(t-1)} = \lambda^{(t-2)}\frac{\int S^{(t-2)}(\mathbf{x})\,d\mathbf{x}}{\int S^{(t-1)}(\mathbf{x})\,d\mathbf{x}} \qquad (9.8.4)$$

where t is the number of iterations. As in the standard numerical method, $S^{(0)}(\mathbf{x})$ is set to an initial guess and $\psi^{(1)}(\mathbf{x})$ is calculated by the Monte Carlo method. Then $\psi^{(t)}$ for $t = 2, 3, \ldots$ are successively calculated until convergence.

Since $S^{(t-1)}(\mathbf{x})$ in Eq. (9.8.2) is always given by the previous iteration, the estimation of $\psi^{(t)}(\mathbf{x})$ by the Monte Carlo method is reduced to the methods discussed in previous sections. With the estimate for the total collision density in the $(t-1)$th iteration $\psi^{(t-1)}$, the first collision sites of fission neutrons in the next iteration are sampled from Eq. (9.8.3) by using random numbers. There are various ways of sampling the first collision sites. The detail of the sampling is left for the reader as an exercise.

A few effective techniques to accelerate the convergence of the eigenvalue are outlined next:

1. The nearer the initial guess to the source distribution, the faster the convergence. The initial guess may be obtained by a few-energy-group diffusion calculations or from the previous result of a similar problem.
2. Linear acceleration methods for neutron diffusion equations, for example, the Chebyshev polynomial method, may be used.
3. Computations until convergence may be done by using a small number of neutrons in one generation. The sampling number may be enlarged at any time.
4. Convergence of iterations may be accelerated by using an approximate Green's function that is obtained from the latest cycle of iteration.

The Green's function method is to some extent similar to coarse-mesh rebalancing for the diffusion equation and has been studied by various investigators.[5,37-40] If we denote the fission source for the tth generation by $S^{(t)}$, the fission sources for two sequential generations can be related as

$$S^{(t)}(\mathbf{x}) = \lambda^{(t-1)} \int G(\mathbf{x}' \to \mathbf{x}) S^{(t-1)}(\mathbf{x}')\, d\mathbf{x}' \qquad (9.8.5)$$

where $G(\mathbf{x}' \to \mathbf{x})$ is the probability that if a fission neutron makes its first collision at $\mathbf{x}'$, then one of the fission neutrons in the next generation makes its first collision in $d\mathbf{x}$ about $\mathbf{x}$. Even though $G(\mathbf{x}' \to \mathbf{x})$ cannot be explicitly written in terms of K and K_ν, the approximation to $G(\mathbf{x} \to \mathbf{x}')$ can be obtained by using two successive source distributions, $S^{(t-1)}(\mathbf{x})$ and $S^{(t)}(\mathbf{x})$, during the Monte Carlo calculations. Once an approximate kernel $\hat{G}(\mathbf{x}' \to \mathbf{x})$ is obtained, the auxiliary eigenvalue problem can be written as

$$\tilde{S}(\mathbf{x}) = \tilde{\lambda} \int \hat{G}(\mathbf{x}' \to \mathbf{x}) \tilde{S}(\mathbf{x}')\, d\mathbf{x}' \qquad (9.8.6)$$

As $\hat{G}$ becomes closer to G, $\tilde{S}$ becomes a better approximation to S.

In order to calculate $\hat{G}$ by using $S^{(t-1)}$ and $S^{(t)}$, we divide the reactor into a number of subregions $\mathscr{D}_i$, $i = 1, 2, \ldots, I$. Suppose that $N_i^{(t-1)}$ is the integrated fission source (more exactly, the first collisions that the fission neutrons make) in region i and that $N_i^{(t-1)}$ will cause the fission source of intensity $N_{i,j}^{(t)}$ for the next generation in region j. $N_i^{(t-1)}$ and $N_{i,j}^{(t)}$ can be calculated through a cycle of Monte Carlo calculation. By using $N_i^{(t-1)}$ and

$N_{i,j}^{(t)}$, the approximate Green's function may be obtained in the form

$$G_{i,j} = \frac{\int_j d\mathbf{x} \int_i d\mathbf{x}' \, G(\mathbf{x}' \to \mathbf{x}) S^{(t-1)}(\mathbf{x}')}{\int_i d\mathbf{x}' \, S^{(t-1)}(\mathbf{x}')} = \frac{N_{i,j}^{(t)}}{N_i^{(t-1)}} \qquad (9.8.7)$$

where integrals are extended over subregion i or j. Using $G_{i,j}$ thus obtained, the approximate eigenvalue problem corresponding to Eq. (9.8.6) is constructed as

$$\tilde{N}_j = \tilde{\lambda} \sum_i G_{i,j} \tilde{N}_i \qquad (9.8.8)$$

Equation (9.8.8) can be solved by the standard power method. By using the solution of Eq. (9.8.8), the fission source for the tth generation is corrected to

$$S^{(t)}(x) \frac{\tilde{N}_i}{N_i} \quad \text{for } \mathbf{x} \text{ in } \mathcal{D}_i \qquad (9.8.9)$$

where we assume $\tilde{N}_i$ is normalized as

$$\sum_i \tilde{N}_i = \sum_i N_i \qquad (9.8.10)$$

The Monte Carlo calculation for the tth generation starts with Eq. (9.8.9) as the source distribution rather than $S^{(t)}(\mathbf{x})$. The Green's function method may be applied once after every few iterations.

9.9. MONTE CARLO METHOD FOR STEADY-STATE HEAT-CONDUCTION EQUATIONS

The idea of solving heat-conduction equations by the Monte Carlo method is not new,[32] but did not attract much practical interest until various further contributions were made. In the past decade, however, several extensions and applications have been performed. The Monte Carlo heat-conduction calculation is not necessarily an economical alternative to the finite difference or finite element method, but it is more versatile and powerful than other methods for three-dimensional and complicated geometries. The Monte Carlo method can calculate the temperature at isolated points by bypassing the necessity of obtaining an entire temperature profile. In this section, we first consider a discrete random-walk model and relate it with the difference approximation to the steady-state heat-conduction equation. Then the continuous random walk to estimate the solution of the steady state is discussed in the framework of Haji-Sheikeh and Sparrow.[33]

SIMPLE ONE-DIMENSIONAL DISCRETE PROBLEM

Suppose a discrete system as shown in Fig. 9.6, where G_0 and G_3 are boundary states, G_1 and G_2 are ordinary states, and S_1 and S_2 are source states. The random walk of the particles is assumed to obey the following rules:

1. Every particle stays in a state for 1 sec before going to another state.
2. A particle generated in the source state S_i goes to the ordinary state G_i with probability 1.
3. The total number of particles in G_0 and G_3 are always kept constant equal to ϕ_0 and ϕ_3 respectively.
4. A particle that entered G_1 or G_2 stays there for 1 sec and then moves to the left or right state with probability 0.5 to each direction.
5. A particle originally at G_0 goes to G_1 with probability 0.5 or terminates its life without going to G_1 with probability 0.5. A similar rule applies to G_3, too.
6. If a particle reaches G_0 from G_1 or G_3 from G_2, the life of the particle is terminated.
7. The number of particles in the source states is always maintained to be Q_i.

Suppose the average number of particles staying at G_i, $i = 1, 2$, is to be estimated. In a straightforward Monte Carlo simulation, we place ϕ_0, ϕ_3, Q_1, and Q_2 particles at G_0, G_3, S_1, and S_2, respectively, at a time and play with those particles until all the particles quit the game. Those particles are played as follows. A particle in S_i goes to G_i in the next second and then to G_{i-1} or G_{i+1}. The direction of motion is decided with a random number. The random walk of each particle is followed until the particle reaches G_0 or G_3.

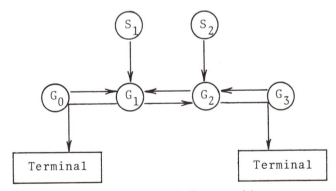

Figure 9.6 A simple discrete model.

Every time a particle visits G_i, the score for G_i is increased by 1. The cumulative score of visits to G_i per game averaged over a number of games gives an estimate for the average number of particles staying in G_i. As the total number of particles played is increased, the answer approaches the true average.

If we denote the true average of the answers for i by ϕ_i, then ϕ_i obviously satisfy

$$
\begin{bmatrix} \phi_0 \\ \phi_1 \\ \phi_2 \\ \phi_3 \end{bmatrix} = \begin{bmatrix} 0 & 0 & 0 & 0 \\ 0.5 & 0 & 0.5 & 0 \\ 0 & 0.5 & 0 & 0.5 \\ 0 & 0 & 0 & 0 \end{bmatrix} \begin{bmatrix} \phi_0 \\ \phi_1 \\ \phi_2 \\ \phi_3 \end{bmatrix} + \begin{bmatrix} \phi_0 \\ Q_1 \\ Q_2 \\ \phi_3 \end{bmatrix} \tag{9.9.1}
$$

It is easily seen that Eq. (9.9.1) is a finite difference equation

$$
\phi_i = \frac{\phi_{i-1} + \phi_{i+1}}{2} + Q_i, \qquad i = 1, 2 \tag{9.9.2}
$$

with prescribed boundary values for ϕ_0 and ϕ_3. Therefore, the Monte Carlo game gives an approximate solution for Eq. (9.9.2).

If the answer at only one grid point, for example, ϕ_1, is wanted, the Monte Carlo estimate can be obtained by the adjoint Monte Carlo game as indicated in Section 9.6 (see Appendix for deriving adjoint equations and their physical meaning). The adjoint equation for Eq. (9.9.1) is obtained by transposing the coefficient matrix as

$$
\begin{bmatrix} \phi_0^* \\ \phi_1^* \\ \phi_2^* \\ \phi_3^* \end{bmatrix} = \begin{bmatrix} 0 & 0.5 & 0 & 0 \\ 0 & 0 & 0.5 & 0 \\ 0 & 0.5 & 0 & 0 \\ 0 & 0 & 0.5 & 0 \end{bmatrix} \begin{bmatrix} \phi_0^* \\ \phi_1^* \\ \phi_2^* \\ \phi_3^* \end{bmatrix} + \begin{bmatrix} Q_1^* \\ Q_2^* \\ Q_3^* \\ Q_4^* \end{bmatrix} \tag{9.9.3}
$$

where ϕ_i^* are the number of adjoint particles at G_i and Q_i^* is the adjoint source. We set $Q_i^* = 0$ except for $Q_1^* = 1$. In each adjoint Monte Carlo game, we assume that only one adjoint particle starts at S_1 and performs a random walk in accordance with the same rule as for normal particles until it reaches a boundary state. The total number of visits of the adjoint particle at each state is scored and denoted as c_i^*. If the adjoint game is repeated a sufficiently large number of times and the average of c_i^* is taken, then it gives an approximation to ϕ_i^*. The answer of the adjoint Monte Carlo for ϕ_1 is then given by

$$\phi_1 = (\phi_0^*, \phi_1^*, \phi_2^*, \phi_3^*) \, \text{col} \, (\phi_0, Q_1, Q_2, \phi_3)$$
$$\approx \langle c_0^* \rangle \phi_0 + \langle c_1^* \rangle Q_1 + \langle c_2^* \rangle Q_2 + \langle c_3^* \rangle \phi_3 \qquad (9.9.4)$$

where $\langle c_i^* \rangle$ denotes the average of c_i^* and ϕ_0, Q_1, Q_2, and ϕ_3 are the prescribed values.

The score of the adjoint game may be counted in a different manner, in which the score as an estimate for ϕ_1 is directly obtained. c_1 is set to zero before an adjoint particle starts a random walk at S_1. Whenever an adjoint particle stops at G_i, c_1 is increased by Q_i. When the adjoint particle enters the boundary state G_0, c_1 is increased by ϕ_0. Similarly, c_1 is increased by ϕ_3 if the adjoint particle enters G_3. The average of c_1 for a large number of adjoint games is an estimate for ϕ_1. The computational time for the adjoint game to satisfy a required accuracy is smaller than the normal game. This difference becomes increasingly more significant as the number of grids increases.

TWO-DIMENSIONAL DISCRETE MODEL

The extension of the discrete random walk to a two-dimensional problem is straightforward. We consider an equally spaced two-dimensional mesh system, for which the discrete heat conduction equation is

$$\phi_{i,j} = \tfrac{1}{4}(\phi_{i-1,j} + \phi_{i+1,j} + \phi_{i,j-1} + \phi_{i,j+1}) + Q_{i,j} \qquad (9.9.5)$$

The boundary values are prescribed as $\phi_{i,j} = \phi_B$, where ϕ_B may vary at each grid on the boundary. Assuming that $\phi_{i,j}$ for $i = u$ and $j = v$ is to be estimated, the adjoint Monte Carlo is played as follows. An adjoint particle starts its random walk from (u, v). It moves to one of $(u-1, v)$, $(u+1, v)$, $(u, v-1)$, and $(u, v+1)$ with probability $1/4$ for each direction, and so on. The score of this adjoint game is initialized as $c_{u,v} = 0$ before the random walk. As the particle stays in (i, j) for a second, $c_{u,v}$ is increased by $Q_{i,j}$. If the adjoint particle reaches a boundary state, $c_{u,v}$ is increased by ϕ_B and the life of the particle is terminated. After a large number of games, the average of $c_{u,v}$ gives an approximation to $\phi_{u,v}$.

CONTINUOUS MODEL

A natural extension of the discrete random walk to the heat-conduction equation

$$-p\nabla^2 \phi(x, y) = Q(x, y) \qquad (9.9.6)$$

for a homogeneous two-dimensional geometry is to start with its integral form because the integral equation is similar to the matrix equation. We assume that the boundary condition is given by $\phi = \phi_B(s)$ along the boundary, where s is the coordinate on the boundary. An integral form of Eq. (9.9.6) may be written as

$$\phi(\mathbf{r}) = \frac{1}{2\pi} \int_{2\pi} \phi(\mathbf{r} + \gamma\boldsymbol{\omega}) \, d\boldsymbol{\omega} + Q(\mathbf{r}) \qquad (9.9.7)$$

where (x, y) is abbreviated by $\mathbf{r}$, $\boldsymbol{\omega}$ is a unit directional vector on the x–y plane and γ is an arbitrary constant that is smaller or equal to the shortest distance between (x, y) and the boundary, as shown in Fig. 9.7. The physical meaning of Eq. (9.9.7) is that when the circle of radius γ within the boundary is drawn with $\mathbf{r}$ as the center, then $\phi(r)$ is the average of $\phi(r')$ on the circle plus $Q(\mathbf{r})$. Assume that the solution of Eq. (9.9.5) for $\mathbf{r} = \mathbf{r}_0$ is wanted. In playing the adjoint Monte Carlo game, a circle with $\mathbf{r}_0$ as the center and the radius γ that is equal to the shortest distance between $\mathbf{r}_0$ and the boundary is drawn as illustrated in Fig. 9.7. The probability distribution on the circle is constant. Then by using a random number ξ, a point on the circle is selected, which is denoted by $\mathbf{r}_1$. The next step of random walk is executed by using $\mathbf{r}_1$ as the center and by constructing a circle whose radius is the shortest distance between $\mathbf{r}_1$ and the boundary. The random walk is terminated if the particle reaches the boundary. The total score of the particle is given by

$$c(\mathbf{r}_0) = \sum_{k=0}^{K} Q(\mathbf{r}_k) + \phi_B(\mathbf{r}_K) \qquad (9.9.8)$$

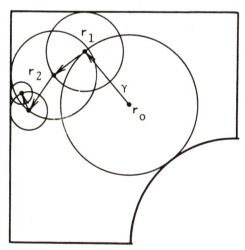

Figure 9.7   Random walk of an adjoint particle in heat transfer calculation.

where r_k is the center of the kth circle, K is the total number of points that the particle stopped, and $\phi_B(r_K)$ is the boundary value at the last stop of the particle.

One difficulty with the present approach is that as a particle approaches boundary, the circles become smaller and smaller and the random walk becomes time consuming. To circumvent this, the following approximate representation of the boundary is employed:

a. If the radius is smaller than a prescribed value, the particle is regarded as having reached the boundary.
b. If a particle moves to an angular position in a circle that deviates by less than a prescribed criterion, then the particle is regarded as having arrived at the boundary.
c. If the radius is smaller than a prescribed criterion, the continuous random walk can be switched to a discrete random walk based on the difference approximation.

Other types of boundary conditions than the Dirichlet type are discussed in Reference 33.

9.10. MONTE CARLO SOLUTION OF LAPLACE EQUATIONS VIA THE SURFACE-DENSITY TECHNIQUE

The continuous random walk described in Section 9.9 allows large walking steps in the beginning of the random walk, but requires smaller and smaller steps as the walker approaches a boundary. Two different approaches to circumvent this inefficiency have been proposed. The first approach is the surface-density approach[34,35] introduced in this section, and the second approach is essentially the same as the approach introduced in Section 9.9 except that other shapes than circles are used. For the second approach, an unsteady-state version of the method is discussed in Section 9.11.

We now consider a three-dimensional region $\mathscr{D}$ with smooth surface where the field function T satisfies the Laplace equation

$$\nabla^2 T(r) = 0, \qquad r \in \mathscr{D} \tag{9.10.1}$$

and the Dirichlet boundary condition

$$T(r) = T_B(r) \quad \text{for} \quad r \in \Gamma \tag{9.10.2}$$

In Eq. (9.10.2), Γ denotes the external surface of $\mathscr{D}$ and T_B is a prescribed function that is continuous on Γ.

It is known that the solution of this Dirichlet problem is given by the integral[41,42]

$$T(\mathbf{r}) = \int_{\Gamma} \frac{\langle \mathbf{r} - \mathbf{s}, \mathbf{n}(\mathbf{s}) \rangle}{4\pi |\mathbf{r} - \mathbf{s}|^3} \mu(\mathbf{s}) \, d\mathbf{s} \qquad (9.10.3)$$

where $\mathbf{s}$ is the coordinate on Γ, $\langle \, , \, \rangle$ denote the scalar product of two vectors, $d\mathbf{s}$ is the surface area on Γ about $\mathbf{s}$, $\mathbf{n}$ is the unit vector inward-normal to Γ at $\mathbf{s}$, and $\mu(\mathbf{s})$ is the double layer potential and satisfies the following integral equation:

$$\mu(\mathbf{s}) = 2T_B(\mathbf{s}) + \int_{\Gamma} \frac{\langle \mathbf{s}' - \mathbf{s}, \mathbf{n}(\mathbf{s}') \rangle}{2\pi |\mathbf{s}' - \mathbf{s}|^3} \mu(\mathbf{s}') \, d\mathbf{s}' \qquad (9.10.4)$$

For simplicity in later descriptions, we express Eq. (9.10.3) and Eq. (9.10.4) in compact forms as

$$T(\mathbf{r}) = \int R(\mathbf{s} \rightarrow \mathbf{r}) \mu(\mathbf{s}) \, d\mathbf{s} \qquad (9.10.5)$$

$$\mu(\mathbf{s}) = B(\mathbf{s}) + \int K(\mathbf{s}' \rightarrow \mathbf{s}) \mu(\mathbf{s}') \, d\mathbf{s}' \qquad (9.10.6)$$

where

$$B(\mathbf{s}) = 2T_B(\mathbf{s})$$

$$R(\mathbf{s} \rightarrow \mathbf{r}) = \frac{\langle \mathbf{r} - \mathbf{s}, \mathbf{n}(\mathbf{s}) \rangle}{4\pi |\mathbf{r} - \mathbf{s}|^3} \qquad (9.10.7)$$

$$K(\mathbf{s}' \rightarrow \mathbf{s}) = \frac{\langle \mathbf{s}' - \mathbf{s}, \mathbf{n}(\mathbf{s}') \rangle}{2\pi |\mathbf{s}' - \mathbf{s}|^3} \qquad (9.10.8)$$

If we can estimate the solution of Eq. (9.10.6) by such a method as Monte Carlo and introduce it into Eq. (9.10.5), the temperature at $\mathbf{r}$ is obtained. We note that R and K have the following properties

$$\int R(\mathbf{s} \rightarrow \mathbf{r}) \, d\mathbf{s} = 1 \qquad (9.10.9)$$

$$K(\mathbf{s}' \rightarrow \mathbf{s}) < 0 \qquad (9.10.10)$$

$$\int K(\mathbf{s}' \rightarrow \mathbf{s}) \, d\mathbf{s} = -1 \qquad (9.10.11)$$

Equation (9.10.6) is a Fredholm integral equation of the second kind and of the same form as Eq. (9.5.9). In Eq. (9.10.6) $|K(\mathbf{s}' \rightarrow \mathbf{s})| \, d\mathbf{s}$ may be interpreted as the probability that an event at $\mathbf{s}'$ will be followed by an event in $d\mathbf{s}$ about $\mathbf{s}$. $R(\mathbf{s} \rightarrow \mathbf{r})$ may be interpreted as the sensitivity distribution of a detector placed at $\mathbf{r}$ that senses the values of $\mu(\mathbf{s})$ at every point on Γ. Therefore, Eqs. (9.10.5) and (9.10.6) can be calculated as a Monte Carlo problem as follows: (1) the site of the initial event $\mathbf{s}_1$ is selected from $B(\mathbf{s})$ provided that $B(\mathbf{s})$ is normalized, (2) the second, third, ..., and the nth sites of events are selected from the probability distribution functions, respectively,

$$|K(\mathbf{s}_1 \to \mathbf{s}_2)|, \; |K(\mathbf{s}_2 \to \mathbf{s}_3)|, \; \ldots, \quad \text{and} \quad |K(\mathbf{s}_{n-1} \to \mathbf{s}_n)|$$

Since events are selected from $|K|$ rather than K, the weight of an event is set to

$$
\begin{aligned}
w_n &= 1 && \text{for } n = 1 \\
&= \frac{K(\mathbf{s}_{n-1} \to \mathbf{s}_n)}{|K(\mathbf{s}_{n-1} \to \mathbf{s}_n)|} w_{n-1} && \text{for } n > 1
\end{aligned}
\tag{9.10.12}
$$

If the above Monte Carlo approach is tried, however, the following two difficulties will soon arise. The first difficulty is that selecting an event $\mathbf{s}$ from $K(\mathbf{s}' \to \mathbf{s})$ with a fixed $\mathbf{s}'$ is very difficult. The second difficulty is that the weights will have plus and minus signs alternatively and the chain of the events is endless. If we express the probability distribution of the first, second, ..., and nth events by $\mu_1, \mu_2, \ldots$, they are given by

$$
\begin{aligned}
\mu_1(\mathbf{s}) &= B(\mathbf{s}) \\
\mu_2(\mathbf{s}) &= \int K(\mathbf{s}' \to \mathbf{s}) \mu_1(\mathbf{s}') \, d\mathbf{s}' \\
&\;\;\vdots \\
\mu_n(\mathbf{s}) &= \int K(\mathbf{s}' \to \mathbf{s}) \mu_{n-1}(\mathbf{s}') \, d\mathbf{s}'
\end{aligned}
\tag{9.10.13}
$$

It can be easily proven that, as n increases, μ_n and μ_{n-1} will satisfy

$$\mu_n(\mathbf{s}) = -\mu_{n-1}(\mathbf{s})$$

Therefore, the summation

$$\lim_{N \to \infty} \sum_{n=1}^{N} \mu_n(\mathbf{s})$$

will not converge to the real solution $\mu(\mathbf{s})$ of Eq. (9.10.6) but oscillates.

To circumvent the first difficulty described above, Hoffmann and Banks[34] proposed to use the adjoint equations of Eqs. (9.10.5) and (9.10.6) given by

$$T(\mathbf{r}) = \int d\mathbf{s} B(\mathbf{s}) \mu^*(\mathbf{s}) \, d\mathbf{s} \tag{9.10.14}$$

$$\mu^*(\mathbf{s}) = R(\mathbf{s} \to \mathbf{r}) + \int K(\mathbf{s} \to \mathbf{s}') \mu^*(\mathbf{s}') \, d\mathbf{s}' \tag{9.10.15}$$

The derivation of Eq. (9.10.15) is given in Appendix II. In playing an adjoint Monte Carlo game, the first evens **s** is selected from $R(\mathbf{s} \rightarrow \mathbf{r})$ with a fixed **r**. The next and the subsequent events are selected from $K(\mathbf{s} \rightarrow \mathbf{s}')$ with $\mathbf{s}'$ as the previous event. As explained next, selecting **s** from $K(\mathbf{s} \rightarrow \mathbf{s}')$ is by far easier than selecting **s** from $K(\mathbf{s}' \rightarrow \mathbf{s})$.

The probability distribution for the first event in the adjoint game may be written as

$$p(\mathbf{s})\, d\mathbf{s} = \frac{\langle \mathbf{r} - \mathbf{s}, \mathbf{n}(\mathbf{s}) \rangle}{4\pi |\mathbf{r} - \mathbf{s}|^3}\, d\mathbf{s} \qquad (9.10.16)$$

where $p(\mathbf{s})\, d\mathbf{s}$ is the probability that the event occurs in the surface element $d\mathbf{s}$. Referring to Fig. 9.8, the solid angle subtended by the surface element $d\mathbf{s}$ when viewed from **r** is given by

$$d\Omega = \frac{\langle \mathbf{r} - \mathbf{s}, \mathbf{n}(\mathbf{s}) \rangle}{|\mathbf{r} - \mathbf{s}|^3}\, d\mathbf{s} \qquad (9.10.17)$$

so Eq. (9.10.16) becomes

$$p(\mathbf{s})\, d\mathbf{s} = \frac{d\Omega}{4\pi} \qquad (9.10.18)$$

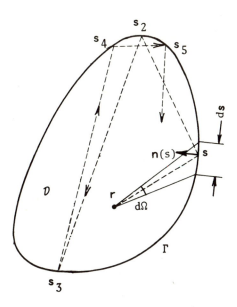

Figure 9.8 Random walk on a surface.

Therefore, the first event can be determined by the selection of an isotropic direction from $\mathbf{r}$, and $\mathbf{s}$ is the point where the straight trajectory strikes Γ.

The probability distribution for the second and the subsequent events of the adjoint particle is given by

$$|K(\mathbf{s}_n \to \mathbf{s}_{n-1})|\, d\mathbf{s}_n = \frac{\langle \mathbf{s}_{n-1} - \mathbf{s}_n, \mathbf{n}(\mathbf{s}_n)\rangle}{2\pi |\mathbf{s}_{n-1} - \mathbf{s}_n|^3}\, d\mathbf{s}_n$$

$$= \frac{d\Omega}{2\pi}, \qquad n \geq 2 \tag{9.10.19}$$

where $\mathbf{s}_{n-1}$ is the site of the previous event and is assumed to be known. Equation (9.10.19) says that the probability that the nth event occurs in $d\mathbf{s}_n$ about $\mathbf{s}_n$ is equal to the solid angle subtended by $d\mathbf{s}_n$ as viewed from $\mathbf{s}_{n-1}$ divided by 2π. Therefore, the nth event is obtained by selecting an isotropic direction from $\mathbf{s}_{n-1}$.

Since the Monte Carlo approach for Eq. (9.10.15) is simulating the series,

$$\mu_1^*(\mathbf{s}) = R\,(\mathbf{s} \to \mathbf{r})$$

$$\mu_n^*(\mathbf{s}) = \int K\,(\mathbf{s} \to \mathbf{s}')\mu_{n-1}^*(\mathbf{s})\, d\mathbf{s}', \qquad n > 1 \tag{9.10.20}$$

μ_n^* will have alternative signs, and the summation of μ_n^* will not converge just as the case of $\mu_n(\mathbf{s})$. So we consider a new series of functions defined by

$$\tilde{\mu}_1^*(\mathbf{s}) = \mu_1^* + \tfrac{1}{2}\mu_1^*$$

$$\tilde{\mu}_2^*(\mathbf{s}) = \tfrac{1}{2}(\mu_2^* + \mu_3^*)$$

$$\vdots \tag{9.10.21}$$

$$\tilde{\mu}_2^*(\mathbf{s}) = \tfrac{1}{2}(\mu_n^* + \mu_{n-1}^*)$$

This is called a *Fredholm series*.[42] We can easily see that $\tilde{\mu}_n^*$ will converge to zero as n increases. The summation of $\tilde{\mu}_n^*$ quickly converges to the solution of Eq. (9.10.15).

Implementation of the surface-density method is illustrated in Fig. 9.8. A walker is started at the point $\mathbf{r}$ where the field value T is desired. Isotropically it leaves $\mathbf{r}$ and moves directly to $\mathbf{s}$ on Γ. Then the boundary value at the point $\mathbf{s}_1$ is saved. The walker leaves $\mathbf{s}_1$ and moves isotropically-inward to another surface point. The walk is continued until the walker reaches four or five different points on the surface. Those sites are designated sequentially as $\mathbf{s}_2, \mathbf{s}_3, \ldots, \mathbf{s}_n$. A straightforward summation of the score at each event will not converge to the solution of Eq. (9.10.15). So referring to Eq. (9.10.21), the score of the game is set to

$$\lambda = [B(\mathbf{s}_1) + \tfrac{1}{2}B(\mathbf{s}_2)] + \tfrac{1}{2}\sum_{n=3}^{N}[B(\mathbf{s}_{n-1}) + B(\mathbf{s}_n)] \tag{9.10.22}$$

where N is the number of events sampled during a game. The average of λ for many games will give an approximation for $T(\mathbf{r})$ of Eq. (9.10.14).

9.11 MONTE CARLO METHOD FOR TRANSIENT HEAT CONDUCTION

We start the derivation of the Monte Carlo method for transient heat transfer problems with the one-dimensional heat conduction equation in the forward difference form as given by Eq. (4.2.12). For simplicity, we assume that the medium is uniform, the mesh interval Δx is constant, and the source term is zero. We assume that the initial condition is given and the temperature at the left and right boundaries are prescribed. If we choose $\Delta t = \Delta x^2/2p$ and set $Q_i = 0$ for simplicity of discussions, the forward difference equation for Eq. (4.1.3) becomes

$$T_i^{(n)} = \tfrac{1}{2}(T_{i-1}^{(n-1)} + T_{i+1}^{(n-1)}) \tag{9.11.1}$$

where $T_i^{(n)}$ is the temperature at $t_n = n\Delta t$ and the spatial grid i. Assuming that $T_j^{(m)}$ is to be found, an adjoint game may be played as follows.[33]

The score of the game is set to $\lambda = 0$ initially. The random walk of an adjoint particle starts at $i = j$ and $n = m$. In the adjoint game the time sequence is reversed. At t_{m-1} the particle goes to $j-1$ or $j+1$ with probability 0.5 for each direction. As a general rule the particle goes to $(i-1, n-1)$ or $(i+1, n-1)$ with the equal probability after a stay at (i, n). If the particle reaches a boundary point at $n > 0$, the score of the particle is set to the prescribed value of the temperature at the boundary. If the particle is still inside of the domain when $n = 0$ is reached, the score of the particle is set to the initial value of the temperature at the point where the particle reaches $n = 0$.

The unsteady Monte Carlo method on the discrete space is now extended to a continuous problem.[36] The unsteady-state heat conduction equation is written as

$$p\nabla^2 T(\mathbf{x}, t) = \frac{\partial T(\mathbf{x}, t)}{\partial t}, \quad t > 0, \; \mathbf{x} \in \mathcal{D} \tag{9.11.2}$$

where $\mathcal{D}$ is the domain of interest. The Dirichlet boundary condition is given by

$$T(\mathbf{x}, t) = T_B(\mathbf{x}, t), \qquad \mathbf{x} \in \Gamma \tag{9.11.3}$$

and the initial condition is

$$T(\mathbf{x}, 0) = T_0(\mathbf{x}), \qquad t = 0 \tag{9.11.4}$$

where T_B and T_0 are prescribed functions and Γ is the external surface of $\mathscr{D}$. We do not consider the heat source in Eq. (9.11.2), although it can be incorporated easily whenever necessary.

In terms of Green's function in Eq. (9.11.2), the solution of this problem may be written as

$$T(\mathbf{x}, t) = \int_{\mathscr{D}} G(\mathbf{x}', 0 \to \mathbf{x}, t) T_0(\mathbf{x}') \, d\mathbf{x}'$$

$$- \int_0^t \int_{\Gamma} T_B(\mathbf{x}', t') p \frac{\partial}{\partial n'} G(\mathbf{x}', t' \to \mathbf{x}, t) \, ds' \, dt'$$

(9.11.5)

where G is Green's function, $\mathbf{s}'$ is the coordinate on Γ and $\partial/\partial n'$ is the derivative outward-normal to Γ with respect to $\mathbf{x}'$. Green's function satisfies the following conditions:

1. $G(\mathbf{x}', t' \to \mathbf{x}, t') = \delta(\mathbf{x} - \mathbf{x}')$

2. $G(\mathbf{x}', t' \to \mathbf{x}, t) = 0$ for $\mathbf{x} \in \Gamma$ and for any $t \geq t'$

3. $\nabla_x^2 G(\mathbf{x}', t' \to \mathbf{x}, t) = \dfrac{\partial}{\partial t} G(\mathbf{x}', t' \to \mathbf{x}, t)$ (9.11.6)

Equation (9.11.5) may be thought as a weighted integral of the initial condition $T_0(\mathbf{x}')$ and the boundary condition $T_B(\mathbf{x}', t)$ with $\mathbf{x}' \in \Gamma$, where G and $-p(\partial/\partial n')G$ are the weighting functions. The following properties of the Green's function are important (see Problem 14):

a. $\displaystyle \int_{\mathscr{D}} G(\mathbf{x}', 0 \to \mathbf{x}, t) \, d\mathbf{x}' - \int_0^t \int_{\Gamma} p \frac{\partial}{\partial n'} G(\mathbf{x}', t' \to \mathbf{x}, t) \, ds' \, dt' = 1$ (9.11.7)

b. If we denote the first term of Eq. (9.11.7) by $\alpha_x(t)$, then it can be easily proven that

$$\alpha_x(0) = 1$$

$$0 < \alpha_x(t) < 1$$

(9.11.8)

$$\alpha_x(\infty) = 0$$

The physical meaning of the first term of Eq. (9.11.7), or equivalently $\alpha_x(t)$, is the total weight for the initial condition $T_0(\mathbf{x}')$, while the second term of Eq. (9.11.7), or equivalently, $1 - \alpha_x(t)$ is the total weight for the boundary condition $T_B(\mathbf{x}', t')$ with $\mathbf{x}' \in \Gamma$. Notice that the second term of Eq. (9.11.7) is positive with the minus sign in front. Property b indicates that, if t is small, $T(\mathbf{x}, t)$ is mainly determined by $T_0(\mathbf{x}')$, while if t is large $T(\mathbf{x}, t)$ is mainly determined by $T_B(\mathbf{x}', t')$.

If the shape of $\mathscr{D}$ is simple, G is analytically obtained, and $T(\mathbf{x}, t)$ for any given $\mathbf{x}$ and t can be calculated by performing integrations in Eq. (9.11.5).

Unfortunately this is not the case for most realistic cases, because Green's function is available only to extremely simple configurations. So we derive a more versatile method based on the Monte Carlo method that fits any shape of $\mathcal{D}$. However, before revising the basic scheme of Eq. (9.11.5), it is worthwhile to study the Monte Carlo method to perform integrations in Eq. (9.11.5). We first show how a Monte Carlo method can be used to calculate a weighted integral of a function by the following example.

Example 9.7

Suppose the following weighted integral of a smooth function $g(x)$ is desired:

$$I = \int_a^b w(x)g(x)\,dx$$

where $w(x)$ is a normalized weighting function:

$$\int_a^b w(x)\,dx = b - a$$

Three numerical approximation methods are suggested:

1. Extended Simpson's rule

$$I = \frac{b-a}{2N}\left[w(a)g(a) + 2\sum_{n=1}^{N-1} w(x_n)g(x_n) + w(b)g(b) \right]$$

where

$$x_n = \frac{n(b-a)}{N}$$

2. Extended Simpson's rule with a transformed coordinate

$$I = \frac{b-a}{2N}\left[g(a) + 2\sum_{n=1}^{N-1} g(x_n) + g(b) \right]$$

where x_n is found by

$$n\frac{b-a}{N} = \int_a^{x_n} w(x)\,dx$$

3. Monte Carlo method

$$I = \frac{b-a}{N}\sum_{n=1}^{N} g(x_n)$$

where x_n is decided with the nth random number ξ by

$$\xi(b-a) = \int_a^{x_n} w(x)\,dx$$

and N is the total number of random numbers used.

Assuming that the Green's function in Eq. (9.11.5) is analytically known, the Monte Carlo game to estimate $T(\mathbf{x}, t)$ for a fixed point $(\mathbf{x}, t)$ can be devised as follows.

Step 1. Draw a random number ξ. If $\xi < \alpha(t)$, we decide to sample the temperature from $T_0(\mathbf{x}')$ and go to Step 2. If $\xi > \alpha(t)$, we decide to sample the temperature from $T_B(\mathbf{x}', t')$ and go to Step 3.

Step 2. Normalize $G(\mathbf{x}', 0 \to \mathbf{x}, t)$ as

$$\tilde{G}_{x,t}(\mathbf{x}') \equiv \frac{G(\mathbf{x}', 0 \to \mathbf{x}, t)}{\int_{\mathscr{D}} G(\mathbf{x}', 0 \to \mathbf{x}, t)\,d\mathbf{x}'}$$

and determine $\mathbf{x}'$ by $\tilde{G}$ and a set of random numbers. Then set the score of this game to $T_0(\mathbf{x}')$ and go to the next game.

Step 3. With the ξ drawn in Step 1, decide t' by

$$\alpha(t') = \xi$$

Then decide $\mathbf{x}'$ by a set of random numbers and the normalized probability distribution of x':

$$F(\mathbf{x}') = \frac{p(\partial/\partial n')G(\mathbf{x}', t' \to \mathbf{x}, t)}{\int_\Gamma p(\partial/\partial n')G(\mathbf{x}', t' \to \mathbf{x}, t)\,ds'}, \quad \mathbf{x}' \in \Gamma$$

Set the score to $T_B(\mathbf{x}', t')$, and go to the next step.

In order to extend the Monte Carlo integration for Eq. (9.11.5) for any arbitrary shaped domain $\mathscr{D}$, we consider a subdomain V of a simple shape contained in $\mathscr{D}$ as shown in Fig. 9.9. We assume that (1) Green's function for V is easily obtained, (2) $T(\mathbf{x}, t)$ for a fixed point $(\mathbf{x}, t)$ is desired, and (3) $(\mathbf{x}, t)$ is contained in $\mathscr{D}$. In terms of the temperature $T(\mathbf{x}', t')$ for $\mathbf{x}'$ on S, which is the boundary of V, $T(\mathbf{x}, t)$ is expressed by Eq. (9.11.5), except that $\mathscr{D}$ and Γ are altered to V and S, respectively. Suppose Vs are selected so that S and Γ coincide as much as possible. If the Monte Carlo game with Green's function for V is started, the score of the game is obtained whenever $t' = 0$ is reached or the part of S overlapping with Γ is reached. If the surface point on S but not on Γ is selected, then $T(\mathbf{x}', t')$ at that point is considered as a new unknown. The Monte Carlo game is continued with another configuration

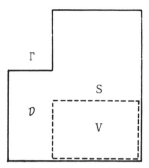

$\mathcal{D}$:   whole domain
V: subdomain for Green's function

Figure 9.9   Illustration of a subdomain.

V' whose surface S' is overlapping with Γ as much as possible. The game continues until $t' = 0$ is reached or x' on Γ is reached. Green's function for two simple shapes of V for a two-dimensional domain $\mathcal{D}$ is illustrated next.

CIRCULAR SUBDOMAINS FOR V

Green's function for a circular (or infinite cylinder) subdomain may be expressed by using the Bessel–Fourier series. If the point $\mathbf{x}$ is the center of the circle, Green's function becomes much simpler and is given by

$$G(\mathbf{x}', t' \to \mathbf{x}, t) = \sum_{k=1}^{\infty} a_k J_0\left(\mu_k \frac{r}{R}\right) \exp\left(\frac{-p\mu_k^2(t-t')}{R^2}\right) \qquad (9.11.9)$$

where $r = |\mathbf{x} - \mathbf{x}'|$, R is the radius of the circle, μ_k is the kth root of the Bessel function $J_0(z)$, and a_k is a coefficient given by

$$a_k = \frac{2}{\pi R^2[J_1(\mu_k)]^2} \qquad (9.11.10)$$

This Green's function has the following properties:

a. $G(\mathbf{x}', t' \to \mathbf{x}, t) = 0$ for $\mathbf{x}' \in S$

b. $G(\mathbf{x}', t' \to \mathbf{x}, t') = \delta(\mathbf{x}' - \mathbf{x})$

c. $\dfrac{\partial}{\partial n'} G(\mathbf{x}', t' \to \mathbf{x}, t)\bigg|_{\mathbf{x}' \in S} = \dfrac{\partial}{\partial r'} G(\mathbf{x}', t' \to \mathbf{x}, t)\bigg|_{r=R}$

$$= -\sum_{k=1}^{\infty} \frac{2\mu_k}{\pi R^3 J_1(\mu_k)} \exp\left(\frac{-p\mu_k^2(t-t')}{R^2}\right)$$

d. $\alpha_{\mathbf{x}}(\tau) = \displaystyle\int_V G(\mathbf{x}', t' \to \mathbf{x}, t)\, d\mathbf{x}' = 2\pi \int_0^R G(\mathbf{x}', t' \to \mathbf{x}, t) r'\, dr'$

$\qquad = \displaystyle\sum_{k=1}^{\infty} \frac{4}{\mu_k J_1(\mu_k)} \exp\left(\frac{-p\mu_k^2 \tau}{R^2}\right)$

In these equations S is the circle of radius R and $\tau = t - t'$.

The Monte Carlo game for unsteady heat transfer with circular subdomains is similar to the continuous model for steady-state heat transfer described in Section 9.9 and may be written as follows:

Step 1. Draw the largest circle contained in $\mathscr{D}$ with $\mathbf{x}$ as the center. The value $\tau = t - t'$ is selected by

$$\alpha(\tau) = \xi$$

where ξ is a random number. If $\tau \leq 0$, go to Step 2; otherwise go to Step 3.

Step 2. Determine $\mathbf{x}'$ in V with the following normalized probability distributions. The cumulative angular probability distribution in terms of polar coordinate with $\mathbf{x}$ at the origin is

$$F(\omega) = \frac{\omega}{2\pi}$$

The cumulative radial probability distribution is obtained by integrating Eq. (9.11.9) with respect to r as

$$H(r) = \frac{\sum_{k=1}^{\infty} \gamma_k r J_1[\mu_k (r/R)]}{\sum_{k=1}^{\infty} \gamma_k R J_1(\mu_k)}$$

where

$$\gamma_k = \frac{a_k}{\mu_k} \exp\left(\frac{-p\mu_k^2 t}{R^2}\right)$$

Set the score of this game to $T_0(\mathbf{x}')$. Go to the next game.

Step 3. Determine $\mathbf{x}'$ on S with the cumulative probability distribution

$$F(\omega) = \frac{\omega}{2\pi}$$

where ω is the angular coordinate on the circle with the center at $\mathbf{x}$. If $\mathbf{x}'$ is on Γ, set the score to $T_B(\mathbf{x}', t')$. If not, restart Step 1 with the largest circle in $\mathscr{D}$

with the center at $\mathbf{x}'$. The values of $\mathbf{x}$ and t in Step 1 are reset to $\mathbf{x}'$ and t', respectively, determined in this step.

The average of a large number of games will give an estimate for $T(\mathbf{x}, t)$. The random walk with circular subdomains has the problems stated in Section 9.9, so some approximate treatment of the boundary is necessary to accelerate termination of the random walks.

RECTANGULAR SUBDOMAINS FOR V

The more the fraction of Γ overlapping with S, the more the chance of terminating the random walk. From this point of view, circular subdomains have poor contact with Γ. If V has straight-line boundaries that are parallel to the x and y axis, rectangular subdomains[36] will have a significantly larger fraction of overlap with Γ. For a rectangular V with length Y and width X, as illustrated in Fig. 9.10, Green's function is given by

$$G(\mathbf{x}', t' \to \mathbf{x}, t) = \sum_{k=1}^{\infty} \sum_{l=1}^{\infty} a_{kl} \sin \frac{k\pi(a_2 - x')}{X} \sin \frac{l\pi(b_2 - y')}{Y} \exp\left[-\gamma_{kl}(t - t')\right]$$

$$(9.11.11)$$

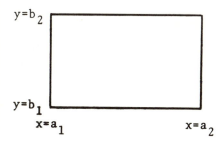

Figure 9.10 Rectangular region for Green's function.

where

$$X = a_2 - a_1, \qquad Y = b_2 - b_1$$

$$\gamma_{kl} = p\left[\left(\frac{k\pi}{X}\right)^2 + \left(\frac{l\pi}{Y}\right)^2\right]$$

$$a_{kl} = \frac{4}{XY} \sin \frac{k\pi(a_2 - x)}{X} \sin \frac{l\pi(b_2 - y)}{Y}$$

Figure 9.11 illustrates a random walk with rectangular subdomains. Notice that the particle reaches a boundary after four paths.

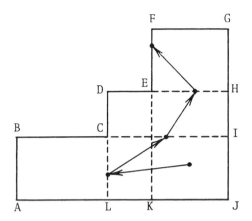

Four rectangular regions used for
Green's function are

V_1:   DHJL V_2:   ABIJ

V_3:   DHJL V_4:   FGJK

Figure 9.11 Illustration of a random walk with
rectangular subregions for Green's
function.

PROBLEMS

1. Considering a binomial event with the probability $p = 0.1$ for *yes* and $q = 1 - p = 0.9$ for *no*, calculate the variance and standard deviations of the number of obtaining *yes* out of 100, 1000, and 10,000 trials, respectively.

2. Consider a homogeneous medium for monoenergetic particles. The probability that a particle travels a distance s and makes a collision in ds about s is $p(s)\, ds = \Sigma e^{-\Sigma s}\, ds$.
   a. Suppose we estimate the mean travel distance $\langle s \rangle$ by the Monte Carlo method; what is the variance of s?
   b. If $\langle s \rangle$ is estimated by the average of N sampling, how large an N is necessary in order that the average s_N has the confidence level of 99% for the confidence interval $|(s_N - \langle s \rangle)/\langle s \rangle| < 0.001$?

3. Suppose a particle walks randomly on a straight line going left and right. The path length between two steps is subject to the probability distribution e^{-s}. The direction after a stop is either left or right with equal

probability. By using the following random numbers sequentially, find how far the fifth stop is from the starting point.

$$0.1452, \quad 0.0337, \quad 0.7835, \quad 0.1816, \quad 0.5389$$

4. Draw the flow chart of the analog Monte Carlo calculation for monoenergetic neutron transport, which involves only elastic scattering and capture (no fission) in a bare homogeneous sphere with radius R_0. Assume there is a point source at the center of the sphere.

5. Suppose a particle moving randomly on a straight line going right and left. When the particle enters a collision, the probability of changing the direction is 0.5 and that of unchanging the direction is 0.5. The probability that the particle travels the distance from a collision to the next collision falls between s and $s + ds$, is $e^{-\Sigma s} \, ds$, where Σ is a constant. If this problem is biassed by $I(x) = e^{+ax} (a > 0)$, then what is the mean distance of the particle going to the left and that of the particle going to the right?

6. The kernel of the nonanalog Monte Carlo game that is biased by the adjoint solution may be written

$$\tilde{K}(\mathbf{x}, \mathbf{x}') = \tilde{\gamma}\tilde{K}_\Omega(\Omega \to \Omega')\tilde{K}_r(\mathbf{r} \to \mathbf{r}')$$

Discuss how $\tilde{\gamma}$, $\tilde{K}_\Omega$, and $\tilde{K}_r$ are related to γ, K_Ω, and K_r, respectively, for the analog Monte Carlo game.

7. If $K(\mathbf{x}, \mathbf{x}') \, d\mathbf{x}'$ is the probability that a particle entering a collision at $\mathbf{x}$ will make the next collision in $d\mathbf{x}'$ about $\mathbf{x}'$, what is the probability that the same particle makes k additional collisions so the $(k + 1)$th collision will be in $d\mathbf{x}'$ about $\mathbf{x}'$?

8. Suppose the monoenergetic particle transport in a cylinder with two concentric regions and a line source of particles at the center. Draw the flow chart of Monte Carlo calculations to estimate the rate of leakage from the cylinder. Assume that there are no other reactions than isotropic scattering and capture.

9. How can the average total flux in a region be estimated by the analog Monte Carlo for monoenergetic neutron transport?

10. Suppose all the sites of collisions have been recorded during a Monte Carlo calculation. Draw a flow chart for estimating the rate of leakage of particles from the system by using the next-event estimator.

11. Assuming that the source is uniformly distributed in a finite cylinder with height H and radius R, draw a flow chart of the subprogram to sample a source particle.

12. Suppose four boxes that are identified by the index $i = 1, 2, 3, 4$ and a ball that is randomly moving from one box to another. Initially a ball is thrown into one of the boxes with probability S_i, $i = 1 \sim 4$. The probability that the ball goes from i to j is given by K_{ij}. If the ball comes to the fourth box ($i = 4$), the random walk is terminated.

a. Denoting the expected number of times that the ball visits the ith box by ψ_i, derive the analytical equation for ψ_i.

b. Assuming we are interested in estimating ψ_1 by the Monte Carlo method, derive the adjoint equation.

c. Derive the analytical formula for the variance of the analog Monte Carlo simulation for ψ_1.

d. If the game is biased by the adjoint solution, what is the probability for the ball going from i to j in the biased game? Investigate also how the random walk in the biased game is terminated.

13. Prove that the expected answer of Monte Carlo calculations is unaffected by adding delta scatterers.

14. a. Prove

$$p\nabla_{x'}^2 G(\mathbf{x}', t' \to \mathbf{x}, t) + \frac{\partial}{\partial t'} G(\mathbf{x}', t' \to \mathbf{x}, t) = 0$$

b. Prove Eq. (9.11.7) by performing integrations in the following equation:

$$\int_0^t \int_{\mathscr{D}} \left[p\nabla_{x'}^2 G(\mathbf{x}', t' \to \mathbf{x}, t) + \frac{\partial}{\partial t'} G(\mathbf{x}', t' \to \mathbf{x}, t) \right] d\mathbf{x}'\, dt' = 0$$

REFERENCES

1. E. D. Cashwell and C. J. Everett, *A practical manual on the Monte Carlo method for random walk problems*, Pergamon, New York (1959).

2. E. A. Straker, *The Morse code—A multigroup neutron and gamma ray transport code*, ORNL-4585, also CCC-127/Morse, Oak Ridge National Laboratory, Tenn. (1970).

3. D. C. Irving, R. M. Freestone, Jr., and F. B. K. Kam, *05R, a general purpose Monte Carlo neutron transport code*, ORNL-3622, also ORNL-3715 and ORNL-TH-1245, Oak Ridge National Laboratory, Tenn. (1965).

4. F. A. R. Schmidt, "Status of the Monte Carlo development," *Proceedings of the IAEA Seminar on Numerical Reactor Calculations*, Vienna, Jan. 17–21 (1972).

5. M. H. Kalos, F. R. Nakache, and J. Celnik, in *Computing methods in reactor physics*, H. Greenspan, C. N. Kelber, and D. Okrent, Eds., Gordon and Breach, New York (1968).

6. H. Rief and H. Kschwendt, "Reactor analysis by Monte Carlo," *Nucl. Sci. Eng.*, **30**, 395–418 (1967).

7. J. Spanier and E. M. Gelbard, *Monte Carlo principles and neutron transport problems*, Addison-Wesley, Reading, Mass. (1969).

8. W. Matthes, *Some applications of the Monte Carlo method at the Euratom Research Center Ispra*, CONF-710302, U.S. Atomic Energy Commission, Division of Technical Information, Springfield, Va. (1971).

9. W. E. Vesely, F. J. Wheeler, and R. S. Marsden, *The raffle Monte Carlo code and its use to obtain correlated estimates of collapsed cross sections*, CONF-710302, U.S. Atomic Energy Commission, Division of Technical Information, Springfield, Va. (1971).

10. N. P. Buslinko, *The Monte Carlo method, the method of statistical trials*, Pergamon, New York (1966).

11. H. A. Meyer, *Symposium on Monte Carlo methods*, Wiley, New York (1956).

12. J. M. Hammersley and D. C. Handscomb, *Monte Carlo methods*, Methuen, London (1964).

13. A. M. Weinberg and E. P. Wigner, *The physical theory of neutron chain reactors*, University of Chicago, Chicago, Ill. (1958).

14. G. I. Bull and S. Glasstone, *Nuclear reactor theory*, Van Nostrand, Princeton, N.J. (1970).

15. K. M. Case and P. F. Zweifel, *Linear transport theory*, Addison-Wesley, Reading, Mass. (1967).

16. W. Feller, *An introduction to probability theory and its applications*, 2nd edit., Wiley, New York (1965).

17. N. A. Frigerio and N. A. Clark, "A random number set for Monte Carlo computations," *Trans. Amer. Nucl. Soc.*, **22**, 238 (1975).

18. R. R. Coveyou, "On random number generation," *Proc. NEACRP meeting of a Monte Carlo study group*, ANL-75-2, Argonne National Laboratory, Argonne, Ill. (1975).

19. T. E. Hull and A. R. Dobell, "Random number generators," *Soc. Ind. App. Math.*, **4**, 230–254 (1962).

20. M. Abramowitz and I. A. Stegun, *Handbook of mathematical functions*, National Bureau of Standards, U.S. Department of Commerce (1970).

21. Rand Corporation, *A million random digits with 100,000 normal deviates*, Free Press, Glencoe, Ill. (1955).

22. H. Honeck, "The calculation of thermal utilization and disadvantage factor in uranium/water lattices," *Nucl. Sci. Eng.*, **18**, 49–68 (1964).

23. A. Leonard, Collision probabilities and response matrices: An overview, CONF-750413, Vol. II, National Technical Information Service, Springfield, Va. (1975).

24. F. A. R. Schmidt, "Biased angle selection in Monte Carlo shielding calculations," *Nucl. Eng. Des.*, **15**, 209 (1971).

25. J. Spanier, *A new multi-stage procedure for systematic variance reduction in Monte Carlo*, CONF-710302, Vol. 2, National Technical Information Center, Springfield, Va. (1971).

26. F. H. Clark, *The exponential transform as an importance sampling device*, ORNL-Radiation Safety Information Center 14, Oak Ridge National Laboratory, Tenn. (1966).

27. H. A. Steinberg, M. H. Kalos, and E. S. Troubetzkoy, *Machine generation of Monte Carlo sampling functions*, CONF-710302, National Technical Information Center, Springfield, Va. (1971).

28. V. R. Cain, "Applications of SN adjoint flux calculations to Monte Carlo biasing, *Trans. Amer. Nucl. Soc.*, **10**, 399 (1967).

29. C. E. Burgart and P. N. Stevens, *A general method of importance sampling the angle of scattering in Monte Carlo calculations*, ORNL-TM-2890, Oak Ridge National Laboratory, Tenn. (1970).

30. R. R. Coveyou, V. R. Cain, and K. J. Yost, "Adjoint and importance in Monte Carlo application," *Nuc. Sci. Eng.*, **27**, 219-234 (1967).

31. C. E. Burgart, *Capabilities of the MORSE multigroup Monte Carlo code in solving reactor eigenvalue problems*, ORNL-TM-3662, Oak Ridge National Laboratory, Tenn.; also in *The Proceedings of the IAEA Seminar on Numerical Reactor Calculations*, Vienna, Jan. 17–21 (1972).

32. M. E. Muller, "Some continuous Monte Carlo methods for the Dirichlet problem," *Ann. Math. Stat.*, **27**, 569–589 (1956).

33. A. Haji-Sheikh and E. M. Sparrow, "The solution of heat conduction problems by probability methods," *J. ASME*, **89**, 121–131 (1967).

34. T. J. Hoffman and N. E. Banks, "Monte Carlo solution to the Dirichlet problem with the double layer potential density," *Trans. Am. Nucl. Soc.*, **18**, 136 (1974).

35. T. J. Hoffman and N. E. Banks, "Extension of the surface-density Monte Carlo technique to heat transfer in non-convex regions," *Trans. Am. Nucl. Soc.*, **19**, 164 (1974).

36. E. S. Troubetzkoy and N. E. Banks, "Solution of heat diffusion equation by Monte Carlo," *Trans. Am. Nucl. Soc.*, **19**, 163 (1974).

37. J. B. Parker and E. R. Woodcock, "Monte Carlo criticality calculations," progress in nuclear energy, Series IV, **4**, 441, Pergamon Press (1961).

38. R. C. Gast and N. R. Candelore," Monte Carlo stretegies and uncertainties," ANL-75-2, Argonne National Laboratory, Argonne, Ill. (1974).

39. L. L. Carter and N. J. McCormick, "Source convergence in Monte Carlo calculations," *Nucl. Sci. Eng.* **36**, 438–441 (1969).

40. T. Asaoka, Y. Nakahara, S. Miyasaka, K. Horikami, T. Suzuki, T. Nishida, Y. Taji, T. Tsuchihashi, H. Gotoh, Y. Seki, M. Nakagawa, and J. Hirota, *Coarse-mesh rebalancing acceleration for eigenvalue problems*, ANL-75-2, Argonne National Laboratory, Argonne, Ill. (1974).

41. S. G. Mikhlin, *Linear integral equations*, Hindustan Publishing Corporation, Delhi, India (1960).

42. M. V. Lovitt, *Linear integral equations*, Dover, New York (1950).

Chapter 10 Computational Fluid Dynamics—II

10.1 SCOPE OF COMPUTER SIMULATIONS FOR AERODYNAMIC FLOW

Progressively improved approximations to the full governing equation for the aerodynamic flow around wings and bodies of aircrafts have been developed as computers have improved. However, even with the most advanced computer of today, it is not yet possible to obtain numerical solutions to the full Navier–Stokes equation for complex configurations in a practical amount of time and cost. Chapman, Mark, and Pirtle[1] divided the past, present, and future development in aerodynamic computations into the following stages.

Stage 1. *Linearized inviscid approximation* (past). Two- and three-dimensional approximations were developed during the 1930s through the 1960s. They are currently used in aircraft design. Viscous and nonlinear inviscid terms are neglected.

Stage 2. *Inviscid nonlinear approximation* (1971–1975). The two- and three-dimensional approximations for transonic flow have been developed. No viscous terms are considered. They cannot be applied to flow-separation problems.

Stage 3. *Viscous, time-averaged Navier–Stokes equations* (1975–1978?). No terms of the Navier–Stokes equation are neglected, but certain terms involving turbulent momentum, and energy and heat transport are modeled. Such codes have been developed already for two-dimensional separated turbulent flows.[2-4] In order to provide practical computational results for three-dimensional problems, the following two developments will be necessary: (1) accurate turbulence models for separated and attached flow and (2) a computer capability two orders of magnitude greater than Illiac IV.[5]

Stage 4. *Complete time-dependent Navier–Stokes equations* (distant future, 1985?). In essence, all of the significant-size turbulent eddies in a given flow would be computed for a sufficiently long time. This stage requires several orders of magnitude more computation than the third stage. With the present speed of computer development as described in Section 5.1, it is not unrealistic to expect a computer that is large enough and fast enough to handle this stage in the next ten years.

It is presumed in accomplishing Stages 3 and 4 that more sophisticated and more efficient computational methods are developed along with the development of new computers.

   Our focus of attention in this chapter is on the finite difference techniques developed in Stage 2. Most of the computational methods for inviscid nonlinear equations have been formulated with the finite-difference approximation, but very little[6,34] of the finite-element methods has been applied. The latter approach has, however, the advantage of conforming to the shape of wings and bodies much more easily than the former. The current trend of fitting irregular shapes of boundary with the finite difference schemes is by means of coordinate transformations,[30–31] although no detail is described in this book. The nonlinear term of the flow equation becomes particularly important in transonic flow. The nonlinear difference equations for transonic flow are basically elliptic but have some hyperbolic domains that represent the local supersonic flow. In solving the difference equations, two kinds of approaches have been used. The first is the successive-over-relaxation (SOR) method for elliptic equations with nonlinear coefficients. The second is the fast-direct-solution (FDS) method. Section 10.2 describes the application of the fast-direct method to inviscid linear problems. Section 10.3 through 10.5 describe the solution for the transonic thin-airfoil perturbation equation by both SOR and FDS. Section 10.6 discusses the transonic flow calculation without approximation to the Euler equation. Sections 10.7 and 10.8 introduce the fast-direct-solution methods.

10.2   FINITE-DIFFERENCE SOLUTION TO SUBSONIC AERODYNAMIC FLOW

A peculiar aspect of the finite-difference solution for lifting airfoil problems is that the external boundary conditions cannot be prescribed but have to be determined as a part of the solution. Even though the incompressible inviscid flow equation is given as a Laplace equation, it must be solved at least several times until all the boundary conditions become consistent. The

Laplace equation for the airfoil flow problem has been solved by the fast-direct-solution (FDS) method[7-12] rather than iteratively. The FDS method is a direct-solution technique that applies to the finite-difference approximation for the Poisson equation in a uniform mesh system for a rectangular geometry. The FDS method works with an external boundary condition of the Dirichlet type or with certain other types of simple boundary conditions. If any internal boundary condition exists, FDS cannot be applied in a straightforward manner but needs multiple applications to solve a single problem. Nevertheless, the computing time of FDS is so short that it has been extensively used for fluid analysis. The objective of this section is to introduce the finite difference solution for incompressible flow problems developed by Martin.[13] The principle of FDS is discussed in Section 10.7. We assume in this section that FDS is available whenever necessary.

We consider the incompressible flow around an airfoil with any arbitrary cross-section shape immersed in a uniform stream. The x coordinate is chosen to be parallel to the chord direction as shown in Fig. 10.1. The direction of flow is specified by α (angle of attack), which is the angle between the direction of the fluid at infinity and the x coordinate. The flow distribution can be calculated in terms of the stream function or by the velocity potential function. In each case, the flow equation is reduced to a Laplace equation.[35] We choose here the stream function $\psi(x, y)$ defined by

$$u = \frac{\partial \psi}{\partial y}, \qquad v = -\frac{\partial \psi}{\partial x} \qquad (10.2.1)$$

where u and v are velocities in the x and y directions, respectively. Then the flow equation becomes

$$\nabla^2 \psi = 0 \qquad (10.2.2)$$

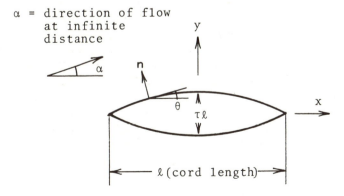

Figure 10.1 A biconvex airfoil.

There are three boundary conditions to be satisfied. (1) Internal boundary condition along the airfoil surface: Since the stream function along a surface is constant, we set $\psi = 0$ along the airfoil surface. (2) Kutta condition:[14] This condition implies the flow leaves the trailing edge smoothly. Mathematically, the Kutta condition is expressed by

$$(u_t - u_b)_{\text{TE}} = 0 \qquad (10.2.3)$$

where u_t is the horizontal velocity on the upper side of the airfoil, u_b is that on the down side, and TE denotes the trailing edge. (3) External boundary condition: The physical boundary conditions at infinity are

$$u = U_\infty \cos \alpha$$
$$v = U_\infty \sin \alpha \qquad (10.2.4)$$

where U_∞ is the fluid velocity at infinity and α is the angle of attack.

The first approximation to be made is that the external boundary must be placed at a finite distance from the airfoil so that the finite difference approximation is applied to Eq. (10.2.2). In order to obtain the boundary condition applicable at a finite distance, we consider the asymptotic solution at far field. The flow around any airfoil may be viewed from a long distance as the combination of doublet and vortex superimposed on a uniform flow. Therefore, the stream function at far field is given in the form

$$\psi_{\text{ext}} = U_\infty(y \cos \alpha - x \sin \alpha) - \frac{\mu y}{2\pi(x^2 + y^2)} - \frac{\Gamma}{2\pi}(\ln \sqrt{x^2 + y^2} + \text{const})$$

$$(10.2.5)$$

where μ is the intensity of doublet and Γ is the circulation. The values of μ and Γ are unknowns to be found through the solution. It should be noticed, however, that as $r^2 = (x^2 + y^2)$ increases, the influence of the second term decreases much more quickly than the third term. So if the distance from the airfoil to the boundary is sufficiently large, the second term of Eq. (10.2.5) can be neglected. Since circulation is caused because of the Kutta condition, Γ is determined (or adjusted) in such a way that the Kutta condition is satisfied. The constant in the third term of Eq. (10.2.5) is dependent on the shape and size of the airfoil. This constant is not arbitrary because the stream function along the airfoil has been set to zero (only one stream line can be set to a prescribed value, see Section 5.5). For thin airfoils, Martin proposed using the constant that is analytically found for a flat plate. The constant for a flat plate of length l is $\ln(4/l)$. Accuracy of this approximation for the constant is discussed in reference 13.

The finite difference solution for an arbitrarily shaped airfoil is developed through the following three cases.

Case 1. *Airfoil surface passing grid points*

In order to illustrate how to apply the FDS method to the airfoil flow problem with an internal boundary condition, the effect of the Kutta condition is neglected and the external boundary condition, Eq. (10.2.5), is assumed to be prescribed including the circulation Γ. We consider the uniform grid system for a finite rectangular domain so that FDS can be applied. It is also important for FDS that the number of grids in one direction is $2^m - 1$, where m is an integer. For simplicity, we also assume that the airfoil surface line coincides with grid points as shown in Fig. 10.2.

Equation (10.2.2) is now approximated by the difference equation

$$\frac{2\psi_{i,j} - \psi_{i-1,j} - \psi_{i+1,j}}{\Delta x^2} + \frac{2\psi_{i,j} - \psi_{i,j-1} - \psi_{i,j+1}}{\Delta y^2} = 0 \qquad (10.2.6)$$

or equivalently

$$L\psi_{i,j} = 0 \qquad (10.2.6a)$$

where L is the difference (or discrete Laplace) operator. The boundary conditions for $\psi_{i,j}$ are

$$\psi_{i,j} = 0 \quad \text{for } (i,j) \in \mathscr{S}_{\text{int}} \qquad (10.2.7)$$

$$\psi_{i,j} = \psi_{\text{ext}}(x_i, y_i) \quad \text{for } (i,j) \in \mathscr{S}_{\text{ext}} \qquad (10.2.8)$$

where $\mathscr{S}_{\text{int}}$ and $\mathscr{S}_{\text{ext}}$, respectively, denote the internal (airfoil surface) and external boundaries, and ψ_{ext} is the value of Eq. (10.2.5) on $\mathscr{S}_{\text{ext}}$ and assumed here to be prescribed.

In applying FDS, the solution of Eq. (10.2.6a) is partitioned into two parts as

$$\psi_{i,j} = \psi_{i,j}^{(a)} + \psi_{i,j}^{(b)} \qquad (10.2.9)$$

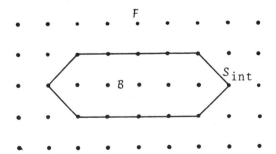

Figure 10.2 A hypothetical airfoil surface passing grid points.

In Eq. (10.2.9) $\psi_{i,j}^{(a)}$ and $\psi_{i,j}^{(b)}$ are the solutions of the following problems:

$$L\psi_{i,j}^{(a)} = 0 \qquad \text{for entire } \mathcal{F} + \mathcal{B}$$
$$\psi_{i,j}^{(a)} = \psi_{\text{ext}}(x_i, y_j) \quad \text{for } (i,j) \in \mathcal{S}_{\text{ext}} \tag{10.2.10}$$

$$L\psi_{i,j}^{(b)} = 0 \qquad \text{for entire } \mathcal{F}$$
$$\psi_{i,j}^{(b)} = 0 \qquad \text{for } (i,j) \in \mathcal{S}_{\text{ext}} \tag{10.2.11}$$
$$\psi_{i,j}^{(b)} = -\psi_{i,j}^{(a)} \qquad \text{for } (i,j) \in \mathcal{S}_{\text{int}}$$

where $\mathcal{F}$ denotes the entire rectangular domain excluding the airfoil and $\mathcal{B}$ is the domain occupied by the airfoil. Equation (10.2.10) is solved by FDS. So that Eq. (10.2.11) can be also solved by FDS, it is replaced by an equivalent problem as

$$L\psi_{i,j}^{(b)} = f_{i,j} \quad \text{for entire } \mathcal{F} \text{ and } \mathcal{B}$$
$$\psi_{i,j}^{(b)} = 0 \quad \text{for } (i,j) \in \mathcal{S}_{\text{ext}} \tag{10.2.12}$$

where $f_{i,j}$ is the fictitious source with the following properties:

1. $f_{i,j} = 0$ for $(i,j) \in \mathcal{F}$
2. $f_{i,j}$ can have nonzero value only on and inside $\mathcal{S}_{\text{int}}$
3. $f_{i,j}$ are adjusted so that $\psi_{i,j}^{(b)}$ satisfy $\psi_{i,j}^{(b)} = -\psi_{i,j}^{(a)}$ for $(i,j) \in \mathcal{S}_{\text{int}}$

   Solution of Eq. (10.2.12) needs further devices because the fictitious source is not explicitly prescribed. Suppose the total number of the grid points on $\mathcal{S}_{\text{int}}$ is K, and they are numbered by $l = 1, 2, \ldots, K$. The discrete function $\psi_{i,j}^{(b)}$ can be expressed in the form,

$$\psi_{i,j}^{(b)} = \sum_{l=1}^{K} c_l u_{i,j}^{(l)} \tag{10.2.13}$$

In the above equation, c_l are undetermined coefficients and $u_{i,j}^{(l)}$ are independent functions satisfying

$$Lu_{i,j}^{(l)} = \delta_{i,j}^{(l)} \quad \text{for } (i,j) \in \mathcal{F} + \mathcal{B}, \qquad l = 1, 2, \ldots, K$$
$$u_{i,j}^{(l)} = 0 \quad \text{for } (i,j) \in \mathcal{S}_{\text{ext}} \tag{10.2.14}$$

where $\mathcal{F} + \mathcal{B}$ denotes the entire domain and

$$\delta_{i,j}^{(l)} = 1 \quad \text{if } (i,j) \text{ is the } l\text{th grid on } \mathcal{S}_{\text{int}}$$
$$= 0 \quad \text{otherwise} \tag{10.2.15}$$

Equation (10.2.13) for each l can be solved by FDS. The undetermined coefficients c_l are determined by introducing Eq. (10.2.13) into the third equation of Eq. (10.2.11) and solving the resulting equation for c_l:

$$\sum_{l=1}^{K} c_l u_{i,j}^{(l)} = -\psi_{i,j}^{(a)} \quad \text{for the } K \text{ grids } (i,j) \text{ on } \mathscr{S}_{\text{int}} \tag{10.2.16}$$

Evidently, the linear combination

$$\psi_{i,j} = \psi_{i,j}^{(a)} + \sum_{l=1}^{K} c_l u_{i,j}^{(l)} \tag{10.2.17}$$

satisfies Eqs. (10.2.6) through (10.2.8). However, the value of $\psi_{i,j}$ for each grid point is calculated by using FDS once again rather than using Eq. (10.2.17). This is because the summation of $u_{i,j}^{(l)}$ according to Eq. (10.2.17) requires a large amount of memory to store $u_{i,j}^{(l)}$ for all i, j, and l. The equation for the final application of FDS to find $\psi_{i,j}$ is given by

$$\begin{aligned} L\psi_{i,j} &= f_{i,j} & \text{for } (i,j) \in \mathscr{F} + \mathscr{B} \\ \psi_{i,j} &= \psi_{\text{ext}}(x_i, y_j) & \text{for } (i,j) \in \mathscr{S}_{\text{ext}} \end{aligned} \tag{10.2.18}$$

with no internal boundary condition, where

$$f_{i,j} = \sum_l c_l \delta_{i,j}^{(l)}$$

In this equation, c_l have been determined by Eq. (10.2.16).

Case 2. *Airfoil Surface Passing Mesh Intervals*

Generally, the airfoil surface does not coincide with the grid points. In this case, interpolations between grid points becomes necessary in order to satisfy the inner boundary condition $\psi = 0$ along the airfoil surface. Here we consider exactly the same problem as Case 1 except the airfoil surface does not coincide with the grid points.

The first step is to determine the special boundary grids r_k, $k = 1, 2, \ldots, K$, along the airfoil surface at which the internal boundary condition $\psi = 0$ is imposed. The total number of boundary grids K is independent of the number of grid points inside the airfoil surface, but it is desirable that K is equal to or less than the number of grids inside the internal boundary. We assign a grid point just internal to the airfoil surface to the boundary grid at r_k. Although the position of r_k is arbitrary in general, we place r_k on one of the mesh lines in order to make the interpolation easy. An example of r_ks and their correspondence to the adjacent internal grid is illustrated in Fig. 10.3.

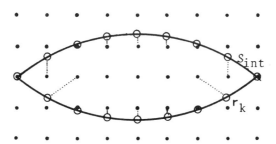

Figure 10.3 An airfoil surface not passing grid
points.

The next step is to calculate $\psi^{(a)}$ and $u^{(l)}$. The calculation procedure is
exactly the same as the previous case except that Eq. (10.2.15) is replaced by

$$\delta_{i,j}^{(l)} = 1 \quad \text{if } (i, j) \text{ is the internal grid assigned to the } l\text{th boundary grid}$$

$$= 0 \quad \text{otherwise}$$

(10.2.18)

The arbitrary coefficients in Eq. (10.2.13) must be evaluated by using the
internal boundary condition at the special boundary grids r_l. Since we do not
have the value of ψ on those special grids, an interpolation formula must be
used. However, since r_l has been located on a mesh line, ψ at r_l can be easily
obtained by interpolating between two adjacent grid points. This interpolation applies to all of $\psi^{(a)}$ and $u^{(l)}$ to evaluate the values at r_l.

Case 3.   Lifting Airfoil

When the Kutta condition is imposed but the circulation Γ is unknown, the
solution is partitioned into three parts:

$$\psi_{i,j} = \psi_{i,j}^{(a)} + \Gamma \psi_{i,j}^{(c)} + \sum_{l=1}^{K} c_l u_{i,j}^{(l)}$$

(10.2.19)

We consider here the same problem as Case 2 except that the airfoil is lifting,
and accordingly the Kutta condition must be satisfied. In Eq. (10.2.19), the
three components satisfy the following conditions:

$$L\psi_{i,j}^{(a)} = 0 \qquad\qquad \text{for entire } \mathcal{F} \text{ and } \mathcal{B}$$

$$\psi_{i,j}^{(a)} = U_\infty(y_i \cos \alpha - x_i \sin \alpha) \text{ for } (i, j) \in \mathcal{S}_{\text{ext}}$$

(10.2.20)

(No internal B.C.)

$$L\psi_{i,j}^{(c)} = 0 \qquad\qquad \text{for entire } \mathcal{F} \text{ and } \mathcal{B}$$

$$\psi_{i,j}^{(b)} = -\frac{1}{2\pi}\left(\ln\sqrt{x_i^2 + y_i^2} + \ln\frac{4}{l}\right) \text{ for } (i,j) \in \mathcal{S}_{\text{ext}} \qquad (10.2.21)$$

(No internal B.C.)

$$L\psi_{i,j}^{(l)} = f_{i,j}^{(l)}, \quad l = 1, 2, \ldots, K \quad \text{for entire } \mathcal{F} \text{ and } \mathcal{B}$$

$$u_{i,j}^{(l)} = 0 \qquad\qquad \text{for } (i,j) \in \mathcal{S}_{\text{ext}} \qquad (10.2.22)$$

(No internal B.C.)

Each of Eqs. (10.2.20) through (10.2.22) is solved by FDS. The $K+1$ unknowns in Eq. (10.2.18), Γ and c_l, $l = 1, 2, \ldots, K$, are determined by the K internal boundary conditions on the airfoil and the Kutta condition. The difference approximation for the Kutta condition, Eq. (10.2.3), is given in terms of ψ by

$$\frac{1}{2\Delta y}[(-3\psi_{i,j} + 4\psi_{i,j+1} - \psi_{i,j+2}) + (3\psi_{i,j} - 4\psi_{i,j+1} + \psi_{i,j-2})] = 0$$

$$(10.2.23)$$

where the upward difference is used for u_t, the downward difference is used for u_b, and (i,j) is the grid point at the trailing edge.

10.3 TRANSONIC THIN AIRFOIL PERTURBATION THEORY

As the Mach number approaches unity, the flow around an airfoil will have supersonic region (see Fig. 10.4). The thin airfoil perturbation theory for transonic region was developed in the late 1950s,[15–17] but the solution was not obtained until recently because of the nonlinearity of the flow equation. The first solution for the transonic thin airfoil equation was by means of a finite difference technique and the relaxation technique. Several other numerical methods are currently being investigated by various investigators. Two numerical methods for the transonic equation are described in Sections 10.4 through 10.6. The purpose of this section is to derive the transonic perturbation equation in the most straightforward manner.

Consider a thin airfoil moving through a compressible fluid to the negative x direction with speed $|U_\infty|$. An observer on the airfoil will see a uniform stream U_∞ on which the perturbation u and v are superimposed. The

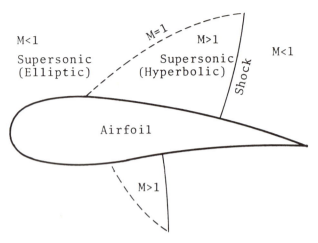

Figure 10.4   Transonic flow around an airfoil (Ms in this figure
are local Mach numbers).

equations of motion and continuity for compressible inviscid fluid may be
written as

$$UU_x + VU_y = -\frac{c^2}{\rho}\rho_x \tag{10.3.1}$$

$$UV_x + VV_y = -\frac{c^2}{\rho}\rho_y \tag{10.3.2}$$

$$\rho(U_x + U_y) + U\rho_x + V\rho_y = 0 \tag{10.3.3}$$

where suffixes x and y mean the partial derivatives with respect to x and y,
respectively. In Eqs. (10.3.1) through (10.3.3), $U \equiv U(x, y)$ and
$V \equiv V(x, y)$ are the velocity of fluid in the x and y directions, respectively,
$\rho \equiv \rho(x, y)$ is the density, and c is the local sonic velocity defined by
$c^2 = dp/d\rho$. Multiplying Eqs. (10.4.1) and (10.4.2) by U and V, respectively,
and adding the two equations, we obtain

$$U^2U_x + UV(U_x + V_y) + V^2V_y = -\frac{c^2}{\rho}(U\rho_x + V\rho_y) \tag{10.3.4}$$

Substituting Eq. (10.3.3) into the right side of Eq. (10.3.4) and dividing by
$-c^2$ yield

$$\left(1 - \frac{U^2}{c^2}\right)U_x - \frac{UV}{c^2}(U_y + V_x) + \left(1 - \frac{V^2}{c^2}\right)V_y = 0 \tag{10.3.5}$$

Assuming that the fluid velocity without the disturbances, U_∞, is in the direction of the x axis (no lifting case is considered in this section: $\alpha = 0$ in Fig. 10.1), the total velocity at any point may be written by

$$U(x, y) = U_\infty + u(x, y)$$
$$V(x, y) = v(x, y) \tag{10.3.6}$$

where u and v are the components of the disturbances (perturbations) caused by the airfoil.

We define the velocity potential $\phi(x, y)$ by

$$\phi_x = u, \qquad \phi_y = v \tag{10.3.7}$$

and assume the irrotational condition

$$v_x = u_y$$

or equivalently

$$\phi_{yx} = \phi_{xy} \tag{10.3.8}$$

Introducing Eq. (10.3.6) into Eq. (10.3.5) and using Eq. (10.3.7) yield

$$\left[1 - \frac{(U_\infty + u)^2}{c^2}\right]\phi_{xx} - \frac{2(U_\infty + u)v}{c^2}\phi_{xy} + \left(1 - \frac{v^2}{c^2}\right)\phi_{yy} = 0 \tag{10.3.9}$$

Assuming that the perturbed velocity components are much smaller than the sonic speed, the second and the third terms of Eq. (10.3.9) are approximated by

$$-\frac{2(U_\infty + u)v}{c^2}\phi_{xy} \approx 0$$
$$\left(1 - \frac{v^2}{c^2}\right)\phi_{yy} \approx \phi_{yy} \tag{10.3.10}$$

When U_∞/c approaches unity, the first term of Eq. (10.3.9) requires careful attention as discussed in the following paragraph. If the first order term is neglected, then Eq. (10.3.9) will approach $\phi_{yy} = 0$ as the Mach number approaches unity, thus leading to an unacceptable result.

The sonic speed c in the adiabatic flow is a function of flow velocity and satisfies the adiabatic condition

$$c^2 + \frac{\gamma - 1}{2}[(U_\infty + u)^2 + v^2] = c_\infty^2 + \frac{\gamma - 1}{2}U_\infty^2 \tag{10.3.11}$$

where c is the local sonic speed and c_∞ is the sonic speed at the infinite distance where the fluid speed is U_∞. The Mach number at infinite distance is defined by

$$M = \frac{U_\infty}{c_\infty} \qquad (10.3.12)$$

By neglecting the second order terms in Eq. (10.3.11) and using Eq. (10.3.12), the sonic ratio is derived as

$$\left(\frac{c}{c_\infty}\right)^2 \approx 1 - (\gamma - 1)M^2 \frac{u}{U_\infty} \qquad (10.3.13)$$

The inverse sonic ratio may also be written as

$$\left(\frac{c_\infty}{c}\right)^2 \approx 1 + (\gamma - 1)M^2 \frac{u}{U_\infty} \qquad (10.3.14)$$

The first term of Eq. (10.3.9) then becomes

$$1 - \frac{(U_\infty + u)^2}{c^2} = 1 - \left(\frac{c_\infty}{c}\right)^2 \left[M^2 \left(1 + 2\frac{u}{U_\infty}\right) + \left(\frac{u}{U_\infty}\right)^2 \right]$$

$$\approx 1 - M^2 \left(1 + \frac{2u}{U_\infty}\right) - M^4(\gamma - 1)\frac{u}{U_\infty} \qquad (10.3.15)$$

where the second order terms are neglected. If we approximate the last term by $M^4 \approx M^2$, Eq. (10.3.15) becomes

$$1 - M^2 - M^2(\gamma + 1)\frac{u}{U_\infty} \qquad (10.3.15a)$$

Thus Eq. (10.3.10) is reduced to the usual form of the transonic thin airfoil equation

$$\left[1 - M^2 - M^2(\gamma + 1)\frac{u}{U_\infty} \right] \frac{\partial^2 \phi}{\partial x^2} + \frac{\partial^2 \phi}{\partial y^2} = 0 \qquad (10.3.16)$$

The approximation $M^4 \approx M^2$ is valid if $M \approx 1$. In order to improve the accuracy of Eq. (10.3.16) and to make the result more consistent with experimental results, other powers such as $M^{3/2}$ or $M^{7/4}$ are also used instead of M^2 in the third term of Eq. (10.3.15a) by some investigators.[15–19] A more elaborate approach for deriving Eq. (10.3.16) is seen in Reference 16. If $M^2 \ll 1$ or $M^2 \gg 1$, the nonlinear term u/U_∞ in Eq. (10.3.16) becomes negligible; hence Eq. (10.3.16) becomes the thin airfoil perturbation equation for subsonic and supersonic flow.

The boundary condition at the airfoil surface is that the velocity component normal to the surface is zero, namely

$$U_{\infty n} + \frac{\partial \phi}{\partial n} = 0 \qquad (10.3.17)$$

where $U_{\infty n}$ is the component of U_∞ normal to the surface and $\partial/\partial n$ is the derivative outward normal to the surface. With the angle θ between the surface and the x coordinate (see Fig. 10.1), $U_{\infty n}$ may be written as

$$U_{\infty n} = \mp U_\infty \sin \theta \qquad (10.3.18)$$

where the minus and plus signs refer to the upper and lower surfaces, respectively. Suppose the shape of the surface between the leading edge to the trailing edge is given by a smooth function $f(x)$. Assuming the airfoil is very thin, $\sin \theta$ is approximated by

$$\sin \theta \approx \tan \theta = f'(x) \qquad (10.3.19)$$

where f' is the first derivative. Therefore, Eq. (10.3.17) becomes

$$\frac{\partial \phi}{\partial n} = \pm U_\infty f'(x) \qquad (10.3.20)$$

Since we consider only very thin airfoils, Eq. (10.3.20) may be approximated by

$$\frac{\partial \phi}{\partial y} = \pm U_\infty f'(x) \qquad (10.3.20a)$$

If, for example, $f(x)$ is given by

$$f(x) = \pm \tau \left(0.5l - \frac{2x^2}{l} \right) \qquad (10.3.21)$$

where l is the chord length and τ is the ratio of the maximum airfoil thickness to the chord length, then the airfoil surface boundary condition becomes

$$\frac{\partial \phi}{\partial y} = \mp 4\tau x \frac{U_\infty}{l} \qquad (10.3.22)$$

In solving the thin airfoil perturbation equation, Eq. (10.3.20a) or Eq. (10.3.22) is applied on the horizontal line, neglecting the thickness of the airfoil, rather than considering the actual shape of the surface.

The external boundary conditions is $u = v = 0$ at infinity. Practically, the numerical solution of Eq. (10.3.16) should be found in a finite domain. There are three approaches to the external boundary condition at a finite distance, as follows:

a. The first approach is to set at a finite distance

$$u = v = 0 \qquad (10.3.23)$$

According to Lomax and Martin,[28] the error due to this crude approximation applied at half a chord length from the airfoil is significant at the external boundary but is small at the airfoil.

b. The second approach is to use the solution of the linearized subsonic flow equation for a doublet. In the far field, the transonic flow around any nonlifting airfoil can be viewed as the doublet superposed on a uniform flow, and the perturbed flow approaches the solution of the linearized equation asymptotically,

$$\beta\phi_{xx} + \phi_{yy} = 0 \qquad (10.3.24)$$

The solution of Eq. (10.3.24) for a doublet can be obtained easily by using the Prandtl–Glauert transformation[16] and is given by

$$\phi(x, y) = \frac{\mu}{2\pi} \frac{x}{(x^2 + \beta^2 y^2)} \qquad (10.3.25)$$

where μ is the intensity of the doublet. The value of μ is not prescribed, but is given in terms of $u = \phi_x$ as[18]

$$\mu = 2 \int_{-l/2}^{l/2} f(x)\, dx + \frac{\gamma+1}{2U_\infty} \int\int_{-\infty}^{\infty} [u(x, y)]^2\, dx\, dy \qquad (10.3.26)$$

The integral in the second term may be approximated by

$$\int\int_{\text{finite}} [u(x, y)]^2\, dx\, dy \qquad (10.3.27)$$

where integrals are extended only over the finite domain. The error involved in this approximation is small since u^2 decays rapidly as the distance from the airfoil increases. In applying Eq. (10.3.25) at a finite distance an iterative procedure is necessary since μ needs the solution of Eq. (10.3.16). Equation (10.3.25) is suitable if Eq. (10.3.16) is solved by a relaxation method, because Eq. (10.3.25) can be evaluated after every few iterations, thus revising the boundary conditions as the iteration proceeds.

c. The third approach is to use the exact solution, if available, for the linearized subsonic flow equation. In the case of the parabolic-arc biconvex airfoil, the exact solution of Eq. (10.3.24) is given in terms of $z = x + i\beta y$ as[23]

$$\beta u - iv = \frac{4}{\pi} U_\infty \tau \left[1 - z \ln \frac{z + 0.5l}{z - 0.5l} \right] \qquad (10.3.28)$$

10.4 ITERATIVE SOLUTION OF THE TRANSONIC THIN AIRFOIL PERTURBATION EQUATION

The solution of Eq. (10.3.16) has not been obtained in any analytical form because of difficulty with the nonlinear term. The solution was first successful by means of the finite difference approximation and an iterative technique based on SLR (the successive-line-relaxation method).† The SOR or SLR method cannot be applied in the classical sense as described in Chapter 3 for the following reasons. First, Eq. (10.3.16) is nonlinear, so the coefficients must be successively revised during the iteration. Second, Eq. (10.3.16) is elliptic in most of the domain but becomes hyperbolic in a few regions around the airfoil. Third, the elliptic region requires over-relaxation, while the hyperbolic region requires under-relaxation. In this section, the numerical method originally developed by Muman and Cole[18,19] is summarized.

If we consider a biconvex airfoil with zero angle of attack, the flow is symmetric about the chord line, so we consider the flow only above the chord line. The left and right external boundaries are set at finite distances from the edge of the air foil, typically, a half to a few chord lengths. The upper boundary is also set at a few chord lengths from the bottom. The mesh space is set as shown in Fig. 10.5. Notice that the line $y = 0$ is located half a mesh space below the bottom grid row $j = 1$. This is for later convenience in dealing with the bottom boundary conditions.

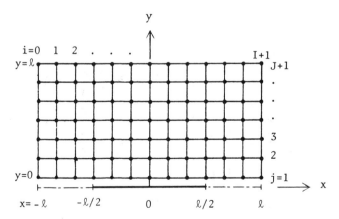

Figure 10.5 The mesh system for an iterative solution of the transonic equation.

† SLR refers to both the successive-line-over-relaxation (SLOR) and the successive-line-under-relaxation (SLUR). SLOR that uses the extrapolation parameter in the range, $1 < \omega < 2$, becomes SLUR if ω is set in the range, $0 < \omega < 1$.

We first rewrite Eq. (10.3.16) as

$$(1 - a - b\phi_x)\phi_{xx} + \phi_{yy} = 0 \qquad (10.4.1)$$

where

$$a = M^2$$

$$b = M^2 \frac{\gamma + 1}{U_\infty}$$

Equation (10.4.1) is elliptic if $a + b\phi_x < 1$ and hyperbolic if $a + b\phi_x > 1$, corresponding to subsonic and supersonic flows, respectively. The central difference scheme with respect to x derivatives is used for subsonic regions, while the backward difference scheme is used for supersonic flow. By doing this, upstream wave propagation is prevented reflecting the supersonic flow, and thus the iterative scheme is stabilized. With the central difference scheme for ϕ_x and ϕ_{xx}, the difference approximation for Eq. (10.4.1) at the grid (i, j) becomes

$$\left[1 - a - b\frac{\phi_{i+1,j} - \phi_{i-1,j}}{2\Delta x} \right] \frac{\phi_{i+1,j} - 2\phi_{i,j} + \phi_{i-1,j}}{\Delta x^2}$$

$$+ \frac{\phi_{i,j+1} - 2\phi_{i,j} + \phi_{i,j-1}}{\Delta y^2} = 0 \qquad (10.4.2)$$

With the backward difference scheme for ϕ_x and ϕ_{xx}, the difference approximation for the supersonic region is

$$\left[1 - a - b\frac{\phi_{i,j} - \phi_{i-2,j}}{2\Delta x} \right] \frac{\phi_{i,j} - 2\phi_{i-1,j} + \phi_{i-2,j}}{\Delta x^2}$$

$$+ \frac{\phi_{i,j+1} - 2\phi_{i,j} + \phi_{i,j-1}}{\Delta y^2} = 0 \qquad (10.4.3)$$

The grid points involved in the above equations are illustrated in Fig. 10.6.

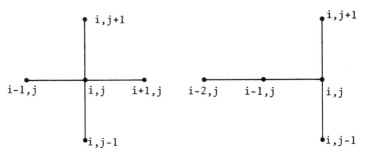

Subsonic region   Supersonic region

Figure 10.6   Difference schemes.

If the external boundary is at about a few chord lengths, the infinite boundary condition $u = v = 0$, or in terms of ϕ, the following difference formulas can be applied without a significant error:

$$\phi_{1,j} = \phi_{2,j} \quad \text{for all } j \quad \text{Left boundary}$$

$$\phi_{I,j} = \phi_{I-1,j} \quad \text{for all } j \quad \text{Right boundary}$$

$$\phi_{i,J} = \phi_{i,J-1} \quad \text{for all } i \quad \text{Top boundary}$$

If the external boundary is set closer to the airfoil, it is desirable to use approach b or c described in Section 10.3. If approach b is adopted, μ is recalculated after every few iterations. The second integral in Eq. (10.3.26) is extended only inside the external boundary since the contribution from outside is negligibly small.

The boundary condition on the bottom line is given by Eq. (10.3.20a) for the airfoil surface, or

$$v = (\phi_y)_{y=0} = 0 \tag{10.4.4}$$

for the nonairfoil bottom boundary in the nonlifting case. The difference approximation for ϕ_{yy} at the lowest grid row, namely, the second term in Eqs. (10.4.2) and (10.4.3) with $j = 1$ is replaced by

$$(\phi_{yy})_{i,1} = \frac{1}{\Delta y}\left[\frac{\phi_{i,2} - \phi_{i,1}}{\Delta y} - (\phi_y)_{y=0}\right] \tag{10.4.5}$$

The second term in the bracket in Eq. (10.4.5) is evaluated by Eq. (10.3.20a) or Eq. (10.4.4).

The selection of Eq. (10.4.2) or Eq. (10.4.3) is determined by checking the sign of the values in the bracket of Eq. (10.4.2). Several approaches to an iterative solution may be developed. In the scheme used by Muman and Cole, two loops of iterations were employed. In the outer loop of iteration, the column line inversion for Eqs. (10.4.2) or (10.4.3) or a combination of both is performed for all the columns of grid points, starting each sweep from the upwind column and completing at the downwind column. The inner iteration loop applies to each column if Eq. (10.4.3) is involved. This loop is necessary because the value of the bracketed term in Eq. (10.4.3) must be recalculated through an iterative procedure for each column. In the first column inversion for a grid column j, the bracketed term of Eq. (10.4.2) is calculated by using the initial guess, and if the sign is negative Eq. (10.4.3) is used for that grid point. The column inversion is to solve a tridiagonal matrix equation and, reduced to using Eqs. (1.7.17) and (1.7.16). The column inversion for the column is repeated until the coefficients of the difference equation converge. When the inner iteration for a line converges, the extrapolation with the SLR iteration parameter ω is applied. If Eq. (10.4.3) is not involved in a grid column, there is no need of inner iteration for that column. The value of ω must be set to $0.9 \sim 1.0$ for the supersonic region and $1 < \omega < 2$ for the subsonic region.

The convergence of the described scheme has not been mathematically proven, but it converges in practice. In a typical case, it took 400 iterations for the outer loop that corresponds to SLR. The inner loop of iteration of the scheme may be eliminated by using the previous iteratives in the brackets in Eqs. (10.4.2) and (10.4.3). This saves overall computing time.[29]

The test previously mentioned to determine the elliptic region and the hyperbolic region was dependent on the bracketed term of Eq. (10.4.2). However, this can lead to the following two inconsistent cases. Case 1: While Eq. (10.4.2) becomes hyperbolic, Eq. (10.4.3) becomes elliptic. Case 2: Both Eqs. (10.4.2) and (10.4.3) are stable because the former remains elliptic while the latter is hyperbolic simultaneously. Muman[19] suggested dealing with Case 1 as the parabolic region and Case 2 as the shock region. The revised test and scheme are used in Section 10.5.

South and Brandt[32] have shown that the computational time can be substantially reduced by using the multigrid method. The transient transonic equation has also been successfully solved by using the alternative direction implicit method.[33]

10.5 SEMIITERATIVE SOLUTION OF THE TRANSONIC EQUATION

As an alternative to the relaxation method for the transonic thin airfoil perturbation equation, FDS has been applied in an attempt to reduce computing time. The method described in this section was developed by Martin and Lomax,[23] in which the linear part of the equation is solved by FDS while the nonlinear part is iteratively solved by the Atkin–Shanks extrapolation formula.[25,26] Since the details of FDS for this particular application is introduced in Section 10.8, we proceed assuming that FDS is available. The present method is considerably faster than relaxation methods, although without further improvement the convergence becomes difficult as the Mach number approaches unity and consequently the size of the supersonic region increases.

A thin symmetrical parabolic-arc biconvex airfoil immersed in a uniform subsonic free stream, which is the same problem as in Section 10.4, is considered here. In terms of the perturbed velocity components, the perturbation equation becomes

$$Au_x + v_y = 0 \qquad (10.5.1a)$$

$$u_y - v_x = 0 \qquad (10.5.1b)$$

where

$$A = 1 - a - bu \qquad (10.5.2)$$

and where Eq. (10.5.1a) is equivalent to Eq. (10.4.1) and Eq. (10.5.1b) to Eq. (10.3.8). Equation (10.5.1) is elliptic if $A > 0$ and hyperbolic if $A < 0$. Since the nonlifting case is considered, the boundary conditions are

1. $u = v = 0$ at infinite distance

2. $v(x, 0) = -4\tau x \dfrac{U_\infty}{l}$ for $-\dfrac{l}{2} < x < \dfrac{l}{2}$

    $= 0 \qquad$ for $x < -\dfrac{l}{2}$ and $\dfrac{l}{2} < x$

where the origin of the coordinate system is set at the center of the airfoil. The first condition is replaced by the approximate condition of Eq. (10.3.28) applied at a finite distance.

FINITE DIFFERENCE FORMULAS

The staggered grid system as shown in Fig. 10.7 is used to formulate the difference approximation for Eq. (10.5.1), whereas the difference equation for Eq. (10.5.1a) is written for the solid box, and that for Eq. (10.5.1b) is

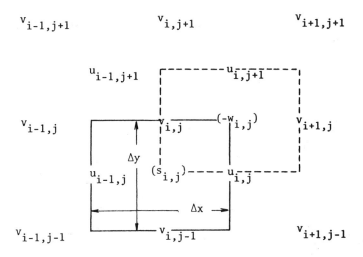

Figure 10.7   Mesh boxes with staggered grid points.

written for the dashed box. The central difference formulas for u_x and v_y in reference to the center of the solid box in Fig. 10.7 are defined by

$$u_{xc} \equiv \frac{u_{i,j} - u_{i-1,j}}{\Delta x} \tag{10.5.3}$$

$$v_{yc} \equiv \frac{v_{i,j} - v_{i,j-1}}{\Delta y} \tag{10.5.4}$$

which are used in the elliptic (subsonic) region. Equation (10.5.3) is replaced in the hyperbolic (supersonic) region by the backward difference formula defined as

$$u_{xb} \equiv \frac{u_{i-1,j} - u_{i-2,j}}{\Delta x} \tag{10.5.5}$$

The central difference formulas for u_y and v_x with respect to the center of the dashed box are defined by

$$u_{yc} \equiv \frac{u_{i,j+1} - u_{i,j}}{\Delta y} \tag{10.5.6}$$

$$v_{xc} \equiv \frac{v_{i+1,j} - v_{i,j}}{\Delta x} \tag{10.5.7}$$

In reference to the solid box in Fig. 10.7, the difference approximation for Eq. (10.5.1) in the elliptic region becomes

$$A_c u_{xc} + v_{yc} = 0 \tag{10.5.8}$$

where

$$A_c = 1 - b - a\frac{u_{i,j} + u_{i-1,j}}{2} \tag{10.5.9}$$

For the hyperbolic region, the corresponding equation is

$$A_b u_{xb} + v_{yc} = 0 \tag{10.5.10}$$

where

$$A_b = 1 - b - a\frac{u_{i-1,j} + u_{i-2,j}}{2} \tag{10.5.11}$$

In Eq. (10.5.11), the "backward" average of u is used consistently with the backward difference formula for u_{xb}. The central difference formulas are used for Eq. (10.5.1b) regardless of the type of region. In reference to the dashed box in Fig. 10.7, the difference approximation for Eq. (10.5.1b) becomes

$$u_{yc} - v_{xc} = 0 \tag{10.5.12}$$

The type of the region of Eq. (10.5.1) is decided by the sign of A in Eq. (10.5.1a). However, when Eq. (10.5.1a) is approximated by Eq. (10.5.8) or Eq. (10.5.10), the signs of A_c and A_b are not always consistent in the vicinity of a boundary between the two types of regions. Muman initially used only A_c for this· test, as described in Section 10.4, but later refined the test as follows:[19]

$$\text{Elliptic if } A_c > 0 \text{ and } A_b > 0$$

$$\text{Hyperbolic if } A_c < 0 \text{ and } A_b < 0$$

The regions not satisfying these tests are further divided into two domains:

$$\text{Parabolic if } A_c < 0 \text{ but } A_b > 0$$

$$\text{Shock if } A_c > 0 \text{ but } A_b < 0$$

For the parabolic region A is set to zero, and the difference equation becomes

$$v_{yc} = 0 \tag{10.5.13}$$

For the shock region the difference equation is deviced to satisfy the shock-jump relations:

$$A_c u_{xc} + A_b u_{xb} + v_{yc} = 0 \tag{10.5.14}$$

In applying FDS to the present problem, the difference equations for any region must be written in the form

$$u_{xc} + v_{yc} = s_{i,j}$$
$$u_{yc} - v_{xc} = 0 \tag{10.5.15}$$

where $s_{i,j}$ is the value of s at the center of the solid-mesh box. Equations (10.5.8), (10.5.10), (10.5.13), and (10.5.14) together with Eq. (10.5.12) are cast in the form of Eq. (15.5.15) by defining s as

ELLIPTIC

$$s_{i,j} = (b + au_c)u_{xc} \tag{10.5.16}$$

HYPERBOLIC

$$s_{i,j} = u_{xc} - (1 - b - au_b)u_{xb} \tag{10.5.17}$$

PARABOLIC

$$s_{i,j} = u_{xc} \tag{10.5.18}$$

SHOCK

$$s_{i,j} = (b + au_c)u_{xc} - (1 - b - au_b)u_{xb} \tag{10.5.19}$$

SUCCESSIVE RELAXATION METHOD

The FDS method solves Eq. (10.5.15) whenever the right side is given. A successive relaxation scheme that successively improves the right side may be written as

$$\mathbf{L}\hat{\mathbf{U}}^{(t)} = \mathbf{S}(\mathbf{U}^{(t-1)})$$
$$\mathbf{U}^{(t)} = \omega\,\hat{\mathbf{U}}^{(t)} + (1-\omega)\mathbf{U}^{(t-1)}$$

(10.5.20)

where $\mathbf{U}$ represents u and v, $\mathbf{L}$ represents the difference operators in Eq. (10.5.15), $\mathbf{S}$ represents the right sides of Eq. (10.5.15), ω is a relaxation parameter $(1 < \omega < 2)$, and t is the number of iterations. It is, however, more advantageous to use the Atkin–Shanks extrapolation until the iterative solution converges reasonably well, and then to switch to the above relaxation method.

ATKIN–SHANKS ALGORITHM

Suppose z_1, z_2, and z_3 are three successive approximations to the correct answer a. If the sequence $\{z_n\}$ is subject to an exponential transient in n in the sense

$$z_n = a + br^n$$

(10.5.21)

where a, b, and r are constant and $|r| < 1$, then any three consecutive values of z_n determine a, b, and r. It is easy to prove that a is given by

$$a = \frac{z_2^2 - z_1 z_3}{2z_2 - z_1 - z_3}$$

(10.5.22)

Equation (10.5.22) is designated as the Atkin–Shanks formula and may be used to estimate the converged value of a sequence that is similar to Eq. (10.5.21). The Atkin–Shanks formula can be also used to accelerate convergence of a function or a vector. Improvement of successive approximation by the Atkins–Shanks formula is the best if the three consecutive approximations are geometric or near geometric. A sequence of approximations is nearly geometric if it is obtained from a power series expansion. In applying the Atkins–Shanks formula, the following steps are repeated until the approximation becomes close to convergence:

Step 1. Prepare an initial guess for the solution namely, $\mathbf{u}^{(0)}$, $\mathbf{v}^{(0)}$, where $\mathbf{u}^{(0)}$ and $\mathbf{v}^{(0)}$ are vectors representing the velocity components of all the grid points.

Step 2. Generate the next three successive approximations, namely, $\mathbf{u}^{(1)}$, $\mathbf{v}^{(1)}$, $\mathbf{u}^{(2)}$, $\mathbf{v}^{(2)}$, and $\mathbf{u}^{(3)}$, $\mathbf{v}^{(3)}$.

Step 3. Apply the Atkin–Shanks formula to each element of **u** and **v** to obtain the estimate.

The result of Step 3 is used as the initial guess in the next cycle of application. The detail of generating the three successive approximations is discussed next.

We extend Eq. (10.5.15) in the compact form to

$$\mathbf{LU} = (1 - \epsilon)\mathbf{S}(\mathbf{U}^{(0)}) + \epsilon\mathbf{S}(\mathbf{U}) \qquad (10.5.23)$$

where the notations are explained after Eq. (10.5.20) except that ϵ is a parameter and $\mathbf{U}^{(0)}$ is an arbitrary initial guess for **U**. Here we express the boundary conditions stated after Eq. (10.5.2) in a compact form as

$$\mathbf{BU} = \mathbf{G} \quad \text{on} \quad \Gamma \qquad (10.5.24)$$

where **B** represents an operator for **U** along the boundary, **G** represents the inhomogeneous terms, and Γ represents the boundaries. The solution of Eq. (10.5.23) is dependent on the value of ϵ, but it becomes identical to the solution of Eq. (10.5.15) when $\epsilon = 1$. Since the solution of Eq. (10.5.23) is a function of ϵ, we expand **U** and **S(U)** in the power of ϵ as

$$\mathbf{U} = \tilde{\mathbf{U}}^{(1)} + \epsilon\tilde{\mathbf{U}}^{(2)} + \epsilon^2\tilde{\mathbf{U}}^{(3)} + \cdots \qquad (10.5.25)$$

$$\mathbf{S}(\mathbf{U}) = \mathbf{S}^{(1)} + \epsilon\mathbf{S}^{(2)} + \epsilon^2\mathbf{S}^{(3)} + \cdots \qquad (10.5.26)$$

By introducing these two expansions into Eq. (10.5.23) and equating the terms having the same power of ϵ, we obtain the series of equations,

$$\mathbf{L}\tilde{\mathbf{U}}^{(1)} = \mathbf{S}(\mathbf{U}^{(0)}) \qquad (10.5.27)$$

$$\mathbf{L}\tilde{\mathbf{U}}^{(2)} = -\mathbf{S}(\mathbf{U}^{(0)}) + \mathbf{S}^{(1)} \qquad (10.5.28)$$

$$\mathbf{L}\tilde{\mathbf{U}}^{(3)} = \mathbf{S}^{(2)} \qquad (10.5.29)$$

The boundary conditions for $\tilde{\mathbf{U}}^{(n)}$, $n = 1, 2, 3$, are

$$\mathbf{B}\tilde{\mathbf{U}}^{(1)} = \mathbf{G}$$

and $\qquad\qquad\qquad\qquad\qquad\qquad\qquad\qquad\qquad (10.5.30)$

$$\mathbf{B}\tilde{\mathbf{U}}^{(2)} = \mathbf{B}\tilde{\mathbf{U}}^{(3)} = \mathbf{0}$$

respectively. By setting $\epsilon = 1$ in Eq. (10.5.25), the solution of Eqs. (10.5.27) through (10.5.29) provides three successive approximations as

$$\mathbf{U}^{(1)} = \tilde{\mathbf{U}}^{(1)} \qquad (10.5.31)$$

$$\mathbf{U}^{(2)} = \tilde{\mathbf{U}}^{(1)} + \epsilon \tilde{\mathbf{U}}^{(2)} \qquad (10.5.32)$$

$$\mathbf{U}^{(3)} = \tilde{\mathbf{U}}^{(1)} + \epsilon \tilde{\mathbf{U}}^{(2)} + \epsilon^2 \tilde{\mathbf{U}}^{(3)} \qquad (10.5.33)$$

Notice that the errors of $\mathbf{U}^{(1)}$, $\mathbf{U}^{(2)}$, and $\mathbf{U}^{(3)}$ are in the order of ϵ, ϵ^2, and ϵ^3, respectively, thus forming a nearly geometric series.

Equation (10.5.22) is applied to each element of Eqs. (10.5.31) through (10.5.33). If there is no initial guess, $\mathbf{U}^{(0)}$ in Eq. (10.5.26) may be set to a null vector. The result of the extrapolation with Eq. (10.5.22) is used as $\mathbf{U}^{(0)}$ in the second cycle of the application of the above procedure. However, as the three successive approximations become too close to convergence, the denominator of Eq. (10.5.22) will lead to a significant round-off error. So, the scheme should be switched to the relaxation scheme.

FINAL FORM OF FINITE DIFFERENCE EQUATIONS

Equation (10.5.23) in the extended form of Eq. (10.5.15) may be explicitly written as

$$u_{xc} + v_{yc} = (1 - \epsilon)s_{i,j}(u^{(0)}) + \epsilon s_{i,j}(u)$$
$$u_{yc} - v_{xc} = 0 \qquad (10.5.34)$$

where $s_{i,j}$ are given by Eqs. (10.5.16) through (10.5.19) for various regions. In accordance with Eq. (10.5.25), both u and v are expanded into the power series in ϵ as

$$u = u^{(1)} + \epsilon(u^{(2)} - u^{(1)}) + \epsilon^2(u^{(3)} - u^{(2)}) + \cdots$$
$$v = v^{(1)} + \epsilon(v^{(2)} - v^{(1)}) + \epsilon^2(v^{(3)} - v^{(2)}) + \cdots \qquad (10.5.35)$$

where $u^{(n)}$ and $v^{(n)}$ are successive approximations. The equation for each successive approximation for the Atkin–Shanks extrapolation is derived by introducing these into Eq. (10.5.34) and collecting the terms of the same order. They may be expressed in a general form as

$$u_{xc}^{(n)} + v_{yc}^{(n)} = s^{(n-1)}$$
$$u_{yc}^{(n)} - v_{xc}^{(n)} = 0$$

for $n = 1, 2$ and 3 where $s^{(n-1)}$ are given for $n = 1$ and 2 by

ELLIPTIC

$$s^{(n-1)} = bu_{xc}^{(n-1)} + au_c^{(n-1)}u_{xc}^{(n-1)}$$

HYPERBOLIC

$$s^{(n-1)} = (b-1)u_{xb}^{(n-1)} + u_{xc}^{(n-1)} + au_b^{(n-1)}u_{xb}^{(n-1)}$$

PARABOLIC

$$s^{(n-1)} = u_{xc}^{(n-1)}$$

SHOCK

$$s^{(n-1)} = (b-1)u_{xb}^{(n-1)} + bu_{xc}^{(n-1)} + a[u_c^{(n-1)}u_{xc}^{(n-1)} + u_{xb}^{(n-1)}]$$

and for $n = 3$ by

ELLIPTIC

$$s^{(2)} = bu_{xc}^{(2)} + a[u_c^{(2)}u_{xc}^{(1)} + u_c^{(1)}(u_{xc}^{(2)} - u_{xc}^{(1)})]$$

HYPERBOLIC

$$s^{(2)} = (b-1)u_{xb}^{(2)} + u_{xc}^{(2)} + a[u_b^{(2)}u_{xb}^{(1)} + u_b^{(1)}(u_{xb}^{(2)} - u_{xb}^{(1)})]$$

PARABOLIC

$$s^{(2)} = u_{xc}^{(2)}$$

SHOCK

$$s^{(2)} = (b-1)u_{xb}^{(2)} + bu_{xc}^{(2)} + a[u_c^{(2)}u_{xc}^{(1)} + u_c^{(1)}(u_{xc}^{(2)} - u_{xc}^{(1)})]$$
$$+ a[u_b^{(2)}u_{xb}^{(1)} + u_b^{(1)}(u_{xb}^{(2)} - u_{xb}^{(1)})]$$

The boundary conditions stated after Eq. (10.5.2) are applied to each of the successive approximations.

10.6 ITERATIVE SOLUTION OF EXACT TRANSONIC EQUATION BY SUCCESSIVE RELAXATION

The iterative solution for the transonic thin foil perturbation equation as discussed in Section 10.4 has been further extended to the exact flow potential equation, Eq. (10.3.9), and applied to blunt-nosed, lifting airfoils in a transonic flow. This section introduces the method developed by Steger and Lomax.[20] The method described here is not restricted by the assumption of thin airfoil theory and applies to almost any shapes of airfoils in inviscid compressible flow.

In order to deal with cambered and blunt-nosed airfoils, a curvilinear orthogonal coordinate system and the corresponding grid system as shown in Fig. 10.8 are used. The thick line ABCDE in Fig. 10.8 is referred to as a control surface, which is close or in many cases exact to the nose of the

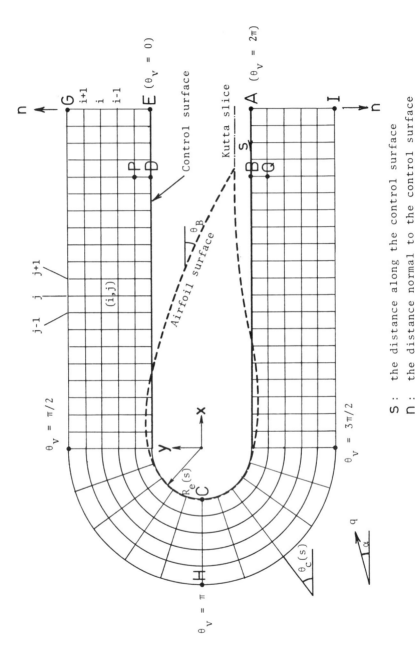

Figure 10.8 The mesh system based on a control surface.

S : the distance along the control surface
n : the distance normal to the control surface

airfoil, while the aft portion of the airfoil lies within the control surface. The flow equation is solved only outside the control surface with appropriate boundary conditions on the control surface. In Fig. 10.8 $\theta_c(s)$ is the angle that the s coordinate makes with the x coordinate of the Cartesian system, and $R_c(s)$ is the radius of curvature of a point s on the control surface. The angle of attack α is the angle between the flow direction at infinity and the x coordinate. The origin of the grid points (i, j) is placed at point A, and uniform mesh space Δs and Δn are used.

The basic equation on the Cartesian coordinate is given by Eq. (10.3.9). On the s–n coordinate system, the basic equation is written as

$$
\frac{c^2 U^2}{H}\left[\frac{1}{H}\phi_{ss} + \left(\frac{1}{H}\right)_s \phi_s\right] + (c^2 - V^2)\phi_{nn} - \frac{2UV}{H}\phi_{sn}
$$
$$
- \frac{1}{HR_c}\left[(c^2 - U^2)\phi_n + \frac{2UV}{H}\phi_s\right] = 0
\tag{10.6.1}
$$

where

$$
c^2 = c_\infty^2 - \left(\frac{\gamma - 1}{2}\right)(U^2 + V^2 - q^2)
\tag{10.6.2}
$$

$$
U(s, n) = q \cos(\theta_c - \alpha) + \frac{1}{H}\phi_s
\tag{10.6.3}
$$

$$
V(s, n) = -q \sin(\theta_c - \alpha) + \phi_n
\tag{10.6.4}
$$

$$
H(s, n) = \frac{R_c(s) + n}{R_c(s)}
\tag{10.6.5}
$$

In Eq. (10.6.1) through Eq. (10.6.5), subscripts s and n denote the partial derivatives, c is the local sonic velocity, c_∞ is the sonic velocity at infinity, U and V are the total velocity components in the s and n directions, respectively, $\phi(s, n)$ is the flow potential for the perturbed flow, and q is the flow velocity at infinity with angle α to the x direction.

The boundary conditions consist of the external boundary condition along EGHIA in Fig. 10.8, the inner boundary condition along ABCDE, and the Kutta condition at the trailing edge. The relation between the Kutta condition and the external boundary condition is essentially the same as discussed in Section 10.2. The external boundary condition for the flow potential in the lifting case ($\alpha \neq 0$) is approximated by the far field solution for a linearized subsonic flow with a vortex center in the airfoil and is given by

$$
\phi_\infty(x, y) = \frac{\Gamma}{2\pi} \tan^{-1} \frac{y\sqrt{1 - M^2}}{x}
\tag{10.6.6}
$$

where Γ is the circulation, M is the Mach number at infinity, and x and y are Cartesian coordinates with the origin located at a reference point in the airfoil as shown in Fig. 10.8. The values of

$$\theta_v = \tan^{-1} \frac{y\sqrt{1-M^2}}{x} \tag{10.6.7}$$

for typical points are shown in Fig. 10.8. Notice that θ_v jumps by 2π from A to E. The boundary condition along the airfoil surface is given by

$$V(n, s) = U(n, s) \tan [\theta_B(s) - \theta_c(s)] \tag{10.6.8}$$

where $\theta_B(s)$ is the angle that the actual surface of the airfoil makes with the x direction. Where the control surface coincides with the actual airfoil surface, the boundary condition is $V(n, s) = 0$. Equation (10.6.8) is evaluated on the control surface, rather than on the actual airfoil surface, just as in the case of thin airfoil perturbation theory. The Kutta condition implies that U becomes continuous across the Kutta slice.

The difference scheme depends on whether a grid point is in the subsonic flow or supersonic flow. If a point (i, j) is subsonic, the central difference formulas applied to the first and second order partial derivatives in Eq. (10.6.1) through Eq. (10.6.4) are

$$(\phi_{ss})_{i,j} = \frac{\phi_{i,j+1} - 2\phi_{i,j} + \phi_{i,j-1}}{(\Delta s)^2} \tag{10.6.9}$$

$$(\phi_s)_{i,j} = \frac{\phi_{i,j+1} - \phi_{i,j-1}}{2\Delta s} \tag{10.6.10}$$

$$(\phi_{sn})_{i,j} = \frac{\phi_{i+1,j+1} + \phi_{i-1,j-1} - \phi_{i+1,j-1} - \phi_{i-1,j+1}}{4\Delta s \, \Delta n} \tag{10.6.11}$$

$$(\phi_{nn})_{i,j} = \frac{\phi_{i+1,j} - 2\phi_{i,j} + \phi_{i-1,j}}{(\Delta n)^2} \tag{10.6.12}$$

$$(\phi_n)_{i,j} = \frac{\phi_{i+1,j} - \phi_{i-1,j}}{2\Delta n} \tag{10.6.13}$$

In the supersonic region Eqs. (10.6.12) and (10.6.13) are used for the n derivatives, but backward difference formulas are used for the s derivatives in the region above the airfoil, and forward difference formulas are used below the airfoil. For example, the backward difference formulas with second order accuracy are

$$(\phi_{ss})_{i,j} = \frac{2\phi_{i,j} - 5\phi_{i,j-1} + 4\phi_{i,j-2} - \phi_{i,j-3}}{(\Delta s)^2} \tag{10.6.14}$$

$$(\phi_s)_{i,j} = \frac{5\phi_{i,j-1} - 8\phi_{i,j-2} + 3\phi_{i,j-3}}{2\Delta s} \qquad (10.6.15)$$

$$(\phi_{sn})_{i,j} = \frac{1}{4\Delta n \Delta s}[3(\phi_{i+1,j} - \phi_{i-1,j})$$

$$-4(\phi_{i+1,j-1} - \phi_{i-1,j-1}) + (\phi_{i+1,j-2} - \phi_{i-1,j-2})] \qquad (10.6.16)$$

The forward difference formulas are similar but extended for the opposite direction.

The airfoil surface boundary condition, Eq. (10.6.8), may be written

$$\phi_n - q \sin(\theta_c - \alpha) = \left[\frac{1}{H}\phi_s + q \cos(\theta_c - \alpha)\right]\tan(\theta_B - \theta_c) \qquad (10.6.17)$$

where Eqs. (10.6.3) and (10.6.4) are used, and $H = 1$ on the control surface. The derivatives are replaced by the difference formulas,

$$(\phi_n)_{1,j} = \frac{-(3+\beta_j)\phi_{1,j} + (4+2\beta_j)\phi_{2,j} - (1+\beta_j)\phi_{3,j}}{2\Delta n} \qquad (10.6.18)$$

$$\left(\frac{\phi_s}{H}\right)_{1,j} = \frac{1}{4\Delta s}\left[(2+\beta_j)(\phi_{1,j+1} - \phi_{1,j-1}) - \frac{\beta_j}{H_{2,j}}(\phi_{2,j+1} - \phi_{2,j-1})\right] \qquad (10.6.19)$$

where $\beta_j \Delta n/2$ is the distance between the control surface and the airfoil surface at s.

Lomax and Steger used the SLR method with the line inversions along n directions to solve the difference equation for Eq. (10.6.1). The sweep in each iteration starts at a horizontal grid line at $\theta_v = \pi$ and proceeds in two directions downstream, one sweeping over the airfoil and another under the airfoil. The matrix for the inversion becomes a tridiagonal matrix regardless of the type of the region. The coefficients of the difference equations were calculated with the previous iteratives. An efficient value of the extrapolation parameter ω for SLR has been found to be close to but less than unity for the supersonic and slightly less than two in the subsonic region. Whether a grid point belongs to the subsonic region or supersonic region is determined by checking the value of the coefficient of the first term of Eq. (10.6.1). The circulation Γ is revised during the iterations on a trial-and-error basis. Steger and Lomax used an interactive cathode ray tube that displayed the value of ω, Γ, the maximum residue, and the pressure distributions as the iteration proceeded. The values of ω and Γ were interactively changed during the iterations. The final value of Γ is the value that makes the velocity continuous across the Kutta slice. However, the use of a cathode ray tube is not essential. Instead, for example, the time-sharing terminal may be used to alter the parameters during the iterative procedure. Even an algorithm that automatically determines these parameters could be developed. As future work, the development of a more efficient method of accelerating the convergence rate is desirable.

10.7 THE FAST DIRECT SOLUTION (FDS) METHODS FOR THE POISSON EQUATION

In computational fluid dynamics, the finite difference analog of the Poisson equation is solved frequently. In fact, the methods of analysis described in previous sections of this chapter all require solutions of the Poisson equation (or Cauchy–Riemann equation). The computational cost for such a calculation depends strongly on efficiency in solving the Poisson equation in the finite difference form. If the geometry or the boundary condition is complicated, the difference equation would have to be solved by an iterative approach as described in Chapter 3. However, when the uniformly spaced mesh system for a rectangular geometry with simple boundary conditions is considered, the direct solution methods,[7–12] much faster than the iterative or the Gaussian elimination methods, are available. Once a fast direct solution program is developed, it can be applied even to a nonrectangular geometry or to a problem with complicated boundary conditions. Among several fast direct methods, probably the cyclic reduction method, which was suggested by Golub, devised by Hockney,[11,12] and later improved by Buneman,[7,10] is most widely used. Lomax and Martin[28] extended the cyclic reduction to the Cauchy–Riemann equation. The objective of this section is to introduce Buneman's improved algorithm. The application to the Cauchy–Riemann equation is described in Section 10.8.

In applying the cyclic-reduction method, the rectangular geometry is divided into a uniformly spaced mesh system. Although the boundary conditions acceptable to the present method include the Dirichlet type, the Neumann type, and a cyclic boundary condition, we consider only the Dirichlet boundary condition (see Problem 10 for other types of boundary conditions). With I columns times J rows of uniformly spaced grids excluding those on the boundaries, the finite difference equations for the Poisson equation with the Dirichlet boundary conditions become a block-tridiagonal matrix as illustrated by Eq. (3.2.25). Because of uniform mesh spacing, the block-tridiagonal matrix becomes particularly simple. The nonnull off-diagonal blocks are all identity matrix, while the diagonal blocks are all identical tridiagonal matrix. So that the cyclic reduction method is efficient, it is important that the number of the blocks, namely, J, are equal to $2^l - 1$, where l is an integer. For simplicity, we assume $l = 3$ in the subsequent discussions.

With $J = 2^3 - 1 = 7$, the difference equation may be written as

$$
\begin{bmatrix}
\mathbf{A} & -\mathbf{I} & & & & & \\
-\mathbf{I} & \mathbf{A} & -\mathbf{I} & & & & \\
& -\mathbf{I} & \mathbf{A} & -\mathbf{I} & & & \\
& & -\mathbf{I} & \mathbf{A} & -\mathbf{I} & & \\
& & & -\mathbf{I} & \mathbf{A} & -\mathbf{I} & \\
& & & & -\mathbf{I} & \mathbf{A} & -\mathbf{I} \\
& & & & & -\mathbf{I} & \mathbf{A}
\end{bmatrix}
\begin{bmatrix}
\mathbf{x}_1 \\ \mathbf{x}_2 \\ \mathbf{x}_3 \\ \mathbf{x}_4 \\ \mathbf{x}_5 \\ \mathbf{x}_6 \\ \mathbf{x}_7
\end{bmatrix}
=
\begin{bmatrix}
\mathbf{y}_1 \\ \mathbf{y}_2 \\ \mathbf{y}_3 \\ \mathbf{y}_4 \\ \mathbf{y}_5 \\ \mathbf{y}_6 \\ \mathbf{y}_7
\end{bmatrix}
\tag{10.7.1}
$$

where $\mathbf{A}$ is a tridiagonal matrix of order I, $\mathbf{I}$ is an identity matrix of the same order, $\mathbf{x}_j$ represents the solution along the jth grid row, and $\mathbf{y}_j$ represents the inhomogeneous terms for the jth grid row. As a step to reduce Eq. (10.7.1) to a lower order, we multiply the even rows by $\mathbf{A}$:

$$
\begin{bmatrix}
\mathbf{A} & -\mathbf{I} & & & & & \\
-\mathbf{A} & \mathbf{A}^2 & -\mathbf{A} & & & & \\
& -\mathbf{I} & \mathbf{A} & -\mathbf{I} & & & \\
& & -\mathbf{A} & \mathbf{A}^2 & -\mathbf{A} & & \\
& & & -\mathbf{I} & \mathbf{A} & -\mathbf{I} & \\
& & & & -\mathbf{A} & \mathbf{A}^2 & -\mathbf{A} \\
& & & & & -\mathbf{I} & \mathbf{A}
\end{bmatrix}
\begin{bmatrix}
\mathbf{x}_1 \\ \mathbf{x}_2 \\ \mathbf{x}_3 \\ \mathbf{x}_4 \\ \mathbf{x}_5 \\ \mathbf{x}_6 \\ \mathbf{x}_7
\end{bmatrix}
=
\begin{bmatrix}
\mathbf{y}_1 \\ \mathbf{A}\mathbf{y}_2 \\ \mathbf{y}_3 \\ \mathbf{A}\mathbf{y}_4 \\ \mathbf{y}_5 \\ \mathbf{A}\mathbf{y}_6 \\ \mathbf{y}_7
\end{bmatrix}
\tag{10.7.2}
$$

By adding the two adjacent rows to each even row, Eq. (10.7.2) becomes

$$
\begin{bmatrix}
\mathbf{A} & -\mathbf{I} & & & & & \\
\mathbf{0} & \mathbf{A}^{(2)} & \mathbf{0} & -\mathbf{I} & & & \\
& -\mathbf{I} & \mathbf{A} & -\mathbf{I} & & & \\
& -\mathbf{I} & \mathbf{0} & \mathbf{A}^{(2)} & \mathbf{0} & -\mathbf{I} & \\
& & & -\mathbf{I} & \mathbf{A} & -\mathbf{I} & \\
& & & -\mathbf{I} & \mathbf{0} & \mathbf{A}^{(2)} & \mathbf{0} \\
& & & & & -\mathbf{I} & \mathbf{A}
\end{bmatrix}
\begin{bmatrix}
\mathbf{x}_1 \\ \mathbf{x}_2 \\ \mathbf{x}_3 \\ \mathbf{x}_4 \\ \mathbf{x}_5 \\ \mathbf{x}_6 \\ \mathbf{x}_7
\end{bmatrix}
=
\begin{bmatrix}
\mathbf{y}_1 \\ \mathbf{y}_2^{(2)} \\ \mathbf{y}_3 \\ \mathbf{y}_4^{(2)} \\ \mathbf{y}_5 \\ \mathbf{y}_6^{(2)} \\ \mathbf{y}_7
\end{bmatrix}
\tag{10.7.3}
$$

where

$$\mathbf{A}^{(2)} = \mathbf{A}^2 - 2\mathbf{I} \tag{10.7.4}$$

$$\mathbf{y}_j^{(2)} = \mathbf{A}\mathbf{y}_j + \mathbf{y}_{j-1} + \mathbf{y}_{j+1}, \qquad j = 2, 4, 6 \tag{10.7.5}$$

Since the even rows of Eq. (10.7.3) are independent of the odd rows, they may be separated as

$$\begin{bmatrix} \mathbf{A}^{(2)} & -\mathbf{I} & \\ -\mathbf{I} & \mathbf{A}^{(2)} & -\mathbf{I} \\ & -\mathbf{I} & \mathbf{A}^{(2)} \end{bmatrix} \begin{bmatrix} \mathbf{x}_2 \\ \mathbf{x}_4 \\ \mathbf{x}_6 \end{bmatrix} = \begin{bmatrix} \mathbf{y}_2^{(2)} \\ \mathbf{y}_4^{(2)} \\ \mathbf{y}_6^{(2)} \end{bmatrix} \tag{10.7.6}$$

In terms of the solution of Eq. (10.7.6) the odd rows of Eq. (10.7.3) may be written as

$$\begin{aligned} \mathbf{A}\mathbf{x}_1 &= \mathbf{y}_1 + \mathbf{x}_2 \\ \mathbf{A}\mathbf{x}_3 &= \mathbf{y}_3 + \mathbf{x}_2 + \mathbf{x}_4 \\ \mathbf{A}\mathbf{x}_5 &= \mathbf{y}_5 + \mathbf{x}_4 + \mathbf{x}_6 \\ \mathbf{A}\mathbf{x}_7 &= \mathbf{y}_7 + \mathbf{x}_6 \end{aligned} \tag{10.7.7}$$

Although Eq. (10.7.6) is easier to solve than Eq. (10.7.1), it may be further reduced. Multiplying the second row of Eq. (10.7.6) by $\mathbf{A}^{(2)}$ and adding the first and third rows, Eq. (10.7.6) becomes

$$\begin{bmatrix} \mathbf{A}^{(2)} & -\mathbf{I} & \\ \mathbf{0} & \mathbf{A}^{(3)} & \mathbf{0} \\ & -\mathbf{I} & \mathbf{A}^{(2)} \end{bmatrix} \begin{bmatrix} \mathbf{x}_2 \\ \mathbf{x}_4 \\ \mathbf{x}_6 \end{bmatrix} = \begin{bmatrix} \mathbf{y}_2^{(2)} \\ \mathbf{y}_4^{(3)} \\ \mathbf{y}_6^{(2)} \end{bmatrix} \tag{10.7.8}$$

where

$$\mathbf{A}^{(3)} = (\mathbf{A}^{(2)})^2 - 2\mathbf{I} \tag{10.7.9}$$

$$\mathbf{y}_4^{(3)} = \mathbf{A}^{(2)}\mathbf{y}_4^{(2)} + \mathbf{y}_2^{(2)} + \mathbf{y}_6^{(2)} \tag{10.7.10}$$

The second row of Eq. (10.7.8) may be separated as

$$\mathbf{A}^{(3)}\mathbf{x}_4 = \mathbf{y}_4^{(3)} \tag{10.7.11}$$

In terms of $\mathbf{x}_4$, the first and third rows of Eq. (10.7.8) may be written as

$$\begin{aligned} \mathbf{A}^{(2)}\mathbf{x}_2 &= \mathbf{y}_2^{(2)} + \mathbf{x}_4 \\ \mathbf{A}^{(2)}\mathbf{x}_6 &= \mathbf{y}_6^{(2)} + \mathbf{x}_4 \end{aligned} \tag{10.7.12}$$

Thus the solution of Eq. (10.7.1) can be obtained by solving the linear equations of order I in the sequence: Eq. (10.7.11), Eq. (10.7.12), and Eq.

(10.7.7). Equation (10.7.7) is easy to solve when the right sides are given, because $\mathbf{A}$ is tridiagonal [see Eqs. (1.7.15) and (1.7.16)]. $\mathbf{A}^{(3)}$ and $\mathbf{A}^{(2)}$ in Eqs. (10.7.11) and (10.7.12), respectively, are not tridiagonal, but the solutions to those equations may be obtained by applying the solution to the tridiagonal problem repeatedly as follows. In general, $\mathbf{A}^{(n)}$ satisfies the recursion relation

$$\mathbf{A}^{(1)} = \mathbf{A}$$
$$\mathbf{A}^{(2)} = (\mathbf{A}^{(1)})^2 - 2\mathbf{I}$$
$$\vdots \qquad\qquad (10.7.13)$$
$$\mathbf{A}^{(n)} = (\mathbf{A}^{(n-1)})^2 - 2\mathbf{I}$$

It is proven in Appendix V that $\mathbf{A}^{(n)}$ can be factorized as

$$\mathbf{A}^{(n)} = \prod_{k=1}^{2^{n-1}} \mathbf{G}_k^{(n)} \qquad (10.7.14)$$

where $\mathbf{G}_k^{(n)}$ is a tridiagonal matrix given by

$$\mathbf{G}_k^{(n)} = \mathbf{A} - \lambda_k^{(n)} \mathbf{I} \qquad (10.7.15)$$

$$\lambda_k^{(n)} = 2 \cos\left[(2k-1)\frac{\pi}{2^n}\right] \qquad (10.7.16)$$

Therefore, Eq. (10.7.11) is solved by applying the tridiagonal solution four times as

$$\mathbf{x}_4 = (\mathbf{G}_4^{(3)})^{-1}(\mathbf{G}_3^{(3)})^{-1}(\mathbf{G}_2^{(3)})^{-1}(\mathbf{G}_1^{(3)})^{-1}\mathbf{y}_4^{(3)} \qquad (10.7.17)$$

The algorithm thus described requires multiplication of matrices to calculate the inhomogeneous terms. This is not favorable because a repeated multiplication of matrices may yield very large numbers among the matrix elements that cause round-off errors (see Problem 10.10). Multiplication of matrices can be avoided by the following modifications devised by Buneman.

Suppose Eqs. (10.7.6) and (10.7.11) are written in the form

$$\begin{bmatrix} \mathbf{A}^{(2)} & -\mathbf{I} & \\ -\mathbf{I} & \mathbf{A}^{(2)} & -\mathbf{I} \\ & -\mathbf{I} & \mathbf{A}^{(2)} \end{bmatrix} \begin{bmatrix} \mathbf{x}_2 \\ \mathbf{x}_4 \\ \mathbf{x}_6 \end{bmatrix} = \begin{bmatrix} \mathbf{A}^{(2)} \ \mathbf{p}_2^{(2)} + \mathbf{q}_2^{(2)} \\ \mathbf{A}^{(2)} \ \mathbf{p}_4^{(2)} + \mathbf{q}_4^{(2)} \\ \mathbf{A}^{(2)} \ \mathbf{p}_6^{(2)} + \mathbf{q}_6^{(2)} \end{bmatrix} \qquad (10.7.18)$$

and

$$\mathbf{A}^{(3)}\mathbf{x}_4 = \mathbf{A}^{(3)}\mathbf{p}_4^{(3)} + \mathbf{q}_4^{(3)} \qquad (10.7.19)$$

respectively. If $\mathbf{p}_k^{(n)}$ and $\mathbf{q}_k^{(n)}$ could be calculated without performing any matrix multiplication, then no multiplication would be required in the whole process, because the multiplication implied by $\mathbf{A}^{(n)}\,\mathbf{p}_k^{(n)}$ is canceled by $\mathbf{A}^{(n)}$ on the left side. The vectors $\mathbf{p}_k^{(n)}$ and $\mathbf{q}_k^{(n)}$ can be determined as follows.

For $n = 2$, we first write

$$\mathbf{y}_j^{(2)} \equiv \mathbf{A}^{(2)}\mathbf{p}_j^{(2)} + \mathbf{q}_j^{(2)} = \mathbf{A}^2\mathbf{p}_j^{(2)} - 2\mathbf{p}_j^{(2)} + \mathbf{q}_j^{(2)} = \mathbf{A}\mathbf{q}_j^{(1)} + \mathbf{q}_{j-1}^{(1)} + \mathbf{q}_{j+1}^{(1)} \qquad (10.7.20)$$

where $\mathbf{q}_j^{(1)} = \mathbf{y}_j$, the second equation of Eq. (10.7.4) is used to obtain the second equality, and Eq. (10.7.5) is used for the third equality. We specify $\mathbf{p}_j^{(2)}$ by

$$\mathbf{A}^2\mathbf{p}_j^{(2)} = \mathbf{A}\mathbf{q}_j^{(1)} \qquad (10.7.21)$$

or equivalently

$$\mathbf{A}\mathbf{p}_j^{(2)} = \mathbf{q}_j^{(1)} \qquad (10.7.22)$$

then

$$\mathbf{q}_j^{(2)} = 2\mathbf{p}_j^{(2)} + \mathbf{q}_{j-1}^{(1)} + \mathbf{q}_{j+1}^{(1)} \qquad (10.7.23)$$

In Eq. (10.7.22), $\mathbf{p}_j^{(2)}$ is determined by solving a tridiagonal matrix equation as $\mathbf{p}_j^{(2)} = \mathbf{A}^{-1}\mathbf{q}_j^{(1)}$. The vectors $\mathbf{p}_j^{(2)}$ and $\mathbf{q}_j^{(2)}$ may be expressed in terms of $\mathbf{q}$s as

$$\mathbf{q}_j^{(2)} = 2\mathbf{A}^{-1}\mathbf{q}_j^{(1)} + \mathbf{q}_{j-1}^{(1)} + \mathbf{q}_{j+1}^{(1)} \qquad (10.7.24)$$

$$\mathbf{p}_j^{(2)} = \tfrac{1}{2}(\mathbf{q}_j^{(2)} - \mathbf{q}_{j-1}^{(1)} - \mathbf{q}_{j+1}^{(1)}) \qquad (10.7.25)$$

For $n = 3$, the vectors $\mathbf{p}_j^{(3)}$ and $\mathbf{q}_j^{(3)}$ can be calculated in a similar procedure. We write

$$\mathbf{y}_j^{(3)} \equiv \mathbf{A}^{(3)}\mathbf{p}_j^{(3)} + \mathbf{q}_j^{(3)} = ((\mathbf{A}^{(2)})^2 - 2\mathbf{I})\mathbf{p}_j^{(3)} + \mathbf{q}_j^{(3)}$$
$$= \mathbf{A}^{(2)}(\mathbf{A}^{(2)}\mathbf{p}_j^{(3)} + \mathbf{q}_j^{(3)}) + \mathbf{A}^{(2)}\mathbf{p}_{j-2}^{(2)} + \mathbf{q}_{j-2}^{(2)} + \mathbf{A}^{(2)}\mathbf{p}_{j+2}^{(2)} + \mathbf{q}_{j+2}^{(2)} \qquad (10.7.26)$$

where Eq. (10.7.9) is used to obtain the second equality and Eq. (10.7.10) is used to obtain the third equality. In order to specify $\mathbf{p}_j^{(3)}$ and $\mathbf{q}_j^{(3)}$, we look at both sides of the last equality of Eq. (10.7.26) and set

$$(\mathbf{A}^{(2)})^2\mathbf{p}_j^{(3)} = \mathbf{A}^{(2)}(\mathbf{A}^{(2)}\mathbf{p}_j^{(2)} + \mathbf{q}_j^{(2)} + \mathbf{p}_{j-2}^{(2)} + \mathbf{p}_{j+2}^{(2)}) \qquad (10.7.27)$$

$$\mathbf{q}_j^{(3)} = 2\mathbf{p}_j^{(3)} + \mathbf{q}_{j-2}^{(2)} + \mathbf{q}_{j+2}^{(2)} \qquad (10.7.28)$$

Premultiplying Eq. (10.7.27) by $(\mathbf{A}^{(2)})^{-2}$ yields

$$\mathbf{p}_j^{(3)} = \mathbf{p}_j^{(2)} + (\mathbf{A}^{(2)})^{-1}(\mathbf{q}_j^{(2)} + \mathbf{p}_{j-2}^{(2)} + \mathbf{p}_{j+2}^{(2)}) \qquad (10.7.29)$$

By introducing Eq. (10.7.29) into Eq. (10.7.28) and using Eq. (10.7.25), we have

$$\mathbf{q}_j^{(3)} = \mathbf{q}_{j-2}^{(2)} - \mathbf{q}_{j-1}^{(1)} + \mathbf{q}_j^{(2)} - \mathbf{q}_{j+1}^{(1)} + \mathbf{q}_{j+2}^{(2)}$$

$$+ (\mathbf{A}^{(2)})^{-1}(-\mathbf{q}_{j-3}^{(1)} + \mathbf{q}_{j-2}^{(2)} - \mathbf{q}_{j-1}^{(1)} + 2\mathbf{q}_j^{(2)} - \mathbf{q}_{j+1}^{(1)} + \mathbf{q}_{j+2}^{(2)} - \mathbf{q}_{j+3}^{(1)})$$

$$(10.7.30)$$

where $j = 4$ (or multiple of 4 in general). Equation (10.7.30) includes only $\mathbf{q}_j$s. The inversion $(\mathbf{A}^{(2)})^{-1}$ can be performed by applying the solution for a tridiagonal equation twice with the factorization of Eq. (10.7.14).

The cyclic reduction method is applicable to a block-tridiagonal matrix of order $2^l - 1$ with any l. The vectors $\mathbf{q}_j^{(n)}$ and $\mathbf{p}_j^{(n)}$ with $n > 3$ are given by

$$\mathbf{q}_j^{(n)} = \mathbf{q}_{j-2s}^{(n-1)} - \mathbf{q}_{j-s}^{(n-2)} + \mathbf{q}_j^{(n-1)} - \mathbf{q}_{j+s}^{(n-2)} + \mathbf{q}_{j+2s}^{(n-1)}$$

$$+ (\mathbf{A}^{(n-1)})^{-1}(-\mathbf{q}_{j-3s}^{(n-2)} + \mathbf{q}_{j-2s}^{(n-1)} - \mathbf{q}_{j-s}^{(n-2)} + 2\mathbf{q}_j^{(n-1)} \qquad (10.7.31)$$

$$- \mathbf{q}_{j+s}^{(n-2)} + \mathbf{q}_{j+2s}^{(n-1)} - \mathbf{q}_{j+3s}^{(n-2)})$$

$$\mathbf{p}_j^{(n)} = \tfrac{1}{2}(\mathbf{q}_j^{(n)} - \mathbf{q}_{j-m}^{(n-1)} - \mathbf{q}_{j+m}^{(n-1)}) \qquad (10.7.32)$$

where $s = 2^{n-3}$, j is a multiple of 2^{n-1}, $m = 2^{n-2}$, and those vectors with indices that are out of the range are set to null vectors. It should be noticed that $\mathbf{q}_j^{(n)}$ and $\mathbf{p}_j^{(n)}$ are all expressed in terms of $\mathbf{q}_j$s. Storing only $\mathbf{q}_j^{(n)}$ during the reduction process is necessary, and $\mathbf{p}_j$s are calculated by using Eq. (10.7.32) whenever needed. The inverse $(\mathbf{A}^{(n-1)})^{-1}$ is performed by applying the tridiagonal solution repeatedly with Eq. (10.7.14).

10.8 THE FAST DIRECT SOLUTION (FDS) METHOD FOR THE CAUCHY–RIEMANN EQUATION

An elliptic equation may be written in the form of coupled first order equations

$$\frac{\partial}{\partial x}u(x, y) + \frac{\partial}{\partial y}v(x, y) = s(x, y)$$

$$(10.8.1)$$

$$\frac{\partial}{\partial y}u(x, y) - \frac{\partial}{\partial x}v(x, y) = -w(x, y)$$

If $s = w = 0$, Eq. (10.8.1) is the Cauchy–Riemann equation. When s or w or both of them are not equal to zero, Eq. (10.8.1) may be called the inhomogeneous Cauchy–Riemann equation. The advantage of using the form of Eq. (10.8.1) in the fluid dynamics is that u and v directly represent velocity components without any transformation. The objective of this section is to describe the direct solution method for the finite difference approximation for Eq. (10.8.1) developed by Lomax and Martin.[28] The approach is based on the cyclic reduction method and is similar to the case for the Poisson equation as described in Section 10.7, so only the essential difference is discussed in the following. It is presumed that a uniform and rectangular geometry with a Dirichlet boundary condition is considered.

Although there are several ways of selecting the grid system, Lomax and Martin proposed using the staggered mesh system as shown in Fig. 10.7, which provides the central difference approximations for Eq. (10.8.1). The finite difference approximations for Eq. (10.8.1) become

$$\frac{u_{i,j} - u_{i-1,j}}{\Delta x} + \frac{v_{i,j} - v_{i,j-1}}{\Delta y} = s_{i,j} \tag{10.8.2}$$

$$\frac{u_{i,j+1} - u_{i,j}}{\Delta y} - \frac{v_{i+1,j} - v_{i,j}}{\Delta x} = -w_{i,j} \tag{10.8.3}$$

where Eq. (10.8.2) is written in reference to the solid box in Fig. 10.7, while Eq. (10.8.3) is in reference to the dotted box.

In discussing the direct solution for the set of Eqs. (10.8.2) and (10.8.3), let us first consider the example of an entire grid system as shown in Fig. 10.9. As the boundary conditions, the values of u are specified along the left and top corners, while the values of v are specified along the right and bottom corners. This combination of boundary conditions is not a unique selection but is a convenient one if the bottom corner is chosen as a chord line as in Section 10.5. Since the top grid row has a special array, Eq. (10.8.3) for $u_{i,j}$ at the top row is replaced by

$$\frac{u_{i,j+1} - u_{i,j}}{\Delta y/2} - \frac{v_{i+1,j} - v_{i,j}}{\Delta x} = -w_{i,j} \tag{10.8.4}$$

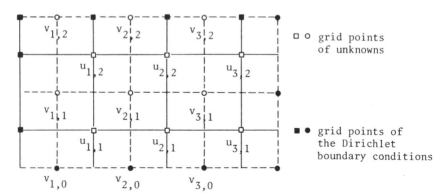

$\square$ $\circ$ grid points of unknowns

$\blacksquare$ $\bullet$ grid points of the Dirichlet boundary conditions

Figure 10.9 A system of staggered grid points.

The difference equations for Fig. 10.9 may be written in the matrix form as

$$
\begin{bmatrix}
\alpha & 0 & 0 & & & & 1 & & \\
-\alpha & \alpha & 0 & & 0 & & & 1 & & 0 \\
0 & -\alpha & \alpha & & & & & & 1 \\
& & & \alpha & 0 & 0 & -1 & & 1 \\
& 0 & & -\alpha & \alpha & 0 & & -1 & & 1 \\
& & & 0 & -\alpha & \alpha & & & -1 & & 1 \\
-1 & & & & 1 & & \alpha & -\alpha & 0 \\
& -1 & & & & 1 & & 0 & \alpha & -\alpha & & 0 \\
& & -1 & & & & 1 & 0 & 0 & \alpha \\
& & & -1 & & & & & & \alpha/2 & -\alpha/2 \\
& & & & -1 & & & 0 & & & \alpha/2 & -\alpha/2 \\
& & & & & -1 & & & & & & \alpha/2
\end{bmatrix}
\begin{bmatrix} u_{11} \\ u_{21} \\ u_{31} \\ u_{12} \\ u_{22} \\ u_{32} \\ v_{11} \\ v_{21} \\ v_{31} \\ v_{12} \\ v_{22} \\ v_{32} \end{bmatrix}
=
\begin{bmatrix} f_{11} \\ f_{21} \\ f_{31} \\ f_{12} \\ f_{22} \\ f_{32} \\ g_{11} \\ g_{21} \\ g_{31} \\ g_{12} \\ g_{22} \\ g_{32} \end{bmatrix}
$$

(10.8.5)

where $\alpha = \Delta y/\Delta x$, and f and g include $s\Delta y$ and $-\omega \Delta y$, respectively, together with the terms carrying the boundary conditions. We express Eq. (10.8.5) more compactly as

$$
\begin{bmatrix}
\mathbf{A} & 0 & \mathbf{I} & 0 \\
0 & \mathbf{A} & -\mathbf{I} & 2\mathbf{I} \\
-\mathbf{I} & \mathbf{I} & \mathbf{B} & 0 \\
0 & -\mathbf{I} & 0 & \mathbf{B}
\end{bmatrix}
\begin{bmatrix} \mathbf{u}_1 \\ \mathbf{u}_2 \\ \mathbf{v}_1 \\ \frac{1}{2}\mathbf{v}_2 \end{bmatrix}
=
\begin{bmatrix} \mathbf{f}_1 \\ \mathbf{f}_2 \\ \mathbf{g}_1 \\ \mathbf{g}_2 \end{bmatrix}
$$

(10.8.5a)

where $\mathbf{A}$ and $\mathbf{B}$ are 3×3 submatrices, $\mathbf{I}$ is the identity matrix, and $\mathbf{u}$, $\mathbf{v}$, $\mathbf{f}$, and $\mathbf{g}$ are subvectors that represent the blocks in the unknown vector and inhomogeneous terms in Eq. (10.8.5).

To reduce the order of Eq. (10.8.5a), we perform the following sequence of operations:

a. Subtract the second row from the first row.
b. Add $\mathbf{A}$ times the third row to the first row.
c. Add $\mathbf{A}$ times the last row to the second row.
d. Add the last row to the third row.

Then Eq. (10.8.5a) becomes

$$
\left[\begin{array}{cc|cc}
\mathbf{0} & \mathbf{0} & 2\mathbf{I}+\mathbf{AB} & -2\mathbf{I} \\
\mathbf{0} & \mathbf{0} & -\mathbf{I} & 2\mathbf{I}+\mathbf{AB} \\
\hline
-\mathbf{I} & \mathbf{0} & \mathbf{B} & \mathbf{B} \\
\mathbf{0} & -\mathbf{I} & \mathbf{0} & \mathbf{B}
\end{array}\right]
\left[\begin{array}{c}
\mathbf{u}_1 \\
\mathbf{u}_2 \\
\hline
\mathbf{v}_1 \\
\frac{1}{2}\mathbf{v}_2
\end{array}\right]
=
\left[\begin{array}{c}
\mathbf{a}_1 \\
\mathbf{a}_2 \\
\hline
\mathbf{b}_1 \\
\mathbf{b}_2
\end{array}\right]
\qquad (10.8.5b)
$$

Notice that the first and second rows have become independent of $\mathbf{u}_1$ and $\mathbf{u}_2$ in Eq. (10.8.5b). Therefore, Eq. (10.8.5b) can be solved by the two steps:

$$
\left[\begin{array}{cc}
2\mathbf{I}+\mathbf{AB} & -2\mathbf{I} \\
-\mathbf{I} & 2\mathbf{I}+\mathbf{AB}
\end{array}\right]
\left[\begin{array}{c}
\mathbf{v}_1 \\
\frac{1}{2}\mathbf{v}_2
\end{array}\right]
=
\left[\begin{array}{c}
\mathbf{a}_1 \\
\mathbf{a}_2
\end{array}\right]
\qquad (10.8.6)
$$

$$
\left[\begin{array}{c}
\mathbf{u}_1 \\
\mathbf{u}_2
\end{array}\right]
=
\left[\begin{array}{cc}
\mathbf{B} & \mathbf{B} \\
\mathbf{0} & \mathbf{B}
\end{array}\right]
\left[\begin{array}{c}
\mathbf{v}_1 \\
\frac{1}{2}\mathbf{v}_2
\end{array}\right]
-
\left[\begin{array}{c}
\mathbf{b}_1 \\
\mathbf{b}_2
\end{array}\right]
\qquad (10.8.7)
$$

In order to extend this method to a larger system, we consider the case that the number of rows of u grids is eight. Then the equation corresponding to Eq. (10.8.6) becomes

$$
\left[\begin{array}{cccccccc}
\mathbf{C} & -\mathbf{I} & & & & & & \\
-\mathbf{I} & \mathbf{C} & -\mathbf{I} & & & & & \\
& -\mathbf{I} & \mathbf{C} & -\mathbf{I} & & & & \\
& & -\mathbf{I} & \mathbf{C} & -\mathbf{I} & & & \\
& & & -\mathbf{I} & \mathbf{C} & -\mathbf{I} & & \\
& & & & -\mathbf{I} & \mathbf{C} & -\mathbf{I} & \\
& & & & & -\mathbf{I} & \mathbf{C} & -2\mathbf{I} \\
& & & & & & -\mathbf{I} & \mathbf{C}
\end{array}\right]
\left[\begin{array}{c}
\mathbf{v}_1 \\
\mathbf{v}_2 \\
\mathbf{v}_3 \\
\mathbf{v}_4 \\
\mathbf{v}_5 \\
\mathbf{v}_6 \\
\mathbf{v}_7 \\
\frac{1}{2}\mathbf{v}_8
\end{array}\right]
\qquad (10.8.8)
$$

where $\mathbf{C} = 2\mathbf{I} + \mathbf{AB}$ and only the left side is shown. The cyclic reduction technique can be applied at this stage. By multiplying the even rows by $\mathbf{C}$ and adding the adjacent odd rows to each even row, Eq. (10.8.8) is successively reduced to

$$
\left[\begin{array}{cccc}
\mathbf{C}^{(2)} & -\mathbf{I} & & \\
-\mathbf{I} & \mathbf{C}^{(2)} & -\mathbf{I} & \\
& -\mathbf{I} & \mathbf{C}^{(2)} & -2\mathbf{I} \\
& & -\mathbf{I} & \mathbf{C}^{(2)}
\end{array}\right]
\left[\begin{array}{c}
\mathbf{v}_2 \\
\mathbf{v}_4 \\
\mathbf{v}_6 \\
\frac{1}{2}\mathbf{v}_8
\end{array}\right]
\qquad (10.8.9)
$$

$$\begin{bmatrix} \mathbf{C}^{(3)} & -2\mathbf{I} \\ -\mathbf{I} & \mathbf{C}^{(3)} \end{bmatrix} \begin{bmatrix} \mathbf{v}_4 \\ \frac{1}{2}\mathbf{v}_8 \end{bmatrix} \qquad (10.8.10)$$

$$[\mathbf{C}^{(4)}][\tfrac{1}{2}\mathbf{v}_8] \qquad (10.8.11)$$

where $\mathbf{C}^{(n)}$ satisfies the recursion formula

$$\begin{aligned} \mathbf{C}^{(1)} &= \mathbf{C} \\ \mathbf{C}^{(n)} &= (\mathbf{C}^{(n-1)})^2 - 2I, \qquad n > 1 \end{aligned} \qquad (10.8.12)$$

and also Eq. (10.7.14) with $\mathbf{A}$ replaced by $\mathbf{C}$. This simplification is due to the particular choice of the grid points for u at the top grid row as well as to the total number of grid rows equal to 2^l, where l is an integer. The rest of the calculations, including the right sides of Eqs. (10.8.8) through (10.8.11) and the solution to the odd blocks, is very similar to the case of the Poisson equation in Section 10.7.

PROBLEMS

1. Write the flow chart of a program to calculate the incompressible fluid flow about an infinitely long rectangular column immersed in a uniform stream, assuming a FDS subroutine is available. Assume also that the boundary of the rectangle passes grid points.

2. Repeat Problem 1 for a rotating cylinder.

3. Explain why Eqs. (10.4.2) and (10.4.3) become unstable in the hyperbolic and elliptic regions, respectively. (Hint: consider the time difference equation for

$$\frac{\partial^2 \phi}{\partial t^2} = a \frac{\partial^2 \phi}{\partial x^2} \qquad a > 0 \text{ (hyperbolic)}$$

$$a < 0 \text{ (elliptic)}$$

and apply the Fourier stability analysis described in Chapter 4.)

4. Referring to Section 10.6, apply the difference approximation directly for Eq. (10.3.10) for a transonic flow, and draw the flow chart of a program based on SLR (successive-over-relaxation for elliptic region and successive-under-relaxation for hyperbolic region).

5. Express Eqs. (10.5.13) and (10.5.14) in terms of the velocity potential ϕ, and discuss how those can be incorporated in the relaxation method described in Section 10.4.

6. Prove Eq. (10.5.22).

7. Apply the Atkin–Shanks extrapolation algorithm to the nonlinear problem in Example 7.11.

8. Consider a uniformly spaced grid system for a rectangular geometry. Assuming the Dirichlet boundary conditions are given for all the four sides, show that the difference equations for the Poisson equation are written in the form of Eq. (10.7.1). (Hint: see Example 1.5.)

9. Write a flow chart for the FDS code for a 15×15 block-tridiagonal matrix.

10. Explain why the round-off error of $\mathbf{A}^n \boldsymbol{\phi}$ increases as n increases, where $\boldsymbol{\phi}$ is an arbitrary vector and $\mathbf{A}$ is the tridiagonal matrix in Eq. (10.7.1). (Hint: expand $\boldsymbol{\phi}$ into the eigenvectors of $\mathbf{A}$, and apply the power method.)

11. Explain why there is no serious round-off error in $\mathbf{A}^{-n} \boldsymbol{\phi}$ for a large n in contrast to Problem 10.

REFERENCES

1. D. R. Chapman, H. Mark, and M. W. Pirtle, "Computers vs. wind tunnels for aerodynamic flow simulations," *Astronaut. Aeronaut.*, **13**, 23–35 (1975).

2. G. S. Deiwert, "Numerical simulation of high Reynolds number transonic flows," AIAA Paper 79-603, AIAA 7th Fluid and Plasma Dynamics Conference, Palo Alto, Calif., June 17–19 (1974).

3. D. C. Wilcox, "Calculations of turbulent boundary layer shock wave interaction," *AIAA J.*, **11**, 1592–1594 (1973).

4. B. S. Baldwin and R. W. MacCormack, "Numerical solution of the interaction of a strong shock wave with a hypersonic turbulent boundary layer," AIAA paper 74–558, AIAA 7th Fluid and Plasma Dynamics Conference, Palo Alto, Calif., June 17–19 (1974).

5. W. J. Bouknight, A. S. Denenberg, D. E. McIntyre, J. M. Randall, A. H. Sameh, and D. L. Slotnick, "The Illicac IV system," *Proc. IEEE*, **60**, 369–388 (1972).

6. T. Brantanow and A. Ecer, "Computational considerations in application of the finite element method for analysis of unsteady flow around airfoils," *Proceedings of the AIAA Computational Fluid Dynamics Conference*, Palm Springs, Calif., July 19–20 (1973).

7. B. L. Buzbee, G. H. Golub, and C. W. Nielson, "On direct methods for solving Poisson's equations," *Siam J. Numer. Anal.*, **7**, 627–656 (1970).

8. F. W. Dorr, "The direct solution of the discrete Poisson equation on a rectangle," *SIAM Review*, **12**, 248–263 (1970).

9. P. Concus and G. H. Golub, "Use of fast direct methods for the efficient numerical solution of nonseparable elliptic equations," *SIAM J. Numer. Anal.*, **10**, 1103–1120 (1973).

10. O. Buneman, *A compact noniterative Poisson solver*, SUIPR Rpt. 294, Institute for Plasma Research, Stanford Univ., Calif. (1969).

11. R. W. Hockney, "A fast direct solution of Poisson's equation using Fourier analysis," *J. Assoc. Comp. Mach.*, **12**, 95–113 (1965).

12. R. W. Hockney, "The potential calculation and some applications," *Methods Computat. Phys.*, **9**, 135–211 (1970).

13. E. D. Martin, "A generalized-capacity-matrix technique for computing aerodynamic flows," *Computers Fluids*, **2**, 79–97 (1974).

14. L. J. Clancy, *Aerodynamics*, Wiley, New York (1975).

15. J. D. Cole and A. F. Messiter, "Expansion procedures and similarity laws for transonic flow, part 1. Slender bodies at zero incidence," *Z. Angew. Math. Phys.*, **8**, 1–25 (1957).

16. H. Ashley and H. Landahl, *Aerodynamics of wings and bodies*, Addison-Wesley, Reading, Mass. (1965).

17. J. D. Cole, *Perturbation methods in applied mathematics*, Blaisdell, Waltham, Mass. (1968).

18. E. M. Murman and J. D. Cole, "Calculation of plane transonic flows," *AIAA J.*, **9**, 114–121 (1971).

19. E. M. Murman, "Analysis of embedded shock waves calculated by relaxation methods," *AIAA J.*, **12**, 626–633 (1974).

20. J. L. Steger and H. Lomax, "Numerical calculations of transonic flow about two-dimensional airfoil by relaxation procedures, *Proceedings of the AIAA 4th fluid and plasma dynamics conference*, June 21–23 (1971).

21. F. R. Bailey and J. L. Steger, "Relaxation techniques for three-dimensional transonic flow about wings," *AIAA J.*, **11**, 318–325 (1973).

22. F. R. Bailey, "On the computation of two- and three-dimensional steady transonic flows by relaxation methods," VKI Lecture Series, *Progress in numerical fluid dynamics*, von Karman Inst. for Fluid Dynamics, Rhode–St. Genese, Belgium, Feb. 11–15 (1974).

23. E. D. Martin and H. Lomax, "Rapid finite-difference computations of subsonic and slightly supercritical aerodynamic flows," *AIAA J.*, **13**, 579–586 (1975).

24. E. D. Martin, "Progress in application of direct elliptic solvers to transonic flow computations," pp. 839–870 in *Aerodynamic Analysis Requiring Advanced Computers*; Part II, NASA SP-347, NASA Ames Research Center, Moffett Field, Calif. (1975).

25. A. C. Aitken, "On Bernoulli's numerical solution of algebraic equations," *Proc. Royal Soc. Edinburgh*, **46**, 289–305 (1926).

26. D. Shanks, "Nonlinear transformations of divergent and slowly convergent sequences," *J. Math. Phys.*, **34**, 1–42 (1955).

27. E. D. Martin, "Direct numerical solution of three-dimensional equations containing elliptic operators," *Int. J. Numer. Methods Eng.*, **6**, 201–212 (1973).

28. H. Lomax and E. D. Martin, "Fast direct numerical solution of the nonhomogeneous Cauchy–Riemann equations," *J. Comp. Phys.*, **15**, 55–80 (1974).

29. J. L. Steger, private communication (1976).

30. F. C. Thames, J. F. Thompson, and W. Mastin, "Numerical solution of the Navier–Stokes equations for arbitrary two-dimensional airfoils," in *Aerodynamic analyses requiring advanced computers*, NASA-SP-347, Proceedings of a conference held at Langley Research Center, Hampton, Va. March 4–6 (1975).

31. J. F. Thompson, F. C. Thames, and C. Wayne, "Automatic numerical generation of body-fitted curvilinear coordinate system for field containing any number of arbitrary two-dimensional bodies," *J. Comp. Phys.*, **15**, 299–319 (1974).

32. J. C. South, Jr. and A. Brandt, "The multigrid method: fast relaxation for transonic flows," *Advances in engineering science*, **4**, NASA CP-2001, National Technical Information Service, Springfield, Va. (1976).

33. W. F. Ballhaus and J. L. Steger, "Implicit-factorization schemes for the low-frequency transonic equation," NASA TM X-73,082, Ames Research Center and U.S. Army Air Mobility R & D Laboratory, Moffett Field, Calif. (1975).

34. S. F. Shen, "An aerodynamicist looks at the finite element method," *Finite elements in fluids*, Vol. 2, R. H. Gallagher, J. T. Oden, C. Taylor, and O. C. Zienkiewicz, Eds., Wiley, New York (1975).

35. V. L. Streeter, *Fluid dynamics*, McGraw-Hill, New York (1948).

Appendix I   Adjoint Equations and Their Physical Meaning

This appendix is intended to supplement the material in Chapters 6 and 9. The adjoint equations for linear equations are derived, and the meanings of the adjoint solutions are discussed here with the neutron transport and diffusion problem in mind. The discussions in this appendix, however, apply to other physical problems whenever the meanings of the basic equations are appropriately interpreted. The adjoint equation for heat conduction in the integral form is discussed in Appendix II.

1.   DERIVATION OF ADJOINT OPERATORS

Linear equations in reactor physics such as multigroup diffusion equations, kinetic equations, neutron transport equations for specific geometries, slowingdown equations, and so on are special cases of the general neutron Boltzmann equation. Neutron transport in reactor systems is described by three space coordinates $\mathbf{r}$, the kinetic energy of the neutron in the laboratory system E, the direction of motion $\boldsymbol{\Omega}$, and time t. Therefore, all the linear equations in reactor physics may be written in the general operator form as

$$L\phi(\mathbf{r}, E, \boldsymbol{\Omega}, t) = S(\mathbf{r}, E, \boldsymbol{\Omega}, t) \tag{1}$$

where L is the operator including integral or differential operators, or both, ϕ is the neutron flux distribution, and S is the external source. Equation (1) is usually defined in a certain domain:

$$\mathbf{r} \in \mathscr{D} \text{ (finite space)}$$

$$E_0 < E < E_1$$

$$t_i < t < t_f$$

where E_0 is the lower limit of energy, E_1 is the upper limit of energy, t_i is the initial time, and t_f is the final time. The solution of Eq. (1) is subject to boundary conditions at the external boundaries of $\mathscr{D}$, and initial conditions at $t = t_i$. If the domain of interest is divided into nonoverlapping subdomains and the equation of the form of Eq. (1) is defined in each of the subdomains, then inner boundary conditions are necessary. Equation (1) is called a normal equation in contrast to the adjoint equation, which is discussed next.

The adjoint equation is defined as

$$L^*\phi^*(\mathbf{r}, E, \mathbf{\Omega}, t) = S^*(\mathbf{r}, E, \mathbf{\Omega}, t) \tag{2}$$

where ϕ^* is the adjoint function, S^* is the adjoint source, and L^* is the adjoint operator defined by

$$\int_{\mathscr{D}} d\mathbf{r} \int_{E_0}^{E_1} dE \int d\mathbf{\Omega} \int_{t_i}^{t_f} dt(\phi^* L\phi) = \int_{\mathscr{D}} d\mathbf{r} \int_{E_0}^{E_1} dE \int d\mathbf{\Omega} \int_{t_i}^{t_f} dt(\phi L^*\phi^*) \tag{3}$$

Equation (3) can be more simply expressed as

$$\langle \phi^*, L\phi \rangle = \langle \phi, L^*\phi^* \rangle \tag{4}$$

where the brackets mean the scalar product. When $L^* = L$ the operator L is called self-adjoint. The definition of the adjoint source S^* is rather arbitrary, and of course, the physical meaning of ϕ^* changes depending on the definition of S^*. The solutions of adjoint equations are subject to the boundary condition at the external surface of the space and the final condition at $t = t_f$. Those conditions are easily found in the process of finding the adjoint operator by Eq. (3).

When S in Eq. (1) is zero and the linear equation is a homogeneous steady-state problem, the operator L must include one undetermined parameter λ as an eigenvalue to ensure the solution. The linear operator for homogeneous problem can be written in the form

$$L = A - \lambda F \tag{5}$$

The adjoint operator can be found in the same way as for Eq. (5) and may be written

$$L^* = A^* - \lambda F^* \tag{6}$$

Before discussing the physical meaning of the adjoint equations, the adjoint equations for specific problems are shown.

Example 1.   One-Group Neutron Diffusion Equation

We first consider the equation

$$-\nabla D \nabla \phi(\mathbf{r}) + \Sigma \phi(\mathbf{r}) - \frac{1}{\lambda} \nu \Sigma_f \phi(\mathbf{r}) = 0 \tag{7}$$

for a multiregion reactor as shown in Fig. A.1, where Γs are inner boundaries and $\mathcal{D}$s are subregions in each of which D and Σ are constant and λ is an eigenvalue. The operator L of this equation is

$$L = -\nabla D\nabla + \Sigma - \frac{1}{\lambda}\nu\Sigma_f \qquad (8)$$

We try to find the adjoint operator L^* by using Eq. (3), assuming ϕ and ϕ^* are continuous in the entire region. The left hand side of Eq. (3) involves the term that can be transformed by partial integration as

$$-\int dv[\phi^*(\nabla D\nabla\phi)] = -\int_s ds\left[\phi^*D\frac{\partial}{\partial n}\phi\right] + \int dv[\nabla\phi^*D\nabla\phi]$$

$$= -\int_s ds\left[\phi^*D\frac{\partial}{\partial n}\phi\right] + \int_s ds\left[\phi D\frac{\partial}{\partial n}\phi^*\right] - \int dv[\phi\nabla D\nabla\phi^*] \qquad (9)$$

In Eq. (9), the surface integral is extended over all the inner and outer boundary surfaces, the inner surfaces are counted twice, that is, once from one side and another time from the other side, and $\partial/\partial n$ is the derivative outward normal to the surface.

If $D(\partial/\partial n)\phi$ and $D(\partial/\partial n)\phi^*$ are all continuous across the inner boundaries, the two surface integrals along any internal boundary cancel each other. Furthermore, if ϕ^* and ϕ both satisfy the outer boundary condition, which are either $\phi^* = \phi = 0$ or $(\partial/\partial n)\phi^* = (\partial/\partial n)\phi = 0$, all the surface

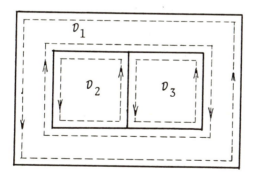

The surface integral along an internal boundary is performed twice, while the surface integral along an outer boundary is performed only once.

Figure A.1 The direction of surface integrals for a multiregion problem.

integrals along the external boundary disappear. Then the adjoint operator is found to be

$$L^* = -\frac{d}{dx}D\frac{d}{dx} + \Sigma_a - \frac{1}{\lambda}\nu\Sigma_f \tag{10}$$

Since L^* is identical to L, this problem is self-adjoint.

Example 2.   Two-Group Neutron Diffusion Equation

The two-group diffusion equation

$$-\nabla D_1\nabla\phi_1(\mathbf{r}) + \Sigma_1\phi(\mathbf{r}) = \frac{1}{\lambda}\nu\Sigma_f\phi(\mathbf{r})$$

$$-\nabla D_2\nabla\phi_2(\mathbf{r}) + \Sigma_2\phi(\mathbf{r}) = \Sigma_{sl}\phi(\mathbf{r}) \tag{11}$$

may be written in the form of Eq. (1), if we define $S = 0$ and

$$L \equiv \mathbf{L} = \begin{bmatrix} -\nabla D_1\nabla + \Sigma_1, & -\frac{1}{\lambda}\nu\Sigma_f \\ -\Sigma_{sl}, & -\nabla D_2\nabla + \Sigma_2 \end{bmatrix} \tag{12}$$

and

$$\phi \equiv \boldsymbol{\phi} = \mathrm{col}\,[\phi_1, \phi_2] \tag{13}$$

The left side of Eq. (4) becomes

$$\langle \boldsymbol{\phi}^*, \mathbf{L}\boldsymbol{\phi} \rangle \tag{14}$$

where

$$\boldsymbol{\phi}^* = \mathrm{col}\,[\phi_1^*, \phi_2^*] \tag{15}$$

We introduce Eqs. (12), (13), and (15) into Eq. (14), and apply integration by part twice to the terms including a differential operator. If the normal and the adjoint solution both satisfy all the internal boundary conditions and either the zero flux or flat flux boundary conditions along the external boundary, then Eq. (4) implies

$$L^* \equiv \mathbf{L}^* = \begin{bmatrix} -\nabla D_1\nabla + \Sigma_1, & -\Sigma_{sl} \\ -\frac{1}{\lambda}\nu\Sigma_f, & -\nabla D_2\nabla + \Sigma_2 \end{bmatrix} \tag{16}$$

It is seen that L^* is the transpose of L.

Example 3. Point Kinetic Equation without Delayed Neutrons

For the point kinetic equation given by

$$\frac{dN(t)}{dt} = \frac{\rho}{l} N(t) + S(t) \tag{17}$$

the operator of Eq. (1) becomes

$$L = \frac{d}{dt} - \frac{\rho}{l} \tag{18}$$

If we consider an arbitrarily long time interval $[t_i, t_f]$ the left side of Eq. (4) becomes

$$\langle N^*, LN \rangle = \int_{t_i}^{t_f} dt\, N^* \left(\frac{d}{dt} - \frac{\rho}{l} \right) N = [N^*N]_{t_i}^{t_f} - \int_{t_i}^{t_f} dt\, N \left(\frac{d}{dt} + \frac{\rho}{l} \right) N^* \tag{19}$$

If the initial condition of N and the final condition of N^* are given by

$$N(t_i) = 0 \tag{20}$$

and

$$N^*(t_f) = 0 \tag{21}$$

respectively, then the adjoint operator becomes

$$L^* = -\frac{d}{dt} - \frac{\rho}{l} \tag{22}$$

It is seen that the sign of the time derivative in Eq. (22) is altered from Eq. (18). Equation (20) can be used as a general initial condition, since $N = 0$ if t_i is selected as a time before the neutron source is introduced into the reactor. Eq. (21) may be also considered as a general final condition if t_f is selected as a time after the period of interest.

Example 4. Time-Dependent Neutron Diffusion Equation

For the space time kinetic equation given by

$$\frac{1}{v} \frac{d}{dt} \phi = +\nabla D \nabla \phi(\mathbf{x}, t) - \Sigma_a \phi(\mathbf{x}, t) + \nu \Sigma_f \phi(\mathbf{x}, t) + S(\mathbf{x}, t) \tag{23}$$

the operator L is written as

$$L = \frac{1}{v} \frac{d}{dt} - \nabla D \nabla + \Sigma_a - \nu \Sigma_f \tag{24}$$

The left hand side of Eq. (4) becomes

$$\langle \phi^*, L\phi \rangle = \int_{t_i}^{t_f} dt \int d\mathbf{x}\, \phi^* \left[\frac{1}{v}\frac{d}{dt} - \nabla D\nabla + \Sigma_a - \nu\Sigma_f \right] \phi \qquad (25)$$

Assuming that ϕ and ϕ^* satisfy inner and outer boundary conditions, the right side of Eq. (25) becomes

$$\langle \phi^*, L\phi \rangle = \int d\mathbf{x} \left[\phi^* \frac{1}{v}\phi \right]_{t_i}^{t_f} + \int_{t_i}^{t_f} dt \int d\mathbf{x}\, \phi \left[-\frac{1}{v}\frac{d}{dt} - \nabla D\nabla + \Sigma_a - \nu\Sigma_f \right] \phi^*$$

$$(26)$$

The first term on the right side of Eq. (26) becomes zero, if the following conditions are satisfied:

$$\phi(\mathbf{x}, t_i) = 0 \qquad (27)$$

and

$$\phi^*(\mathbf{x}, t_f) = 0 \qquad (28)$$

therefore, the adjoint operator is found to be

$$L^* = -\frac{1}{v}\frac{d}{dt} - \nabla D\nabla + \Sigma_a - \nu\Sigma_f \qquad (29)$$

It is seen again that the sign of the time derivative of Eq. (29) is altered from Eq. (24).

Example 5.   Monoenergetic Neutron Transport Equation

Consider the transport equation for slab geometries

$$L\phi(z, \boldsymbol{\Omega}) = S(z)$$

where

$$L = \mu\frac{d}{dz} + \Sigma - \int d\boldsymbol{\Omega}' \Sigma_s (\boldsymbol{\Omega}' \to \boldsymbol{\Omega}) \qquad (30)$$

The left side of Eq. (4) becomes

$$\langle \phi^*, L\phi \rangle = \int d\boldsymbol{\Omega} \int dz\, \phi^*(z, \boldsymbol{\Omega}) \left[\mu\frac{d}{dz}\phi + \Sigma\phi - \int d\boldsymbol{\Omega}' \Sigma_s (\boldsymbol{\Omega}' \to \boldsymbol{\Omega})\phi(z, \boldsymbol{\Omega}') \right]$$

$$(31)$$

The first term of the right side becomes

$$\int d\boldsymbol{\Omega} \int_{z_0}^{z_1} dz \left[\phi^* \mu\frac{d}{dz}\phi \right] = \int d\boldsymbol{\Omega} \left[\phi^* \mu\phi \right]_{z_0}^{z_1} - \int d\boldsymbol{\Omega} \int_{z_0}^{z_1} dz \left[\mu\frac{d}{dz}\phi^* \cdot \phi \right]$$

$$(32)$$

If the normal angular flux satisfies the external boundary conditions,

$$\phi(z, \mathbf{\Omega}) = 0 \quad \text{for } n \cdot \mu > 0 \tag{33}$$

where $n = 1$ for $z = z_0$ and $n = -1$ for $z = z_1$, and also if

$$\phi^*(z, \mathbf{\Omega}) = 0 \quad \text{for } n \cdot \mu < 0 \tag{34}$$

the first term on the right side of Eq. (32) becomes zero. Using this result, Eq. (31) becomes

$$\langle \phi^*, L\phi \rangle = \langle \phi, L^*\phi^* \rangle \tag{35}$$

where the adjoint operator is defined as

$$L^* = -\mu \frac{d}{dz} + \Sigma - \int d\mathbf{\Omega}' \, \Sigma_s(\mathbf{\Omega} \to \mathbf{\Omega}') \tag{36}$$

Comparing Eq. (36) with Eq. (30) it is seen that the sign of the first order differential operator is altered and positions of $\mathbf{\Omega}$ and $\mathbf{\Omega}'$ in Σ_s are exchanged.

Example 6. Neutron Slowing-Down Equation

The neutron slowing-down equation in an infinite monoatomic medium is

$$\phi(E)\Sigma(E) = \int_E^{E/\alpha} \Sigma_s(E')\phi(E') \frac{dE'}{(1-\alpha)E'} + Q(E) \tag{37}$$

or equivalently

$$\phi(E)\Sigma(E) - \int_0^\infty G(E' \to E)\phi(E') \, dE' = Q(E) \tag{38}$$

where

$$G(E' \to E) = \Sigma_s(E') \frac{1}{(1-\alpha)E'}, \quad E < E' < \frac{E}{\alpha}$$
$$= 0, \qquad\qquad\qquad E > E' \quad \text{or} \quad \frac{E}{\alpha} < E' \tag{39}$$

The left side of Eq. (4) for (38) becomes

$$\langle \phi^*(E), L\phi(E) \rangle = \int_0^\infty dE \phi^*(E)\phi(E)\Sigma(E)$$
$$- \int_0^\infty dE \phi^*(E) \int_0^\infty dE' G(E' \to E)\phi(E') \tag{40}$$
$$= \int_0^\infty dE' \phi(E') \left[\phi^*(E')\Sigma(E') - \int_0^\infty dE G(E' \to E)\phi^*(E) \right]$$

Therefore, the adjoint operator is found to be

$$L^* = \Sigma(E') - \int_0^\infty dE\, G(E' \rightarrow E) = \Sigma(E') - \Sigma_s(E') \frac{1}{(1-\alpha)E'} \int_{\alpha E'}^{E'} dE \tag{41}$$

Example 7. Integral Transport Equation

The integral transport equation for slab geometry may be written

$$\int_a^b dx'\, G(x' \rightarrow x)\phi(x') = S(x) \tag{42}$$

where $G(x' \rightarrow x)$ is the transport kernel. The normal operator is

$$L = \int_a^b dx'\, G(x' \rightarrow x) \tag{43}$$

Therefore,

$$\langle \phi^*(x), L\phi(x) \rangle = \int_a^b dx\, \phi^*(x) \int_a^b dx'\, G(x' \rightarrow x)\phi(x')$$
$$= \int_a^b dx'\, \phi(x') \int_a^b dx\, G(x' \rightarrow x)\phi^*(x) \tag{44}$$

The adjoint operator becomes

$$L^* = \int_a^b dx'\, G(x \rightarrow x') \tag{45}$$

In Eq. (45), x and x' in G are exchanged compared with Eq. (43).

2. PHYSICAL MEANING OF ADJOINT EQUATIONS

Let us consider the steady-state equation

$$L\phi(\mathbf{x}) = S(\mathbf{x}) \tag{46}$$

and the corresponding adjoint equation

$$L^*\phi^*(\mathbf{x}) = S^*(\mathbf{x}) \tag{47}$$

where $\mathbf{x}$ is the generic phase-space coordinate including space, energy and direction. In the case of the one-group diffusion equation for slab geometries, $\mathbf{x}$ is simply reduced to the one-dimensional space coordinate. In order to find the physical meanings, we set $S^*(\mathbf{x})$ equal to a specific cross section, Σ_α, which is included in the normal operator L. For example, Σ_α

may be the absorption cross section of some particular region or total cross section for some specific materials.

The solutions of Eqs. (46) and (47) may be expressed as

$$\phi(\mathbf{x}) = \int d\mathbf{x}' G(\mathbf{x}' \to \mathbf{x}) S(\mathbf{x}') \tag{48}$$

$$\phi^*(\mathbf{x}) = \int d\mathbf{x}'' G^*(\mathbf{x}'' \to \mathbf{x}) S^*(\mathbf{x}'') \tag{49}$$

where G and G^* are the Green's functions defined by

$$L G(\mathbf{x}' \to \mathbf{x}) = \delta(\mathbf{x} - \mathbf{x}') \tag{50}$$

$$L^* G^*(\mathbf{x}'' \to \mathbf{x}) = \delta(\mathbf{x} - \mathbf{x}'') \tag{51}$$

Premultiplying Eqs. (50) and (51), respectively, by $G^*(\mathbf{x}'' \to \mathbf{x})$ and $G(\mathbf{x}' - \mathbf{x})$ and integrating over $\mathbf{x}$ yield

$$\int d\mathbf{x} G^*(\mathbf{x}'' \to \mathbf{x}) L G(\mathbf{x}' \to \mathbf{x}) = G^*(\mathbf{x}'' \to \mathbf{x}') \tag{52}$$

and

$$\int d\mathbf{x} G(\mathbf{x}' \to \mathbf{x}) L^* G^*(\mathbf{x}'' \to \mathbf{x}) = G(\mathbf{x}' \to \mathbf{x}'') \tag{53}$$

By definition of the adjoint operator, as discussed in Section 1, the left sides of Eq. (52) and (53) are equal. Therefore, we have

$$G^*(\mathbf{x}'' \to \mathbf{x}') = G(\mathbf{x}' \to \mathbf{x}'') \tag{54}$$

It is seen that $\mathbf{x}'$ and $\mathbf{x}''$ are exchanged between G^* and G. Using Eq. (54) and the definition of the adjoint source as $S^* = \Sigma_\alpha(\mathbf{x})$, Eq. (49) becomes

$$\phi^*(\mathbf{x}) = \int d\mathbf{x}'' G(\mathbf{x} \to \mathbf{x}'') \Sigma_\alpha(\mathbf{x}'') \tag{55}$$

Since $G(\mathbf{x} \to \mathbf{x}'')$ is the neutron flux at $\mathbf{x}''$ due to the delta function neutron source at $\mathbf{x}$, ϕ^* is the total reaction rate with respect to Σ_α when a delta function source is placed at $\mathbf{x}$.

Suppose one desires to calculate an integral

$$I = \int d\mathbf{x} \Sigma_\alpha(\mathbf{x}) \phi(\mathbf{x}) \tag{56a}$$

A straightforward approach is to solve Eq. (46) and introduce the solution to Eq. (56a). However, an alternative approach is to solve Eq. (47) with $S^* = \Sigma_\alpha(\mathbf{x})$ and then calculate

$$I = \int S(\mathbf{x}) \phi^*(\mathbf{x}) \, d\mathbf{x} \tag{56b}$$

Equivalence of Eq. (56a) to Eq. (56b) may be shown by introducing Eq. (49) into Eq. (56b) as

$$I = \int d\mathbf{x} S(\mathbf{x}) \int d\mathbf{x}'' G^*(\mathbf{x}'' \rightarrow \mathbf{x}) \Sigma_\alpha(\mathbf{x}'')$$
$$= \int d\mathbf{x}'' \Sigma_\alpha(\mathbf{x}'') \int d\mathbf{x} G(\mathbf{x} \rightarrow \mathbf{x}'') S(\mathbf{x}) \, d\mathbf{x} = \int d\mathbf{x}'' \Sigma_\alpha(\mathbf{x}'') \phi(\mathbf{x}'').$$

where Eqs. (48) and (54) are also used. This equivalence can be proven more simply if we premultiply Eqs. (46) and (47) by $\phi^*(\mathbf{x})$ and $\phi(\mathbf{x})$, respectively, integrate the resulting equations, and then use Eq. (4). Thus the reaction rate such as Eq. (56a) can be computed by solving the adjoint equation rather than the normal equation. In fact, this technique is often used in the Monte Carlo calculations.

There is no essential change in the physical interpretation of time-dependent adjoint solution. Consider time-dependent equations

$$L\phi(\mathbf{x}, t) = S(\mathbf{x}, t) \tag{57}$$

$$L^*\phi^*(\mathbf{x}, t) = \Sigma_\alpha(\mathbf{x}, t) \tag{58}$$

where we assume the initial conditions, $\phi(\mathbf{x}, 0) = 0$ for Eq. (57) and final condition, $\phi(\mathbf{x}, t_f) = 0$ for Eq. (58). The Green's functions are defined as

$$LG(\mathbf{x}', t' \rightarrow \mathbf{x}, t) = \delta(\mathbf{x} - \mathbf{x}')\delta(t - t') \tag{59}$$

$$L^*G^*(\mathbf{x}'', t'' \rightarrow \mathbf{x}, t) = \delta(\mathbf{x} - \mathbf{x}'')\delta(t - t'') \tag{60}$$

Multiplying Eqs. (59) and Eq. (60) by $G^*(\mathbf{x}'', t'' \rightarrow \mathbf{x}, t)$ and $G(\mathbf{x}', t \rightarrow \mathbf{x}, t)$, respectively, subtracting the second equation from the first equation and integrating the resulting equation over $\mathbf{x}$ and t yield

$$G^*(\mathbf{x}'', t'' \rightarrow \mathbf{x}', t') = G(\mathbf{x}', t' \rightarrow \mathbf{x}'', t'') \tag{61}$$

We again see that the arguments in G^* and G are interchanged. By using G^*, the solution of Eq. (58) for the final condition $\phi(\mathbf{x}, t_f) = 0$ becomes

$$\phi^*(\mathbf{x}, t) = \iint G^*(\mathbf{x}', t' \rightarrow \mathbf{x}, t)\Sigma_\alpha(\mathbf{x}', t') \, d\mathbf{x}' \, dt'$$
$$= \iint G(\mathbf{x}, t \rightarrow \mathbf{x}', t')\Sigma_\alpha(\mathbf{x}', t') \, d\mathbf{x}' \, dt' \tag{62}$$

Since $G(\mathbf{x}, t \rightarrow \mathbf{x}', t')$ is the flux at $\mathbf{x}'$ and t' when the delta function source is at $\mathbf{x}$ and t, $\phi^*(\mathbf{x}, t)$ is the total reaction rate associated with $\Sigma_\alpha(\mathbf{x}, t)$ due to the point impulse neutron source at $\mathbf{x}$ and t.

The previous discussions do not apply to steady state of homogeneous equations because of $S = S^* = 0$. Physically, a homogeneous steady-state equation applies only to a critical reactor. The physical meaning of the adjoint equation for the homogeneous equation may be clarified by considering the response of a critical reactor to a point impulse neutron source. If we consider the one-group model for simplicity, the critical flux distribution satisfies

$$L\phi(x) \equiv \frac{d}{dx} D \frac{d}{dx} \phi(x) - (\Sigma_a - \nu\Sigma_f)\phi(x) = 0 \tag{63}$$

The corresponding adjoint equation is

$$L^*\phi^*(x) \equiv \frac{d}{dx} D \frac{d}{dx} \phi^*(x) - (\Sigma_a - \nu\Sigma_f)\phi^*(x) = 0 \tag{64}$$

Equations (63) and (64) are identical because L is self-adjoint, although the remainder of this appendix applies to non-self-adjoint problems with appropriate modifications. Suppose Eq. (63) has hold until $t = 0$. If a point-impulse neutron source is introduced into the reactor at $t = 0$, the neutron flux after $t = 0$ is subject to a transient and Eq. (63) does not hold until the flux again settles to a steady state. We denote the flux before $t = 0$ by $\psi_1(x)$ and the flux long after the transient by $\psi_2(x)$. Both ψ_1 and ψ_2 satisfy Eq. (63).

The neutron flux during the transient satisfies

$$\frac{1}{v} \frac{\partial}{\partial t} \phi(x, t) - \frac{d}{dx} D \frac{d}{dx} \phi(x, t) + (\Sigma_a - \nu\Sigma_f)\phi(x, t) = \delta(x - x_0)\delta(t) \tag{65}$$

where x_0 is the location of the point source and δ is a Dirac delta function. Equation (65) is, of course, reduced to Eq. (63) at steady state. The initial condition for Eq. (65) is

$$\phi(x, 0) = \psi_1(x) \tag{66}$$

Equations (63), (64), and (65) are subject to the same boundary conditions. For the current discussion we may assume the zero-flux external boundary conditions.

We expand the solution of Eq. (65) in the eigenfunction series of Eq. (63). They are defined by

$$L\phi_n = \omega_n\phi_n, \qquad n = 0, 1, 2, \ldots, \infty \tag{67}$$

where ω_n are eigenvalues and numbered as

$$\omega_0 > \omega_1 > \omega_2 > \cdots \tag{68}$$

Notice that $\omega_0 = 0$ and ϕ_0 corresponds to the physically meaningful solution of Eq. (63). The adjoint eigenfunctions may be defined also by

$$\phi_n^* = \phi_n \tag{69}$$

Equation (67) is a Sturm–Liouville equation introduced in Section 2.1. Its eigensolutions are complete and orthogonal:

$$\int \phi_m^* \phi_n \, dx = \delta_{mn}$$

where δ_{mn} is the Kronecker delta and we assumed that the eigenfunctions are normalized. The eigenvalue problem Eq. (67) is somewhat different from Eq. (7). The former is called the ω *mode* while the latter is called the λ *mode*. The eigenvectors used in Chapter 2 are all the λ mode. The ω mode eigenfunctions may be used to expand any flux distribution, while the λ-mode eigenfunction may be used to expand any fission source distribution but not the flux distribution. The fundamental eigenfunctions of both modes are identical if the reactor is just critical. For a transient problem the ω mode is more useful than the λ mode.

In terms of ϕ_n, the solution of Eq. (65) may be expressed as

$$\phi(x, t) = \sum_{n=0}^{\infty} a_n(t)\phi_n(x) \tag{70}$$

where $a_n(t)$ are coefficients to be determined. For $t < 0$ we obviously have

$$\left.\begin{aligned} a_0(t) &= \frac{\psi_1(x)}{\phi_0(x)} = b \\ a_n(t) &= 0 \quad \text{for } n > 0 \end{aligned}\right\} \quad t < 0 \tag{71}$$

which are the initial conditions for $a(t)$. In order to derive equations for $a_n(t)$, we introduce Eq. (70) into Eq. (65). Premultiplying Eq. (65) by $\phi_n^*(x)$ and integrating over x, we obtain

$$\frac{1}{v}\frac{d}{dt}a_n(t) - \omega_n a_n(t) = \delta(t)\phi_n^*(x_0) \tag{72}$$

The solution of Eq. (72) for $t > 0$ with the initial condition, Eq. (71), becomes

$$\begin{aligned} a_0(t) &= b + \phi_0^*(x_0)v \\ a_n(t) &= \phi_n^*(x_0) \exp(\omega_n v t)v, \qquad n > 0 \end{aligned} \tag{73}$$

Since $\omega_n < 0$ for $n > 0$, all of $a_n(t)$ with $n > 0$ approach zero after a sufficiently long time. Then $\phi(x, t)$ becomes

$$\phi(x, \infty) \equiv \psi_2(x) = [b + \phi_0^*(x_0)v]\phi_0(x) \tag{74}$$

The second term in the bracket of Eq. (74) is the effect of the point-impulse source. In other words, a point-impulse source increases the flux (or power) level of the critical reactor in proportion to the value of the adjoint eigenfunction at the location of the source. The adjoint eigenfunction for the reactor equation is also called *the importance function* because $\phi^*(x)$ represents the importance of the point impulse source at x in increasing the power level.

Appendix II   Adjoint Equation of Surface Density Equation

(Supplement to Section 9.10)

The Laplace equation for the unknown function $T(\mathbf{r})$ with the Dirichlet boundary condition may be transformed to an integral equation

$$T(\mathbf{r}) = \int d\mathbf{s}\mu(\mathbf{s})R(\mathbf{s}\to\mathbf{r}) \tag{1}$$

$$\mu(\mathbf{s}) - \int d\mathbf{s}'\mu(\mathbf{s}')K(\mathbf{s}'\to\mathbf{s}) = B(\mathbf{s}) \tag{2}$$

where $\mathbf{r}$ is the coordinate in the volume of interest, $\mathbf{s}$ is the coordinate on the surface, μ is the surface density, and R and K are integral kernels. In Eqs. (1) and (2) $B(\mathbf{s})$ is a prescribed function on the boundary. If we fix $\mathbf{r}$, Eq. (1) may be thought of as a weighted integral of $\mu(\mathbf{s})$ with weight $R(\mathbf{s}\to\mathbf{r})$. In an analogy to particle transport, $\mu(\mathbf{s})$ may be viewed as the particle density distribution and $R(\mathbf{s}\to\mathbf{r})$ as a detector sensitivity distribution on the surface.

If we define the normal operator of Eq. (2) as

$$L = 1 - \int d\mathbf{s}'K(\mathbf{s}'\to\mathbf{s}) \tag{3}$$

then the adjoint operator becomes

$$L^* = 1 - \int d\mathbf{s}'K(\mathbf{s}\to\mathbf{s}') \tag{4}$$

Using the "detector" as the adjoint source, an adjoint equation for Eq. (2) is written as

$$\mu^*(\mathbf{s}) = R(\mathbf{s}\to\mathbf{r}) + \int d\mathbf{s}'\mu^*(\mathbf{s}')K(\mathbf{s}\to\mathbf{s}') \tag{5}$$

Once $\mu^*(\mathbf{s})$ is obtained, $T(\mathbf{r})$ can be calculated according to Eq. (56b) of Appendix I by

$$T(\mathbf{r}) = \int d\mathbf{s} B(\mathbf{s})\mu^*(\mathbf{s}) \qquad (6)$$

Appendix III   Integral for Triangular Finite Elements

This appendix is the proof of Eq. (7.3.11):

$$\int_e \eta_j \eta_i \, dV = \frac{\Delta^{(e)}}{12}(1+\delta_{ij}) \tag{1}$$

We first consider a right-angled triangle as shown in Fig. A.2. The shape functions η_m and η_n in the right-angled triangle are

$$\eta_m = \frac{x}{a}, \qquad \eta_n = \frac{y}{b}$$

In the case of $i = j = n$, the left side of Eq. (7.3.11) becomes

$$\int_e \eta_n^2 \, dV = \int_0^b \left[\int_0^{a(1-y/b)} \left(\frac{y}{b}\right)^2 dx \right] dy$$

$$= \int_0^b a \left(\frac{y}{b}\right)^2 \left(1 - \frac{y}{b}\right) dy = \frac{ab}{12} = \frac{\Delta^{(e)}}{6} \tag{2}$$

where

$$\Delta^{(e)} = \text{area of the triangle} = \frac{ab}{2}$$

is used. In the case of $i = n$ and $j = m$, we have

$$\int_e \eta_n \eta_m \, dV = \int_0^b \left[\int_0^{a(1-y/b)} \frac{xy}{ab} \, dx \right] dy$$

$$= \frac{1}{ab} \int_a^b y \frac{a^2(1-y/b)^2}{2} \, dy = \frac{ab}{24} = \frac{\Delta^{(e)}}{12} \tag{3}$$

Thus Eq. (7.3.11) is proven for a right-angled triangle.

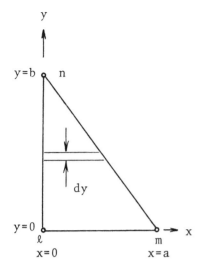

Figure A.2.

In order to prove Eq. (7.3.11) for an arbitrary triangle, we consider another triangle as shown in Fig. A.3, where l and m are at the same position as the previous triangle and the height is the same, while c is arbitrary: the triangle in Fig. A.3 is obtained by shifting the top apex horizontally. The two triangles obviously have the same area.

If we consider the volume element with width dy at y, η_n is constant in the volume element in each triangle, while η_m is a linear function of x in the volume elements. Furthermore, η_m at the right end of the volume element is identical for both triangles; $\eta_m = 0$ at the left end. Therefore, the integral of $\eta_n \cdot \eta_m$ in the volume element is identical for the two triangles. This proves that Eq. (7.3.11) is valid for the second triangle, also.

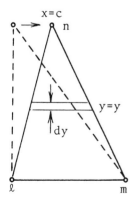

Figure A.3.

Appendix IV   Iterative Schemes for Implicit Methods of Characteristics

The implicit characteristic method for fluid flow described in Section 5.4 requires, at every time step, solution of a large linear system whose coefficient matrix is not always band-diagonal. The computer time, therefore, becomes longer as the fluid network becomes more complicated. This appendix describes fast-convergent iterative schemes that save computer time for the implicit characteristic method.

Considering a simple uniform fluid conduit, the characteristic equations, Eq. (5.2.21) and Eq. (5.2.23), may be written as

$$\frac{\partial \chi(x, t)}{\partial t} + (c + v) \frac{\partial \chi(x, t)}{\partial x} = \frac{s}{\rho c} \tag{1}$$

$$\frac{\partial \xi(x, t)}{\partial t} + (v - c) \frac{\partial \xi(x, t)}{\partial x} = -\frac{s}{\rho c} \tag{2}$$

$$\left. \begin{array}{c} \chi \\ \xi \end{array} \right\} = v(x, y) \pm \frac{1}{\rho c} p(x, y) \tag{3}$$

$$s = -\frac{Q''' c^2}{\rho (\partial h / \partial \rho)} \tag{4}$$

where the external forces are assumed to be zero and χ and ξ represent the sonic waves traveling forward and backward, respectively, and ρc is assumed to be piecewise constant.

Equations (1) and (2) are stable if integrated in the direction of wave motion.[1] Based on this property, two numerical schemes are derived as follows:

SCHEME A

Denoting the time step by n and applying the backward difference approximation to the time derivatives, Eqs. (1) and (2) become

448

$$\frac{d\chi_n(x)}{dx} = -\gamma[\chi_n(x) - \chi_{n-1}(x)] + \frac{\gamma s \Delta t}{\rho c} \tag{5}$$

$$\frac{d\xi_n(x)}{dx} = \tilde{\gamma}[\xi_n(x) - \xi_{n-1}(x)] - \frac{\tilde{\gamma} s \Delta t}{\rho c} \tag{6}$$

where $\gamma = 1/(c + v)\Delta t$ and $\tilde{\gamma} = 1/(c - v)\Delta t$. Assuming χ_{n-1} and ξ_{n-1} are known at the spatial grids, $x_j, j = 1, 2, \dots, J$, and using an appropriate initial condition, Eq. (5) is integrated forward. When $\chi_{n,j-1} \equiv \chi_n(x_{j-1})$ is known, an analytical integration of Eq. (5) in $[x_{j-1}, x_j]$ yields

$$\chi_{n,j} = \chi_{n,j-1} e^{-\gamma \Delta x} + \left[\chi_{n-1,j-1} + \frac{s \Delta t}{\rho c}\right](1 - e^{-\gamma \Delta x})$$

$$+ \frac{1}{\gamma \Delta x}[\chi_{n-1,j} - \chi_{n-1,j-1}][1 - (1 + \gamma \Delta x) e^{-\gamma \Delta x}] \tag{7}$$

where c and v are assumed to be known from the previous iteration, and a linear interpolation for $\chi_{n-1}(x)$ in $x_{j-1} < x < x_j$ is used.

Numerical integration such as a low order Runge–Kutta method may be used instead of analytical integration. Once the final value of $\chi_n(x)$ is obtained, the initial value of $\xi_n(x)$ for the backward integration of Eq. (6) is estimated, and vice versa. Equations (5) and (6) are alternatively integrated until convergence for each time step. This scheme is unconditionally stable.

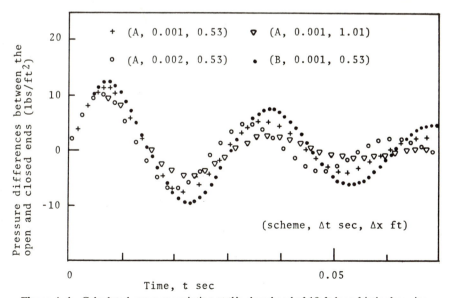

Figure A.4   Calculated pressure variation at the closed end of 10-ft-long frictionless pipe.

SCHEME B

Along the characteristic lines passing (t_n, x_j), difference approximations for Eqs. (1) and (2) may be written as

$$\chi_{n,j} = \chi_{n,j-1} + \gamma \Delta x \left(-\chi_{n,j-1} + \chi_{n-1,j-1} + \frac{s \Delta t}{\rho c} \right) \tag{8}$$

$$\xi_{n,j} = \xi_{n,j+1} + \tilde{\gamma} \Delta x \left(-\xi_{n,j+1} + \xi_{n-1,j+1} - \frac{s \Delta t}{\rho c} \right) \tag{9}$$

Equations (8) and (9) are calculated forward and backward, respectively, in a manner similar to Scheme A. Upon convergence, this scheme gives exactly the same result as in Eqs. (5.4.11) and (5.4.12). The semiimplicit technique can be easily incorporated, thus providing unconditional stability.

Neither scheme takes more than two iterations to satisfy a very tight convergence criterion for all the problems tested. Figure A.4 illustrates the numerical results for the same test problem as in Example 5.1.

REFERENCE

1. J. F. O'Brien and B. L. Edge, "A hybrid computer approach to the solution of water quality in an estuary," *Advances in Computer Methods for Partial Differential Equations*, AICA, Rutgers University, Department of Computer Science, New Brunswick, New Jersey (1975).

Appendix V   Factorization Formula for the FDS Method

(Proof of Eq. (10.7.14))

From Eq. (10.7.13), $\mathbf{A}^{(n)}$ is obviously a polynomial of degree 2^{n-1} in the matrix $\mathbf{A}$. Therefore, $\mathbf{A}^{(n)}$ may be written

$$\mathbf{A}^{(n)} = \sum_{j=0}^{2^{n-1}} c_j^{(n)} \mathbf{A}^j, \qquad n > 1 \tag{1}$$

where

$$c_j^{(n)} = 1 \quad \text{for } j = 2^n \tag{2}$$

Consider the polynomial of a variable having the same coefficients as Eq. (1):

$$a^{(n)} = \sum_{j=0}^{2^{n-1}} c_j^{(n)} a^j \tag{3}$$

Because of Eq. (10.7.13), $a^{(n)}$ satisfies the recursion formula,

$$a^{(n)} = (a^{(n-1)})^2 - 2 \tag{4}$$

If we set

$$a^{(1)} = a = 2 \cos \theta \tag{5}$$

it is easily proven that Eq. (4) is satisfied by

$$a^{(n)} = 2 \cos (2^{n-1}\theta) \tag{6}$$

Equation (6) becomes zero when

$$\theta = \frac{2j-1}{2^n}\pi, \qquad j = 1, 2, \ldots, 2^{n-1} \tag{7}$$

Using Eq. (5), the roots of $a^{(n)} = 0$ are

$$a = 2 \cos \left(\frac{2j-1}{2^n}\pi\right), \qquad j = 1, 2, \ldots, 2^{n-1} \tag{8}$$

Therefore, $a^{(n)}$ must be in the form

$$a^{(n)} = \prod_{j=1}^{2^{n-1}} \left[a - 2 \cos \left(\frac{2j-1}{2^n} \pi \right) \right] \tag{9}$$

where Eq. (2) is used to normalize the polynomial. By replacing a by $\mathbf{A}$ and 1 by $\mathbf{I}$, we obtain

$$\mathbf{A}^{(n)} = \prod_{j=1}^{2^{n-1}} \left[\mathbf{A} - 2 \cos \left(\frac{2j-1}{2^n} \pi \right) \mathbf{I} \right] \tag{10}$$

Thus Eq. (10.7.14) is proven.

Index